Design of Mechanical and Electrical Systems in Buildings

J. Trost

Ifte Choudhury

Texas A&M University

Upper Saddle River, New Jersey
Columbus, Ohio

Library of Congress Cataloging-in-Publication Data
Trost, J.
  Design of mechanical and electrical systems in buildings / J. Trost, Ifte Choudhury,
    p. cm.
  Includes index.
  ISBN 0-13-097235-5
  1. Buildings—Environmental engineering. 2. Buildings—Mechanical equipment—Design
and construction. 3. Buildings—Electric equipment—Design and construction. I.
Choudhury, Ifte. II. Title.
TH6031.T76 2004
696—dc22   2003059664

Editor in Chief: Stephen Helba
Executive Editor: Ed Francis
Editorial Assistant: Jennifer Day
Production Editor: Holly Shufeldt
Production Coordination: UG / GGS Information Services, Inc.
Design Coordinator: Diane Ernsberger
Cover Designer: Jim Hunter
Production Manager: Matt Ottenweller
Marketing Manager: Mark Marsden

This book was set in Cochin by UG / GGS Information Services, Inc. It was printed and
bound by Courier Kendallville, Inc. The cover was printed by Coral Graphic Services, Inc.

Pearson Education Ltd.   Pearson Education—Japan
Pearson Education Singapore Pte. Ltd.   Pearson Education North Asia Ltd.
Pearson Education Canada, Ltd..   Pearson Educación de Mexico, S.A. de C.V.
Pearson Education Australia Pty. Limited   Pearson Education Malaysia Pte. Ltd.

10 9 8 7 6 5 4 3
ISBN 0-13-097235-5

PREFACE

*The snow melts on the mountain
And the water runs down to the spring,
And the spring in turbulent fountain,
With a song of youth to sing,
Runs down to the riotous river,
And the river flows to the sea,
And the water again
Goes back in rain
To the hill where it is used to be.*

W. R. Hearst, 1941

Efficient buildings provide shelter and comfort by selectively using or resisting natural energy flows. The fuel and electrical energy required to operate a building during its useful life will usually cost more than the building's construction. This book deals with efficient building design in terms of (1) lighting and electrical installations, (2) heating, ventilation, and air-conditioning, and (3) water supply and drainage sub-systems of a building.

This book is a primer for students, architects, constructors, managers, occupants, and owners who wish to refine and improve their understanding of efficiency in building operation. Committed readers can develop a working knowledge of the design decisions, equipment options, and operations of different building sub-systems.

Readers who study the text and complete review problems will be able to:

+ Design, size, and detail the different sub-systems installations.
+ Select fixtures and components.
+ Integrate all the building sub-systems with site, building, foundations, structure, materials, and finishes.

A secondary goal is to respect the reader's time, talent, and perception by presenting the materials in a concise, lucid format. Illustrations are included in the text to expand and reinforce the information presented, and actual building applications are emphasized for each topic covered. Study problems follow each chapter so that readers can develop confidence in their abilities to apply knowledge and skills.

Fundamentals of lighting, perceptions, design, and examples are presented in Part I. Part II explains electrical energy that lights, heats, cools, and powers buildings.

Chapters 14 through 18 in Part III will allow the reader to build a knowledge base concerning comfort, heat flow, building heating-cooling equipment, and system selection. Chapter 19 presents a method for predicting annual building heating-cooling costs, and Chapters 20 and 21 outline opportunities for designing efficient buildings based on a working knowledge of heat flow.

Chapters 22 to 24 in Part IV cover general discussions about water supply, building and site drainage, site irrigation, waterscape, and methods and principles of building plumbing. Chapter 25 details plumbing installations in example residence and office occupancies, and Chapter 26 outlines plumbing work for fire and HVAC applications.

ACKNOWLEDGMENTS

Particular thanks are due to Charles W. Berryman, University of Nebraska, Lincoln; Daphene E. Cyr, IUPUI, Indianapolis; Jere C. Hamilton, University of Wyoming; Steven R. Hultin, Colorado State University; Richard M. Kelso, University of Tennessee, Knoxville; and Robert H. Murphy, Wentworth Institute of Technology (retired), for their assistance with the text review.

CONTENTS

Contents

PART I
LIGHTING

□

CHAPTER 1
Light and Perception

What you see is what you see.

Stella

This short chapter discusses sight, the most powerful human sense. We seek meaning in the image's light forms, and responses to light are a logical starting point for lighting-design studies. Read the text, enjoy the images, and note four visual stimuli that command attention. Don't be dismayed by words like *luminance* and *illuminance*. Luminance is visible light leaving a surface, and illuminance is the invisible radiant energy that causes luminance. Both are discussed at length in the next chapter. Light influences perception, and designers use patterns created by light, shade, and shadow to enhance surfaces and spaces.

1.0 LIGHTING TERMS

Professions develop jargon to describe the concepts and equipment they work with. This special language also protects future income. To work and communicate effectively with lighting professionals, you should speak their language. Skim this list of terms before you begin Part I and review it after each chapter to build a language foundation that supports effective communication with the individuals who light your buildings.

Acuity   Fine detail vision.

Adaptation   Weighted average of light entering the eye.

Ballast   Device that starts and regulates current in fluorescent and HID lamps.

Beam candlepower   Spot light rating. Beam luminous intensity.

Beam spread   Vertex angle in degrees that defines a cone of light leaving a lamp.

Brightness   A comparative sensation. Ranges from dim to dazzling.

Candlepower or **Candela**   Unit of luminous intensity. One candlepower or candela = 12.57 lumens.

CBCP   Center beam candlepower.

Cd/m²   Candela per square meter. Metric for luminance.

CIE   Commission International de l'Eclairage.

Color   A comparative sensation.

Color temperature   A source color metric based on Kelvin temperature.

CR   Cavity ratio. A numerical index of space proportions used in lighting calculations.

CRF   Contrast rendition factor. A numerical index for diffuse lighting.

CRI   Color rendering index for lamps.

CU   Coefficient of utilization. The percent of lamp lumens that reach the work plane.

DF   Daylight factor. The percent of outdoor light available inside a building.

Efficacy   An efficiency index for lamps. Measured in lumens per watt.

ESI   Equivalent sphere illumination. Diffuse shadow-free light.

Fluorescent   Lamps that emit light by exciting fluorescent materials with an electric arc.

Footcandle   Unit of illuminance. Density of light incident on a plane or surface. One footcandle = 1 lumen per square foot.

Glare   Uncomfortable brightness contrast.

Halogen   An incandescent lamp in which the filament is surrounded by iodine gas.

HID lamps   High-intensity discharge. Arc discharge lamps (see *mercury, metal halide,* and *sodium*).

Highlight   Interesting brightness contrast.

Illuminance   Density of light incident on a plane or surface. Measured in footcandles or lumens per square foot.

Incandescent   Light emission caused by heating.

Lamp life   Total operating hours when 50% of a group of lamps have failed.

Lightness   A comparative sensation of reflectance or transmission. Ranges from black to white or black to clear.

LLF   Light loss factor. The percent of lamp output lost over time.

Low E   Glass coatings used to limit heat and light transmission.

Lumen   Unit of luminous flux, light quantity. Lamp output is measured in lumens.

Luminaire   A lamp or light fixture.

Luminance or **luminous**   Emitting light, giving off light.

Lux   Metric unit of illuminance. Density of light incident on a plane or surface. One lux = 1 lumen per square meter.

mA   Amperes ÷ 1,000. Unit of electric current for fluorescent lamps.

Mercury lamp   An HID lamp that uses mercury vapor to conduct an electric arc.

Metal halide lamp   An HID lamp that uses a mixture of several conductive vapors to carry an electric arc.

Munsell   A system for defining colors.

Ostwald   A system for defining colors.

Photopic vision   Cone vision, color vision, detail vision.

Scotopic vision   Rod vision. Black and white, peripheral night vision.

Sodium lamp   An HID lamp that uses sodium vapor to conduct an electric arc.

VCP   Visual comfort probability. Numerical index for fixture glare.

1.1 BRIGHTNESS CONTRAST

SIGHT

Vision is the most powerful human sense. Visual talent, the ability to use the eye-brain interface effectively, is a unique ability comparable to athletic skill or musical aptitude. Light is the essence of art and design. Artists use the phrase "understanding light" to describe their craft, and master artists refine and perfect their visual talents by creating images that amplify perception.

Appraisal

The eye-brain interface is much more sophisticated than a fine camera (see Figure 1.1). The eye's range exceeds most camera-film combinations, and the brain adds assessment and judgment to each visual image. Mental appraisals range from simple tasks like locating a missing object to scholarly pursuits interpreting fine art or archaeological artifacts.

Contrasts Draw Attention

Drawings illustrate familiar examples of the evaluation and interpretation that occurs as the eye-brain interface sees. Four human responses to visual information—*brightness*, *pattern*, *motion*, and *color*—are emphasized because they form the foundations of lighting design. Many lighting designers believe all four are actually responses to brightness and lightness* contrasts (see Figure 1.2).

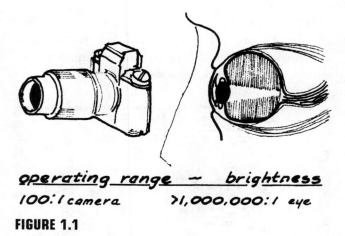

operating range ~ brightness

100:1 camera   >1,000,000:1 eye

FIGURE 1.1

---

*Lightness and brightness are different sensations. This text uses "brightness" for both.

FIGURE 1.2

Brightness contrast attracts attention. When an object in the visual field is brighter than its surroundings, eyes instinctively focus on that object. This eye-brain driven reflex is well documented in merchandising where the sales of impulse items can be directly related to brightness contrast. People are more likely to buy merchandise when it is displayed considerably brighter than the immediate surroundings.

Increasing overall brightness in a space does *not* increase brightness contrast, so lighting designers exploit the visual response to brightness contrast by varying lighting levels. The eye-brain response to a brightness contrast of 5:1 is the same as its response to a brightness contrast of 50:10, but the latter requires much more energy.

The eye adapts quickly to surrounding brightness, and the adapted eye can comfortably function in brightnesses ten times or on one-tenth its adaptation level. Museums exploit both the eye's range and the brightness contrast response when illuminating delicate art. Curators keep ambient lighting dim and use small light sources to illuminate paintings.

Highlight, *sparkle*, *flash*, *radiance*, *gleam*, and *brilliance* are terms used to describe brightness contrast. Lighting designers add words such as *punch*, *focus*, *dramatic*, and *noble* to describe the visual impact of their work. Jargon notwithstanding, brightness contrast is usually accomplished by spotlighting surfaces or by displaying light-reflecting objects on light-absorbing backgrounds.

1.2 PATTERN, FORM, AND SIZE

The eye-brain looks for meaning in all visual information. When incoming images are new or incomplete, testing and analysis are typical responses. Judgments about object dimensions, form, and location are affected by experience and a desire to make sense of visual signals (see Figure 1.3).

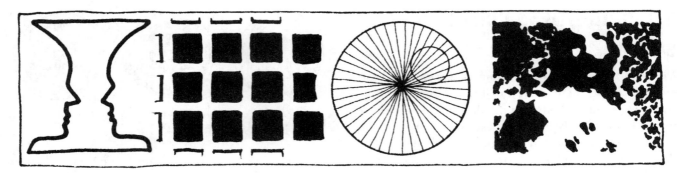

FIGURE 1.3

The eye-brain attempts to interpret each visual input, but object size seems to vary with the background perspective, and straight lines appear to curve when viewed on a radial background. An array of lines on a flat surface becomes a cube, and the front and back faces of the cube exchange positions. Some patterns seem to cycle, while others reveal static images. The desire to seek meaningful information from visual input seems to be perpetual, and oscillations suggest a continuing eye-brain feedback and testing loop. Pattern recognition examples confirm continual test cycles in the seeing process (see Figure 1.4).

Architectural designs exploit form and pattern to create visual impressions. Lines, planes, and volumes define spaces, and lighting can enhance space perceptions. Light can make a wall seem more distant, making a room appear more spacious (see Figure 1.5). Lighting designs use viewers' continuing search for information to make space look longer, wider, or taller. Changing the mix and intensity of light in a room changes viewer space perceptions.

1.3 MOTION AND COLOR

MOTION

Moving objects—fluids, animals, people, or light sources—all command immediate visual attention. Motion, like brightness contrast, draws the human eye (see Figures 1.6 and 1.7). The reflex is probably defensive. It helps people avoid danger, such as speeding cars.

Designers exploit motion's attraction with all manner of signs and devices, from flickering candles and throbbing lava lamps to the rotating mirror spheres used to enhance dancing. Moving lights

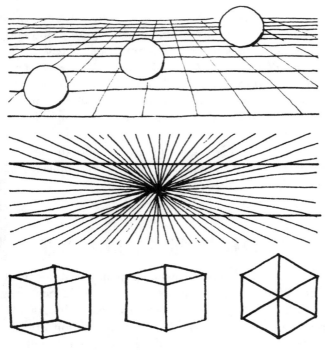

FIGURE 1.4

FIGURE 1.5

FIGURE 1.6

FIGURE 1.7

maximize visual attraction by combining motion and brightness contrast.

Motion, brightness contrast, pattern, and color are the essential components of Disneyland rides, video games, Las Vegas slot machines, and university visualization laboratories.

COLOR

Color is a simple word for a complex mental sensation. Color vision is a comparative process. White snow remains white when seen in deep shadow, and yellow and blue retain their identity in sunlight or shade.

Although colors maintain identity when lighting varies, the appearance of a color seems to change when viewed with other colors. Selecting colors that complement one another is an artistic opportunity built on comparative vision.

Color, like brightness contrast, pattern, and motion, has demonstrable effects on sight and perception. Blue and green are described as "cool," while red, yellow, and orange are labeled "warm." Controlled tests studying behavioral responses to color reinforce cool-warm concepts. Volunteers who performed the same tasks in identical red and blue rooms consistently overestimated air temperature, background noise levels, and time spent in the red room. Room size was overestimated in the blue room.

Warm and cool colors are further described as advancing and receding. An alignment test confirms these descriptors; when volunteers are asked to align a blue and a red sphere of equal size, a significant majority place the blue sphere nearer the eye.

Warm colors are "active and stimulating," while cool colors are "relaxing and restful." Fast-food establishments use warm interior colors to encourage rapid dining. At your next fast-food experience, notice the effects of brightness contrast, color, and music on the way people eat.

Color Terms

Because color vision is a comparative mental process, the terms that describe color perception are not precise. Use the following jargon only when it confirms your visual impressions.

White is a balanced mixture of all light wavelengths, black is the absence of light, and grays are blends of black and white.

Color is a perception caused by an unbalanced mixture of light wavelengths. Green light has more energy in the wavelengths we describe as green and less in other wavelengths. Green objects reflect more light in the green wavelengths.

Hue is a pure color. Blue and red are hues.

Chroma, *purity*, and *saturation* describe a color's freedom from white or gray.

Value is the relative lightness or darkness of a color. Yellow is lighter than blue.

Tints, *tones*, and *shades* are colors diluted with white, gray, and black (see Figure 1.8).

Brightness

Color vision ceases in moonlight. A blue light appears brighter than an equal red light at night, but the two reverse in daylight. As brightness increases, warm colors fade or weaken before cool colors, and intense illumination weakens all colors.

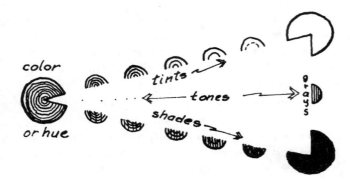

FIGURE 1.8

What you measure   **What you see**

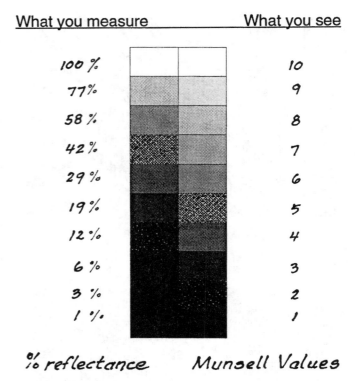

% reflectance   *Munsell Values*

FIGURE 1.9

Color constancy is a term used to describe the eye-brain's ability to recognize colors illuminated by unusual light sources. Color constancy is not absolute, and good designers are careful to select colors using the type of light source specified for the intended location. This means selecting exterior house paint colors outdoors, not in a paint store.

Complement seeking is an eye-brain effort to substitute the complement of a single color that dominates a scene. Evidence of complement seeking is the after-image people report following concentration on a monochromatic pattern. Red meats seem more attractive when displayed in green surroundings, on a green background.

Defining Colors

The visible spectrum spans a range of wavelengths, and several systems have been developed to define color. This black and white format is inadequate, and serious lighting-design students must commit to further reading and investigation.

    CIE The Commission Internationale de l'Eclairage (illumination) defines a color with three numerical values that represent percentages of red, green, and blue. CIE values are also used to describe lamp output.

    Ostwald The Ostwald System catalogs colors in triangles of progressive tints, tones, and shades.

    Munsell The Munsell Color System classifies colors by surface reflectance in ten steps

perceived as equal increments of surface lightness. Black approaches Munsell Value (MV) 1, and white approaches MV 10. Each MV step is equal to the square root of a color's reflectance. Munsell Value differences are excellent indicators of lightness contrast. A difference of 4 MV is a moderate contrast, whereas a difference of 8 MV is a strong contrast. Figure 1.9 lists Munsell reflectances on the left and approximates ten equal perception steps on the right.

1.4 RESPONSES AND DESIGN PALETTE

Brightness contrast, pattern, motion, and color can be used to inform and influence viewers. These four variables are the lighting designer's palette.

    In marketing, visual attraction is the first step toward a sale. In architecture, light is used to inform and influence viewer perception, mood, and behavior. Light informs and influences on several levels. At a building entry, light may show the route to a door,

A

B

C

D

FIGURE 1.10

but light can also communicate sensations of safety, space, warmth, and welcome. Light in buildings defines spaces and surfaces. It enhances the shapes and forms that enclose space, and it commands viewer attention. Light can stimulate, excite, and entice as it complements art and architecture. The first step in lighting design is visualizing brightness patterns that make objects or spaces attractive and interesting (see Figures 1.10A–D).

Because the eye-brain continually searches for meaning in visual information, lighting designers have an opportunity to communicate beyond just reflex reactions to brightness, pattern, motion, and color. For example, people:

- Make more purchases when merchandise displays are notably brighter than their surroundings. Increased sales suggest they may prefer the variety that brightness contrast adds to a retail environment.
- Interpret a line as longer when arrowheads reverse, and they perceive rooms or spaces as longer, wider, or taller when light is used to emphasize one dimension or plane.
- Are attracted by motion in the visual field and find some surroundings more pleasing when motion is part of the lighting scheme.

LIGHTING DESIGN

Perception is the foundation of intellectual activity, and sight is our strongest perceptive sense. Talented lighting designers know sight is an instinctive mental process that searches for information and meaning; they use light to interest, inform, and influence people.

A common thread ties the Kimbell Museum in Texas, the Seagram Building in New York, and Philip Johnson's Glass House in Connecticut. Richard Kelly designed lighting for all three. Kelly's trinity of lighting design perceptions are focal glow, ambient luminescence, and play of the brilliants.

Focal glow is the campfire of all time, . . . the sunburst through the clouds, and the shaft of sunshine that warms the far end of the valley. Focal glow commands attention and interest. It fixes the gaze, concentrates the mind and tells people what to look at. It separates the important from the unimportant.

Ambient luminescence is a snowy morning in open country. It is underwater in the sunshine, or inside a white tent at high noon. Ambient luminescence minimizes the importance of all things

and all people. It fills people with a sense of freedom of space and can suggest infinity.

Play of the brilliants is the aurora borealis, . . . the Versailles hall of mirrors with its thousands of candle flames. Play of the brilliants is Times Square at night, . . . the magic of a Christmas tree, Fourth of July skyrockets. It quickens the appetite and heightens all sensation. It can be distracting or entertaining.*

Exciting and rewarding careers await individuals who understand the eye's range and the mind's responses to light. Sight is a comparative evaluation of lightness and brightness, which designers exploit to attract attention and influence perception.

1.5 PRODUCTIVITY

Lighting affects human perception, and light can stimulate behavioral responses. Customers eat more rapidly in a fast-food environment with bright lighting and warm colors than in a candlelit, white tablecloth, gourmet restaurant. Shoppers make more impulse purchases when selected merchandise is noticeably brighter than its surroundings. People perceive space changes when a room's lighting is changed from concealed source to exposed source.

However, while light influences perception and behavior, lighting will not control a complex individual behavior like productivity. Productivity is influenced by a mix of visual inputs and mental impulses that can include pleasure, power, lust, envy, delight, love, and memories of a recent meal. Lighting is an important contributor to human perception, and light affects simple behaviors like impulse buying or circulation paths. Light may influence a complex behavior such as productivity but lighting will **not** reliably improve productivity over time.

RESEARCH

Many studies confirm severely limited relationships between lighting and behavior, but a 1975 lighting research project discovered a *10% increase* in employee productivity when illuminance values were increased from 50 to 150.** Another study found *no improvement*

---

*"The Great Illuminator" by Philip Cialdella & Clara D. Powell, *Lighting Design and Application*, May 1993, pp. 58–65.
**"Productivity," *Electrical World*, June 1977.

in visual performance when illuminance values exceeded 50 and very little improvement from 30 to 50.* Review this research and read further if lighting numbers stimulate your interest. Always check sponsors before accepting research findings. Disposable diaper manufacturers and cloth diaper makers funded separate studies to judge *the environmental impact of diapers*. Findings were contradictory, both at a 99% confidence interval.

---

*Donald Ross, *The Limitations of Illumination as a Determinant of Task Performance*, IEEE annual meeting, October 1977, Los Angeles.

MORE

Readers finding this chapter's coverage of sight, pattern, and color too brief will enjoy: *Image Object and Illusion*, a collection of *Scientific American* articles published in 1974; and Edwin H. Land, "The Retinex Theory of Color Vision," *Scientific American*, December 1977. "Retinex" is Land's word for the retina-cortex sensation called vision.

REVIEW QUESTIONS

1. Lamps produce luminance and illuminance. Which is visible?
2. Name the phenomena that can attract the eye.
3. Explain "color."
4. Tones are colors or hues mixed with what?
5. Name three color cataloging systems. Which one is used to describe lamp output?

ANSWERS

1. Luminance is visible.
2. Brightness contrast, pattern/form/size, motion, and color.
3. Color is an unbalanced light mixture with dominant wavelengths.
4. Gray.
5. Munsell, Ostwald, and CIE. CIE is used to describe lamp output.

CHAPTER 2
Lighting Metrics

Light as energy is measured by photometry and follows its laws. Light for vision has no such simple rules or procedures . . . The eye is the only instrument that can evaluate light for vision . . .

Louis Erhardt

Develop lighting knowledge by reading and applying the photometric terms explained in the text. Luminance values broaden your lighting vocabulary, but the numbers don't represent visual impressions. Brightness is a sensation, and visual impressions are comparative evaluations of lightness and brightness contrasts.

Work through the problems following the text and then use a light meter to develop an understanding of illuminance, luminance, and reflectance values.

2.0 LIGHT AND SIGHT

LIGHT

Light is:

- Radiant energy that excites the human retina and creates visual sensation.
- An energy form the human eye can see.

These definitions provide some useful information but offer less than complete revelation. The next paragraph offers a little more detail, but a lucid explanation of light is difficult.

In 1900 Max Karl Ernst Ludwig Planck (1858–1947) postulated that radiant energy could be described by units called *quanta* and that the energy content of quanta increases as wavelength decreases. Albert Einstein (1879–1955) developed the concept of wave-particle dualism and defined the *photon* with wave and particle properties. Dual properties permit wave theory to explain color, while particle theory explains transmission and absorption.

Understanding light is a worthy aspiration. Happily, lighting design is possible without complete comprehension. In this text, light is defined simply as visible electromagnetic energy.

Wavelength

The electromagnetic spectrum catalogs radiant energy by wavelength and frequency. Light occupies a small visible portion of the spectrum with wavelengths ranging from 380 to 770 nanometers. A nanometer is very small (10^{-9} meters), so the wavelengths of light fall between 15 and 30 millionths of an inch (see Figure 2.1).

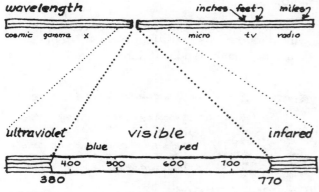

visible spectrum wavelength-nanometers

FIGURE 2.1

Light with wavelengths longer than 610 nanometers is called "red," and light with wavelengths between 440 and 500 nanometers is "blue." As wavelength decreases, frequency and energy increase. Blue light is more energetic than red light.

Most light sources also emit longer and shorter invisible wavelengths. Longer wave *infrared* radiation is perceived as heat, but when visible wavelengths are absorbed, they also degrade to heat. Shorter wave *ultraviolet* can cause photoelectric, fluorescent, and photochemical effects. Ultraviolet radiation also causes suntanning and can be used to kill germs.

Color Temperature

A colored light source has a large part of its energy concentrated around a specific wavelength. A colored object reflects most incident light in a narrow wavelength band. White light includes all visible wavelengths, but wavelength content varies for specific light sources.

A blackbody is one that absorbs all incident radiation at all temperatures, and also radiates a maximum at all wavelengths for a given temperature. The color of a light source can be quantified using *Kelvin* temperature to describe light emitted by a blackbody radiator. When blackbody temperature reaches 800 K, red light is emitted. The light becomes yellow at 3,000 K, white at 5,000 K, pale blue at 8,000 K, and bright blue at 60,000 K. Zero K is absolute zero or minus 273°C. The following table gives approximate color temperatures for a few familiar light sources.

Source	Color Temperature K
Sunlight at sunset	1,800 K
Sunlight at noon	4,800 K
Incandescent lamps	2,600–3,100 K
Fluorescent lamp (cool white)	4,200 K
Northwest sky	Up to 25,000 K

Build color knowledge by studying the CIE chromaticity diagram where "good" light sources lie along a blackbody locus based on color temperature.

Velocity

In a vacuum all light wavelengths travel 300,000 kilometers per second. In glass or water, the velocity of light is reduced, and each wavelength travels at a different speed.

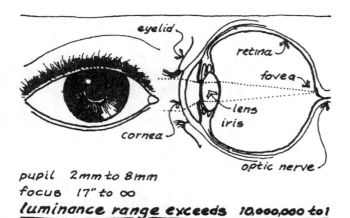

pupil   2mm to 8mm
focus   17" to ∞
luminance range exceeds   10,000,000 to 1

FIGURE 2.2

FIGURE 2.3

SIGHT

Light enters the eye through a protective cornea and aqueous humor (see Figure 2.2). Then it passes through an opening in the iris called the pupil. The size of the pupil controls the amount of light that will be passed on by the lens. Ciliary muscles change lens curvature so that light from sources at differing distances is focused on the surface of the retina. The retina is covered with sensors called *rods* that provide night vision and peripheral vision.

Detail vision and color vision occur near the center of the retina in a tiny area called the fovea. Sensors called *cones* dominate in the fovea and respond to red, green, or blue light. Cones do not respond to low light levels. Most colors are perceived as gray tones in moonlight.

The optic nerve carries electrical signals from the retina to the brain for interpretation. About 1 million optic nerve fibers serve more than 100 million rods and cones, so the retina also processes visual information.*

Adaptation

The range of adapted vision extends a log step above or below the adaptation level without discomfort. When moving from indoors to outdoors, notice how easily and quickly your eyes adapt to large changes in visual field brightness (see Figure 2.3). Adaptation is

---

*An excellent discussion of color vision can be read in "The Retinex Theory of Color Vision" by Edwin H. Land, *Scientific American*, December 1977.

instant for a tenfold change, 100:1 takes a second or two, and 1,000:1 can take 30 seconds or more.

Numerical luminance value changes do *not* indicate the eye's sensitivity to change; a luminance change from 1 to 2 is perceived as one-tenth the change from 1 to 1,000.

2.1 PHOTOMETRIC TERMS

Numbers used to describe visual experiences are taken from meters that cannot see. Study the following terms and numbers to develop a working lighting vocabulary, but *don't* believe the numbers represent visual sensations.

HISTORY

A *standard candle* was the historical starting point for photometric terms and measurements (see Figure 2.4). The candle was carefully constructed with detailed specifications for wax, wick, and burning rate, and was assigned a luminous intensity rating of *1 candlepower*.

Three candle-related terms were defined to describe light quantity, light density, and surface brightness:

- The *lumen* was defined as the quantity of light incident on a 1-square-foot surface 1-foot distant from a standard candle.

1 candle power ~ 12.57 lumens

illuminance 1 foot away ~ 1 fc
1 footcandle = 1 lumen per sqft.

if the surface facing the candle reflects 1 lumen per square foot its luminance is 1 footlambert

standard candle

FIGURE 2.4

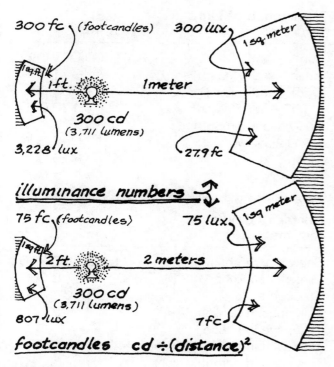

FIGURE 2.5

- The *footcandle* was defined as the density of illumination (illuminance) 1 foot distant from a standard candle (see Figure 2.5).
- The *footlambert* was defined as a surface luminance of 1 lumen per square foot.

Luminance and brightness are not equivalent because brightness is a sensation that includes contrast in the field of view and adaptation, while luminance is a photometric unit. A luminous candle seems bright on a dark night but dull on a sunny day.

HISTORICAL TERMS

Lighting terms based on the standard candle include:

Candlepower—Luminous intensity. The light source intensity of a standard candle.

Lumen—Luminous flux. Quantity of light. A standard candle emits 12.57 lumens. A sphere with a 1-foot radius has a surface area of 12.57 $(4\pi r^2)$ square feet, so the illuminance 1 foot distant from a standard candle is 1 lumen per square foot.

Footcandle—Illuminance. Density of illumination incident on a surface or plane. One

footcandle is an illuminance of 1 lumen per square foot.

Footlambert—Luminance. A surface that emits 1 lumen per square foot, *measured in the direction being viewed*, has a luminance of 1 footlambert (see Figure 2.6).

CURRENT TERMS

The standard candle burns no more. It has been replaced by the *candela*, which is defined as 1/60 of the intensity of a square centimeter of a blackbody radiator at 2,045 K. Fortunately, the output of a standard candle is equal to a candela, and most historical terms are unchanged.

Candela (cd) or **candlepower**—Unit of *luminous intensity* for a light source. One candela (cd) equals 1 candlepower (cp), and 1 candela yields 12.57 lumens.

Lumen—*Luminous flux* or quantity of light.

Footcandle (fc) or **lux**—*Illuminance units*. The density of illumination incident at a point on a surface, measured in lumens per square foot. One lumen per square foot is 1 footcandle. Lux is the SI unit of illuminance. One lux is 1 lumen per square meter, so 1 fc is about 10 lux (10.76).

cd/m^2 or **nit**—Unit of luminance. Luminous intensity per unit of projected area, for a reflecting or an emitting surface. One cd/m^2 or 1 nit is a luminance of 1 lumen per square meter *in the direction being viewed*. If the reflecting or emitting surface diffuses light, 1 lumen per square meter creates a luminance of $1/\pi$ cd/m^2 (Figure 2.6).

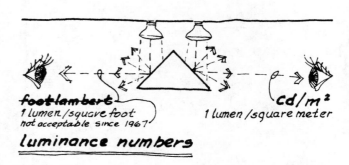

FIGURE 2.6

BEAM CANDLEPOWER

Illuminance decreases as distance to a light source increases (fc or lux = cp/d^2). Manufacturers specify *beam candlepower* and beam spread for spotlights and floodlights to aid illuminance calculations (see Figure 2.7).

Assume a 300 candela point source to verify the photometric numbers shown in Figure. 2.5 on the facing page. Then complete the end of chapter problems to ensure understanding.

NUMBERS

The following tables illustrate familiar viewing experiences quantified with approximate photometric numbers.

Intensity Values	cd
Sun (clear day)	670,000,000
75 watt spot lamp (average)	4,500
75 watt flood lamp (average)	1,800
Candle flame (1 foot distant)	1

Illuminance Values	fc
Sunlight + skylight (clear day)	10,000
Foggy day (dense fog)	1,000
Commercial offices	10–100
Residence interior (at night)	10
Candle flame (1 foot distant)	1
Moonlight (clear, full moon)	0.01

Luminance of Light Sources	cd/m²
Sun	+/− 2,000,000,000
Incandescent lamps	
500 watt (clear)	12,000,000
100 watt (frosted)	170,000
Fluorescent lamp	
(40 watt cw)	7,000
Moon	5,000
Candle flame	5,000

Values are for brightest spot on each source.

Surface Luminance (estimated)	cd/m²
Snow in sunlight	20,000–30,000
Grass in sunlight	2,000
Grass in shade	100
This page in sunlight	13,000
This page in shade	1,000
This page in library at night	150
This page in home at night	20

Converting Photometric Units

Multiply	By	To Get
Candela	1	Candlepower
Candela	12.57	Lumens
Footcandle	0.09	Lux
Lux	10.76	Footcandles
cd/m²	0.29	Footlamberts
Footlambert	3.42	cd/m²

2.2 VISUAL PERFORMANCE
HOW MUCH LIGHT?

Five seeing variables define visual tasks:

- Luminance
- Size
- Contrast
- Time or speed
- Accuracy

When a task is difficult, increased luminance can help maintain *speed and accuracy* (see Figure 2.8).

Professor H. R. Blackwell studied visual performance with students at Ohio State University. His laboratory tests in 1958 defined the luminance and

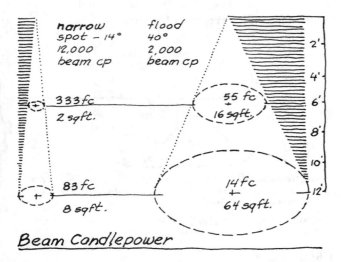

Beam Candlepower

FIGURE 2.7

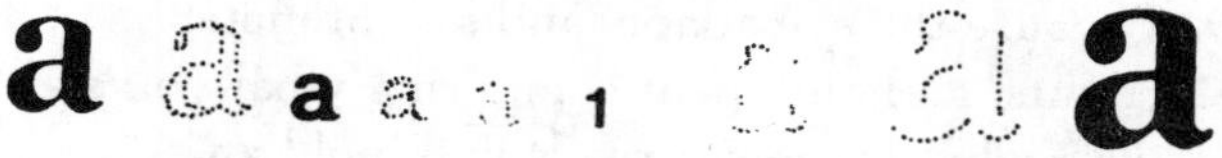

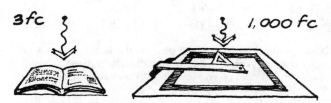

FIGURE 2.8

the related illuminance needed to correctly perform a variety of visual tasks. Reading high-contrast type like this text was found to require about 2 footcandles, and tracing a blueprint required about 1,000 footcandles (see Figure 2.9).

When reading 8 point type, to achieve 99% seeing accuracy, the minimum page illuminance falls between 1 and 2 footcandles, for a viewer under 40 years of age with a visual assimilation time of one-fifth of a second."

3fc   1,000 fc

FIGURE 2.9

Age erodes acuity, and older viewers may need as much as 50% more illuminance for a given seeing task.

RECOMMENDATIONS

Visual performance research findings don't directly apply to workplace seeing tasks because the motion, contrast, and glare typical in the workplace visual field are absent in laboratory tests.

Lighting manufacturers, government agencies, and the Illuminating Engineering Society (IES) recommend illuminance for various visual tasks. The following small example is from the 1981 IES Handbook.*

Recommended Illuminance	fc
Dining	5–10
Sewing, light to medium fabrics	50–100
Jewelry and watch manufacturing	200–500

*Chapter 7 offers more illuminance recommendations and calculations.

Logic, n. —The art of thinking and reasoning . . .
Major Premise: Sixty men can do a piece of work sixty times as quickly as one man.
Minor Premise: One man can dig a post-hole in sixty seconds; therefore—
Conclusion: Sixty men can dig a post-hole in one second.

This may be called syllogism arithmetical, in which, by combining logic and mathematics, we obtain a double certainty and are twice blessed.

Ambrose Bierce
The Devil's Dictionary

EXAMPLE PROBLEMS

Complete the following problems and activities to build your lighting vocabulary; then check the answers that follow.

1. Find the total luminous flux emitted by a lamp rated at 100 candelas.

2. Find the illuminance (lux) 1 meter distant from a point light source rated at 100 candelas.
3. Find the illuminance (lux) 2 meters distant from a point light source rated at 100 candelas.
4. Find the illuminance (fc) 1 foot distant from a point light source rated at 100 candelas.
5. Find the illuminance (fc) 6.56 feet distant from a point light source rated at 100 candelas.
6. Find the luminance of a diffusing surface that reflects 70% of incident light if the illuminance density is 25 lux.
7. Find the luminance of a diffusing surface that reflects 70% of incident light if the illuminance density is 25 fc.
8. Which two preceding problems (1–7) have visible answers?
9. Use a light meter to measure several indoor and outdoor illuminance levels. Then try to estimate illuminance without the meter.

ANSWERS

1. **1,257 lumens** 1 cd = 12.57 lumens, (100)(12.57) = 1,257 lumens.
2. **100 lux** 100 cd = 1,257 lumens = 100 lumens per square meter at 1 meter distant. One lumen per square meter = 1 lux.
3. **25 lux** Lux = cd/d^2 = $100/2^2$ = 25 lux. The density of light decreases as distance to the light source increases. For a *point source* of light each doubling of distance reduces lux or footcandles by a factor of 4. (This calculation can be used to determine illuminance under spotlights when beam candlepower is known. A method for calculating overall illuminance in buildings will be developed in Chapter 7.)
4. **100 fc** 100 cd = 1,257 lumens = 100 lumens per square foot 1 foot away.
   One lumen per square foot = 1 fc.
5. **2.3 fc** $100/6.56^2$ = 2.32 fc (see problem 3). 6.56 feet = 2 meters, and 2.32 fc = 25 lux.
6. **5.6 nits (cd/m^2)** Illuminance = 25 lux = 25 lumens per square meter. If reflectance is 70%, surface luminance is (25)(70%) = 17.5 lumens per square meter = 17.5/π or 5.57 cd/m^2.
7. **59.9 nits (cd/m^2)** Illuminance = 25 fc = 25 lumens per square foot. If reflectance is 70%, surface luminance is (25)(70%) = 17.5 lumens per square foot or (17.5)(10.76) = 188.3 lumens per square meter = 188.3/π or 59.9 cd/m^2.
8. Only luminance values (see problems 6 and 7) are visible, but the numbers do not describe visual sensations. The diffusing surface in problem 7 is about ten times more luminous than the surface in problem 6, but it does *not* appear ten times as bright. Luminance values can be assigned to lightness (reflectance) and brightness, but these values don't duplicate visual sensations. The eye-brain sees lightness from black to white, and brightness across a broad scale from dim to dazzling.
9. Because the eye adapts and a light meter counts, an hour spent comparing what you see with what the meter reads will build an appreciation of illuminance and luminance values. GE sells a light meter that displays illuminance values for about $45. If that price strains your budget, a handheld photographic light meter can be used. Verify and/or adjust the following table by comparing photographic and light meter readings.

fc	f stop	Time	ASA
2	1.4	1/30	400
10	2.8	1/30	400
20	4	1/30	400
50	5.6–8	1/30	400
100	8–11	1/30	400
200	11–16	1/30	400

You can also use a camera's through the lens light meter, but you'll have to experiment with a bit of tracing paper taped over the lens and develop your own conversion scale.

To estimate reflectance with a light meter, first measure incident footcandles and then rotate the meter 180° to measure reflected footcandles. Divide reflected fc by incident fc to estimate % reflectance.

The following chart offers some approximate reflectance values you may verify.

Surface %	Reflectance
Mirror	90–95
White paint	60–90
Concrete	10–60
Birch (clear finish)	20–40
Walnut (oil finish)	5–15
Grass	5–15
Asphalt	1–10

CHAPTER 3
Lamps

Lighthouse, n. A tall building on the seashore in which the government maintains a lamp and the friend of a politician.

Ambrose Bierce

This chapter describes incandescent, fluorescent, and high-intensity discharge lamps—three lamp types that comprise most building lighting. As you read about each, try to develop convictions about appropriate lamps for specific applications. The text doesn't suggest "best" lamps for a restaurant, church, hospital, or school because lamp selection embraces design intent and experience. When you see an attractive lighting installation try to understand its pleasing features in terms of your impressions about brightness contrast and color. Then note lamp types and wattages used to attract your interest.

Lamp manufacturers quote "efficacy" numbers to compare lamps, and luminaire manufacturers note lighting "watts per square foot" for building installations, but there's more to lighting design than lamp lumens and electrical costs.

3.0 LAMP EFFICACY, LIFE, AND COLOR

The human eye's sensitivity to light peaks in the yellow-green range (near 555 nanometers) and the lumen is defined by the human eye. A perfectly efficient light source would emit 680 lumens of yellow-green light. A perfectly efficient white light source might emit from 200 to 400 lumens per watt, depending on the definition of "white light."

EFFICACY

Efficacy is a lighting term for the number of lumens a lamp produces from each watt of electrical input energy. A standard 100 watt incandescent lamp with a clear glass bulb radiates about 1,750 lumens and has an efficacy of 17.5.

LIFE

Lamp life is rated in hours; a rating of 1,000 hours means that after 1,000 hours, half of a representative sample of lamps are still operable and the other half have burned out. Lamp life ratings are based on typical conditions. Abnormal supply voltage, vibration, ambient temperature, or frequent on-off cycles can reduce lamp life.

COLOR

Color temperature is a term used to describe the color of light. Standard incandescent lamps operate at color temperatures from 2,600 to 3,000 K. Higher color temperatures indicate more blue energy, and lower color temperatures mean more red energy.

Color temperature is *not* an indicator of the appearance of colored objects; instead, a *color rendering index* (CRI) is used to compare the appearance of colored objects illuminated by different light sources. Lamp CRIs are comparative numbers that vary with color temperature. An incandescent lamp is assigned a CRI of 100 at 2,900 K, and north sky daylight is assigned a CRI of 100 at 7,500 K. Manufacturers recommend lamps with CRIs of 90 or more when color rendition is an important consideration.

Efficacy, life, color, and operating characteristics are discussed for each of three principal lamp types: incandescent, fluorescent, and high-intensity discharge.

3.1 INCANDESCENT LAMPS

Electric arc lights were developed in the early 1800s, but Thomas Edison is credited with developing the first marketable incandescent lamp in 1879. Edison's lamp, an evacuated bulb with a filament of carbonized sewing thread, delivered 1.4 lumens per watt. Tungsten filaments were first used in 1907, and in 1913 lamps filled with inert gas replaced the evacuated bulb. Incandescents are not very efficient. A standard incandescent lamp converts most of the electrical input to heat and less than 10% to light, but low-cost lamps and fixtures make incandescent lamps a popular choice for many commercial applications and most residential lighting.

EFFICACY

The efficacy of standard incandescent lamps extends from 10 to 30 lumens per watt. Smaller lamps rate near 10, while larger lamps are more efficient because their filaments can operate at higher temperatures. Halogen incandescent lamps offer still higher efficacy, but decorative and long-life lamps yield fewer lumens per watt (see Figure 3.1).

LIFE

Incandescent lamps usually fail when switched on. A 100 watt tungsten filament operates near 4,800°F (2,900 K) and weakens as it ages. When a cold lamp is turned on, the initial current may be ten times normal operating current, and this current surge can break a weak filament (see Figure 3.2). Incandescent lamps that are not cycled on and off may burn for many years.

Standard incandescent lamps have a rated life of 750 or 1,000 hours. Light output decays 10 to 20% by the time a lamp fails. Decorative lamps are rated from 1,500 to 4,000 hours, and extended service lamps are available with 2,000 to 8,000 hour ratings.

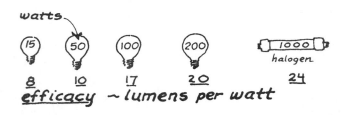

FIGURE 3.1

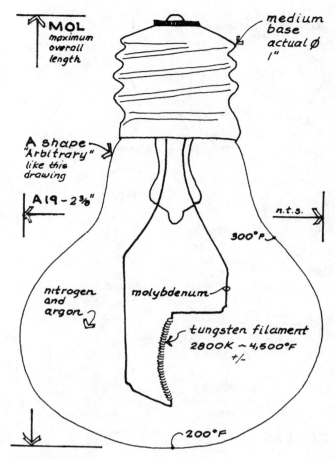

FIGURE 3.2

CRI value of 100, are considered "good" color rendering light sources.

OPERATION

Reduced input *voltage* reduces incandescent lamp efficacy and color temperature, but increases lamp life. Increasing lamp voltage has the opposite effect. As the table shows, a small change in input voltage can have a large impact on lamp life. See also Figure 3.4.

Voltage	% Light	% Life
Design (120 volts)	100	100
More (130 volts)	130	40
Less (110 volts)	75	300

Effects are approximate for a standard 1,000 hour, 120 volt lamp.

Dimming controls reduce lamp voltage and can be good investments because they extend lamp life. However, dimmers are *not* used for museum lighting or spaces where color rendition is an important concern. Incandescent dimmers are inexpensive, and solid state incandescent dimmers conserve electrical energy.

SPECIAL LAMPS

Halogen incandescent lamps provide longer life (2,000 to 4,000 hours), less decay, and more light than standard lamps. Halogen lamps include a quartz tube that surrounds the lamp filament with iodine gas (see Figure 3.5). Iodine vapor in the quartz tube limits filament evaporation and minimizes light loss over time.

These lamps offer longer life but yield less light (see Figure 3.3).

COLOR

The color temperature of light emitted by standard incandescent lamps ranges from 2,600 to 3,000 K, photographic lamps operate near 3,400 K, and tungsten melts at 3,665 K (Figure 3.2). Incandescent light is called "warm" because most visible energy is emitted in the yellow-orange-red wavelengths that flatter human skin tones. Incandescents, with an assigned

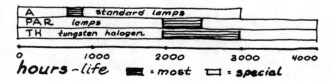

FIGURE 3.3

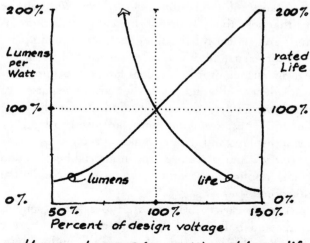

FIGURE 3.4

21

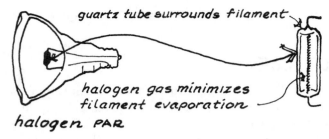

quartz tube surrounds filament

halogen gas minimizes filament evaporation

halogen PAR

FIGURE 3.5

Halogen lamps are available in multi-reflector (MR), parabolic aluminized reflector (PAR), and tubular (T) shapes.

Low-voltage incandescent lamps have compact filaments that permit improved beam control. Twelve or 24 volt lamps can deliver a brighter (and smaller) beam of light than 120 volt lamps of equal wattage. Remote transformers can serve several lamps, or a transformer can be included with each light fixture.

Low-voltage building lighting applications include spot lighting and decorative lighting; low-voltage halogen lamps combine the advantages of both types. In garden and landscape lighting, low-voltage systems offer easy installation and reduced shock risk.

Incandescent lamps designed for *high-voltage* (240 or 277 volt) operation are used to reduce wiring costs in branch lighting circuits.

Long-life lamps are actually designed for 130 or 140 volt power. Operated at 120 volts, they last a long time and cut replacement labor, but emit less light than standard lamps.

Quality colored lamps use *interference filters* on the lens that selectively transmit one color (wavelength-band) and reflect other colors back into the lamp.

Interference filters can also be applied to the reflecting surface of incandescent lamps. Certain coatings reflect visible light, but they are transparent to longer-wave infrared energy radiating from the lamp filament. These *cool beam* lamps are used to light ice cream displays and theater makeup rooms.

Infrared lamps are used for industrial heating and drying applications, and for warming residential bathrooms.

A great variety of *decorative* types, sizes, shapes, colors, and bulb coatings are available (see Figure 3.6). Globular lamps are often used to hide a low-budget porcelain socket, and silver bowl lamps, developed years ago for use in indirect fixtures, are

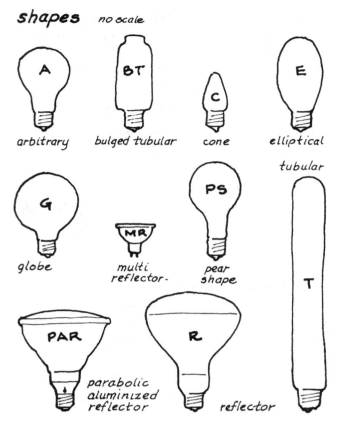

shapes no scale

A — arbitrary

BT — bulged tubular

C — cone

E — elliptical

G — globe

MR — multi reflector

PS — pear shape

tubular

T

PAR — parabolic aluminized reflector

R — reflector

FIGURE 3.6

"discovered" again by lighting designers every 10 years. There may be more lamp types than beer brands; if you cannot find the "right" lamp at your local lighting dealer, try theater, auto, or electronics suppliers.

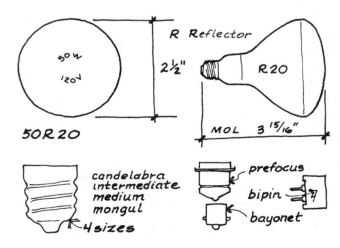

50 W 120 V

R Reflector

2½"

R20

50R20

MOL 3 15/16"

candelabra
intermediate
medium
mongul
4 sizes

prefocus

bipin

bayonet

FIGURE 3.7

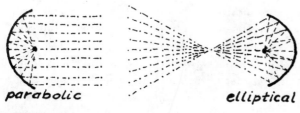

parabolic   elliptical

FIGURE 3.8

ER

FIGURE 3.10

LAMP SHAPE AND SIZE

A code composed of watts plus a letter-number is used to specify lamp size and shape. The letter designates shape, and the number is the lamp's largest diameter measured in eighths of an inch. For example, 50 watt R 20 describes a reflector lamp 2.5 inches in diameter (see Figure 3.7).

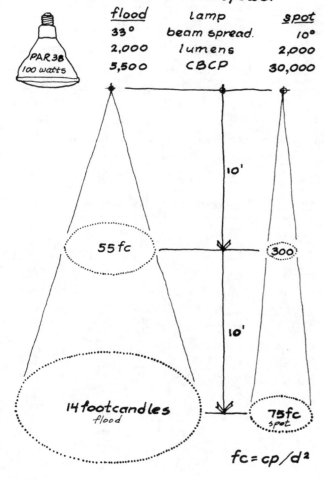

CBCP center beam candlepower

PAR 38
100 watts

flood	lamp	spot
33°	beam spread.	10°
2,000	lumens	2,000
5,500	CBCP	30,000

10'

55 fc

300

10'

14 footcandles
flood

75 fc
spot

$$fc = cp/d^2$$

FIGURE 3.9

Several lamp base designs are used to accommodate differing lamp sizes and operating watts. A medium screw base is used for most standard lamps from 15 to 300 watts. A larger mogul screw base is used for large lamps. Other base designs are used with special purpose lamps or fixtures.

Reflector lamps are designed to concentrate and aim light output. Their efficacy is lower than that of standard lamps, but reflector lamp performance is measured in beam candlepower and beam spread instead of lumens per watt. 100 watt spot and flood lamps produce equal lumens, but the spot's beam is much brighter and smaller (Figure 3.9). Reflector (R), multi-reflector (MR), and parabolic aluminized reflector (PAR) lamps are designed for spot (narrow beam) or flood (wide beam) lighting applications. R lamps are less expensive than PAR lamps, but PAR lamps offer better beam control and they can survive raindrops. Small MR lamps offer superior beam control for highlighting applications.

Elliptical reflector (ER) lamps are designed for use in pinhole fixtures because their light output converges outside the lamp (see Figures 3.8 and 3.10). Recently they have been touted as energy-conserving lamps, but incandescent lamps are a poor choice when energy conservation is a primary goal.

3.2 FLUORESCENT LAMPS

Thomas Edison entered a fluorescent lamp patent in 1896, but production of practical commercial fluorescent lamps only began in the mid-1930s.

Fluorescent lamps are arc discharge light sources. An electric current flows through mercury vapor inside an evacuated glass tube. The arc radiates energy, and this radiation excites fluorescent materials that convert arc radiation to visible light. Arc current in standard 4-foot lamps is about 430 mA (milli-amperes), less than half an ampere. A phosphor mixture coats the tube and sets light color. Flu-

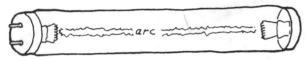

cathode ~ electron source for arc
vacuum ~ low mercury vapor pressure

phosphors 2-3 grams fluoresce when
excited by arc

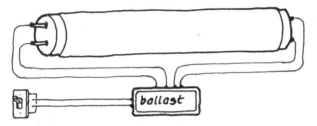

FIGURE 3.11

orescent lamps are controlled by a *ballast* that starts
the arc and limits arc current (see Figure 3.11).

EFFICACY

Fluorescent lamp-ballast efficacy ranges from 60 to
nearly 100 lumens per watt. New lamp output depre-
ciates before stabilizing after 100 hours of operation,
so "initial" lumens are measured after a 100 hour burn-
in. Efficacy is initial lumens divided by input watts.
Electronic ballasts increase efficacy by increasing the
electrical power supply frequency from 60 Hz to
25,000 Hz.

Ambient temperature affects efficacy. As temperature drops, mercury vapor condenses and lamps
flicker and dim. Outdoor winter operation of fluorescent lamps requires special low-temperature ballasts,
and glass lamp enclosures are also required in cold
climates.

LIFE

Compact fluorescent lamps have life expectancies
around 10,000 hours. Larger standard lamps have
rated lives near 20,000 hours.

Because fluorescent lamps are long-lived, their
light output is significantly reduced as they approach
failure. Manufacturers quote "mean lumens" at 40%
of life when light output has dropped about 15%
below the initial lumen rating of a new lamp. Lamp
life estimates assume 3 hours of operation per start.
If lamps operate 12 hours per start, their life ex-

pectancy increases by about 20%. Years ago a cun-
ning advertisement advocating 24-hour-a-day lighting touted: *"they're never turned off so they last longer."*
Thinking readers grasped the trick in this apparent
truth, but many less thoughtful designers embraced
24-hour lighting for buildings constructed in the
1950s and 1960s.

Recall that at rated life half of a group of lamps
have failed. Some building operators replace individual lamps when they fail, but others prefer group replacement at 70% of rated life when few failures have
occurred.

After 20,000+ hours lamp cathodes are worn,
tube ends are blackened, mercury may be contaminated, phosphors take less pleasure in arc stimulation, and lamps fail. Lamp disposal is an environmental concern because a 1995 vintage 4-foot lamp
contains about 23 milligrams of mercury. Histori-
cally, discarded lamps have accounted for 3.8%[*] of
the total mercury found in municipal solid waste.

CRI

As mentioned earlier, the CRI is a number index used
to compare the color of surfaces illuminated by a lamp
to the color of the same surfaces illuminated by incandescent light or daylight. A CRI value of 100 is assigned to daylight and incandescent light. Fluorescent
lamps with CRI values above 70 approximate daylight
or incandescent light (at the color temperature of the
lamp being tested).

Color Rendering Index

Source (Kelvin Temperature)	CRI
Incandescent (2,900 K)	100
CW fluorescent (4,400 K)	66
WW fluorescent (3,100 K)	55
WWX fluorescent (3,020 K)	77
Improved color fluorescent (3,000 to 4,000 K)	To 90
North sky daylight (7,500 K)	100

[*]National Electrical Manufacturers Association. "Environmental
Impact Analysis: Spent Mercury-Containing Lamps." Rosslyn, VA,
2000.

COLOR

Careful fluorescent lamp selection can create a lighting environment that complements any interior color scheme. Most fluorescent lamps emit light with higher color temperature than incandescent lamps—they produce more blue-green energy and less red-orange-yellow energy. However, a great variety of "white" fluorescent lamps are available; "triphosphor" improved color lamps use a mix of three phosphors to produce spikes of blue, green, and red spectral energy that imitate incandescent light.

OPERATION

Voltage

The operating voltage of a fluorescent lamp is controlled by a ballast. In large buildings, fluorescent lighting circuits are usually served at 277 volts (347 volts in Canada) to reduce wiring costs.

Dimming

Dimming fluorescent lamps is more expensive than dimming incandescent lighting. Extra dimming ballasts and/or electronic ballasts and special controls are required. An alternate approach is to switch half of the lamps in a fluorescent installation, permitting two illuminance choices.

LAMP TYPES AND BASES

Four fluorescent lamp types are used for most building lighting applications (see Figure 3.12).

Preheat lamps require starters and take a few seconds to light. Preheat applications include some compact lamps and residential under-cabinet fixtures. Preheat lamps have bipin bases. Older preheat fixtures had replaceable starters.

Instant start or "slimline" lamps light immediately. They have single-pin bases.

Rapid start lamps light after a slight delay; most use bipin bases.

High-output and *very-high-output* lamps have recessed contact bases. These lamps use increased arc current to produce more light, but they have lower efficacy than standard lamps.

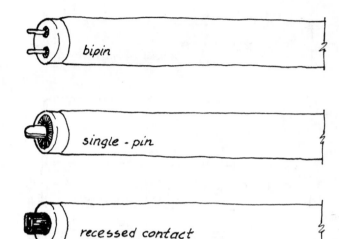

FIGURE 3.12

High-output lamps are rated at 800 mA and very-high-output lamps operate at 1,500 mA.

LAMP DESIGNATIONS

Compact fluorescent lamps use small tubes, but standard fluorescent lighting installations use 1.5-inch (T12) and 1-inch (T8) lamps. T indicates tubular shape, and the following number is the tube diameter in eighths of an inch. T8 lamps offer "better" color and higher efficacy than T12 lamps.

Fluorescent lamps are manufactured in a great variety of shapes and lengths, but 4-foot and 8-foot tubes offer more light per lamp dollar. Actual tube length is slightly less than the nominal dimension. Fluorescent fixtures accommodate 4-foot lamps in a total fixture length of 48 inches. Square fluorescent ceiling fixtures use bent "U" tubes or "compact" lamps instead of straight lamps.

Most fluorescent lamp designations begin with the letter F. Review the following generic examples to understand the code, but use *only* current manufacturers' catalogs for lamp selection. See also Figure 3.13.

F20T12/CW/RS is a 20 watt, 2' long, 1.5" diameter, cool white, rapid start lamp.

F40T12/WW/RS is a 40 watt, 4' long, 1.5" diameter, warm white, rapid start lamp.

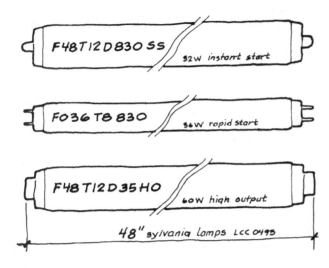

FIGURE 3.13

F40T12/WWX/U is a 40 watt, 1.5″ diameter, deluxe warm white, "U" shaped lamp. *This lamp fits a 2′ square fixture. It is a 4′ lamp bent into a "U" shape.*

Instant start (slimline) lamps use lamp length descriptors instead of watts.

F42T6/CW is a 42″ long, 3/4″ diameter, cool white (slimline) instant start, 25 watt lamp.

F48T12/W is a 48″ long, 1.5″ diameter, white (slimline) instant start, 40 watt lamp.

F96T12/CW is a 96″ long, 1.5″ diameter, cool white (slimline) instant start, 75 watt lamp.

F96T12/W/HO is a 96″ long, 1.5″ diameter, white (800 mA) high-output, 110 watt lamp.

F96T6/D/VHO is a 96″ long, 3/4″ diameter, "daylight" color (1,500 mA) very-high-output, 215 watt lamp.

Danger! The preceding descriptors are generic. The following four lamps from two recent manufacturers' catalogs illustrate variations.

FO96T8 is a 96″ long, 1″ diameter, "Octron" (265 mA) 59 watt lamp manufactured by Osram/Sylvania.

CF5DS/827 is a 1/2″ diameter, 4″ long, 5 watt, compact fluorescent lamp, with a CRI of 82 and a color temperature of 2,700 K, manufactured by Osram/Sylvania.

F40T12SP30/RS/WMP is a 48″ long, 1.5″ diameter, 3,000 K, rapid start, "Watt-Mizer Plus," 32 watt lamp, manufactured by General Electric.

F30T12SP41/RS/WM is a 36″ long, 1.5″ diameter, 4,100 K, rapid start, "Watt-Mizer," 25 watt lamp, manufactured by General Electric.

Older standard fluorescent lamps used about 10 watts per foot of length, but new lamp watts are best found in manufacturers' catalogs.

Newer energy-efficient fluorescent lamps are advertised for their color and watts saved, but the lamp codes *don't* indicate efficacy. The following examples for 48-inch GE lamps illustrate:

Description	CRI	Efficacy
F40T12CWX	89	53
F40T12CW/RS/WM	62	78
F32T8SPX30	84	92

Note: F32T8SPX30 above indicates watts for a 48″ lamp. This makes lamp identification even more challenging.

BALLASTS

Fluorescent and HID lamps must be connected through ballasts that start the arc and limit operating current (see Figure 3.14). Actual lighting watts are determined by the lamp-ballast combination.

An aged 40 watt lamp controlled by an obsolete magnetic ballast will use nearly 50 watts, but a new 32 watt lamp served by an electronic ballast can consume *less than* 32 watts. Three types of fluorescent ballasts—*preheat*, *rapid start*, and *instant start*—serve corresponding lamp types. Some T8 high-efficacy lamps operate on rapid start or instant start ballasts but instant start lamps and ballasts are most efficient.

Magnetic or Electronic?

Magnetic ballasts offer low initial cost but they are noisy and less efficient than electronic ballasts. Ballasts are rated A through F for noise output. An A

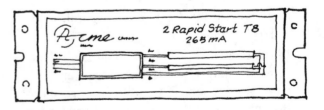

FIGURE 3.14

rated ballast is least noisy when two standard 4-foot lamps are served by each ballast. Longer lamps, or one lamp per ballast, will make more noise.

In renovation work, you may find older fixtures that use preheat lamps and ballasts. Some rapid start lamps will work on preheat ballasts, but matched lamp-ballast combinations will be more efficient.

Electronic ballasts are selected for many new lighting installations because they produce more light, less heat, and less noise than magnetic ballasts. They offer power factors approaching 1, dimming capability, and low harmonic distortion.* Manufacturers quote "ballast factor"—lamp lumen output as a percent of rated lumens compared to an ANSI "reference ballast."

SPECIAL FLUORESCENT LAMPS
Rapid Start or Instant Start

Rapid start (RS) fluorescent lamps dominate the market, but they start poorly in cold weather. Instant start (slimline) lamps are more expensive and have less life expectancy, but they start at lower ambient temperatures than RS and are frequently used in outdoor applications.

HO and VHO

Standard loading for a 4-foot fluorescent lamp is a bit less than half an ampere (see Figure 3.15). High-output (800 mA) and very-high-output (1,500 mA) rapid start lamps are available. HO/VHO lamps produce more light, but their efficacy is lower than standard lamps. When high illuminance is required, HO and VHO lamps offer lower first costs than standard lamps because fewer fixtures are required.

Compact

Compact fluorescent lamps' wattages range from 5 to 50 and lengths from 5 to 24 inches. Compact circular lamps have been available for many years, but recent compact lamps are small "U" shape tubes or clusters of tubes. The smaller (4 to 8 inches) compact lamps were developed to replace incandescents. Some include attached ballasts and screw bases to permit installation in incandescent sockets; others use detached ballasts (see Figure 3.16).

Compact fluorescents can provide real electrical cost savings, but they are not very effective in spot lighting applications. With life ratings up to 10,000 hours, these lamps can cut incandescent energy costs by as much as 70%.

Larger (up to 24 inches) compact fluorescent lamps rated up to 50 watts compete with standard and "U" lamps in 2-foot square fixtures.

Energy-Efficient Fluorescent Lamps

Manufacturers have developed a number of new lamp-ballast combinations that offer energy savings when compared to standard lamps and ballasts. Note that *both* lamps and ballasts must be used to realize the advertised savings. Installing an energy-efficient lamp in an old fixture with an antique ballast will not produce great savings (and some lamps will not light).

One efficient lamp-ballast combination uses triphosphor lamps with electronic ballasts to achieve efficacies as high as 100 lumens per watt. 48-inch lamps operate at 265 mA and offer good color rendering characteristics.

Cold Cathode

Cold cathode fluorescent lamps use an internal soft iron cylinder as a source of arc electrons. Although cold cathode lamps are less efficient than standard fluorescents, they last for many *years*. Cold cathode lamp tubes can be shaped to fit curves or angles, so they are often used in decorative light coves.

FIGURE 3.15

*Harmonic distortion caused by magnetic ballasts on three-phase lighting circuits can overload the common neutral conductor.

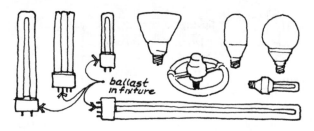

FIGURE 3.16

Neon

Cold cathodes are also used as an electron source in neon lamps. Neon emits an orange-red light when excited by a high-voltage arc. Neon lamps operate at high voltage with currents near 30 mA and consume about 7 watts per foot of length. Krypton and argon emit other colors. Phosphors or tinted glass are used to modify color.[*]

3.3 HIGH-INTENSITY DISCHARGE LAMPS

High-intensity discharge (HID) lamps are intense light sources that produce light by passing an electric arc through a conducting vapor. The arc radiates visible light, and most HID arcs operate at high temperature and pressure. Light color is determined by the elements used in the conducting vapor.

It takes time to start, heat, vaporize, and stabilize HID arcs, and warm-up times as long as 5 minutes are not unusual. Ceramic or quartz arc tubes are used to withstand temperatures that can exceed 1,000°C, and an outer glass bulb is designed to retain lamp materials in the event of a violent failure (see Figure 3.17).

All HID lamps require ballasts to strike the arc and to limit arc current. Magnetic ballasts add about 15% to lamp watts; electronic ballasts permit dimming, but they don't increase light output.

Most lamps will *not* light for some minutes after a power interruption because temperatures and pressures in the arc tube exceed ballast capacity. Restart time can vary from 1 to more than 10 minutes; supplementary lighting or special hot-start lamps are required when HID sources are used for sports or assembly lighting.

HID lamps have average lives as long as 40,000 hours. Manufacturers use a great variety of lamp descriptors: ANSI codes H for mercury, S for sodium, and M for multiple arc metals. Although other shapes are available, E (elliptical), B (bulged), and T (tubular) shapes are used extensively for these lamps (see Figure 3.18). Medium and mogul bases are typical, and available wattages range from 35 to more than 2,000. Arc tubes are designed to operate in a specific burning position. Tilting lamps reduce efficiency and life.

HID EFFICACY AND LIFE

Mercury and sodium HID lamps were developed in the 1930s and significantly improved by the mid-1960s. Sodium lamps are available in low-pressure (SOX) and high-pressure (HPS) versions. Vaporized metals carry the visible arc, mercury lamps produce green light, and sodium lamps yield yellow light. Metal halide lamps use mercury, sodium, scandium, and other halide conductors to produce "white" light.

Mercury

Clear mercury lamps produce green-blue light (5,710 K). "Deluxe" mercury lamps add a phosphor coating on the outer bulb to "whiten" light output.

Efficacy of mercury lamps ranges from 30 to 60 lumens per watt, starting time is 3 to 7 minutes, and life can exceed 24,000 hours with considerable decay as lamps age. Mercury sources are being replaced with more efficient metal halide or sodium lamps, but some landscape lighting designers prefer clear mercury to enhance foliage.

Sodium

Sodium lamps are very efficient sources of yellow-orange light. Low-pressure sodium looks slightly orange (1,740 K), and high-pressure sodium ap-

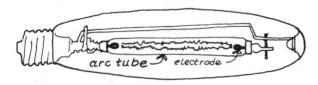

FIGURE 3.17

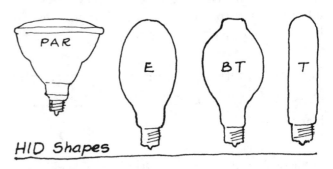

HID Shapes

FIGURE 3.18

[*]For more information, read "Neon: Argon to Xenon" by Michael Cohen, *Lighting Design and Application*, April 1995.

pears yellow (2,100 K) or "golden" in advertising descriptions.

Low-pressure sodium (SOX) lamps produce monochromatic light near the human eye's peak sensitivity. Initial efficacy can be as high as 180, but color rendering is dreadful. Lamp life ranges from 10,000 to 20,000 hours, and light output drops less than 10% as lamps age. Low-pressure sodium is often used for roadway lighting where economical operation is more important than color rendition.

High-pressure sodium lamps have initial efficacy values as high as 140 and life ratings from 16,000 to 40,000 hours. Light output decays about 20% as lamps age. Improved color lamps are available, but CRI increases come at the expense of lamp life and efficacy.

Metal Halide

Metal halide HID lamps use mixtures of mercury, sodium, thallium, scandium, and additional conducting materials in the arc tube to obtain "white" light output. Efficacy approaches 100 lumens per watt, and life ratings range from 10,000 to 20,000 hours. MH lamps are extensively used for sports lighting and increasingly in commercial and industrial applications where an intense light source is desired.

Color Rendering Index

Source (Kelvin Temperature)	CRI
Mercury (5,710 K)	15
"Improved" mercury (3,900 K)	To 50
MH (3,600 K)	To 70
Sodium—low psi (1,740 K)	—
Sodium—high psi (2,100 K)	22
"Improved" sodium (2,200 K)	To 70

Source: GE Lamp Catalog #9200/21.

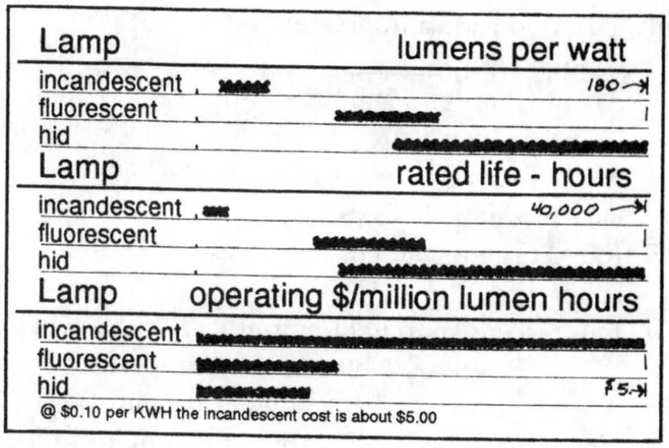

Lamp	lumens per watt
incandescent	
fluorescent	1
hid	

Lamp	rated life - hours
incandescent	40,000 →
fluorescent	1
hid	

Lamp	operating $/million lumen hours
incandescent	
fluorescent	1
hid	5 →

@ $0.10 per KWH the incandescent cost is about $5.00

FIGURE 3.19

COMPARISONS

See Figure 3.19 for comparisons between incandescent, fluorescent, and HID lamps.

OTHER LAMPS

The "E" lamp is an inductive lamp that uses radio waves to excite a light-emitting plasma. Development of this lamp has been well publicized. Currently, it is produced by a number of manufacturers.

Electroluminescence is a term for light emitted by certain materials when exposed to an alternating electric field. An example is the small night light that plugs into a duplex outlet. Attempts to develop large marketable electroluminescent sources have not been successful.

Lasers, light pipes, liquid crystals, and light-emitting diodes begin just the "L" *list* of other available lamps.

REVIEW QUESTIONS

1. Efficacy is measured in _____.
2. An ideal light source would produce about _____ lumens from each watt of input energy.
3. Low-pressure sodium HID lamps have efficacies as high as _____.
4. Which lamp type—incandescent, fluorescent, or HID—has the best CRI?
5. Rated lamp life is the number of operating hours until _____.
6. Are tungsten-halogen lamps ballasted?
7. An easy way to double the life of an incandescent lamp is _____.
8. Which fluorescent lamp-ballast type offers the highest efficacy?
9. What is the length and diameter of an F42T8 fluorescent lamp?

10. The arc inside a fluorescent lamp tube is conducted by _____.
11. Arc current for a standard 48-inch fluorescent lamp is about _____.
12. Name four advantages of electronic ballasts over magnetic ballasts.
13. Rated life for a standard fluorescent lamp is about _____.
14. Fluorescent lamp light output will drop about _____% after 8,000 hours of operation.
15. A spotlight is rated 100,000 beam candlepower. How many footcandles are in the center of the beam 100 feet distant?
16. A lamp designated H400E28 is _____.
17. How many watts should be allowed on the electrical circuit that serves the H400 lamp in question 16?
18. Which HID lamp offers the highest efficacy?
19. Which HID lamp offers the longest life?
20. Which HID lamp offers the highest CRI?
21. Why do large buildings serve fluorescent and HID lighting at 277 volts?
22. Name two functions of a ballast.

ANSWERS

1. lumens per watt
2. 680
3. 180
4. incandescent, CRI = 100
5. 50% of a representative lamp group fail
6. No. They are incandescent lamps.
7. reduce supply voltage slightly with a dimmer
8. instant start (slimline)
9. 42 inches long and 1-inch diameter
10. mercury vapor
11. half an ampere—actually 425 or 430 mA (energy-conserving lamps operate at 265 mA or less)
12. higher power factor, reduced harmonic distortion, higher efficacy, and less noise
13. 20,000 hours
14. 15
15. 10
16. a 400 watt, elliptical, mercury lamp, 3.5 inches in diameter
17. 460 watts—ballast adds 15%
18. low-pressure sodium (SOX)
19. high-pressure sodium
20. Metal halide. Manufacturers claim CRI values of 90 for "improved" MH and sodium lamps, but the sodium CRI is at a very low color temperature.
21. Higher voltage allows more lamps on each 20 ampere circuit. This reduces electrical system costs.
22. start the arc and limit arc current

CHAPTER 4
Lighting Design

Hobbes—How's your snow art progressing?

Calvin—I've moved into abstraction.

Hobbes—Ah!

Calvin—This piece is about the inadequacy of traditional imagery and symbols to convey meaning in today's world.
By abandoning representationalism, I'm free to express myself with pure form, specific interpretation gives way to a more visceral response.

Hobbes—I notice your oeuvre is monochromatic.

Calvin—Well C'mon, it's just snow.

Bill Watterson '94

Lighting design is a two-step process. Designers visualize desired brightness contrasts and patterns, and then select lamps and luminaires to fulfill their visual intentions.

This chapter begins with a design-oriented discussion of luminaires that illuminate and/or decorate. Read to develop convictions about brightness, shade, and shadow patterns that enhance rooms or spaces by emphasizing architectural features or by adding bright focal points.

Lighting shapes viewer perceptions, and the palette of artistic lighting design includes brightness contrast, pattern, and color.

Three luminaire types using concealed lamps are described as cone, band, or uniform light sources. Each luminaire is discussed emphasizing design possibilities, and opinions are offered concerning appropriate applications.

Decorative luminaires that illuminate and add visual interest are briefly discussed; the chapter ending review pages include design sketch templates.

4.0 SELECT LUMINAIRES

Luminaire is the proper word for a lighting assembly that connects one or more lamps to a power source and distributes their light. Luminaire components can include diffusers, reflectors, lenses, prisms, shields, louvers, airways, transformers, ballasts, and electrical wiring. *Fixture* is a less descriptive word, often substituted for luminaire in common usage, and occasionally in this text.

Because brightness contrast attracts viewer attention, lighting designers emphasize objects or surfaces when they select luminaires with concealed lamps. Selecting luminaires with visible lamps or diffusers emphasizes the luminaire.

Two simple observations are the basis of luminaire selection. Beyond providing comfortable lighting that accommodates activities expected to take place in the room or space, **luminaires are selected (1) to emphasize objects or surfaces**, *or* **(2) to provide decorative focal points in a space.**

Decorative luminaires illuminate surfaces, and concealed source luminaires draw viewer attention to bright surfaces near lamps. Talented lighting professionals select luminaires only after choosing decoration *or* surface illumination as the primary design intent.

A plaster sculpture is illustrated in diffuse and directional light (see Figures 4.1A and B). Designers prefer directional luminaires when shade and shadow patterns enhance the illuminated object.

4.1 ILLUMINATE OR DECORATE

A luminaire recessed in the ceiling can be an excellent choice when the design intent is to illuminate a beautiful table. A crystal chandelier can be an equally fine choice above the same table when design intent is an added focal point. Fixtures with concealed lamps emphasize architectural features, objects, openings, sur-

A

B

FIGURE 4.1

faces, and details. Decorative fixtures add focal points, variety, pattern, and interest.

Two dining room scenes compare a concealed source fixture recessed in the ceiling with a decorative chandelier (see Figures 4.2 and 4.3). Both provide adequate light for dining and conversation, and both can be examples of "good" lighting design.

Admirers of the concealed fixture claim that it better illuminates gourmet dishes, and its bright specular reflections in china, crystal, and silverware provide variety and interest. They allege that the soft light reflected from the table to the diners stimulates intimate dinner conversation.

Chandelier admirers believe that a stepped cone of flashing prisms is the perfect complement for a sumptuous table. They also claim that dinner conversation is more spirited when diners see each other covered with a lace of specular reflections.

Intimate conversation is usually associated with less light, and spirited conversation with bright light,

so a dimmer can mitigate contentions about conversation. Perceptive readers will agree, however, that each fixture creates a very different dining ambience.

4.2 DESIGNS

Design is the creative visualization that precedes fixture selection. Accomplished designers first visualize the space, surfaces, and objects to be illuminated. Then they invent and evaluate mental images of brightness contrast and patterns that can make the space visually inviting, comfortable, and interesting. Finally, designers choose from their mental images the brightness patterns that best enhance the space. Intelligent fixture selection happens only after designers mentally *invent*, *visualize*, and *evaluate* brightness patterns.

Comparative room sketches illustrate a variety of design approaches using fixtures with concealed lamps that illuminate objects or surfaces. Creative lighting designers visualize such scenes *before* selecting fixtures. As you study the sketches, think about the fixture type and location that could produce the brightness patterns shown; then review each sketch a second time and pick a favorite design. Finally, discuss the design elements that attracted your attention.

Figure 4.4A compares designs that "wash" walls with light. Illuminating a wall can make it seem more distant so one room should appear "longer" and the other "wider." The left scene increases apparent ceiling height by illuminating beams, and it adds contrast by spotlighting the vase and the tabletop. The right scene uses a dark window and tabletop to provide variety and contrast in the field of view.

Figure 4.4B is very similar, but it shows a two-way valence light source. Uplight coves make the ceiling appear higher, and downlight wall washers make the wall seem more distant. The dark valence face yields dramatic contrast. A lighter face or a perforated face will soften this effect. Installing or constructing valences or coves demands true edges and careful joining because the lamps highlight all faults.

Figure 4.4C paints walls with arches of light. The brightness patterns are more dramatic and can be more interesting than uniform surface luminance. Careful spacing is important when using brightness patterns. Fixtures too close together

FIGURE 4.2

FIGURE 4.3

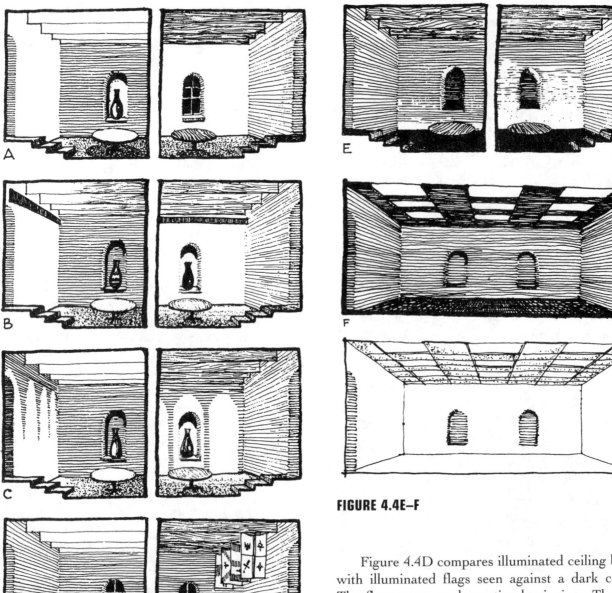

FIGURE 4.4A–D

FIGURE 4.4E–F

yield an expensive wall "wash," and fixtures too far apart can create trivial patterns. Manufacturers offer graphic data on arch shape and dimensions for specific lamp-fixture combinations.

A designer's palette just begins with bright arches on walls. The heart shape in Figure 4.5 is just a spotlight aimed into a corner. What sources and surfaces will produce the other shapes?

Figure 4.4D compares illuminated ceiling beams with illuminated flags seen against a dark ceiling. The flags serve as decorative luminaires. The light they reflect adds focal points and modifies room ambience. Visualize the fixture type, location, and mounting details that you would use to best illuminate the beams or the flags.

Figure 4.4E illustrates a low light source with dim and bright intensity. The low line of light suggests the horizon at sea; some designers claim a floating effect. Compare this figure with Figures 4.4A through 4.4D and notice how different the room appears when light is introduced below eye level.

Visualize a concealed vertical fluorescent in the corner of a room. Do you think you'll like the visual effect? If so, build and test one behind an opaque cove.

Figure 4.4F is included as an example of uniform lighting, an approach that spreads lumens equally

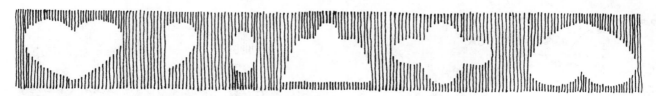

FIGURE 4.5

throughout a space. Why is uniform lighting used in many schools and offices, but in few homes, restaurants, or retail stores?

Proponents claim that uniform lighting provides an ideal environment for performing visual tasks, but critics contend it creates dull and boring spaces that fail to recognize the eye's range and the stimulation brightness contrast can offer. Figure 4.4F illustrates a typical uniform lighting scheme in large rooms with dark or light interior surfaces. Luminaires in the lower view include diffusers designed to minimize ceiling brightness contrast.

4.3 RECESSED DOWNLIGHTS

ARCHITECTURAL LIGHTING?

The next few pages emphasize fixtures with shielded lamps used to illuminate surfaces (see Figure 4.6). "Architectural lighting" is a possible descriptor for these built-in luminaires. Three fixture types provide *cones* of light, *bands* of light, or *uniform* lighting.

CONES OF LIGHT

Recessed can downlights can be effective luminaires when used to emphasize objects or surfaces. Manufacturers detail these fixtures to project a cone of illumination, and designers locate and aim them to accent or silhouette. Lamps or reflectors designed to produce a narrow cone provide dramatic brightness patterns that can enhance a painting, a plant, or a table. However, using these fixtures to illuminate large surfaces or an entire room is an aesthetic null that wastes money and energy.

Clearance

Recessed fixtures typically include a housing, support arms, and visible trim. If fluorescent or HID lamps are used, they'll also include a ballast. When installed in insulated ceilings, an extra housing is required.

Recessed can luminaires shield the lamp to minimize glare, so considerable space above the ceiling is required. Where space is limited, specify smaller cans with R-20 or MR-16 lamps.

Trims

A black step baffle cuts glare but reduces light output. Quality reflective cone trim allows you to specify an economical A lamp, but the housing requires more vertical clearance to conceal the source. The pinhole must use an ER lamp, so check the light distribution profile to determine the most effective mounting height.

The opal diffuser is not an effective accent light because the opal is a bright spot on the ceiling and it provides a wider (and less intense) cone of illumination.

Wall washers can increase visual interest where their brightness patterns complement arches in an arcade (see Figure 4.7). However, unless you sell fixtures for a living, there are better ways to illuminate wall surfaces. Compact fluorescent wall washers or fluorescent coves are an effective alternative where scallop patterns are not essential design features.

A single luminaire yields a more dramatic brightness contrast pattern than a group of closely spaced fixtures, and it will do a better job accenting an object or emphasizing a surface.

Framing spots use reflectors and a mask to outline a painting or sculpture with an exact brightness envelope. A wall washer or a small luminaire mounted on the picture frame can provide almost the same dramatic effect because the eye is drawn by brightness, but responds to the painting's contrast instead of brightness pattern (see Figure 4.8).

OPINIONS

Recessed cans are elements of quality lighting design when they are used to accent, emphasize, focus, or direct viewer attention to a specific object or surface.

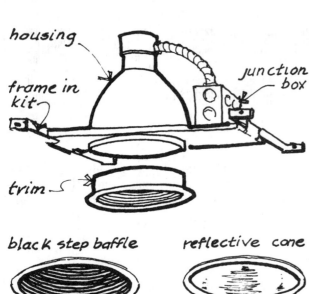

housing

frame in kit

junction box

trim

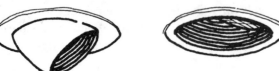

black step baffle

reflective cone

pinhole

opal diffuser

eyeball

adjustable

lens

ugg!

visible lamp

slot

wall washer

FIGURE 4.6

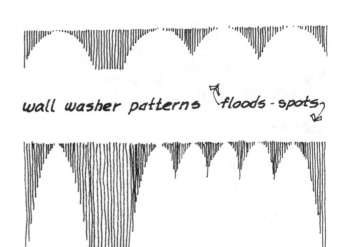

wall washer patterns   floods-spots

horizontal surface patterns

FIGURE 4.7

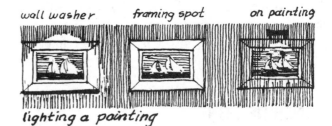

wall washer   framing spot   on painting

lighting a painting

FIGURE 4.8

They are a poor choice for illuminating large surfaces or areas.

Since recessed cans are not efficient area lights, don't specify them with compact fluorescent lamps. Compact fluorescents save energy, but they won't deliver a tight beam of accent light. If you need intense brightness contrast, try DC or HID lamps (see Figure 4.9). Translucent colored trims add a decorative edge to recessed cans, but this bright accent rim can compete with the illuminated object for viewer attention. Develop your opinions about such fixtures by careful study of installations you find appealing.

FIGURE 4.9

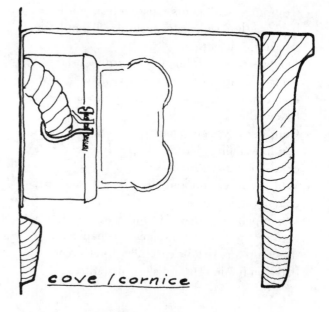

cove / cornice

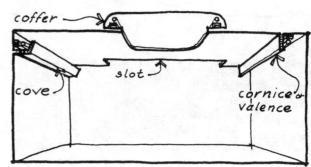

FIGURE 4.10

4.4 COVE LIGHTING

BANDS OF LIGHT

Coves are appropriate fixtures where the brightness patterns they produce make a room more interesting or inviting. Here *cove** refers to a linear luminaire with hidden lamps. Concealed lamps and a long axis make cove fixtures ideal for surface illumination. They can be used to increase apparent ceiling height, expand room proportions, or dramatize surface textures. Talented designers use coves to create focal planes, embellish architectural lines, define circulation paths, or enhance attractive surfaces or objects. The bands of brightness contrast they create can grace an appealing space or make a bland interior more inviting (see Figure 4.10).

The following are recommended **details** for built-in cove luminaires:

- Use economical 48-inch or 96-inch fluorescent strip fixtures. Specify electronic ballasts or "A" rated magnetic ballasts where quiet operation is important.

---

*A cove luminaire is a shielded uplight, mounted high on a wall, to illuminate upper wall and ceiling. Here *cove* describes any linear fixture with concealed lamps regardless of illuminance direction or fixture location. Coffer, valence, cornice, and slot luminaires are described as coves on these pages.

- Strip fixtures must be mounted on noncombustible surfaces. "Greenfield" (flexible conduit) is used between the box and the fixture to limit ballast noise.
- Mounting brackets and lamp sockets create minor shadow lines. Overlap lamp ends and revise bracket details if shadows are objectionable.
- Set cove dimensions so lamps and fixtures are *not* visible and check sight lines to be sure lamps are *not* mirrored in windows. Allow adequate clearance for re-lamping.
- In up and down coves, try a cool lamp on top and a warm lamp below. Also try switching top and bottom lamps separately. If an installation is too bright, paint the back of the cove black; if it is too dark, add a reflector. If the visible cove face looks too dark, consider perforations.

◆ Cove edges and joints must be true and level because the bright background will emphasize any tilt or defect.

◆ Ducts or structural members can be used to conceal lamps instead of coves *only* when careful erection, painting, and finishing procedures are followed.

◆ Consider coves suspended with wire or chain if you like sailing. They sway gently when the air conditioning system operates.

◆ Where curves and long life are required, specify cold cathode lamps.

◆ When planning a cove installation, check architectural, structural, and mechanical drawings and details carefully. Interference between luminaires, finishes, structure,

ductwork, and piping should be resolved before construction.

The eye "sees" bands of brightness and accommodates gradual brightness change. The sketches in Figure 4.11 illustrate some cove lighting effects.

Study the sketches in Figures 4.11A and 4.11B *before* reading on. Imagine people, furnishings, color, and activity in each space. Which lighting scheme would you prefer if the room was used as an office, a church, or a restaurant? Why?

When ceilings and walls are white, coves (uplight) increase apparent ceiling height. Cornice luminaries (downlight) appear to increase room size and lower the ceiling. What happens when dark finishes or strong advancing colors replace white surfaces?

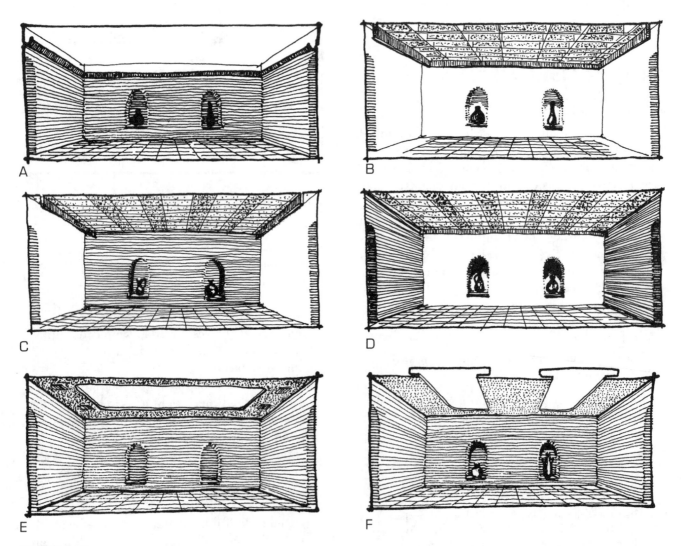

FIGURE 4.11

Cornice lighting on the side walls can make a room seem wider (Figure 4.11C); slot or cornice luminaires on the far wall increase apparent room length (Figure 4.11D). Would either lighting installation be better for church use than the two shown in Figures 4.11A and 4.11B?

Indirect lighting describes an installation with concealed lamps where general room illumination is first reflected by ceilings or walls. *Direct* lighting aims lamp light at objects or work surfaces.

The coffered ceiling shown in Figures 4.11E and 4.11F can increase apparent room height. Coffers are used extensively on cruise ships to make spaces with low ceilings appear more spacious.

Because only surfaces receive direct light, "architectural lighting" might describe the sketches in Figure 4.11. Look at each sketch again and develop your design opinions. Which lighting scheme would you prefer if the room was used as a restaurant, an office, a retail store, or a church? Why?

COVES AND DESIGN

Study each cove lighting illustration and then draw in your own lighting design for a restaurant, church, or office using the top view in Figure 4.12. Add furnishings, people, and at least *two* bright focal points illuminated with recessed cans. Experiment with color to add visual interest.

Surface brightness decreases as the distance from a cove increases, but the eye "sees" just a bright surface. Cove lighting can be especially effective when the illuminated surface is interesting. Interior designers use art, carving, color, and ornament to grace the ceiling surface.

Coves draw attention to the ceiling because brightness contrast attracts the eye. If the ceiling is just an unadorned white plane, it will seem higher when cove lighting is used, but lighting design can offer more than an increase in perceived ceiling height.

Talented lighting professionals view the ceiling plane as a design opportunity.* Cove lighting emphasizes the ceiling surface. Texture, line, or depth can make a ceiling more interesting. The illustrations

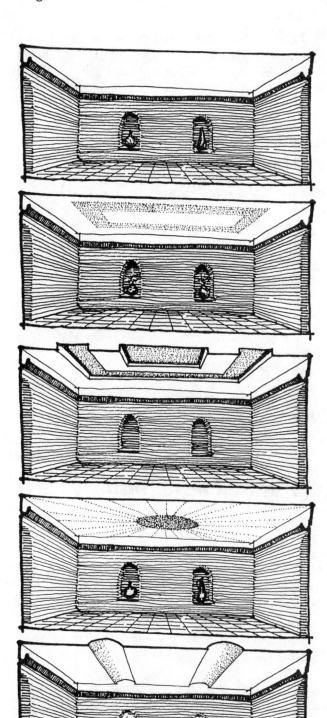

FIGURE 4.12

---

*When working with residential ceilings (sheet rock on wood joists), curved patterns or color accents are more successful than straight-line patterns because ceilings are rarely true and level.

in Figure 4.12 suggest brightness contrast effects using texture, shade, and shadow. The bottom sketch repeats the far wall arches and adds light to silhouette vases and flowers; it is the beginning of a three-dimensional lighting design.

Visual perceptions change as brightness, textures, shades, shadows, and colors change. Lighting is the essence of design. Study the illustrations again after completing the top view. Notice how your perception of the room changes as you consider each sketch. Categorize each view using the following words:

Strong

Weak

Interesting

Dull

Inviting

Attractive

Best

Then select three or four other adjectives that further describe the "best" illustration.

4.5 UNIFORM LIGHTING

Almost any group of luminaires can be spaced to obtain uniform lighting, but fluorescent troffers in a lay in ceiling grid are an economical option. In high-bay applications, HID lamps compete with fluorescents; incandescent sources are seldom used because of high operating costs. Illumination decreases near walls because they absorb light, so spacing luminaires for uniformity requires a centered symmetrical fixture layout (see Figure 4.13).

SPACING

S/MH (spacing to mounting height) is a number used to determine fixture spacing for uniform illumination (see Figure 4.14). Spacing is the center-to-center distance between small fixtures, or the edge-to-edge spacing for 24-inch fluorescents, and mounting height is the fixture-to-desktop distance (or the fixture-to-lane distance in a bowling alley).

Luminaire manufacturers provide S/MH values and photometric diagrams showing candlepower distribution for each fixture. Luminaires with wide light distribution patterns can be spaced further apart, saving fixtures and dollars, but wider light distribution means less shielding and more glare.

GLARE

Luminaires with wide distribution patterns can produce uncomfortable glare. Glare is defined as uncomfortable brightness contrast. A classic glare example is approaching auto headlights at night. Headlights don't produce glare in daylight, and lighting designers minimize glare problems by increasing background brightness, reducing source brightness, or shielding the source.

The eyebrow shields light sources above a 45-degree angle, so bright lamps overhead are not usually glare sources. As the fixture shielding angle decreases, fixture spacing increases, but glare problems are likely (see Figures 4.15 and 4.16).

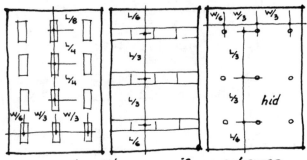

reflected ceilings · uniform schemes

FIGURE 4.13

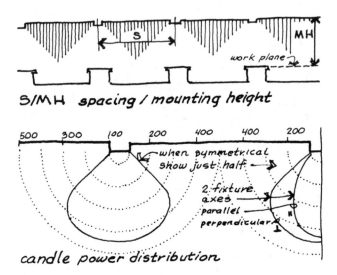

S/MH spacing / mounting height

candle power distribution

FIGURE 4.14

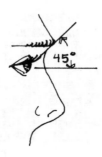

FIGURE 4.15

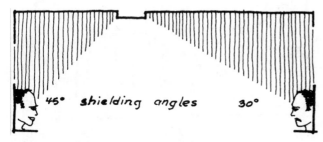

45° shielding angles 30°

FIGURE 4.16

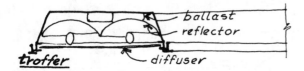

ballast
reflector
troffer
diffuser

FIGURE 4.17

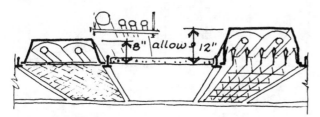

8" allow 12"

FIGURE 4.18

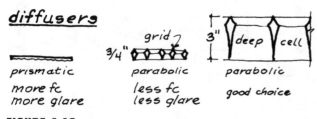

diffusers

prismatic
more fc
more glare

grid
3/4"
parabolic
less fc
less glare

3" deep cell
parabolic
good choice

FIGURE 4.19

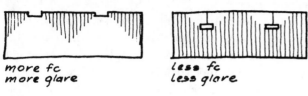

direct or indirect ?

more fc
more glare

less fc
less glare

FIGURE 4.20

Chapter 7 includes a glare index and a calculation method used to determine the number of fixtures required for a given illumination level.

FIXTURES

Suspended ceiling systems with lay in fluorescent troffers rule the school and office lighting market. Many sizes and shapes are available, but when initial cost drives selection, 24″ × 48″ troffers are usually the economical choice (see Figure 4.17).

Shallow troffers are desirable because the space above suspended ceilings is usually crowded with piping, conduit, and ducts. However, troffer depth increases when deep parabolic diffusers are used. Where fire rated ceilings are required, an insulated fixture enclosure takes more space (see Figure 4.18).

Troffer Options

Troffers include ballasts, reflectors, and diffusers; some troffers also have duct connections and dampers for air conditioning. Parabolic diffusers minimize light in the glare zone, and specular reflectors can maximize light output and reduce lighting costs (see Figure 4.19). Removing lamps and installing specular reflectors in existing fixtures _may_ be a good investment.

Other Luminaires

In high-bay or high ceiling retail installations, fixture glare is not usually a problem. HID fixtures or 8-foot fluorescent strips are used extensively because they can deliver uniform illumination economically. Fixture selection is usually based on estimated life cycle costs, and talented designers add brightness contrast with color and spotlights to stimulate sales. Indirect fixtures minimize glare, but direct fixtures deliver more illumination per watt of electrical input (see Figure 4.20).

CU

CU (coefficient of utilization) is an index of fixture efficiency. It is the percent of lamp lumens that reach the work plane. A fixture's CU varies with the size, proportions, and reflectance of the illuminated room (see Figure 4.21). Fixtures in a large room will have a higher CU than the same fixtures in a small room because walls absorb light. Large rooms have proportionally less wall area. White surfaces reflect more

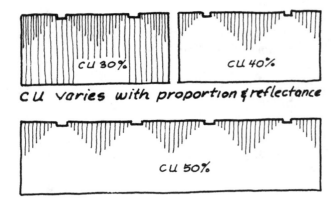

FIGURE 4.21

light than black surfaces, so fixture CU will be higher in a white room.

UNIFORM LIGHTING DESIGN

Many building lighting installations are planned and detailed to provide a constant level of interior illumination. Advocates of uniform lighting believe "good" lighting is assured when equal footcandles are provided in similar rooms and spaces. Uniform lighting "design" is a three-step process:

1. Determine an appropriate illuminance for the intended occupant activity.
2. Select fixtures and calculate the number required.
3. Space the fixtures for uniformity.

The resulting scheme is a grid or rows of evenly spaced luminaires; a calculation method is covered in Chapter 7.

The sketches in Figure 4.22 illustrate the worst uniform lighting installations. Rows of wraparound fixtures cause glare, and the luminous ceiling is uncomfortably bright. The sketches in Figure 4.23 show better uniform lighting using recessed or shielded fixtures to minimize glare.

OPINIONS

Critics describe uniform lighting schemes as boring at best and wasteful at worst. Uniform schemes are boring because they minimize the brightness patterns and contrasts that people rely on for visual reference and information. The fixture rows paint a meaningless, repetitive glare pattern on the ceiling that degrades visual comfort. Uniform schemes waste electrical energy by spreading light like fertilizer on a lawn, instead

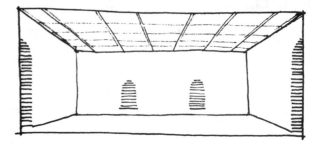

FIGURE 4.22

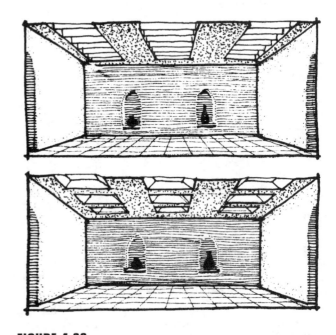

FIGURE 4.23

aiming *less* illumination at specific objects, surfaces, or activity areas.

Interior designers visualize materials, fabrics, and color selections in a given room or space. Talented lighting professionals complement interior colors, textures, and surfaces with light. They use brightness contrast to define and grace attractive objects, details, surfaces, and activity areas. Because uniform schemes minimize brightness patterns and contrasts, they are used infrequently, and only in conjunction with brighter accent areas or surfaces that create visual interest.

Danger! Fixture rows in uniform lighting schemes should reinforce the structural grid. *Don't* lay out fluorescent fixtures in patterns that cross or conflict with the pattern of beams, columns, and walls that define a room.

4.6 DECORATIVE LUMINAIRES

ILLUMINATE AND DECORATE?

Richard Kelly's trinity of lighting-driven perceptions given in Chapter 1 is repeated as an introduction to decorative luminaires.

Focal glow is the campfire of all time, . . . the sunburst through the clouds, and the shaft of sunshine that warms the far end of the valley. Focal glow commands attention and interest. It fixes the gaze, concentrates the mind and tells people what to look at. It separates the important from the unimportant.

Ambient luminescence is a snowy morning in open country. It is underwater in the sunshine, or inside a white tent at high noon. Ambient luminescence minimizes the importance of all things

and all people. It fills people with a sense of freedom of space and can suggest infinity.

Play of the brilliants is the aurora borealis, . . . the Versailles hall of mirrors with its thousands of candle flames. Play of the brilliants is Times Square at night, . . . the magic of a Christmas tree, Fourth of July skyrockets. It quickens the appetite and heightens all sensation. It can be distracting or entertaining.

Artistic lighting begins when designers visualize the brightness patterns that make rooms or objects attractive and interesting. The sketches in Figure 4.24 picture two elegant rooms enhanced and graced by good lighting. Illumination in the domed room comes from shielded lamps that make ceiling surfaces the brightest elements in the field of view. The ornate ceiling acts as a luminaire reflecting and diffusing light throughout the space.

In the rectangular room, a central luminaire provides a bright decorative focal point *and* illuminates the ceiling. Both rooms are ideal settings for effective lighting because of the brightness, shade, and shadow patterns. Light adds to their interesting surfaces and rich architectural details.

Lighting design is easy when illuminated surfaces are interesting. These sketches show rooms with elaborate surfaces where any illumination will create a variety of attractive luminance contrasts. The rectangular room with its decorative luminaire

FIGURE 4.24

43

and the domed room with its indirect fixtures are both examples of good lighting design.

Less exciting rooms demand more from lighting designers who must add brightness patterns, decorative fixtures, and task lighting to create visual interest.

"Decorative" is a poor descriptor for the visual beauty an elegant luminaire can create. Imagine the colors, patterns, and contrasts in a sunset and compare them to the experience of viewing a fine stained glass lamp (see Figure 4.25). The scale varies, but the visual sensations are nearly identical. Windows, sunsets, candles, prisms, lamps, reflectors, sunrise, daylight, and stained glass—all are potential luminaires, and all are potentially beautiful.

The preceding pages emphasized "architectural" lighting. Fixtures with concealed lamps illuminated large surfaces, and the visual results were as attractive and interesting as the illuminated architecture. Decorative luminaires use exposed lamps, reflectors, diffusers, shapes, and color in the field of view to illuminate and *beautify* (see Figure 4.26). Such fixtures add focal points and brightness patterns; they can grace elegant rooms or make ordinary rooms more attractive. Most successful lighting designs include both shielded and decorative luminaires.

Selecting the best fixture for a specific room is a talent built on experience. Artistic designers study lighting installations like professional golfers study courses. They continually create and save memory images of attractive lighting scenes and the fixtures that help create such scenes. Brightness contrast, pattern, color, and motion are design tools; but memory images of lighting effects are a designer's reference library.

The source of a memory image affects its quality; the best images are created by observation of an actual lighting installation. Lesser images may be taken from a lighting showroom, a manufacturer's catalog, and this achromatic text.

Build your memory library of lighting images by continual observation as you shop, dine, travel, and recreate.

add color

FIGURE 4.25

FIGURE 4.26

luminaires by Spring City Electrical

some look better with lamps off ~ all look good with dim lamps.

REVIEW QUESTIONS AND TEMPLATES

1. Name two possible design intentions that can aid luminaire selection.

2. A lighting designer who wants to emphasize a crystal vase of flowers will consider a(n) _____ fixture.

3. A lighting designer who wants to enhance an ornate ceiling will consider a(n) _____ fixture.

A

B

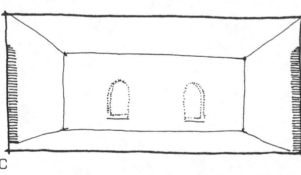

C

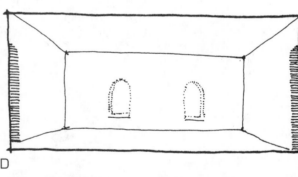

D

E

F

FIGURE 4.27

4. Template sketches are repeated in Figures 4.27A through 4.27F to encourage illustration.* Experiment with pencils, pens, or markers to add accents, shade, shadow, and color.

---

*The black and white illustrations in Figure 4.27 are only a beginning for lighting study. Serious students will enjoy *Detailing Light* by Jean Gorman, © 1995, Whitney Library of Design, New York; and *Perception and Lighting as Formgivers for Architecture* by William Lam, © 1977, McGraw-Hill, New York.

ANSWERS

1. illuminate surfaces and objects *or* create decorative focal points (and illumination)
2. adjustable recessed can with a narrow beam spot lamp
3. cove

CHAPTER 5
Sunlight and Daylight

Architecture is the masterly, correct and magnificent play of forms of light.

Le Corbusier

Sunlighting designs trust the same tools used for lamplighting. People respond to:

- brightness contrast,
- form and pattern,
- color, and
- motion

regardless of the light source.

Sunlight and lamplight designs seek identical goals and use identical techniques. The sun replaces lamps. Windows or apertures serve as luminaires. Sunlighting designs use light to create decorative focal points or to illuminate objects and surfaces.

Architects consider views, site features, and sun angles before locating windows that decorate and illuminate interior spaces. Talented architects create exciting interior patterns and contrasts by exploiting sunlight intensity and its changing angle of incidence. Lesser architects measure and estimate sunlight to model potential lamplight cost savings. Accomplished architects and designers detail sunlighting apertures to minimize summer heat gain and maximize winter heat gain.

5.0 **SUNLIGHT OR DAYLIGHT?**

Sunlight design is emphasized on the following pages. Daylight designs are appropriate in climates where the usual sky condition is overcast. Sunlight offers a moving, directional source five to ten times more intense than daylight (see Figure 5.1). Designs that exploit sunlight are effective in overcast climates, but designs based on daylight will overheat in warm sunny climates.

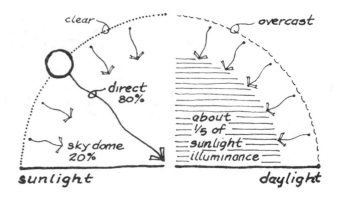

FIGURE 5.1

SUNLIGHT DESIGNS

The eye cannot distinguish sunlight from lamplight, but designers using sunlight for indoor illumination exploit the sun's changing intensity and angle of incidence. Because most light is invisible until reflected, the two principles of luminaire selection are appropriate for sunlighting designs. **Use sunlight to (1) create decorative focal points,** *or* **(2) illuminate and emphasize objects or surfaces** (see Figure 5.2).

This choice is an important starting point for each sunlighting aperture because it shapes occupant perceptions and suggests details for windows or clerestories. Designing and detailing sunlight apertures is more complex than selecting light fixtures because more variables are involved. Size, shape, proportions, location, light control, glazing, mullion pattern, view, and orientation are only the beginning of an extensive list of design decisions.

Focal point apertures include stained glass windows or windows that capture beautiful views. Apertures used to illuminate surfaces or objects include skylights, clerestories, and light shelves (see Figure 5.3). Brightness contrast is the primary tool of lighting design. Bright views, bright windows, and bright surfaces command visual attention.

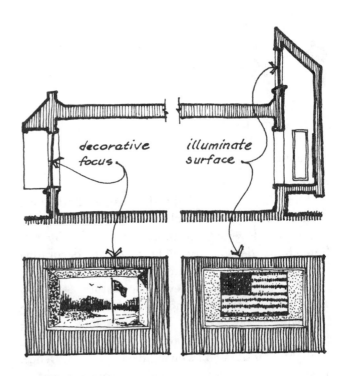

FIGURE 5.2

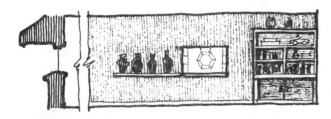

FIGURE 5.3

5.1 FOCAL POINT SUNLIGHTING

A design intent to use sunlight as a focal point *or* to illuminate surfaces helps locate, shape, and size windows. Intent is not an absolute because all sunlight will illuminate surfaces and bright surfaces command attention, but choosing a primary purpose for each sunlighting aperture is the first step in a creative design effort.

Beveled, etched, or stained glass windows are classic examples of focal point sunlighting (see Figure 5.4). Window pattern and color are beyond

the scope of this text, but window size and orientation are not. In homes, *small* openings are appropriate, and orientation should be suited to the occupants' schedule. A decorative bedroom window will probably face east for morning brightness, but a living room window may face west if entertaining hours fall near sunset.

In churches, architects select larger stained glass areas to suit sanctuary size. Window location will be chosen carefully so that colors and brightness contrast complement (but don't overwhelm) services. Stained glass facing southeast will ensure maximum brightness contrast during morning services, and it may even improve a dull Sunday sermon!

View windows are interior focal points (see Figure 5.5). They command visual attention like stained glass when an outdoor scene is bright and stimulating. Well-designed view windows frame scenes and can grace landscape with mullion pattern (see Figure 5.6). Talented architects consider exterior appearance, landscape, and interior light when they shape, locate, size, and detail view windows.

Because inviting views are found at all compass orientations, exterior shading that excludes direct summer sun is an important design consideration.

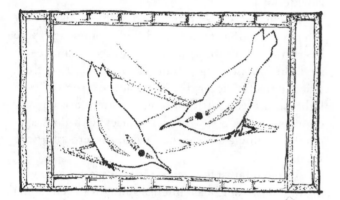

FIGURE 5.4

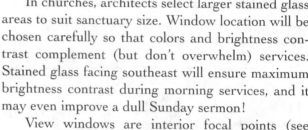

FIGURE 5.6

FIGURE 5.5

recessed apertures frame views _and_ illuminate surfaces.

day ↑ night ↓

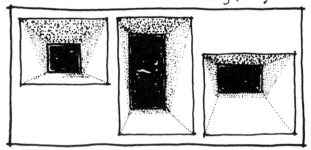

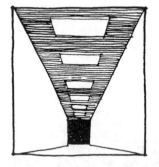 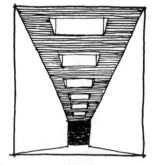

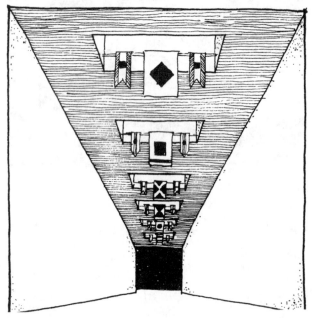

FIGURE 5.7

5.2 SUNLIGHTING SURFACES

When an exterior view is unpleasant or when privacy is desired, sunlighting designs configure openings to illuminate interior surfaces. A most important design consideration for successful surface lighting designs is that _the surface to be illuminated should be interesting._

The Pantheon in Rome is a classic surface-lighting design. On clear days sunlight passing through the dome aperture paints the coffered dome and rotunda walls with a moving bright ellipse. On cloudy days diffuse daylight softens interior details.

Skylights, clerestories, and light wells do _not_ offer interesting views, and light entering these apertures is invisible until it illuminates a surface. A skylight illuminating a room is as boring as a fluorescent troffer doing the same job. A row of skylights and a row of fluorescent troffers are equally depressing, but a row of clerestories with flags suspended below each can be visually interesting (see Figure 5.7).

Visual interest is essential for successful sunlighting of interior surfaces. Intense sunlight (or daylight) allows designers to use illuminated surfaces as focal planes. An aperture designed to wash a wall with sunlight can be enhanced by changing the wall texture or color. A tapestry, a plant, or a piece of pottery displayed on the bright wall will make the room more interesting and inviting.

Surface sunlighting effects can be imitated using lamplight. Architects detail lamps in skylight and clerestory apertures to minimize a "black hole" appearance at night (see Figure 5.8).

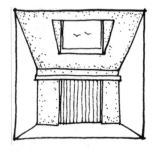 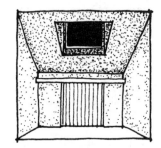

FIGURE 5.8

50

5.3 ORIENTATION

The advantages and disadvantages of selected sunlight orientations between 26 and 50 degrees north latitude follow.

North. Most uniform light intensity during daylight hours, but north-facing windows lose heat all winter. In cold climates, north-facing windows are a poor choice.

East. Great sunrise views, but overheating is likely on summer mornings. Heat gain can be limited with well-placed deciduous trees or vertical exterior shading devices. Interior blinds will also reduce summer heat gain; but with interior blinds closed for most summer mornings, light and view are lost.

South. More light and heat in winter, less light and heat in summer—*good!*

South is the best orientation for heat control. Overhangs can make south-facing windows even more effective by excluding most summer heat without limiting winter heat gain. South-facing windows can provide "free heat" during winter months, and south-facing windows with double glass and insulated night curtains can save heating dollars.

West. Great for watching sunsets but will cause afternoon overheating in summer. Excess summer heat can be controlled with well-placed deciduous trees or vertical exterior shading devices. Interior blinds will also reduce summer heat gain; but with interior blinds closed for most summer afternoons, light and view are lost.

Clerestories. Clerestories and skylights don't offer interesting views, so heat control is a primary design consideration. Clerestory windows facing south ensure easy control of summer heat and maximize winter heat and light. South-facing clerestories with night insulation can cut winter heating costs.

Skylights. Horizontal or sloping roof openings are poor sunlighting apertures because they magnify summer heat gain and maximize winter heat loss. Thoughtful architects use clerestories instead of skylights.

CONTROL HEAT

Because direct sun is a blessing in winter and a curse in summer, effective sunlighting designs *invite heat in winter and exclude it in summer*. Such heat control is easy with south-facing windows in the northern hemisphere. On a clear winter day, south-facing windows receive three times the solar heat as east or west windows, and on a sunny summer afternoon, south-facing glass will gain only one-third as much heat as west-facing glass.

Windows and clerestories combined with overhangs, reflectors, louvers, diffusers, and special glazing are the stuff of heat control. Their design is not difficult, requiring only an intent to control heat as sunlight is used for surface illumination *or* to create bright focal points. Direct winter sunlight will make interior surfaces more inviting on a cold day and add "free" heat. Summer sun can be used to create *small* bright interior focal points without overheating. Carefully detailed windows will accommodate summer views as they exclude direct sun.

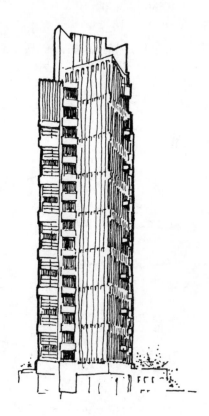

FIGURE 5.9
Frank Lloyd Wright Price Tower Bartlesville, OK 1953

Buildings designed for effective external sun control use different shading forms and details at each compass orientation (see Figure 5.9). Well-designed sun-control elements catch summer heat outside the building without blocking views. Interior blinds can reduce building heat gains, but they limit views.

5.4 SUNLIGHT CONTROL

The size, shape, orientation, and shading of a sunlight aperture should be based on a design intent to use the opening as a focal point or to illuminate interior objects and surfaces. The sun charts and the section angle overlay (in back of text) will help you plan sunlight control details and interior brightness patterns. Study the examples and then solve the review problems on page 62.

Using only the 30°N sun chart, find the time and location of sunrise on 21 April.

time = 5:30 AM
location = 15° north of east

Find the sun's bearing (compass direction from observer) at 2 PM, 21 September.

bearing = 50° west of south

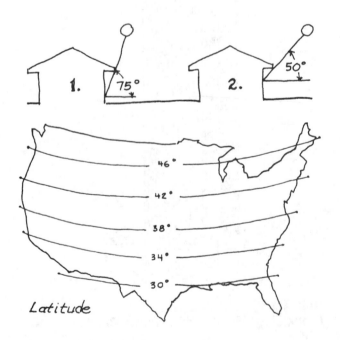

FIGURE 5.10

Using the section angle overlay, pin through the center marks on the overlay and the 30°N sun chart. Set the window faces arrow by rotating the overlay. Read the red section angle directly above the time-date point on the sun chart.

Find the section angle at 3 PM, 21 August, for a window facing south.

answer = 75°

Find the section angle at 3 PM, 21 August, for a window facing 35° west of south.

answer = 50°

See Figure 5.10.

EXAMPLES

Draw the shadow cast by a pole projecting from a wall. Figure 5.11 shows pole and wall dimensions. All examples use the 30°N sun chart. Use an adjustable triangle or a protractor.

Example 1

Wall faces south. Draw the pole's shadows at noon, 21 March; and 3 PM, 22 December. Study the following solution and verify all angles used to locate the shadows. Then overlay the solution example with tracing paper and draw shadows for examples 2, 3, 4, and 5 before looking at the answers in Figure 5.12.

Example 1 Solution (Figure 5.11)

◆ At noon the sun bears due south: 0°.
◆ Draw a line through 3 PM, 22 December, and read the sun's bearing: 45° west of south.

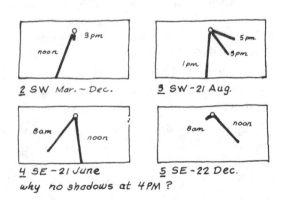

FIGURE 5.12

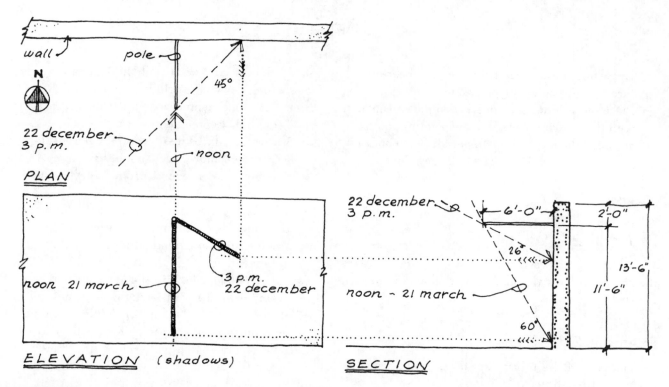

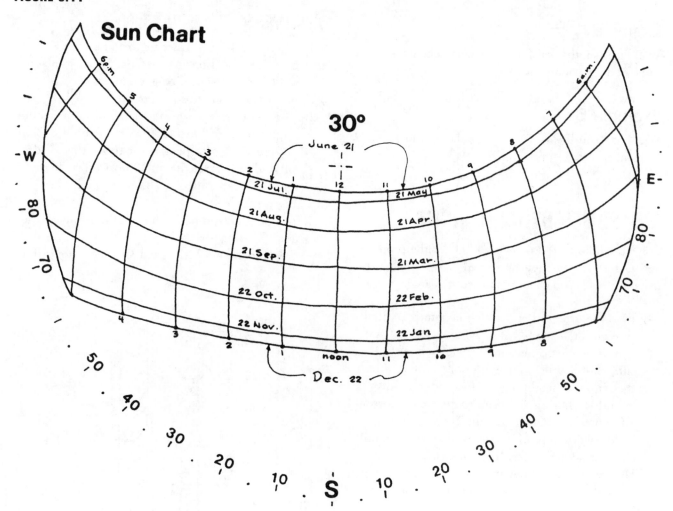

FIGURE 5.11

Sun Chart

30°

- Draw lines on the *plan* from the sun's bearing through the end of the pole to the wall.
- Place the section angle overlay on the sun chart with the "window faces" arrow pointing south and read the red section angle values. Noon, 21 March: 60°; and 3 PM, 22 December: 26°.
- Plot the section angles on the *section* and project.

Example 2

Wall faces south-west. (45° west of south). Draw the pole's shadows at noon, 21 March; and 3 PM, 22 December.

Example 3

Wall faces south-west. Draw the pole's shadows on 21 August at 1 PM, 3 PM, and 5 PM.

Example 4

Wall faces south-east. Draw the pole's shadows on 21 June at 8 AM, noon, and 4 PM.

Example 5

Wall faces south-east. Draw the pole's shadows on 22 December at 8 AM, noon, and 4 PM.

5.5 SUNLIGHTING DESIGNS

Solve the following problems and check the proposed answers. Size the apertures and shading elements using the 30°N sun chart and the section angle overlay.

1. Find the orientation that will maximize annual sunlight on a stained glass window at 10 AM.
2. A new luncheon restaurant wants winter sunlight on a decorative interior fountain. Find the required height of a vertical slot window intended to bring noon sun to the fountain from 1 October through 1 March. The window will be located in a south-facing wall. The fountain is 4 feet tall and 8 feet from the wall.
3. How wide should the window in the preceding example be to ensure sunlight on the fountain from noon until 1 PM?

4. A window faces west (ugh!). Detail an exterior shading device to exclude summer sun after 4 PM. Assume summer extends from 1 May through 30 September.
5. A clerestory faces west (why?). Set the sill height to exclude direct summer sun after 4 PM. Assume summer extends from 1 May through 30 September.
6. A clerestory faces southwest. Set the sill height to exclude direct summer sun after 4 PM. Assume summer extends from 1 May through 30 September.
7. A horizontal trellis is to shade a courtyard located behind a west-facing wall from 4 to 7 PM, between 21 May and 21 September. Find the center-to-center spacing for 2×6 wood trellis members that will exclude direct sunlight during this time.

ANSWERS

1. Study the 10 AM bearings through the year. Sun chart bearings range from 30° east of south to 80° east of south. Select a midrange orientation: 55° *east of south*.
2. The red overlay shows a range of noon section angles from 35° in December, to about 55° on 1 October. Use a protractor to set the window sill and head. Set the sill almost 6 feet high and the head at 16 feet 0 inches high (see Figure 5.13).
3. At noon the sun bears due south. Compare the sun's bearing at 1 PM on 1 October, 22 December, and 1 March. 1 October shows the

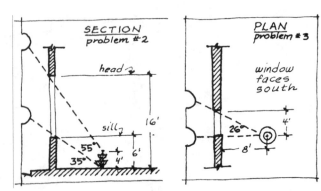

FIGURE 5.13

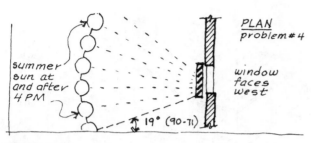

FIGURE 5.14

greatest bearing change—about 26° west of south. Use a protractor on the plan view to set the window width at 4 feet.

4. The sun is low after 4 PM so horizontal shades will not be effective. Detail vertical shades to exclude summer sun and allow winter sun. On 30 September at 4 PM the sun bears about 71° west of south. This is the most southerly

bearing during the summer months. Detail the vertical shades to exclude all sun bearings exceeding 71° west of south. Use a protractor on a plan view of the window to aim and space vertical shades (see Figure 5.14).

5. The highest section angle is 35° at 4 PM on 21 June. Use a protractor to set the sill height (see Figure 5.15).

6. The highest section angle is 50° at 4 PM on 21 June. Use a protractor to set the sill height (see Figure 5.16).

7. The highest section angle is 35° at 4 PM on 21 June. Space members 8 inches apart (9.5" centers). Remember: a 2 × 6 is actually 1.5" × 5.5".

Solar times used on the sun chart must be adjusted for location and daylight savings time.

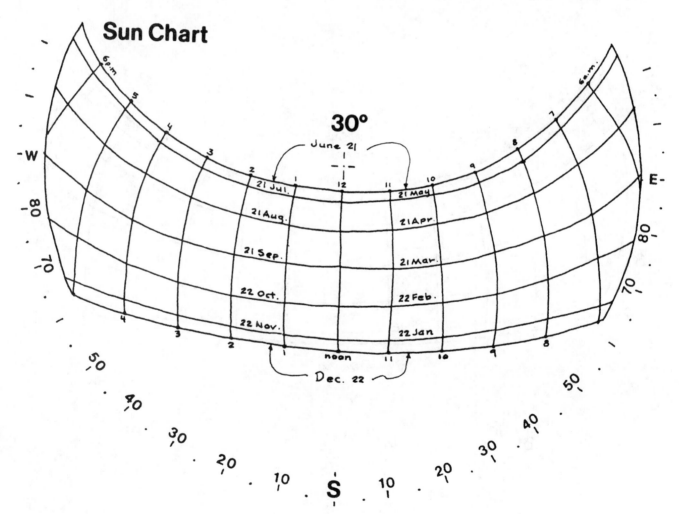

Developed with permission from material copyrighted by LIBBEY-OWENS-FORD CO.

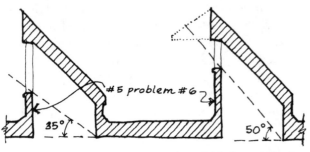

a thoughtful architect
would increase head
overhang (dotted) #6
to increase winter light.

FIGURE 5.15

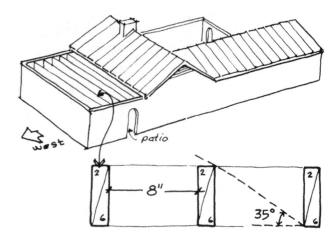

FIGURE 5.16

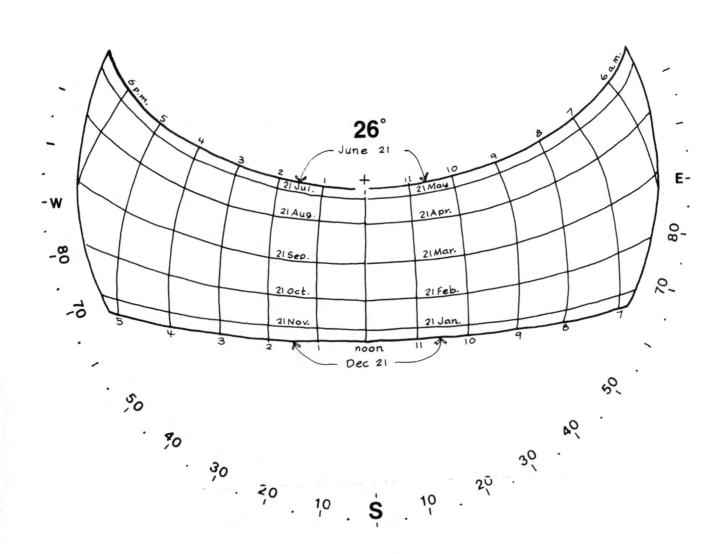

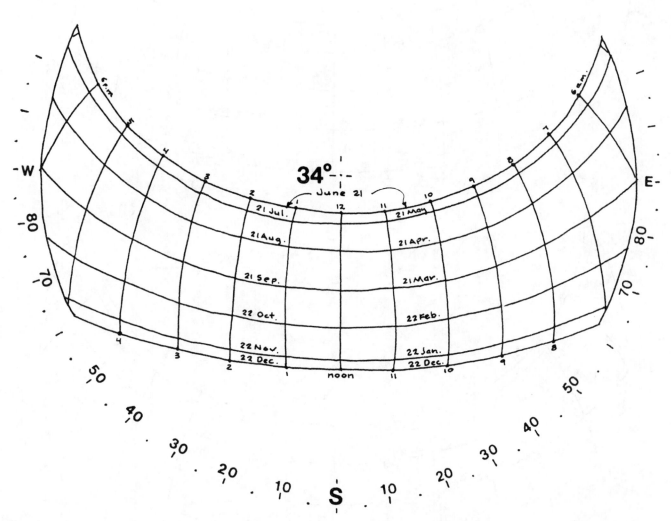

Developed with permission from material copyrighted by LIBBEY-OWENS-FORD CO.

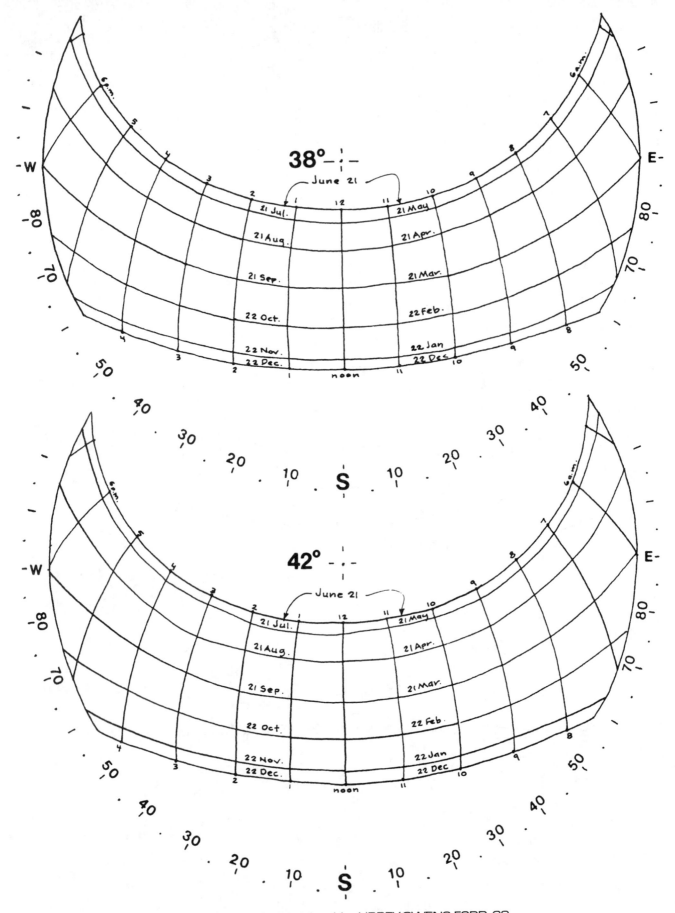

Developed with permission from material copyrighted by LIBBEY-OWENS-FORD CO.

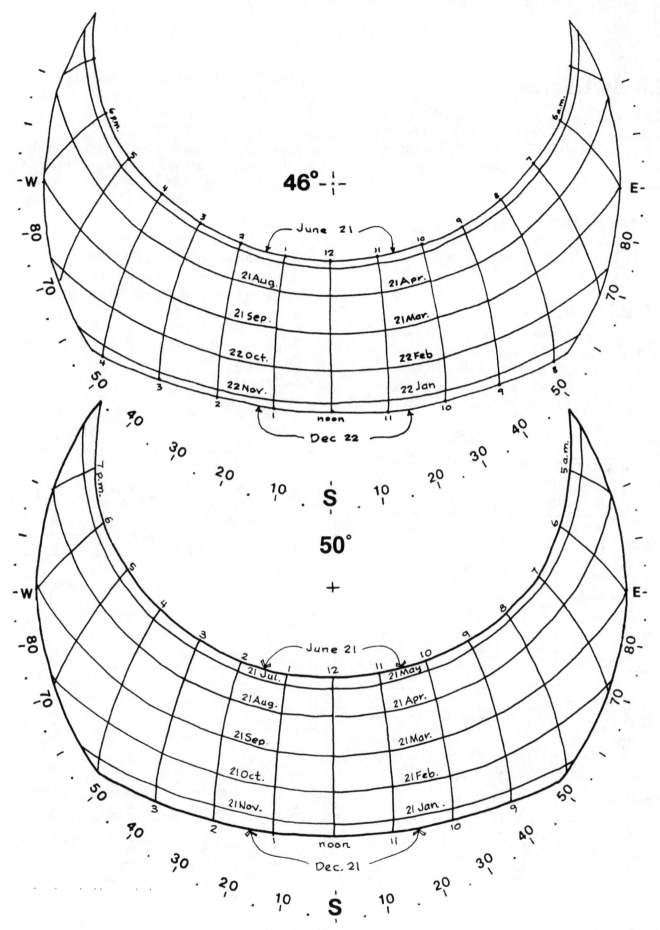

46° - | -

June 21

21 Aug.   21 Apr.

21 Sep.   21 Mar.

22 Oct.   22 Feb

22 Nov.   22 Jan

noon

Dec 22

S

50°

+

June 21

21 Jul.   21 May

21 Aug.   21 Apr.

21 Sep.   21 Mar.

21 Oct.   21 Feb.

21 Nov.   21 Jan.

noon

Dec. 21

S

5.6 DAYLIGHTING

Daylighting considerations shaped industrial America in the early 1900s when sawtooth clerestories illuminated immense factories. Direct sunlight is too intense for most activities where the viewer's eye is adapted to interior lighting levels, so daylighting designs use reflected sunlight or diffuse skylight to replace interior lamplight (see Figure 5.17).

Outdoor clear day illumination can exceed 11,000 footcandles (2,000 from the sky and 9,000 from direct sunlight).

DAYLIGHT FACTOR

A daylight factor (DF) is an estimate of the percent of outdoor horizontal illumination available at an indoor location. A DF of 7 means 7% of the outdoor horizontal illumination is available at the location noted. DFs are calculated by dividing indoor footcandles by out-

door footcandles. Many factors affect daylight penetration into buildings, and the DFs illustrated are maximum values based on ideal conditions (see Figures 5.18 and 5.19).

Some lighting designers enjoy calculating daylight illumination levels in buildings and projecting utility cost savings for replaced lamplight. Many variables are considered in these calculations, including:

- **Climate.** Clear or overcast? number of annual sunshine hours?
- **Illumination.** Sky components? direct component? reflected components?
- **Aperture orientation.** Glass transmission? adjacent buildings or trees? reflectance of surrounding surfaces?
- **Proportions.** Room sizes and heights? aperture sizes and locations?
- **Reflectance.** Interior surfaces? screens? light shelves?

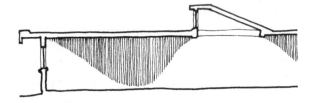

FIGURE 5.17

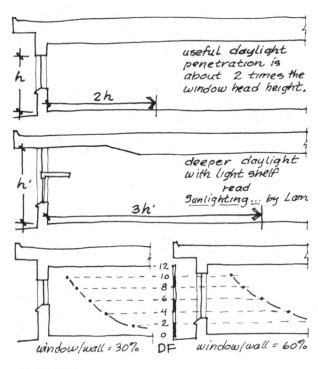

FIGURE 5.18

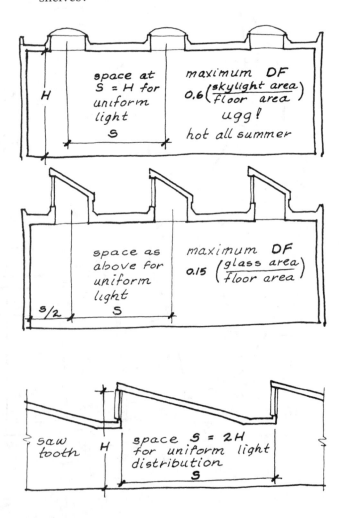

FIGURE 5.19

Readers who enjoy complex estimates can find detailed calculation methods in CIE (Commission Internationale de l'Eclairage) and IES (Illuminating Engineering Society of North America) daylighting publications.

5.7 GLAZING AND VIEWS

GLASS

Glass transmission of heat and light can be quantified with three factors:

- ◆ **U** value—conducted heat transmission
- ◆ **SC**—shading coefficient, radiant heat transmission
- ◆ **VT**—visible transmission (light)

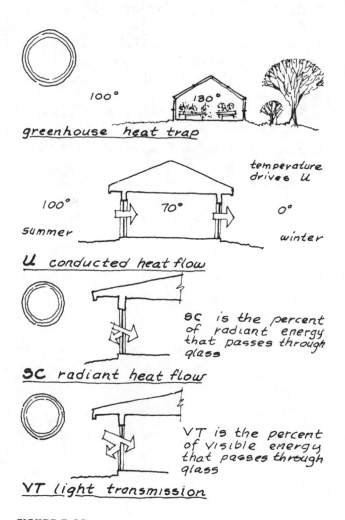

greenhouse heat trap

temperature drives U

100°   70°   0°

summer   winter

U conducted heat flow

SC is the percent of radiant energy that passes through glass

SC radiant heat flow

VT is the percent of visible energy that passes through glass

VT light transmission

FIGURE 5.20

The "greenhouse" effect is a term used to describe the tendency of glass to act as a heat trap (see Figure 5.20). Clear window glass passes 80 to 90% of incoming short-wave solar radiation, but it's opaque to the long-wave infrared radiation emitted by warm indoor surfaces and objects.

Glass manufacturers offer tinted, reflective, and low E products to limit overheating and provide privacy. Tinted and reflective glass transmit less light and heat than clear glass. Low E glass coatings are thin films that selectively reflect or transmit radiation. Low E glass will transmit light with less accompanying heat than other glasses.

VIEWS AND HEAT CONTROL

Because an elegant view offers much more pleasure than rows of architectural sun-control surfaces, the shape, size, orientation, and location of a view window should be selected to frame a beautiful scene.

Shading devices should *not* limit the pleasure viewers take from the colors, contrasts, brightness patterns, and motions in a fine landscape scene. Where architectural shading devices would limit a wonderful west view, delete them and substitute operable interior blinds (see Figure 5.21).

Remember the initial lighting design choice: use light to create decorative focal points *or* to illuminate objects and surfaces. View windows are decorative focal points, while skylights and clerestories usually illuminate interior surfaces or objects.

Heat and light control are primary design considerations for clerestories that don't offer views. Is a clerestory hung with flags and pennants a luminaire or a decorative focal point?

FIGURE 5.21

REVIEW QUESTIONS

Solve the following problems and verify the proposed answers. Size the apertures and shading elements using the 42°N sun chart and the section angle overlay. If you need help, review the examples on page 52–55.

1. Find the orientation that will maximize annual sunlight on a stained glass window at 10 AM.

2. A new luncheon restaurant wants winter sunlight on a decorative interior fountain. Find the required height of a vertical slot window intended to bring noon sun to the fountain from 1 October through 1 March. The window will be located in a south-facing wall. The fountain is 4 feet tall and 8 feet from the wall.

3. How wide should the window in the preceding example be to ensure sunlight on the fountain from noon until 1 PM?

4. A window faces west (ugh!). Detail an exterior shading device to exclude summer sun after 4 PM. Assume summer extends from 1 June through 31 August.

5. A clerestory faces west (why?). Set the sill height to exclude direct summer sun after 4 PM. Assume summer extends from 1 June through 31 August.

6. A clerestory faces south-west. Set the sill height to exclude direct summer sun after 4 PM. Assume summer extends from 1 June through 31 August.

7. A horizontal trellis is to shade a courtyard located behind a west-facing wall from 4 to 7 PM, between 21 May and 21 September. Find the center-to-center spacing for 2 × 6 wood trellis members that will exclude direct sunlight during this time.

8. A north-facing roof slopes 4 in 12. The opening for a 36-inch-square skylight in this roof is 16 inches above the opening in the interior ceiling. Will *direct* summer sun penetrate into the room below this skylight?

ANSWERS

1. Sun's bearing range is 30° to 65° east of south. Face the window 47° east of south.

2. Section angle range is 23° to about 45°. Set the sill 40 inches high and the head 12 feet high.

3. Bearing range is about 20°. Make the window 3 feet wide.

4. Detail vertical shades to exclude all sun bearings exceeding 77° west of south.

5. Set the sill height to exclude section angles less than 35°.

6. Set the sill height to exclude section angles less than 45°.

7. Space 2 × 6 members 8 inches apart to exclude section angles less than 35°.

8. Draw the skylight in section and use a protractor to show the noon section angle in June. Check the 3 PM June section angle.

CHAPTER 6
Design Examples

Space and light and order. These are the things that men need just as much as they need bread or a place to sleep.

Le Corbusier, 1965

The grand white buildings at the World Columbian Exposition (Chicago, 1893) were labeled "second eclectic phase" architectural style. Historians speculate about the social factors that led to public understanding and appreciation of their classic forms and details.

The public may have been less taken by neoclassic style than by the opportunity for an evening of social interaction surrounded by dramatic brightness contrast.

> As the lights went on, the massed human beings below uttered a great sigh. Then in the seats reserved for them, the Cabinet officers, and the Duke and Duchess of Veragua, and other foreign dignitaries began to cheer. The crowd lustily joined in while tightly corseted women fainted and fell like soldiers in battle*

In this chapter, black and white sketches illustrate example lighting designs for two rooms. Partial electrical plans are included; more extensive lighting, switching, and circuiting plans are presented in Chapters 10 and 12. Achromic sketches suggest the brightness contrasts caused by light sources, and the text speculates about perception.

*Margaret Cheney, *Tesla, Man Out of Time*, © 1981, Dorset Press.

6.0 FAMILY ROOM

PERCEPTIONS

The "mood" of a room will change as lighting changes. Just as outdoor scenes on a sunny day "look and feel" more interesting and stimulating than on an overcast day, an indoor space will change as lighting shifts from diffuse to direct, and as the luminance of interior surfaces changes. The effects of light on perception and mood are real design opportunities. They allow lighting design to *influence* viewers.

Light is invisible until it interacts with reflective surfaces or objects, and words describing this interaction are less effective than visual images. Here sketches show contrasts and patterns but lack depth, shade, shadow, highlights, motion, and color.

Study a favorite painting on a dark wood wall and then move it to a white wall. Notice the changes in the painting's colors and your perception of the scene. As your color experience grows, you'll find that all whites are not equal. "White" is the product of many colors and a given white background will enhance some objects and fade others.

These achromic pages offer only a starting point for lighting design. Maturing designers build a library of mental images of the "good lighting" sensations driven by color, highlights, textures, shade, and shadows missing here. The thoughtful use of light to influence perception, mood, or behavior distinguishes quality lighting design.

The example family room is 13 feet wide and 27 feet long (see Figure 6.1A). A window is centered in the north wall. Sliding glass doors and a clerestory face south. The room accommodates a variety of activities including reading, studying, TV viewing, eating, and entertaining. Entertaining may include dancing, singing, gaming, drinking, or quiet conversation. The number of occupants can vary from 1 to

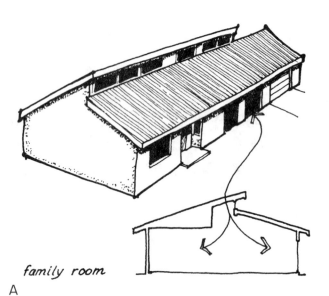

family room

A

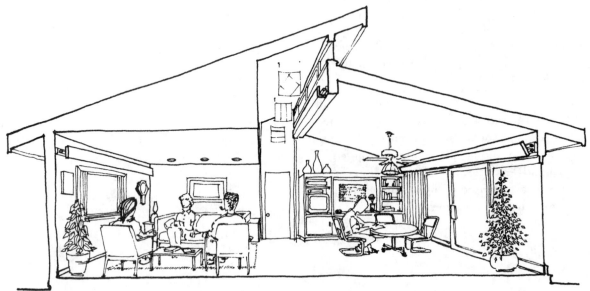

B

FIGURE 6.1

20 or more. The line perspective view in Figure 6.1B will be used to illustrate family room brightness contrasts on the following pages.

DESIGN

Because perception is driven by contrast, sketches emphasize the brightness contrasts created by locating and controlling light sources. Brightness contrast attracts the eye, but lighting design is much more. It is the thoughtful use of *brightness contrast, form-pattern, motion, and color*, to inform, interest, and influence.

Many design decisions precede fixture selection and location (see Figure 6.2). Good lighting and good design begin with thoughtful selection of visible materials, textures, furnishings, and colors. The light-reflecting, light-absorbing, and light-transmitting characteristics of ceilings, walls, doors, windows, and floors generate viewer perceptions, so talented designers visualize day and night lighting opportunities as they select interior colors, textures, and materials.

In Figure 6.3A, dark walls appear closer than light walls. They seem to reduce the apparent room depth. Light walls make the room seem wider and more spacious.

A dark ceiling in Figure 6.3B reduces apparent room height and size. Ceiling board edges or floor tile joints can make the room seem deeper or wider.

Interior furnishings, shapes, colors, and patterns create visual interest (see Figure 6.3C). Some architects and interior designers use an "all white except the floors" formula for interior surfaces to emphasize colors in art and accessories.

A

B

C

FIGURE 6.3

Coolney Playhouse clerestories are elegant daylight luminaires (see Figure 6.4). Heads, jambs, and sills frame the playhouse sky, and Wright's stained glass patterns pleasure children at play. Light, pattern, and color delight the mind and excite the imagination.

On clear days, windows are a room's largest and most intense light sources. They illuminate interior surfaces, *and* they can grace a room by framing and accenting bright scenes. Family room forms, colors, textures, and furnishings are initial lighting-design considerations. Sunlight and daylight deserve equal attention.

The clerestory and sliding doors face south, so eaves exclude direct summer sun but welcome winter sunlight (see Figure 6.5A). South is the only window orientation that allows free winter heat and easy

FIGURE 6.2

F. L. Wright 1912

FIGURE 6.4

control of summer heat. Capable architects consider solar angles and sunlight penetration when they orient a home on a site. Decisions about the location, size, and orientation of windows should support daylighting, view, and interior design concepts.

On a clear winter day, brilliant sunlight fills the clerestory slot and the south end of the family room. Summer daylight will be less intense than winter sunlight, but the room's summer and winter appearance and mood will be nearly identical due to adaptation.

At night the clerestory becomes a dark slot and window views go black. Privacy is lost to passersby, and the windows are dark planes that detract from the comfortable ambience of the room (see Figure 6.5B). Family room blinds are drawn, and lamps recreate the sensation of enclosure and shelter lost to the darkness (see Figure 6.5C).

Because daylight is such a dominant light source, luminaires should be selected *after* visualizing the room in daylight and sunlight. *Luminaires should be selected to emphasize objects or surfaces or to be decorative focal points.*

Perception and mood are set by light sources and by the light-reflecting characteristics of interior surfaces, colors, and furnishings. Family room luminaires were selected and located to complement, emphasize, and enhance these surfaces and furnishings.

Cove fixtures were located to illuminate the clerestory and to wash walls and drapes with bands of brightness. Flags add color and visual interest in the clerestory, and bright drapes make the room seem more spacious. Interior settings that complement a variety of family activities can be achieved using one, two, or all three coves (see Figure 6.6).

Three recessed can fixtures with adjustable apertures emphasize painting(s) on the east wall. A fan light kit is the only decorative luminaire illustrated. It was selected to illuminate the work table, and its stained glass shade provides a decorative color accent. (Be careful selecting fan light kits. Visible lamps will be glare sources with ceilings lower than 8 feet.)

Look back at the sketches and notice the variety of scenes light can paint. The coves and cans increase construction costs, but they add value for owners who take pride in the sensations their homes create. Colors, patterns, shades, shadows, textures, and highlights paint viewer perceptions. Add just a single color to any of these sketches and it will command at-

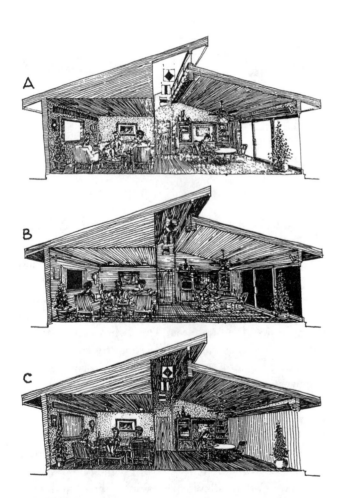

FIGURE 6.5

clerestory

south

north

all

FIGURE 6.6

tention at the expense of the surrounding illustrations.

Lighting variety makes the family room interesting, comfortable, and responsive to the variety of likely family activities. Additional decorative fixtures will add focal points and make the room even more attractive.

Panel Schedule

100 Amp. Two-pole Main Circuit Breaker

1. air conditioner	2. water heater
3. d.o.	4. d.o.
5. dryer	6. oven
7. d.o.	8. d.o.
9. rangetop	10. furnace fan
11. d.o.	12. garbage disp.
13. dishwasher	14. lighting & duplex
15. lighting & duplex	16. d.o.
17. d.o.	18. d.o.
19. d.o.	20. d.o.
21. d.o.	22. d.o.
23. elect. heat	24. elect. heat
25. d.o.	26. d.o.
27. landscape ltg	28–30. spares

ELECTRICAL PLAN

The schematic plan in Figure 6.7 locates fixtures, outlets, and switches for electrical rough-in. Some code authorities require schematic plans for *residential* electrical permits, but many cities require complete circuiting. A circuit plan is shown on page 164 and detailed switching and circuiting are illustrated on page 161 in Chapter 12.

Curved lines connect switches to the fixtures they control:

$ indicates a single-pole switch.

$_d$ is a dimmer switch.

$_3$ is a three-pole switch that allows control of a fixture from two locations.

Where control from more than two locations is required, a four-pole switch must be used. Two switches serve the ceiling fan to allow separate control of the fan and the light kit.

Wiring runs through the fluorescent strip fixtures used as cove lights, and the code requires flexible "Greenfield" conduit where the power supply enters these fixtures. Numbers on the plan refer to circuits noted in the panel schedule. A circle with two lines is the symbol for a duplex outlet to be installed 12 inches above the floor unless another height is

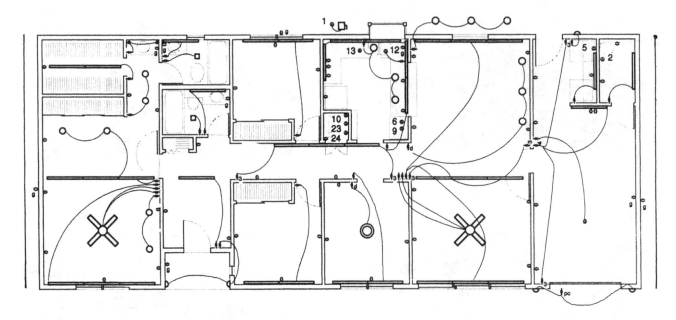

FIGURE 6.7

specified. A circle with an inscribed triangle is the symbol for a special purpose electrical outlet.

6.1 LOBBY

The example office building (Figure 6.8) lobby is used for lighting-design studies. Since it's a first visual experience for residents and guests, the lobby defines "building image." Planters and display cases flank the entry doors, and two large tapestries suspended below a stained glass ceiling add visual interest in the tall space. A bridge, accessed by elevators and a decorative circular stair, carries second-floor resident and guest traffic between the east and west offices. A reception desk, building directory, and small waiting area share the space beneath the bridge (see plan in Figure 6.10).

The perspective sketch in Figure 6.9 and 6.10 is a design template used to invent, visualize, and evaluate brightness patterns and contrasts.

STAINED GLASS CEILING

Lobby lighting design begins with the decorative stained glass ceiling. Opaque walls surround the ceiling at the bridge level. Ground-floor glazing admits only 5% of ambient illumination, so window brightness will not compete with the ceiling for visual attention.

Because sunlight could make the lobby uncomfortably bright, the lighting designer and the stained glass artist agree to limit maximum lobby illumination to 10% of outdoor levels (see Figure 6.11). Outdoor luminance values can range from zero to 10,000 and electric lighting in the office spaces will produce luminance magnitudes from 1 to 100, so lobby brightness in the 10 to 1,000 range should allow comfortable visual adaptation for visitors or residents. If the stained glass uses deep hues that transmit 20% of ambient light the skylight should pass about 50%, but if the stained glass design uses subtle tints that pass 50% of incident light, the skylight should exclude 80%.

At night, industrial light fixtures mounted in the skylight illuminate the ceiling. These fluorescent fixtures are spaced at 1.5 times their height above the stained glass to eliminate lamp image lines in the ceiling. Ceiling luminance is reduced at night, but the stained glass is equally attractive day or night because of viewer adaptation (see Figure 6.12).

Because daylight varies so much from clear to cloudy and from day to night, a scene controller will dim lobby lamps as outdoor light intensity changes.

Ceiling stained glass colors and patterns may include deep hues or subtle tints. The sketches in Figures 6.13 and 6.14 illustrate both possibilities. With light tints the lobby seems taller, but with dark

FIGURE 6.8

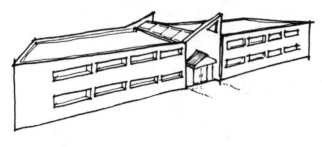

FIGURE 6.9

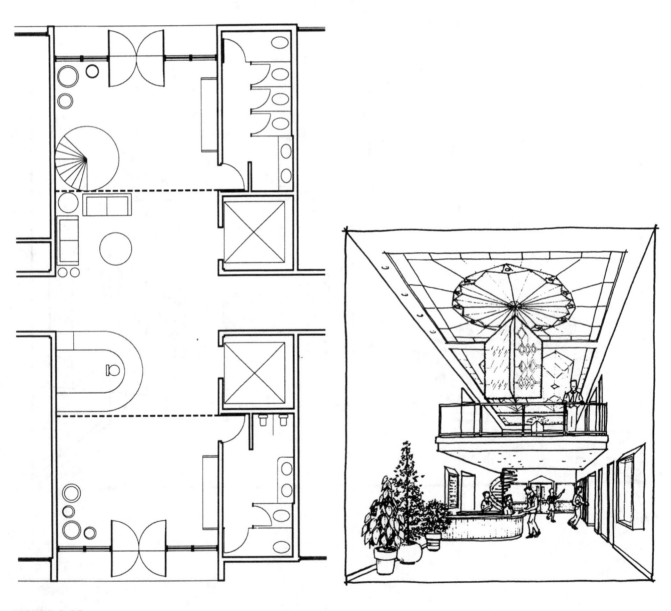

FIGURE 6.10

FIGURE 6.11

FIGURE 6.13

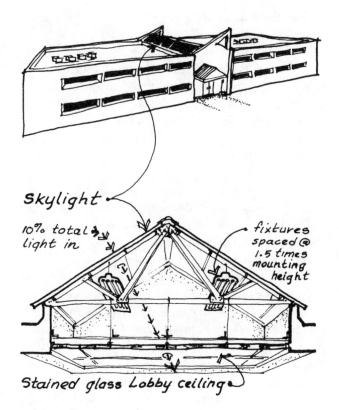

skylight

10% total light in

fixtures spaced @ 1.5 times mounting height

Stained glass Lobby ceiling

FIGURE 6.12

FIGURE 6.14

FIGURE 6.15

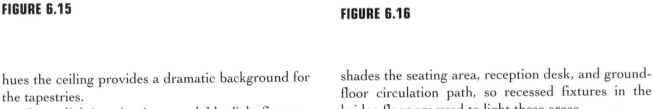

FIGURE 6.16

hues the ceiling provides a dramatic background for the tapestries.

Seven lighting circuits serve lobby light fixtures. Two circuits illuminate the stained glass ceiling, and single circuits light the tapestries, stair, plants, display cases, and porches through dimmer controls that adjust lighting in response to time and daylight changes.

Sketches isolate the ceiling, but perception relates all visible surfaces. Dark brick walls change the lobby's ambience (see Figure 6.15).

An exterior photocell keeps the stained glass backlights off until daylight intensity drops below 200 footcandles. During the evening, ceiling luminance is far below sunny day levels, but the ceiling's stained glass pattern creates a similar visual sensation day or night because of viewer adaptation. After midnight, dimmers cut ceiling backlights to 20% of rated output. When daylight triggers the morning scene, backlights are extinguished.

The spiral stair and the second-floor bridge are attractive lobby structures. Transparent bridge railing panels open the lobby volume and allow visitors to enjoy the complete ceiling pattern. The bridge

shades the seating area, reception desk, and ground-floor circulation path, so recessed fixtures in the bridge floor are used to light these areas.

During daylight hours, a ring of spotlights illuminates the tapestries so that they are brighter than the stained glass ceiling. Narrow beam lamps are aimed to minimize spill light on the floor below, and the ring dims to 60% of rated output for the evening scene. Ring lamps are off from midnight until morning.

Valence lighting accents the vertical faces of the lobby bridge, and a 400 watt recessed HID lamp lights the spiral stair treads from above. These lamps run at full voltage from morning to evening and dim to 60% of rated output from evening to morning.

The building manager plans changing lobby floral displays each month. Seven recessed ceiling fixtures with narrow beam lamps will illuminate the plants and the wall behind them. The floral circuit is photocell controlled to enhance the plants and flowers during evening hours and on overcast days. These lights are off from midnight until noon.

The sketches in Figures 6.16, 6.17, and 6.18 suggest brightness contrast with differing light source

FIGURE 6.17

FIGURE 6.18

settings and surface reflectances. They are only beginning images of lighting design without color, shade, shadow, highlights, and source luminance. Even so, the lobby seems a dramatically different space in each illustration.

Many design decisions precede fixture selection and location. Good lighting and good design begin with thoughtful selection of the materials, textures, furnishings, and colors that shape viewer perceptions. The Fixture Schedule includes all office building fixtures.

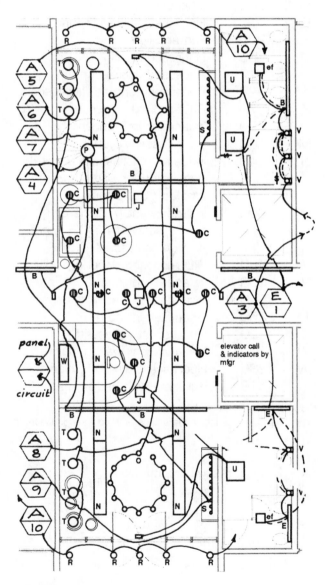

FIGURE 6.19
Lobby Lighting Plan

In the plan shown in Figure 6.19, solid lines denote circuits and dotted lines show switching. Switches are not used on circuits A-4 through A-10. Illumination levels are controlled by timer and photocell signals that adjust seven dimmers.

Fixture Schedule

Key	Watts	Fixture
A	110	Fluorescent strip
B*	130	Fluorescent strip
C*	50	Incandescent can
D	500	Incandescent—custom
E*	72	Fluorescent—wall mount
F	200	Fan and light
G	55	Fluorescent strip
H	72	Recessed fluorescent
J*	28	Recessed compact fluorescent
K	70	Built-in display case
L	250	Incandescent pendant
M	50	Incandescent eyeball
N*	240	Industrial fluorescent (*locate between the skylight and the stained glass ceiling*)
O*	600	Incandescent (*a circle of twelve 50 watt MR-16 spots above each tapestry*)
P*	460	HID recessed in ceiling
R*	100	Incandescent can
S*	500	Incandescent—display case
T*	250	Incandescent can
U*	56	Fluorescent recessed
V*	100	Incandescent wall mount
W*	288	Fluorescent directory sign
=	18	exit lamp with battery pack

*Fixtures used in the lobby.
Circuit loads for the lobby and offices are tabulated in Chapter 12.

Six lighting scenes are programmed:

Morning

Clear winter day

Clear summer day

Cloudy or overcast

Evening

Midnight to sunrise

Scene Circuit Summary

Key	Watts	Fixtures illuminate
A-4	1,138	Stair, bridge, and directory
A-5	1,200	Tapestry (2 @ 600 watts each)
A-6	1,750	Plants (7 @ 250 watts each)
A-7	1,440	Stained glass backlights
A-8	1,440	Stained glass backlights
A-9	1,750	Display cases and under bridge
A-10	1,000	Exterior (10 @ 100 watts)

6.2 DESIGN OPPORTUNITIES

Family room and lobby example sketches are lightness studies lacking shades, shadows, highlights, and color. They are incomplete lighting-design examples, but they serve to illustrate a designer's use of lightness contrasts to inform, interest, and influence people.

When more lighting effects like

brightness,

color,

shade,

shadows,

highlights, and

motion

are added, creative design opportunities multiply.

Design is only part of architectural practice. Vitruvius's design triad:

firmness

commodity

delight

and recent architectural authors:

form

function

economy

include just a single word for design.

Generous authors view architectural design as a pursuit of grace, beauty, and elegance in buildings, but designs evolve in a framework of constraints. Site, program, budget, codes, construction documents, and program changes shape buildings and compete with design considerations in architectural practice.

Design emphasis is greater with lighting. Enhancing a landscape, a building, or a room with light is a simpler task than creating these spaces. Lighting designers pursue beauty by visualizing spaces where light illuminates objects or surfaces *or* creates decorative focal points. They select or design luminaires to accomplish their visual images.

The following information is an attempt to describe lighting-design opportunities. Use what you find helpful.

Begin by identifying interesting objects, details, or features that can pleasure the eye. For landscape lighting the emphasis might be a tree, a pool, a fountain, or a gate; for building lighting, entries are obvious focal points. In interior spaces, light serves occupant activities *and* adds visual interest by emphasizing surfaces, furnishings, art, and architectural features.

View or visualize the space or object to be illuminated, and identify interesting colors, textures, surfaces, and details. *Use light to illuminate objects or surfaces, or to create decorative focal points.*

Visualize opportunities to enhance spaces, rooms, and interesting features with light by:

- Using brightness contrast to accent interesting objects and details. Visualize the patterns that can be created by light, shade, and shadow until you find one that best complements the object or detail.
- Locating luminaires so that grazing light paints surface textures and architectural moldings with light, shade, and shadow.
- Using luminaires to create brightness patterns that enhance occupant perceptions of space proportions and architectural details.
- Selecting light sources that complement space color schemes, and using colored luminaires or luminaires that decorate colored surfaces.

"Decorative" is a poor descriptor for the visual beauty an elegant luminaire can create. Imagine the colors, patterns, and contrasts in a sunset and compare them to the experience of viewing a fine stained glass lamp.

Selecting decorative luminaires for a specific project is a talent built on experience. Artistic designers study lighting installations like professional golfers study courses. They continually create and save memory images of attractive lighting scenes and the fixtures that help create such scenes. Brightness contrast, pattern, color, and motion are design tools; but memory images of lighting effects are a designer's reference library.

TEMPLATES

Make a few copies of Figures 6.20 and 6.21 and use color to illustrate lighting designs for the family room and lobby. Begin with just a warm earth tone and a single accent color. Compare your work to the preceding black and white sketches and see how much just two colors can add to lighting design.

When you're comfortable with two tones, experiment with complement and split complement color schemes.

FIGURE 6.20

FIGURE 6.21

CHAPTER 7
Lighting Numbers

Calvin—I'm not going to do my math homework . . .

Look at these unsolved problems. Here's a number in mortal combat with another.

One of them's going to get subtracted. But why? How? What will be left of him? . . .

If I answered these it would kill the suspense. It would resolve the conflict and turn intriguing possibilities into boring old facts . . .

Hobbes—I never really thought about the literary qualities of math.

Calvin—I prefer to savor the mystery.

Bill Watterson, 13 October 1992

This chapter presents photometric numbers and calculations. Accent lighting uses illumination to create bright objects or surfaces that draw attention. Accent footcandles are easy to estimate when candlepower, beam spread, and distance are known.

Few lighting authorities recommend uniform lighting, but it's a lighting scheme used in too many buildings. Unfortunately, "task" lighting recommendations are often spread across large rooms as uniform illumination, instead of directing light to the smaller areas where a task is performed.

The ability to calculate uniform footcandles is a skill expected of lighting professionals, but these calculations are *not* lighting design. Selecting and spacing fixtures to provide uniform footcandles are not design judgments. They're merely skills that disregard adaptation, and the sensations talented lighting designers rely on to shape perception.

7.0 TERMS AND ACCENTS

REVIEW TERMS

Lumen (lm): Unit of *luminous flux*. The lumen is a measure of light quantity based on the eye's sensitivity. By definition, 1 watt can yield a maximum of 683 lumens. The following units are all related to the lumen.

Candela (cd) or **candlepower:** Unit of *luminous intensity*. Defined as 1/60 of the intensity of a square centimeter of a blackbody radiator at 2,045 K. One candela = 1 lumen per steradian (sr). A steradian is a solid angle at the center of a sphere which subtends an area on the surface of the sphere equal to the radius squared.

Footcandle (fc) or **lux (lx):** Units of *illuminance*. Density of luminous flux uniformly incident upon a surface. One lumen upon a 1-square-foot area is 1 fc. One lumen upon a square meter is 1 lux. One fc = 10.76 lux.

cd/m^2 or **nit:** Unit of *luminance*. Luminous flux leaving a surface. When 1 lumen is reflected or emitted from a 1-square-meter surface, luminance is **1 cd/m^2** *in the direction being viewed*. When 1 lumen is reflected or emitted from a 1-square-meter *diffusing* surface, luminance is **1/π cd/m^2**.

Consider a sphere with a radius of 1 foot surrounding a 1 *candela* point source of light (see Figure 7.1). By definition, *illuminance* at the sphere surface is 1 footcandle or 1 lumen per square foot or 10.76 lux. The sphere's surface area is 12.57 square feet ($4\pi r^2$), so 1 candela emits 12.57 *lumens*. If the sphere trans-

mits and diffuses 50% of incident lumens, its surface *luminance* will be 1.7 cd/m^2.*

ACCENT LIGHTING

Accent lighting rains lumens on interesting objects or surfaces because brightness contrast commands attention. Designers begin by visualizing highlights, shades, shadows, colors, and textures. They select luminaires that enhance the objects or surfaces. A strong brightness contrast sensation usually occurs when the illuminance at an object or surface is at least five times surrounding illuminance. Specifying lamps for a five-fold increase is easy when distance, candlepower, and beam spread are known.

Candlepower sets illuminance at a given distance. Designers seeking brightness contrast use the formula:

$$fc \text{ or } lx = candlepower \div distance^2$$

to estimate illuminance.

Beam spread and the fixture aperture determine the shape of the illuminance cone, and designers use a protractor or trigonometry to estimate the size of the illuminated area (see Figure 7.2).

Lamp manufacturers provide beam candlepower** and beam spread data for accent lighting applications on the facing pages. Lamp and fixture catalogs also provide data on illuminance and size of the illuminated area for typical applications. Study

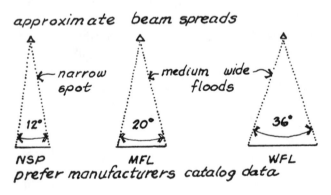

FIGURE 7.2

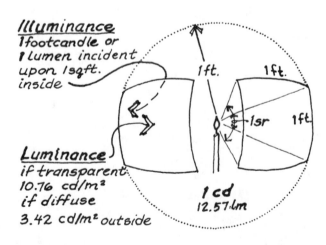

FIGURE 7.1

---

*Sphere radius is 1 foot. Illuminance is 1 fc = 1 lm/ft^2 = 10.76 lm/m^2. Luminance = (50%)(10.76)(1/π) cd/m^2 = 1.7 cd/m^2.
**Manufacturers provide MBCP (maximum beam candlepower) or CBCP (center beam candlepower). Lamps with long filaments produce an elliptical beam instead of a circular beam.

the following examples and work the sample problems at the end of the chapter so you can estimate illuminance values with confidence.

Lamp Table	Beam Spread	Beam cp
A. 25W-R14 SP (12v)	32°	1,040
B. 35W-PAR36 MFL (12v)	28°	1,400
C. 75W-PAR 38 SP (120v)	30°	3,800
D. 25W-PAR36 NSP (12v)	18°	4,500
E. 50W-PAR36 NSP (12v)	18°	8,000
F. 25W-PAR36 VNSP (12v)	10°	15,000
G. 50W-PAR36 VNSP (12v)	8°	23,500
H. 25W-PAR36 (5.5v)	5°	30,000

Use the information in the two tables to select lamps for the following examples, but use current lamp manufacturer data for real applications.

EXAMPLES

1. Select a lamp that will deliver 200+ fc when aimed at a wall 12 feet away.
   Answer: fc = cp ÷ d^2, 200 = (cp) ÷ (12)^2, cp needed is 28,800. Select *lamp H.*
2. How big will the illuminated circle be in example 1? (See Figure 7.3.)

Answer: Lamp H has a 5° beam spread, so the illuminated circle will be about *12 inches in diameter* at a distance of 12 feet. If the lamp axis is not perpendicular to the wall, the illuminated area will be an ellipse instead of a circle.

3. You plan to use two lamps to illuminate a small sculpture displayed on a 20-inch-diameter pedestal. You can mount two H lamps 20 feet away, or two F lamps 10 feet away. Which lamps will deliver maximum illuminance?
   Answer: fc = cp ÷ d^2, two *F lamps* will deliver a total of 300 fc.
4. Average illumination on two intersecting walls is 11 fc. You want to illuminate a heart-shaped area in the corner to accent a floral arrangement (see Figure 7.4). Select the lowest wattage lamp that will increase illuminance by at least 55 fc and produce a heart about 30 inches tall. The lamp will be located 4 feet away from the corner and 3 feet above the center of the accent area.
   Answer: Select 35 watt lamp B. It will deliver 56 fc and produce a heart about 30 inches high. The 25 watt lamps with less beam spread don't produce a tall enough heart. Lamp A only adds 42 fc.

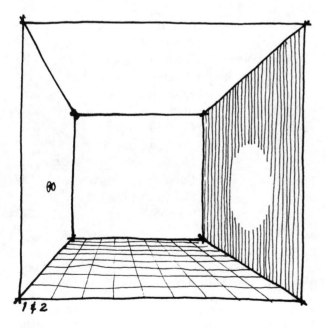

FIGURE 7.3

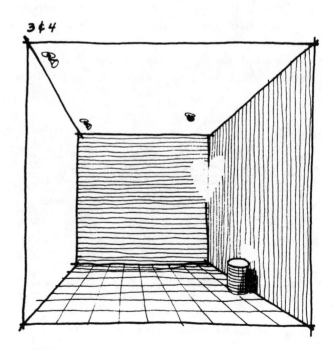

FIGURE 7.4

Illuminated Area Table

Beam Spread	5°	10°	20°	30°	40°
5' distance	5"	10"	22"	33"	44"
10' distance	10"	20"	44"	66"	88"
20' distance	20"	40"	88"	136"	176"

Inches are diameters of illuminated circles.
Note: Most lamps have elliptical beam cones. Use manufacturers' catalog data instead of these example values.

7.1 HOW MUCH LIGHT?

Because the eye adapts, most people can comfortably read this page at any luminance between 4 cd/m^2 and 4,000+ cd/m^2. Illuminance is easier to quantify than luminance. The corresponding illuminance range for reading this page would be 2 to 2,000 footcandles.

Some argue that adaptation and the adapted eye's range make most illuminance recommendations frivolous, but arguments have not discouraged lamp manufacturers, engineers, fixture manufacturers, academics, or government agencies from offering recommendations for lighting visual "tasks." Unfortunately, task lighting is frequently spread over large areas as uniform lighting, instead of emphasizing smaller areas where the task in question is performed.

The graph in Figure 7.5 plots the Davis Reading Test, "speed of comprehension" versus illumination. Two age groups were tested using high- and low-contrast documents.

The graph shows no change in performance between 1 and 500 footcandles, except low contrast for

the 60-year age group showed no change from 10 to 500 footcandles.[*]

The IESNA Handbook defines seven types of visual tasks ranging from easy to difficult, and recommends seven appropriate illumination levels. *General* (uniform) lighting, up to 10 footcandles, is recommended *only* for simple tasks. When illumination exceeding 10 footcandles is recommended, the IESNA specifies "illuminance on task."

Activity		fc
Reading	CRT screen	3
	8 and 10 point type	30
	Glossy magazines	30
	Maps	50
Sewing, hand sewing	Occasional sewing, high contrast	30
	Light to medium fabrics	50
	Dark fabrics, low contrast	100
Jewelry and watch manufacturing		300–1000

Spaces		fc
Auditoriums, assembly		10
Barber shops		50
Court rooms	Seating	10
	Activity areas	50
Food service	Dining	10
	Cashier	30
	Kitchen	50
Hospitals	Lobby	10
	Corridor, night	5
	Corridor, day	10
	Nurse desk	50
	Surgery scrub room	100
	Surgery operating room	100

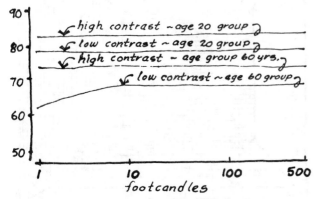

Speed of Comprehension Scores

FIGURE 7.5

[*]For more detail and findings read: S. W. Smith, and M. S. Rea, *Performance of Office-Type Tasks Under Different Types and Levels of Illumination*, The Institute for Research in Vision, Ohio State University, October 1976; and D. K. Ross, *The Limitations of Illumination as a Determinant of Task Performance*, Ross & Baruzzini Inc., St. Louis, March 1978.

Spaces		fc
	Surgery operating table	300–1000
Machine shops	Rough work	30
	Medium work	50
	Fine work	100
	Extra fine work	300–1000
Residence	General	5
	Kitchen counter	50

Outdoors		fc
Construction	Excavation	3
	General construction	5
Ice hockey	Recreational	20
	No special provision for spectators	30
	Facilities for spectators 5000 or less	50
Service station	Approach	15
	Pump area	3

Source: IESNA (2000). *Lighting Handbook: Reference and application.*

Because different activities may require different illumination levels, and because accent lighting is effective when it is at least five times the general illumination level, designers should be able to calculate the number of footcandles delivered by a specific lighting installation.

7.2 LUMINAIRE PHOTOMETRICS

Calculating the number of fixtures needed to provide general lighting in a room is more complex than selecting accent lighting. Fixture manufacturers provide photometric data for uniform lighting, and designers use this data to calculate the number of fixtures required to yield a given illuminance (see Figure 7.6).

CANDLEPOWER PLOTS

Candlepower plots illustrate fixture light output. The plots show candlepower in a circular field, and two or three plots may be given for fixtures with asymmetrical distribution patterns. An ideal distribution for a uniform lighting application spreads light as far as

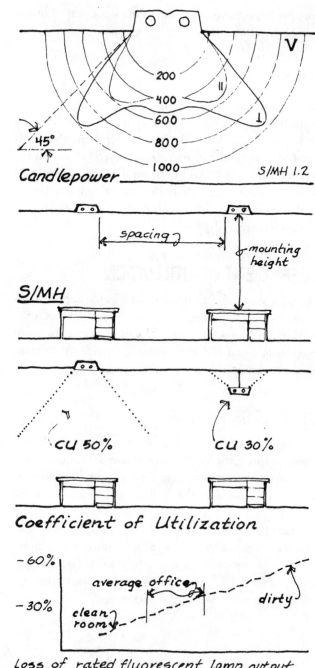

FIGURE 7.6

possible from the fixture, but limits light output in the glare zone above 45°.

S/MH

In uniform lighting applications, the spacing to mounting height ratio (S/MH) indicates a fixture's ability to spread light. As S/MH increases, the number (and cost) of fixtures decreases. Fixture manufacturers include S/MH with candlepower plots, and when fixture light distribution is asymmetrical, S/MH values are given both parallel and perpendicular to fixture.

COEFFICIENT OF UTILIZATION

Coefficient of utilization (CU) is an index of fixture efficiency; it is the percent of lamp lumens that reach the work plane *in a given room*. CU values vary with room proportions and reflectances, so preparatory calculations are required before selecting a CU.

LIGHT LOSS FACTOR

The light loss factor (LLF) is an estimate of illuminance depreciation. Operating considerations such as lamp temperature, lamp age, lamp cleanliness, supply voltage, and the ballast factor affect lumen output.

Danger! Approximate LLFs are used in this text. Readers seeking greater precision must use a more comprehensive reference to study LOF (lamp operating factor), LLD (lamp lumen depreciation), and LDD (luminaire dirt depreciation).

7.3 FOOTCANDLE CALCULATIONS

This section and the Footcandle Calculation form describe a zonal cavity illumination calculation. Read the instructions here first if you're not familiar with the calculation method; then study the example calculation that follows.

Begin by entering the project name, your initials, and the date; then fill in the room dimensions. Next select the fixture and lamp(s) you intend to use, complete the fixture information box,

and attach a copy of the manufacturer's photometric data to the form. Note your lamp lumens and total fixture lumens in the right-hand summary column.

1. Room size, proportions, and reflectances affect footcandles; enter three *cavity height* dimensions in the room view box and the right column summary. The ceiling cavity height is the distance from the ceiling to the fixture. The room cavity extends from the fixture to the work plane. The floor cavity extends from the work plane to the floor. When recessed fixtures are used, the ceiling cavity height is zero. The work plane is usually at desk or countertop height, but in a lobby or bowling alley the work plane is the floor.

   White rooms or large rooms welcome more work plane footcandles than black rooms or small rooms. Enter three *reflectances*: ceiling (pc), wall (pw), and floor (pf).

   Ideally, reflectances are measured with a light meter for specific room finishes, but they are often estimated.

2. Calculate three cavity ratios—CCR, RCR, and FCR—using the formula:

$$CR = (5hc) (L + W) \div (L \times W)$$

   where:
   CR = cavity ratio, hc = height of cavity,
   L = room length, and W = room width.

3. Look up pcc (effective ceiling cavity reflectance) in the table using CCR, pc, and pw. This approximate calculation does not calculate effective floor cavity reflectance because its CU impact is small. Manufacturers' CU tables assume an effective floor cavity reflectance of 20%.

4. Look up the coefficient of utilization (CU) on the manufacturer's data sheet using pcc, pw, and RCR. CU is the percent of lamp lumens expected at this room's work plane using the selected fixture.

5. Use the LLF (light loss factor) table to estimate lumen depreciation.

6. Select required footcandles from the IES recommendations given earlier in the chapter.

Footcandle Calculation project

Fixture Information

Manufacturer - model #
_____

option(s) _____

lamp(s) _____

Fixture Lumens
FL

ballast _____

Room View

1, 2, 3

ceiling reflectance — pc

ceiling cavity height

room cavity height

floor cavity height

average wall reflectance — pw

floor reflectance — pf

Room Plan & Fixture Layout

---

by _____ date _____

Room _____
dimensions _____ x _____
area _____ sqft.

Selected Fixture
lamp lumens _____
# of lamps _____
Fixture lumens _____
FL

1
cavity height & reflectance
ceiling _____' pc _____%
wall _____' pw _____%
floor _____' pf _____%

2
Cavity Ratios
CR = (5 hc) (L+W) ÷ (LxW)
ceiling ~ CCR _____
room ~ RCR _____
floor ~ FCR _____

3
effective ceiling cavity
reflectance ~ pcc _____

4
Coefficient of Utilization
_____ CU

5
Light Loss Factor _____
LLF

6
Footcandles required _____
FC

7
Number of Fixtures =

(FC)(area)÷(FL)(CU)(LLF)

_____

83

7. Calculate the number of fixtures required to ensure a given illumination level using the formula:

$$\text{\# fixtures} = \frac{(\text{footcandles})(\text{room area})}{(\text{FL})(\text{CU\%})(\text{LLF\%})}$$

where:
FL = fixture lumens (total)
CU = coefficient of utilization
LLF = light loss factor

pcc Table (pcc's in field, CCR at left)

pc	80%			70%			50%	
pw	70	50	30	70	50	30	50	30
0	80	80	80	70	70	70	50	50
1	71	66	61	63	58	53	42	39
2	64	56	48	56	48	41	37	30
3	58	47	38	51	40	32	32	24
4	52	40	30	46	35	26	29	20
5	48	35	25	43	32	22	26	17

Example CUs (% CUs in field, RCR left)

pcc	80%			70%			50%	
pw	70	50	30	70	50	30	50	30
0	74	74	74	72	72	72	69	69
1	70	68	67	69	67	66	65	63
2	66	63	61	65	62	59	60	58
3	62	58	55	61	57	54	65	53
4	58	54	50	58	53	49	51	48
5	55	49	45	54	49	45	47	44

LLF Values

	%*
Good	75–80
Average	60–75
Poor	40–60

*Best—85% electronics assembly clean room.
Average—70% classroom or office.
Worst—40% wood cabinet shop.
Direct fixtures up to 85%, indirect to 70%.

EXAMPLE CALCULATION

Fluorescent fixtures will be used to illuminate a classroom measuring 20′ × 30′ × 9′. How many fixtures should be installed to yield 50 footcandles at desktops 30 inches above the floor?

Use a manufacturer's catalog to select a fixture and fill in the fixture information box of the Footcandle Calculation form. The fixture in this example uses three F40 (34 watt CW) fluorescent lamps rated at 2,900 lumens each.

1. Complete the room view box information by entering cavity heights and reflectances. White ceiling reflectance (pc) = 80%, but the average wall reflectance (pw) = 50% because of blackboards and windows.
2. Calculate cavity ratios using the formula:

$$CR = 5hc \, (L + W) \div (L \times W)$$

where:
CR = cavity ratio
hc = height of cavity
L   = room length
W = room width

The ceiling cavity height is zero, so CCR is 0. Room cavity height is 6′-6″, so the RCR is 2.7.

$$(5)(6.5)(20 + 30) \div (600) = 2.7$$

3. Look up pcc in the table. With a CCR (ceiling cavity ratio) of zero, pcc = 80%.
4. Enter the CU table using pcc = 80%, pw = 50%, and RCR = 2.7. Interpolate to select CU = 60% (actually 59.5%).
5. Use the LLF table to estimate LLF = 70%.
6. Required horizontal illumination for this example is 50 footcandles.
7. Calculate the number of fixtures required using the formula:

$$\text{number of fixtures} = \frac{(\text{footcandles})(\text{room area})}{(\text{FL})(\text{CU\%})(\text{LLF\%})}$$

$$8.21 = \frac{(50)(600)}{(8,700)(60\%)(70\%)}$$

8. Lay out fixtures. Eight or nine fixtures will work. Use nine for this room plan—best fit for the ceiling grid, seating arrangement, and

blackboard location. With nine fixtures, the calculated illumination increases to about 55 footcandles.

Lamp Output (initial*)

Lamp Output (initial*)	Lumens
Fluorescent Lamps	
F 40 CW	3,200
F 40 BX (compact)	3,150
F 40 CW (34 watt)	2,900
F 48 CW/HO (60 watt)	4,300
F 48 CW/VHO (110 watt)	6,900
F 96 CW (75 watt)	6,300
F 96 CW/HO (110 watt)	9,200
F 96 CW/VHO (215 watt)	16,000

*Values are approximate. Consult lamp manufacturers for exact lamp lumens.

pcc Values (pcc's in field, CCR at left)

pc	80%			70%			50%	
pw	70	50	30	70	50	30	50	30
CCR								
0	80	80	80	70	70	70	50	50
1	71	66	61	63	58	53	42	39
2	64	56	48	56	48	41	37	30
3	58	47	38	51	40	32	32	24
4	52	40	30	46	35	26	29	20
5	48	35	25	43	32	22	26	17

Fixture CU Table for This Example

CU Values (% CU in field, RCR at left)*

pcc	80%			70%			50%	
pw	70	50	30	70	50	30	50	30
RCR								
0	74	74	74	72	72	72	69	69
1	70	68	67	69	67	66	65	63
2	66	63	61	65	62	59	60	58
3	62	58	55	61	57	54	55	53
4	58	54	50	58	53	49	51	48
5	55	49	45	54	49	45	47	44

*All field CU values are given as percents.
The example fixture is fixture #2 in the following CU tables. Manufacturers' CU tables assume an effective floor cavity reflectance (pfc) of 20%.

LLF Values

LLF Values	%*
Good	75–80
Average	60–75
Poor	40–60

Best—85% electronics assembly clean room.
Average—70% classroom or office.
Worst—40% wood cabinet shop.
Direct fixtures up to 85%, indirect to 70%.

FOOTCANDLE TABLES

Use the tables in this section and the Footcandle Calculation form with manufacturers' CU tables to estimate illumination in a specific room using a particular fixture.

Lamp Output	Lumens*
Incandescent	
50 watt A-19	800
100 watt A-19	1,700
200 watt PS-30	3,700
300 watt PS-30	6,100
Fluorescent	
F 40 CW	3,200
F 40 BX (compact)	3,150
F 40 CW (34 watt)	2,900
F 48 CW/HO (60 watt)	4,300
F 48 CW/VHO (110 watt)	6,900
F 96 CW (75 watt)	6,300
F 96 CW/HO (110 watt)	9,200
F 96 CW/VHO (215 watt)	16,000
MH-HID	
175 watt MH	15,000
250 watt MH	22,000
400 watt MH	40,000
1,000 watt MH	120,000

*Values are approximate. Consult lamp manufacturers' catalogs for exact lumens.

Reflectances

A footcandle meter can provide accurate surface reflectance values. Measure the light reflected by a surface and then rotate the light meter to measure incident light. The reflectance is equal to the reflected value divided by incident value.

Footcandle Calculation

project *Anson Elem. School*

by *A.P.* date *2-7-99*

Room *Classroom 112*
dimensions *20'* x *30'*
area *600* sqft.

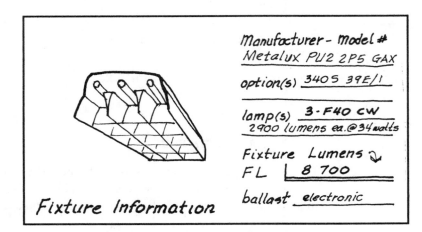

Manufacturer - model #
Metalux PU2 2P5 GAX

option(s) *3405 39E/1*

lamp(s) *3 - F40 CW*
2900 lumens ea. @34 watts

Fixture Lumens ↓
FL *8 700*

ballast *electronic*

Fixture Information

Selected Fixture
lamp lumens *2,900*
# of lamps *3*
Fixture lumens *8,700*

FL

1
cavity height & reflectance
ceiling *0'* pc *80* %
wall *6.5'* pw *50* %
floor *2.5'* pf *20* %

2
Cavity Ratios
$CR = (5 \, hc)(L+W) \div (L \times W)$
ceiling ~ CCR *0*
room ~ RCR *2.7*
floor ~ FCR *1.0*

3
effective ceiling cavity
reflectance ~ pcc *80%*

4
Coefficient of Utilization
60% CU

5
Light Loss Factor *70%*
LLF

6
Footcandles required *50*
FC

7
Number of Fixtures =
$(FC)(area) \div (FL)(CU)(LLF)$

*8.2 use 9**

* 55 footcandles with 9

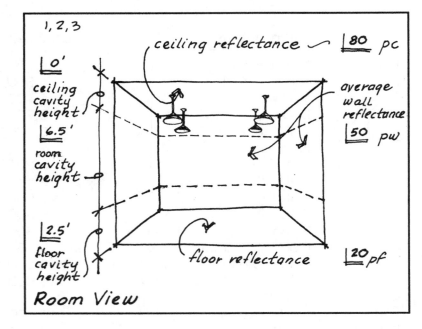

1, 2, 3

ceiling reflectance ~ *80* pc

0'
ceiling
cavity
height

6.5'
room
cavity
height

2.5'
floor
cavity
height

average
wall
reflectance
50 pw

floor reflectance *20* pf

Room View

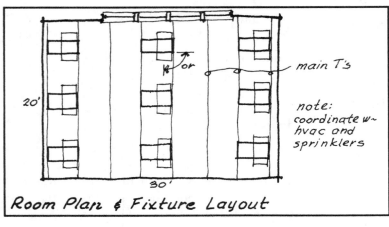

20'

30'

main T's

note:
coordinate w~
hvac and
sprinklers

Room Plan & Fixture Layout

The following approximate values are an imperfect starting point for estimates:

Surface	% Reflectance
Mirror	90–95
White (paint)	80–90
Off white (paint)	60–85
White (ceiling tiles)	70–90
Colored paint (Munsell values)*	
Light value colors (7–9)	40–80
Medium value colors (4–6)	10–40
Dark value colors (1–3)	1–10
Concrete	10–60
Birch (clear finish)	20–40
Walnut (oil finish)	5–15
Grass (or similar carpet)	5–15
Slate blackboard	1–10

*Vary with gloss, texture, age, etc.

pcc Values (pcc's in field, CCR at left)

pc	80%			70%			50%	
pw	70	50	30	70	50	30	50	30
pcc								
0	80	80	80	70	70	70	50	50
1	71	66	61	63	58	53	42	39
2	64	56	48	56	48	41	37	30
3	58	47	38	51	40	32	32	24
4	52	40	30	46	35	26	29	20
5	48	35	25	43	32	22	26	17

LLF Values %*

Good	75–80
Average	60–75
Poor	40–60

*Best—85% electronics assembly clean room.
Average—70% classroom or office.
Worst—40% wood cabinet shop.
Direct fixtures up to 85%, indirect to 70%.

Footcandle Recommendations

Activity		fc
Reading	CRT screen	3
	8 and 10 point size	30
	Glossy magazines	30
	Maps	50
Jewelry and watch manufacturing		300–1000

Spaces		fc
Auditoriums, assembly		10
Barber shops		50
Court rooms	Seating	10
	Activity areas	50
Food service	Dining	10
	Cashier	30
	Kitchen	50
Hospitals	Lobby	10
	Corridor, night	5
	Corridor, day	10
	Nurse desk	50
	Surgery scrub room	100
	Surgery operating room	100
	Surgery operating table	300–1000
Machine shops	Rough work	30
	Medium work	50
	Fine work	100
	Extra fine work	300–1000
Residence	General	5
	Kitchen counter	50

Source: *IESNA Lighting Handbook: Reference & Application.* 2000.

Footcandle Calculation   project _____

by _____ date _____

Room _____
dimensions _____ x ____
area _____ sqft.

Fixture Information

Manufacturer - model #
_____
option(s) _____
lamp(s) _____
Fixture Lumens ⤸
FL |_____
ballast _____

Selected Fixture
lamp lumens _____
# of lamps _____
Fixture lumens _____
                                    FL

1
cavity height & reflectance
ceiling ____' pc _____%
wall ____' pw _____%
floor ____' pf _____%

2
Cavity Ratios
CR = (5 hc) (L+W) ÷ (LxW)
ceiling ~ CCR _____
room ~ RCR _____
floor ~ FCR _____

Room View

1, 2, 3

ceiling reflectance ⌐ |__ pc

ceiling cavity height

average wall reflectance |__ pw

room cavity height

floor cavity height

floor reflectance |__ pf

3
effective ceiling cavity
reflectance ~ pcc _____

4
Coefficient of Utilization
_____ CU

5
Light Loss Factor _____
                              LLF

6
Footcandles required _____
                              FC

7
Number of Fixtures =

(FC)(area)÷(FL)(CU)(LLF)

_____

Room Plan & Fixture Layout

7.4 **COEFFICIENT OF UTILIZATION**

EXAMPLE CUS

Danger! These CU values are generic and approximate! Use manufacturers' CUs for precise calculations.

1.

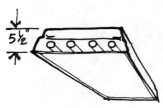
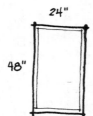

2′ × 4′, 4 F40 Fluorescent Lamps, Flat Prismatic Lens, S/MH 1.5, VCP 55-65

CU Values (% CU in field, RCR left) *

pcc	80%			70%			50%	
pw	70	50	30	70	50	30	50	30
RCR								
0	71	71	71	69	69	69	66	66
1	64	62	60	63	61	59	59	57
2	59	55	51	56	53	50	51	48
3	52	48	43	50	47	43	45	41
4	47	42	37	45	41	37	40	36
5	42	37	32	41	36	31	35	30

*CU values are in percent. pfc = .20%.

2.

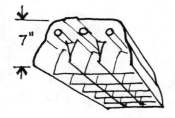
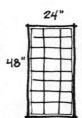

2′ × 4′, 3 F40 Fluorescent Lamps, Parabolic Lens, S/MH 1.3, VCP 94-99

CU Values (% CU in field, RCR left) *

pcc	80%			70%			50%	
pw	70	50	30	70	50	30	50	30
RCR								
0	74	74	74	72	72	72	69	69
1	70	68	67	69	67	66	65	63
2	66	63	61	65	62	60	60	58
3	62	58	55	61	57	54	55	53
4	58	54	50	58	53	49	51	48
5	55	49	45	54	49	45	47	44

*CU values are in percent. pfc = 20%.

3.

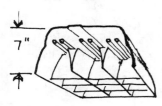

2′ × 2′, 3 F40 Fluorescent Lamps (compact), Parabolic Lens, S/MH 1.3, VCP 80-90

CU Values (% CU in field, RCR left) *

pcc	80%			70%			50%	
pw	70	50	30	70	50	30	50	30
RCR								
0	77	77	77	75	75	75	72	72
1	73	71	70	72	70	68	67	66
2	69	66	63	68	65	62	63	61
3	65	61	57	64	60	57	58	55
4	61	56	52	60	55	51	54	50
5	57	51	47	56	55	47	49	46

*CU values are in percent. pfc = 20%.

1' × 4', 1 F40 Fluorescent Lamp, Parabolic Lens, S/MH 1.3, VCP 94-97.

CU Values (% CU in field, RCR left)*

pcc	80%			70%			50%	
pw	70	50	30	70	50	30	50	30
RCR								
0	73	73	73	72	72	72	68	68
1	69	67	66	68	66	64	63	62
2	65	62	59	64	61	58	59	57
3	61	56	53	60	56	52	54	51
4	57	52	48	56	51	47	50	46
5	53	47	43	52	46	42	45	42

*CU values are in percent. pfc = 20%.

4.

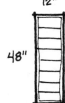

12"

8"

48"

2' × 2' with MH-HID Lamp, 175-250 or 400 Watt, S/MH 1.4, 12 Mounting Height.

CU Values (% CU in field, RCR left)*

pcc	80%			70%			50%	
pw	70	50	30	70	50	30	50	30
RCR								
1	81	77	74	79	75	72	71	69
2	74	69	64	72	69	62	63	60
3	68	61	55	66	62	53	56	52
4	63	54	48	62	55	46	51	46
5	58	49	42	57	50	41	45	40

*All CU values are in percent. pfc = 20%.

5.

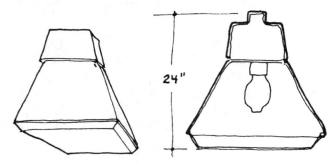

24"

MH-HID 1,000 Watt Lamp, S/MH 1.5, Mounting Height 25'.

CU Values (% CU in field, RCR left)*

pcc	80%			70%			50%	
pw	70	50	30	70	50	30	50	30
RCR								
1	86	84	82	85	82	79	78	77
2	81	79	76	80	77	74	74	72
3	76	74	70	75	73	70	70	68
4	72	70	66	70	68	65	66	64
5	67	65	61	66	64	61	63	60

*All CU values are in percent. pfc = 20%.

6.

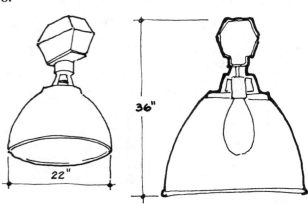

36"

22"

Indirect MH-HID 400 Watt Lamp, Suspend Fixture 3' Below Ceiling and 8' Above Floor

7.

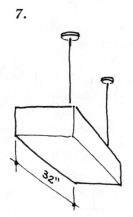

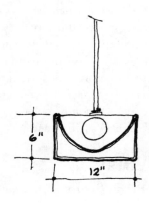

CU Values (CUs in field, RCR left)*

pcc	80%			70%			50%	
pw	70	50	30	70	50	30	50	30
RCR								
1	63	60	57	53	51	49	35	34
2	54	52	48	46	44	41	30	28
3	48	46	41	41	39	35	27	25
4	42	40	35	33	31	26	24	21
5	37	35	30	33	31	26	21	18

*All CU values are in percent. pfc = 20%.

Indirect Fluorescent, 2 F40 Lamps, Suspend 2' Below Ceiling

8.

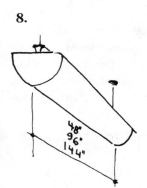

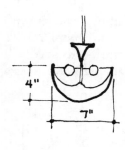

CU Values (CUs in field, RCR left)*

pcc	80%			70%			50%	
pw	70	50	30	70	50	30	50	30
RCR								
1	56	54	52	51	48	46	35	33
2	50	48	46	47	43	41	47	43
3	45	43	40	42	38	36	41	36
4	40	38	34	37	34	31	36	30
5	35	33	28	32	29	25	20	18

*All CU values are in percent. pfc = 20%.

Danger! These CU values are generic and approximate! Use manufacturers' CUs for precise calculations.

7.5 LIGHTING QUALITY NUMBERS

Glare is defined as an excessive or uncomfortable brightness contrast. Auto headlights glare at night but not in daylight. The reflected image of a lamp or window on this page can make reading difficult. Lighting engineers describe such images as "veiling reflections."

Three numerical indices—VCP, ESI, and CRF—attempt to quantify lighting quality.

VCP

Visual comfort probability (VCP) is a fixture glare rating for the worst viewing position in a test room (see Figure 7.7). VCP = 100 means that 100% of observers will be visually "comfortable" (untroubled by glare) in a room illuminated by rated fixtures. VCPs above 80 are considered comfortable.

ESI

Equivalent sphere illumination (ESI) is a term used to describe diffuse shadow-free lighting inside a uniformly illuminated white sphere. Laboratory spheres are used to compare and evaluate lighting installations because an observer looking into such a sphere will *not* see reflected glare or veiling reflections.

The extent to which a given lighting installation duplicates spherical lighting is sometimes quantified in *ESI footcandles*. Individuals who take pleasure in such activity calculate ESI footcandles and compare them with the "raw" horizontal footcandles a light meter measures. A dark room with shielded recessed fixtures might measure 50 "raw" footcandles at the work plane, but calculations could show only 20 spherical ESI footcandles. Raw footcandles and ESI footcandles will be nearly equal in rooms with lumi-nous ceilings, white walls, white furnishings, and white floors.

CRF

Contrast rendition factor (CRF) is a numerical index that compares the uniformity of a specific lighting installation for various viewer locations. Where ESI footcandles and raw footcandles are nearly equal, the CRF value approaches 1. Lighting installations with CRF values near 1.0 are theoretically glare-free, but they can be visually boring—lacking shade, shadows, and brightness contrast.

VDTs

Bright fixtures can create annoying veiling reflections on the specular viewing surface of TVs, computer monitors, and other *visual display terminals* (VDTs) (see Figure 7.8). Suffering users overcome such screen reflections by moving to change the reflection geometry. Lighting designers select indirect or well-shielded fixtures to minimize veiling reflections in VDT workspaces (see Figure 7.9).

Typical VDT screen brightness is about 70 cd/m^2, and many authorities recommend a maximum 5 to 1 ratio for brightness contrasts in VDT workspaces. A 5:1 contrast draws attention, but the adapted eye sees comfortably when field bright-

FIGURE 7.8

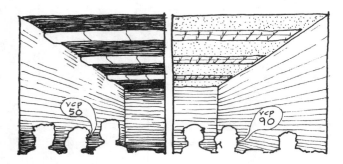

FIGURE 7.7

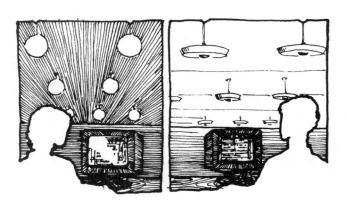

FIGURE 7.9

nesses range from 10 times to one-tenth the adaptation level. A window can easily be 100 times brighter than a computer screen, but some VDT users prefer the visual interest that windows provide in otherwise uniformly dim and bland VDT workspaces.

7.6 ESTIMATES

ESTIMATE WITH WATTS

Limiting the lighting load can reduce electrical costs. New schools and office buildings usually install lighting systems that use less than 2 *watts per square foot*. Review the following example and notice that lighting *watts/sq. ft.* is related to footcandles for a given lamp/fixture combination.

Example

A 20′ × 30′ classroom is illuminated at 55 footcandles by nine 3-lamp fluorescent fixtures. Each lamp is rated at 34 watts including ballast load.

Calculate:

* Watts/sq. ft.
* Footcandles per watt/sq. ft.
* Utility $/year to light the classroom

Answers

* 1.5 watts/sq. ft.

$$(9)(3)(34) \div (20)(30) = 1.5$$

* 36 footcandles per watt/sq. ft.

$$55 \div 1.5 = 36$$

* $184 utility $/year. Assumes 2,500 lighting hours per year and an average electrical cost of $0.08 per kWh including demand charges.

$$(9)(3)(34)(2,500)(\$0.08) = \$183.60$$

Do more calculations to compare alternate lamp-fixture combinations. Be sure to include ballast watts in such comparisons. Efficient electronic fluorescent ballasts may operate lamps at rated watts, but magnetic HID ballasts can add up to 15% to lamp watts.

FOOTCANDLES PER WATT/SQ. FT.

Rough estimates of the *maximum* expected footcandles in large rooms using the most efficient fluores-cent or MH-HID fixtures can be made using the following table.

Footcandles per Watt (per Square Foot)

Fixture	Space Reflectances	fc watt/ sq. ft.
Direct	High (white walls and ceiling)	50
	Low (dark surfaces and shelves)	40
Indirect	High (white walls and ceiling)	30
	Low (dark surfaces and shelves)	15

Example

Estimate lighting watts/sq. ft. when using direct fluorescent fixtures to provide 90 footcandles in an open office area with white walls and ceilings.

Answer

Allow 1.8 watts/sq. ft.

$$(90 \div 50 = 1.8)$$

FOOTCANDLES OR CD/M²

Because illuminance is invisible until reflected, Louis Erhardt proposes luminance (cd/m^2) as a better lighting index than footcandles. His lighting recommendations recognize the eye's ability to adapt to a great range of luminance values and see comfortably at luminances a log step above or below the adapted level.

Erhardt's Lighting Design Recommendations for Interior Activities

Activity	cd/m^2
Casual	1
Normal	10
Demanding	100

Louis Erhardt, *Lighting Design and Application*, January 1993.

Footcandles are easier to calculate than cd/m^2 because reflectances vary for each viewing position, but there is an approximate way to relate illuminance and luminance.

$$luminance = (illuminance)(reflectance) \div \pi$$

When reflectance is 32%, footcandles and cd/m^2 will be numerically equal. Interior scene reflectances range

FIGURE 7.10

from 16% to 64%, and 32% is probably typical for many residential interiors. Dark interiors such as a paneled library with bookshelves on all four walls might have an average reflectance near 16%. Light interiors, for example, a white tiled restroom, may approach 64%.

With 100 fc, the dark room = 50 cd/m^2 and the white room = 200 cd/m^2 (see Figure 7.10).

Convert footcandles to approximate cd/m^2 using the following ratios.

Reflectance	Luminance ÷ Illuminance
16% Low (dark walls and floors)	0.5
32% Average	1
64% High (most interior surfaces white)	2

If illuminance is 10 footcandles, luminance will be 5 cd/m^2 at 16%, 10 cd/m^2 at 32%, and 20 cd/m^2 at 64%.

7.7 LAMP TABLES

INCANDESCENT LAMPS

The following lamp data are provided for academic spotlighting calculations (see Figure 7.11). The tables are *not* inclusive. Many more lamps are available, and new lamps are introduced frequently. Use a current GE, Philips, or Osram Sylvania catalog instead of these tables for designs that will be bid and built.

- *Center beam candlepower* (CBCP) is the intensity in candelas at the center of the beam.
- *Beam spread* is the approximate total angle of the directed beam to where the beam intensity falls to 50% of the maximum value. Two beam spread angles are given for elliptical beams.

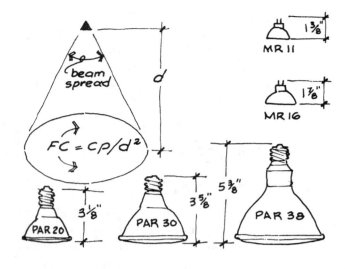

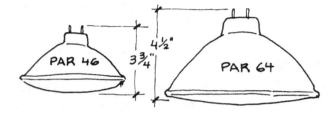

FIGURE 7.11

12 Volt Halogen Spot- and Floodlights*

Watts	Lamp	Beam Spread°	CBCP
20	MR11	30°	600
	MR16	11°	4,500
	MR16	24°	900
	MR16	36°	450
35	MR11	20°	3,000
	MR11	30°	1,300
	MR16	8°	8,100
	MR16	18°	3,240
	MR16	38°	870

12 Volt Halogen Spot- and Floodlights*

Watts	Lamp	Beam Spread°	CBCP
50	MR16	10°	10,800
	MR16	21°	3,330
	MR16	38°	1,395
	MR16	60°	630

120 Volt Halogen Spot- and Floodlights*

Watts	Lamp	Beam Spread°	CBCP
50	PAR20	8°	6,000
	PAR20	27°	1,500
50	PAR30	8°	17,000
	PAR30	26°	3,000
50	PAR38	9°	14,000
	PAR38	27°	3,000
100	PAR38	10°	29,000
	PAR38	27°	7,500

*Federal Energy Legislation bars 50, 75, and 100 watt R30 lamps, and 75, 100, and 150 watt R40 and PAR38 lamps. Halogen lamps are approved replacements for the PARs, and reduced wattage or halogen lamps replace R30s and R40s.

Standard A and PS incandescent lamps can be used with reflectors in floodlighting applications. Reflector design determines beam candlepower. Fixture manufacturers will provide such information.

120 Volt Incandescent Spot- and Floodlights*

Watts	Lamp	Beam Spread	CBCP
200	PAR46	12°×8°	31,000
	PAR46	27°×13°	11,500
300	PAR56	10°×8°	68,000
	PAR56	23°×11°	24,000
	PAR56	37°×18°	11,000
500	PAR64	12°×7°	110,000
	PAR64	23°×11°	37,000
	PAR64	42°×20°	13,000

*Lamp data produced with permission from *GE Lighting's Spectrum 9200 Lamp Catalog*, 21st edition, 1993.

FLUORESCENT LAMPS

Length, watts, and initial lumens are given for selected fluorescent lamps (see Figure 7.12). The following tables are *not* inclusive. Many more lamps are available, and new lamps are introduced frequently. Use these tables for academic calculations, but get a current GE, Philips, or Osram Sylvania catalog for designs that will be bid and built.

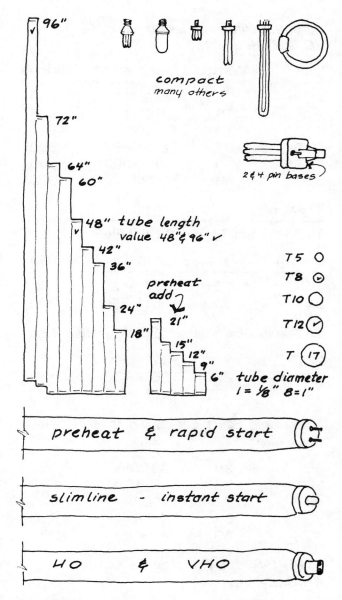

FIGURE 7.12

Use lamp watts for preliminary circuit load estimates, but remember ballast efficiency can increase watts. Use lumens in zonal cavity lighting calculations.

Compact

Compact Lamp and Ballast, 10,000 Hour Life

Length	Watts	Lamp	Lumens
5.4"	15	FLE15TBX	825
6"	20	FLE20TBX	1,200
6.6"	30	FLE30QBX	1,750

Compact Pin Base Lamp, 10,000 Hour Life

Length	Watts	Lamp	Lumens
4.2"	5	F5BX	250
7.5"	3	F13BX	825
10.5"	18	F18BX	1,250
12.8"	27	F27/24BX	1,800
16.5"	39	F39/36BX	2,850
22.5"	50	F50BX	4,000

Tubes

T12 Preheat Lamp, 9,000 Hour Life

Length	Watts	Lamp	Lumens
18"	15	F15T12CW	760
24"	20	F20T12SPX	1,300

T8 Rapid Start Lamp, 20,000 Hour Life

24"	17	F17T8SPX	1375
36"	25	F25T8SPX	2150
48"	32	F32T8SPX	2950

T12 Rapid Start Lamp, 18 to 20,000 Hour Life

36"	25	F30T12SP	2,025
36"	30	F30T12SPX	2,375
48"	32	F40SP	2,650
48"	34	F40SPX	2,900
48"	40	F40SPX	3,350

T12U U-Shape Rapid Start Lamp, 18,000 Hour

22.5"	35	F40LW-U	2,500
22.5"	40	F40SPX-U	3,100

T12 Slimline (Instant Start) Lamp, 12,000 Hour

60"	50	F60T12SP	3,750
72"	55	F72T12SPX	4,800
96"	60	F96T12SPX	6,000
96"	75	F96T12SPX	6,800

T8 Rapid Start Lamp, 15,000 Hour Life

96"	59	F96T8SPX	5,950

T12-HO 800 mA, High-Output Lamp, 12kWh

48"	60	F48T12SPX	4,350
72"	85	F72T12SPX	6,800
96"	110	F96T12SPX	9,350

T12-VHO 1,500 mA, Very-High-Output Lamp, 10,000 Hour Life

48"	110	F48T12CW	6,200
72"	165	F72T12CW	9,700
96"	215	F96T12CW	13,500

Reproduced with permission from *GE Lighting's Spectrum 9200 Lamp Catalog*, 21st edition © 1993.

Metal Halide

Spot- and Flood Applications

Watts	Lamp	Beam Spread	CBCP
70°	PAR38	12	50,000
70°	PAR38	40	6,500
100°	PAR38	12	54,000
100°	PAR38	40	10,000
150°	PAR64	3	300,000
150°	PAR64	13	50,000

The lamps below can also be used with reflectors in spot and flood applications.

Watts	Lamp	Life	CRI	Lumens
32	ED17	10k	70	2,500
50	ED17	5k	70	3,500
70	ED17	12k	70	5,500
100	BD17	15k	70	9,000
150	BD17	15k	70	13,000
175	BD17	15k	65	14,000
250	ED28	10k	65	21,500
400	ED37	20k	65	36,000
1,000	BT56	12k	65	110,000
1,500	BT56	14k	65	155,000
2,000	T7	3k	65	200,000

HID LAMPS

High-intensity discharge lamps are used for sports lighting, roadway lighting, and interior lighting (see Figure 7.13). When these lamps are used with reflectors in floodlighting applications, the reflector design determines beam intensity.

These tables are *not* inclusive. Many more lamps are available, and new lamps are introduced frequently. Use them for academic calculations, but get a current GE, Philips, or Osram Sylvania catalog for designs that will be bid and built.

Remember all HID lamps are ballasted. The ballast usually adds 15% to the rated lamp wattage. Also remember many HID lamps must be used in enclosed fixtures.

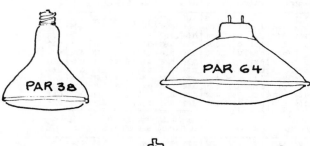

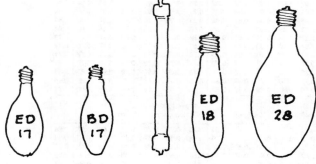

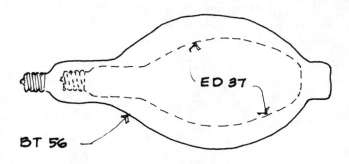

FIGURE 7.13

Sodium

HPS High-Pressure Sodium Clear

Watts	Lamp	Life	CRI	Lumens
250	ED18	24k	22	28,000
400	ED18	24k	22	51,000
1,000	ED18	24k	22	140,000

HPS High-Pressure Sodium Coated

Watts	Lamp	Life	CRI	Lumens
70	ED18	10k	65	3,800
150	ED18	15k	65	10,500
250	ED18	15k	65	22,500
400	ED18	15k	6	37,400

SOX Low-Pressure Sodium

Watts	Lamp	Life	CRI	Lumens
55	SOX35	18k	–	7,650
90	SOX90	16k	–	12,750
135	SOX135	16k	–	22,000
180	SOX180	16k	–	33,000

Mercury

Watts	Lamp	Life	CRI	Lumens
175	ED28	24k	15	7,950
250	ED28	24k	15	11,200
400	BT37	24k	15	22,100

Replacement sodium lamps are available, less watts, more lumens.
Lamp data reproduced with permission from *GE Lighting's Spectrum 9200 Lamp Catalog*, 21st edition.
Lamp length is not given in the illustrations because it varies with lamp wattage.

7.8 CU TABLES

Manufacturers provide CU tables for specific fixtures. The following selected generic CU tables are included for readers who find Section 7.4 too brief.

Typical Intensity Distribution / Typical Luminaire

ρcc →	80			70			50			30			10			0
ρw →	70	50	30	70	50	30	50	30	10	50	30	10	50	30	10	0

1 — Pendant diffusing sphere with incandescent lamp
EFF = 80.5% %DN = 55.9% %UP = 44.1% Lamp = 150A21IF SC (along, across, 45°) = 1.5, 1.5, 1.5

RCR	70	50	30	70	50	30	50	30	10	50	30	10	50	30	10	0
0	0.87	0.87	0.87	0.81	0.81	0.81	0.70	0.70	0.70	0.59	0.59	0.59	0.49	0.49	0.49	0.45
1	0.76	0.71	0.66	0.70	0.66	0.61	0.56	0.52	0.50	0.48	0.44	0.42	0.38	0.36	0.34	0.30
2	0.68	0.60	0.53	0.62	0.55	0.50	0.47	0.42	0.38	0.39	0.35	0.32	0.31	0.29	0.26	0.23
3	0.61	0.52	0.44	0.56	0.48	0.41	0.40	0.35	0.31	0.33	0.29	0.26	0.27	0.24	0.21	0.18
4	0.55	0.45	0.37	0.51	0.42	0.35	0.35	0.30	0.25	0.29	0.25	0.21	0.23	0.20	0.17	0.14
5	0.51	0.40	0.32	0.46	0.37	0.30	0.31	0.25	0.21	0.26	0.21	0.18	0.21	0.17	0.14	0.12
6	0.46	0.35	0.28	0.42	0.33	0.26	0.28	0.22	0.18	0.23	0.19	0.15	0.19	0.15	0.12	0.10
7	0.43	0.32	0.25	0.39	0.29	0.23	0.25	0.20	0.16	0.21	0.16	0.13	0.17	0.13	0.11	0.09
8	0.40	0.29	0.22	0.36	0.27	0.20	0.23	0.17	0.14	0.19	0.15	0.12	0.15	0.12	0.09	0.07
9	0.37	0.26	0.19	0.34	0.24	0.18	0.21	0.16	0.12	0.17	0.13	0.10	0.14	0.11	0.08	0.07
10	0.34	0.24	0.17	0.32	0.22	0.16	0.19	0.14	0.11	0.16	0.12	0.09	0.13	0.10	0.08	0.06

2 — Porcelain-enameled ventilated standard dome with inc. lamp
EFF = 86.5% %DN = 4.0% %UP = 96.0% Lamp = 100A21/5B SC (along, across, 45°) = N/A, N/A, N/A

RCR	70	50	30	70	50	30	50	30	10	50	30	10	50	30	10	0
0	0.99	0.99	0.99	0.97	0.97	0.97	0.93	0.93	0.93	0.89	0.89	0.89	0.85	0.85	0.85	0.83
1	0.91	0.87	0.84	0.89	0.85	0.82	0.82	0.79	0.77	0.79	0.76	0.74	0.76	0.74	0.72	0.71
2	0.83	0.76	0.70	0.80	0.74	0.69	0.71	0.67	0.63	0.69	0.65	0.62	0.66	0.63	0.60	0.59
3	0.75	0.66	0.59	0.73	0.65	0.59	0.62	0.57	0.53	0.60	0.55	0.52	0.58	0.54	0.51	0.49
4	0.69	0.58	0.51	0.67	0.57	0.50	0.55	0.49	0.44	0.53	0.48	0.44	0.51	0.47	0.43	0.41
5	0.63	0.52	0.44	0.61	0.51	0.44	0.49	0.43	0.38	0.47	0.42	0.37	0.46	0.41	0.37	0.35
6	0.58	0.46	0.39	0.56	0.46	0.38	0.44	0.38	0.33	0.43	0.37	0.33	0.41	0.36	0.32	0.31
7	0.53	0.42	0.34	0.52	0.41	0.34	0.40	0.33	0.29	0.39	0.33	0.29	0.38	0.32	0.28	0.27
8	0.50	0.38	0.31	0.48	0.37	0.31	0.36	0.30	0.26	0.35	0.30	0.25	0.34	0.29	0.25	0.24
9	0.46	0.35	0.28	0.45	0.34	0.28	0.33	0.27	0.23	0.32	0.27	0.23	0.32	0.26	0.23	0.21
10	0.43	0.32	0.25	0.42	0.32	0.25	0.31	0.25	0.21	0.30	0.24	0.21	0.29	0.24	0.20	0.19

3 — Bare lamp PAR-38 flood
EFF = 100% %DN = 100% %UP = 0% Lamp = 150PAR38FL SC (along, across, 45°) = 0.6, 0.6, 0.6

RCR	70	50	30	70	50	30	50	30	10	50	30	10	50	30	10	0
0	1.20	1.20	1.20	1.17	1.17	1.17	1.12	1.12	1.12	1.07	1.07	1.07	1.03	1.03	1.03	1.00
1	1.14	1.11	1.08	1.11	1.08	1.06	1.04	1.02	1.00	1.01	0.99	0.97	0.97	0.96	0.95	0.93
2	1.08	1.02	0.98	1.05	1.01	0.97	0.97	0.94	0.91	0.94	0.92	0.89	0.92	0.89	0.87	0.86
3	1.02	0.95	0.90	1.00	0.94	0.89	0.91	0.87	0.84	0.89	0.85	0.82	0.86	0.84	0.81	0.79
4	0.97	0.89	0.83	0.95	0.88	0.83	0.86	0.81	0.78	0.84	0.80	0.77	0.82	0.79	0.76	0.74
5	0.92	0.84	0.78	0.91	0.83	0.77	0.81	0.76	0.72	0.79	0.75	0.72	0.78	0.74	0.71	0.70
6	0.88	0.79	0.73	0.87	0.78	0.73	0.77	0.72	0.68	0.75	0.71	0.68	0.74	0.70	0.67	0.66
7	0.84	0.75	0.69	0.83	0.74	0.69	0.73	0.68	0.64	0.72	0.67	0.64	0.71	0.67	0.64	0.62
8	0.81	0.71	0.66	0.80	0.71	0.65	0.70	0.65	0.61	0.69	0.64	0.61	0.68	0.64	0.61	0.59
9	0.78	0.68	0.62	0.77	0.68	0.62	0.67	0.62	0.58	0.66	0.61	0.58	0.65	0.61	0.58	0.57
10	0.75	0.65	0.60	0.74	0.65	0.59	0.64	0.59	0.56	0.63	0.59	0.56	0.63	0.58	0.55	0.54

4 — PAR-38 Flood with spec. anodized reflector (45 deg. cutoff)
EFF = 66.2% %DN = 100 %UP = 0 Lamp = 150PAR38FL* SC (along, across, 45°) = 0.6, 0.6, 0.6

RCR	70	50	30	70	50	30	50	30	10	50	30	10	50	30	10	0
0	1.10	1.10	1.10	1.07	1.07	1.07	1.02	1.02	1.02	0.98	0.98	0.98	0.94	0.94	0.94	0.92
1	1.06	1.03	1.02	1.03	1.01	1.00	0.98	0.96	0.95	0.94	0.93	0.92	0.91	0.90	0.90	0.88
2	1.02	0.98	0.95	1.00	0.96	0.94	0.94	0.91	0.89	0.91	0.89	0.88	0.88	0.87	0.86	0.84
3	0.98	0.93	0.90	0.96	0.92	0.89	0.90	0.87	0.85	0.88	0.85	0.83	0.86	0.84	0.82	0.81
4	0.94	0.89	0.85	0.93	0.88	0.84	0.86	0.83	0.80	0.84	0.82	0.79	0.83	0.81	0.79	0.77
5	0.91	0.85	0.81	0.90	0.84	0.80	0.83	0.79	0.77	0.81	0.78	0.76	0.80	0.77	0.75	0.74
6	0.88	0.81	0.77	0.87	0.81	0.77	0.80	0.76	0.73	0.78	0.75	0.73	0.77	0.75	0.72	0.71
7	0.85	0.78	0.74	0.84	0.78	0.74	0.77	0.73	0.70	0.76	0.72	0.70	0.75	0.72	0.70	0.69
8	0.82	0.75	0.71	0.81	0.75	0.71	0.74	0.70	0.68	0.73	0.70	0.67	0.72	0.69	0.67	0.66
9	0.79	0.72	0.68	0.78	0.72	0.68	0.71	0.68	0.65	0.70	0.67	0.65	0.70	0.67	0.65	0.64
10	0.77	0.70	0.66	0.76	0.69	0.65	0.69	0.65	0.63	0.68	0.65	0.62	0.68	0.65	0.62	0.61

5 — PAR-38 Flood with black baffle
EFF = 66.2% %DN = 100 %UP = 0 Lamp = 150PAR38FL* SC (along, across, 45°) = 0.6, 0.6, 0.6

RCR	70	50	30	70	50	30	50	30	10	50	30	10	50	30	10	0
0	0.79	0.79	0.79	0.77	0.77	0.77	0.74	0.74	0.74	0.70	0.70	0.70	0.68	0.68	0.68	0.66
1	0.76	0.75	0.73	0.75	0.73	0.72	0.71	0.70	0.69	0.68	0.68	0.67	0.66	0.66	0.65	0.64
2	0.74	0.71	0.69	0.72	0.70	0.68	0.68	0.67	0.65	0.66	0.65	0.64	0.64	0.63	0.63	0.62
3	0.71	0.68	0.66	0.70	0.67	0.65	0.66	0.64	0.62	0.64	0.63	0.61	0.63	0.61	0.60	0.60
4	0.69	0.65	0.63	0.68	0.65	0.62	0.63	0.61	0.60	0.62	0.60	0.59	0.61	0.60	0.58	0.58
5	0.67	0.63	0.60	0.66	0.62	0.60	0.61	0.59	0.57	0.60	0.58	0.57	0.59	0.58	0.56	0.56
6	0.65	0.61	0.58	0.64	0.60	0.58	0.59	0.57	0.55	0.59	0.57	0.55	0.58	0.56	0.55	0.54
7	0.63	0.59	0.56	0.62	0.58	0.56	0.57	0.55	0.53	0.57	0.55	0.53	0.56	0.54	0.53	0.52
8	0.61	0.57	0.54	0.60	0.56	0.54	0.56	0.53	0.52	0.55	0.53	0.52	0.55	0.53	0.51	0.51
9	0.59	0.55	0.52	0.59	0.55	0.52	0.54	0.52	0.50	0.54	0.51	0.50	0.53	0.51	0.50	0.49
10	0.58	0.53	0.51	0.57	0.53	0.50	0.53	0.50	0.49	0.52	0.50	0.48	0.52	0.50	0.48	0.48

Reproduced from the *IESNA Lighting Handbook, 9th Edition,*. 2000, courtesy of the Illuminating Engineering Society of North America.

Typical Luminaire	Typical Intensity Distribution	ρcc →	80			70			50			30			10			0
		ρw →	70	50	30	70	50	30	50	30	10	50	30	10	50	30	10	0

6 — A-lamp downlight with spec. anodized reflector
EFF = 96.2% % DN = 100% % UP = 0% Lamp = 150A21IF SC (along, across, 45°) = 1.2, 1.2, 1.1

RCR ↓	70	50	30	70	50	30	50	30	10	50	30	10	50	30	10	0
0	0.82	0.82	0.82	0.81	0.81	0.81	0.77	0.77	0.77	0.74	0.74	0.74	0.71	0.71	0.71	0.69
1	0.78	0.76	0.75	0.77	0.75	0.73	0.72	0.71	0.70	0.70	0.68	0.68	0.67	0.66	0.66	0.64
2	0.74	0.71	0.68	0.73	0.70	0.67	0.67	0.65	0.63	0.65	0.63	0.62	0.63	0.62	0.61	0.59
3	0.70	0.66	0.62	0.69	0.65	0.61	0.63	0.60	0.58	0.61	0.59	0.57	0.59	0.57	0.56	0.55
4	0.66	0.61	0.57	0.65	0.60	0.56	0.58	0.55	0.53	0.57	0.54	0.52	0.56	0.53	0.51	0.50
5	0.63	0.56	0.52	0.61	0.56	0.52	0.55	0.51	0.48	0.53	0.50	0.48	0.52	0.50	0.48	0.46
6	0.59	0.53	0.48	0.58	0.52	0.48	0.51	0.47	0.45	0.50	0.47	0.44	0.49	0.46	0.44	0.43
7	0.56	0.49	0.45	0.55	0.49	0.44	0.48	0.44	0.41	0.47	0.43	0.41	0.46	0.43	0.41	0.40
8	0.53	0.46	0.41	0.52	0.45	0.41	0.45	0.41	0.38	0.44	0.40	0.38	0.43	0.40	0.38	0.37
9	0.50	0.43	0.39	0.49	0.43	0.38	0.42	0.38	0.35	0.41	0.38	0.35	0.41	0.37	0.35	0.34
10	0.47	0.40	0.36	0.47	0.40	0.36	0.39	0.36	0.33	0.39	0.35	0.33	0.38	0.35	0.33	0.32

7 — 8" Open reflector downlight (32W CFL)
EFF = 64.9% % DN = 100% % UP = 0% Lamp = CF6/32 SC (along, across, 45°) = 1.1, 1.1, 1.1

RCR ↓	70	50	30	70	50	30	50	30	10	50	30	10	50	30	10	0
0	0.77	0.77	0.77	0.75	0.75	0.75	0.72	0.72	0.72	0.69	0.69	0.69	0.66	0.66	0.66	0.65
1	0.73	0.72	0.70	0.72	0.70	0.69	0.67	0.66	0.65	0.65	0.64	0.63	0.63	0.62	0.61	0.60
2	0.69	0.66	0.63	0.68	0.65	0.62	0.63	0.61	0.59	0.61	0.59	0.58	0.59	0.58	0.57	0.55
3	0.66	0.61	0.58	0.64	0.60	0.57	0.59	0.56	0.54	0.57	0.55	0.53	0.55	0.54	0.52	0.51
4	0.62	0.57	0.53	0.61	0.56	0.52	0.55	0.51	0.49	0.53	0.51	0.48	0.52	0.50	0.48	0.47
5	0.58	0.53	0.49	0.57	0.52	0.48	0.51	0.48	0.45	0.50	0.47	0.45	0.49	0.46	0.44	0.43
6	0.55	0.49	0.45	0.54	0.48	0.45	0.47	0.44	0.41	0.46	0.43	0.41	0.46	0.43	0.41	0.40
7	0.52	0.46	0.41	0.51	0.45	0.41	0.44	0.41	0.38	0.44	0.40	0.38	0.43	0.40	0.38	0.37
8	0.49	0.43	0.38	0.48	0.42	0.38	0.42	0.38	0.35	0.41	0.38	0.35	0.40	0.37	0.35	0.34
9	0.47	0.40	0.36	0.46	0.40	0.36	0.39	0.35	0.33	0.38	0.35	0.33	0.38	0.35	0.33	0.32
10	0.44	0.37	0.33	0.44	0.37	0.33	0.37	0.33	0.31	0.36	0.33	0.30	0.36	0.33	0.30	0.29

8 — 8" Open reflector downlight (2-26W CFL)
EFF = 61.6% % DN = 100% % UP = 0% Lamp = (2) CFQ26 SC (along, across, 45°) = 1.5, 1.6, 1.5

RCR ↓	70	50	30	70	50	30	50	30	10	50	30	10	50	30	10	0
0	0.73	0.73	0.73	0.72	0.72	0.72	0.68	0.68	0.68	0.66	0.66	0.66	0.63	0.63	0.63	0.62
1	0.69	0.67	0.66	0.68	0.66	0.64	0.63	0.62	0.61	0.61	0.60	0.59	0.59	0.58	0.57	0.56
2	0.65	0.61	0.59	0.64	0.60	0.58	0.58	0.56	0.54	0.56	0.55	0.53	0.55	0.53	0.52	0.51
3	0.61	0.56	0.52	0.59	0.55	0.52	0.53	0.51	0.48	0.52	0.49	0.47	0.50	0.48	0.47	0.46
4	0.57	0.51	0.47	0.56	0.50	0.47	0.49	0.46	0.43	0.48	0.45	0.43	0.47	0.44	0.42	0.41
5	0.53	0.47	0.42	0.52	0.46	0.42	0.45	0.41	0.39	0.44	0.41	0.38	0.43	0.40	0.38	0.37
6	0.50	0.43	0.38	0.48	0.42	0.38	0.41	0.38	0.35	0.40	0.37	0.35	0.40	0.37	0.34	0.33
7	0.46	0.39	0.35	0.45	0.39	0.35	0.38	0.34	0.31	0.37	0.34	0.31	0.36	0.33	0.31	0.30
8	0.43	0.36	0.32	0.42	0.36	0.32	0.35	0.31	0.29	0.34	0.31	0.28	0.34	0.31	0.28	0.27
9	0.41	0.33	0.29	0.40	0.33	0.29	0.33	0.29	0.26	0.32	0.28	0.26	0.31	0.28	0.26	0.25
10	0.38	0.31	0.27	0.37	0.31	0.27	0.30	0.26	0.24	0.30	0.26	0.24	0.29	0.26	0.24	0.23

9 — 8" Round with cross baffles
EFF = 40.2% % DN = 100 % UP = 0 Lamp = (2) CFQ26 SC (along, across, 45°) = 1.2, 1.2, 1.1

RCR ↓	70	50	30	70	50	30	50	30	10	50	30	10	50	30	10	0
0	0.47	0.47	0.47	0.46	0.46	0.46	0.44	0.44	0.44	0.42	0.42	0.42	0.41	0.41	0.41	0.40
1	0.45	0.44	0.43	0.44	0.43	0.42	0.41	0.41	0.40	0.40	0.39	0.39	0.39	0.38	0.38	0.37
2	0.43	0.41	0.39	0.42	0.40	0.38	0.39	0.37	0.36	0.37	0.36	0.35	0.36	0.35	0.35	0.34
3	0.40	0.37	0.35	0.39	0.37	0.35	0.36	0.34	0.33	0.35	0.33	0.32	0.34	0.33	0.32	0.31
4	0.38	0.35	0.32	0.37	0.34	0.32	0.33	0.31	0.30	0.32	0.31	0.29	0.32	0.30	0.29	0.28
5	0.36	0.32	0.29	0.35	0.32	0.29	0.31	0.29	0.27	0.30	0.28	0.27	0.30	0.28	0.27	0.26
6	0.34	0.30	0.27	0.33	0.29	0.27	0.29	0.27	0.25	0.28	0.26	0.25	0.28	0.26	0.25	0.24
7	0.32	0.28	0.25	0.31	0.27	0.25	0.27	0.25	0.23	0.26	0.24	0.23	0.26	0.24	0.23	0.22
8	0.30	0.26	0.23	0.29	0.25	0.23	0.25	0.23	0.21	0.25	0.23	0.21	0.24	0.22	0.21	0.20
9	0.28	0.24	0.21	0.28	0.24	0.21	0.23	0.21	0.20	0.23	0.21	0.19	0.23	0.21	0.19	0.19
10	0.27	0.22	0.20	0.26	0.22	0.20	0.22	0.20	0.18	0.22	0.20	0.18	0.21	0.19	0.18	0.17

10 — Metal halide downlight
EFF = 63.8% % DN = 78.4 % UP = 21.6 Lamp = M100/C/U SC (along, across, 45°) = 1.2, 1.2, 1.1

RCR ↓	70	50	30	70	50	30	50	30	10	50	30	10	50	30	10	0
0	0.76	0.76	0.76	0.74	0.74	0.74	0.71	0.71	0.71	0.68	0.68	0.68	0.65	0.65	0.65	0.64
1	0.72	0.70	0.68	0.70	0.69	0.67	0.66	0.65	0.64	0.64	0.63	0.62	0.61	0.61	0.60	0.59
2	0.68	0.64	0.62	0.66	0.63	0.61	0.61	0.59	0.57	0.59	0.57	0.56	0.57	0.56	0.55	0.54
3	0.64	0.59	0.56	0.63	0.58	0.55	0.57	0.54	0.52	0.55	0.53	0.51	0.54	0.52	0.50	0.49
4	0.60	0.55	0.51	0.59	0.54	0.50	0.52	0.49	0.47	0.51	0.48	0.46	0.50	0.48	0.46	0.45
5	0.57	0.51	0.46	0.55	0.50	0.46	0.49	0.45	0.43	0.48	0.45	0.42	0.47	0.44	0.42	0.41
6	0.53	0.47	0.43	0.52	0.46	0.42	0.45	0.42	0.39	0.44	0.41	0.39	0.43	0.41	0.38	0.37
7	0.50	0.43	0.39	0.49	0.43	0.39	0.42	0.39	0.36	0.41	0.38	0.36	0.41	0.38	0.35	0.34
8	0.47	0.41	0.36	0.46	0.40	0.36	0.39	0.36	0.33	0.39	0.35	0.33	0.38	0.35	0.33	0.32
9	0.45	0.38	0.34	0.44	0.37	0.33	0.37	0.33	0.31	0.36	0.33	0.31	0.36	0.33	0.30	0.29
10	0.42	0.35	0.31	0.42	0.35	0.31	0.35	0.31	0.28	0.34	0.31	0.28	0.34	0.30	0.28	0.27

Reproduced from the *IESNA Lighting Handbook, 9th Edition,*. 2000, courtesy of the Illuminating Engineering Society of North America.

Typical Luminaire	Typical Intensity Distribution	ρcc →	80			70			50			30			10			0
		ρw →	70	50	30	70	50	30	50	30	10	50	30	10	50	30	10	0
		RCR ↓	EFF = 55.9%			% DN = 76.2%			% UP = 23.8%			Lamp = 22 & 32W circ.* SC (along, across, 45°) = 1.3, 1.3, 1.5						

11 — CFL surface-mounted disk

RCR	70	50	30	70	50	30	50	30	10	50	30	10	50	30	10	0
0	0.63	0.63	0.63	0.60	0.60	0.60	0.55	0.55	0.55	0.50	0.50	0.50	0.45	0.45	0.45	0.43
1	0.57	0.54	0.51	0.54	0.51	0.48	0.46	0.44	0.42	0.42	0.40	0.39	0.38	0.36	0.35	0.33
2	0.51	0.46	0.42	0.48	0.44	0.40	0.40	0.37	0.34	0.36	0.33	0.31	0.32	0.30	0.29	0.27
3	0.46	0.40	0.35	0.44	0.38	0.34	0.35	0.31	0.28	0.31	0.28	0.26	0.28	0.26	0.24	0.22
4	0.42	0.35	0.30	0.40	0.34	0.29	0.31	0.27	0.24	0.28	0.24	0.22	0.25	0.22	0.20	0.19
5	0.39	0.31	0.28	0.37	0.30	0.25	0.27	0.23	0.20	0.25	0.21	0.19	0.22	0.20	0.17	0.16
6	0.36	0.28	0.23	0.34	0.27	0.22	0.24	0.20	0.18	0.22	0.19	0.16	0.20	0.17	0.15	0.14
7	0.33	0.25	0.20	0.31	0.24	0.20	0.22	0.18	0.15	0.20	0.17	0.14	0.18	0.16	0.13	0.12
8	0.31	0.23	0.18	0.29	0.22	0.18	0.20	0.16	0.14	0.18	0.15	0.13	0.17	0.14	0.12	0.11
9	0.28	0.21	0.16	0.27	0.20	0.16	0.18	0.15	0.12	0.17	0.14	0.11	0.16	0.13	0.11	0.10
10	0.27	0.19	0.15	0.25	0.19	0.14	0.17	0.13	0.11	0.16	0.13	0.10	0.14	0.12	0.10	0.09

12 — High bay, open metal reflector, narrow
EFF = 87.5% % DN = 85.9% % UP = 1.6% Lamp = M400/C/U SC (along, across, 45°) = 1.1, 1.1, 1

RCR	70	50	30	70	50	30	50	30	10	50	30	10	50	30	10	0
0	1.04	1.04	1.04	1.01	1.01	1.01	0.96	0.96	0.96	0.92	0.92	0.92	0.88	0.88	0.88	0.86
1	0.98	0.95	0.93	0.96	0.93	0.91	0.89	0.87	0.86	0.86	0.84	0.83	0.82	0.81	0.80	0.78
2	0.92	0.87	0.83	0.90	0.85	0.82	0.82	0.79	0.76	0.79	0.77	0.74	0.77	0.75	0.73	0.71
3	0.86	0.80	0.75	0.84	0.78	0.74	0.76	0.72	0.69	0.73	0.70	0.67	0.71	0.68	0.66	0.64
4	0.81	0.73	0.68	0.79	0.72	0.67	0.70	0.66	0.62	0.68	0.64	0.61	0.66	0.63	0.60	0.59
5	0.76	0.68	0.62	0.74	0.67	0.61	0.65	0.60	0.56	0.63	0.59	0.56	0.62	0.58	0.55	0.54
6	0.72	0.63	0.57	0.70	0.62	0.56	0.60	0.55	0.52	0.59	0.54	0.51	0.57	0.54	0.51	0.49
7	0.67	0.58	0.52	0.66	0.58	0.52	0.56	0.51	0.47	0.55	0.50	0.47	0.54	0.50	0.47	0.45
8	0.64	0.54	0.48	0.62	0.54	0.48	0.53	0.47	0.44	0.51	0.47	0.43	0.50	0.46	0.43	0.42
9	0.60	0.51	0.45	0.59	0.50	0.45	0.49	0.44	0.41	0.48	0.44	0.40	0.47	0.43	0.40	0.39
10	0.57	0.48	0.42	0.56	0.47	0.42	0.46	0.41	0.38	0.45	0.41	0.38	0.45	0.40	0.37	0.36

13 — High bay, open metal reflector, medium
EFF = 83.9% % DN = 95.2% % UP = 4.8% Lamp = M400/C/U SC (along, across, 45°) = 1.6, 1.6, 1.4

RCR	70	50	30	70	50	30	50	30	10	50	30	10	50	30	10	0
0	0.99	0.99	0.99	0.96	0.96	0.96	0.91	0.91	0.91	0.88	0.86	0.86	0.82	0.82	0.82	0.80
1	0.93	0.90	0.87	0.90	0.88	0.85	0.83	0.81	0.80	0.80	0.78	0.77	0.76	0.75	0.74	0.72
2	0.86	0.81	0.77	0.84	0.79	0.75	0.76	0.73	0.70	0.73	0.70	0.68	0.70	0.68	0.66	0.64
3	0.80	0.73	0.68	0.78	0.72	0.67	0.70	0.65	0.61	0.66	0.63	0.60	0.64	0.61	0.58	0.57
4	0.75	0.67	0.61	0.73	0.65	0.60	0.63	0.58	0.54	0.61	0.57	0.53	0.58	0.55	0.52	0.51
5	0.70	0.61	0.54	0.68	0.59	0.54	0.57	0.52	0.48	0.55	0.51	0.48	0.54	0.50	0.47	0.45
6	0.65	0.55	0.49	0.63	0.54	0.48	0.53	0.47	0.43	0.51	0.46	0.43	0.49	0.45	0.42	0.41
7	0.60	0.51	0.44	0.59	0.50	0.44	0.48	0.43	0.39	0.47	0.42	0.39	0.45	0.41	0.38	0.37
8	0.56	0.47	0.40	0.55	0.46	0.40	0.44	0.39	0.35	0.43	0.38	0.35	0.42	0.38	0.34	0.33
9	0.53	0.43	0.37	0.52	0.42	0.36	0.41	0.36	0.32	0.40	0.35	0.32	0.39	0.35	0.31	0.30
10	0.50	0.40	0.34	0.48	0.39	0.33	0.38	0.33	0.29	0.37	0.32	0.29	0.36	0.32	0.29	0.27

14 — High bay, open metal reflector, wide
EFF = 83.8% % DN = 97 % UP = 3 Lamp = M400/C/U SC (along, across, 45°) = 1.9, 1.9, 1.7

RCR	70	50	30	70	50	30	50	30	10	50	30	10	50	30	10	0
0	0.99	0.99	0.99	0.97	0.97	0.97	0.92	0.92	0.92	0.87	0.87	0.87	0.83	0.83	0.83	0.81
1	0.92	0.89	0.86	0.90	0.87	0.84	0.83	0.81	0.79	0.80	0.78	0.76	0.76	0.75	0.74	0.72
2	0.85	0.80	0.75	0.83	0.78	0.73	0.75	0.71	0.68	0.72	0.69	0.66	0.69	0.66	0.64	0.62
3	0.79	0.71	0.65	0.76	0.70	0.64	0.67	0.62	0.58	0.64	0.60	0.57	0.62	0.59	0.56	0.54
4	0.72	0.63	0.57	0.70	0.62	0.56	0.60	0.55	0.51	0.58	0.53	0.50	0.56	0.52	0.49	0.47
5	0.67	0.57	0.50	0.65	0.56	0.50	0.54	0.48	0.44	0.52	0.47	0.43	0.50	0.46	0.43	0.41
6	0.62	0.51	0.44	0.60	0.50	0.44	0.49	0.43	0.39	0.47	0.42	0.38	0.46	0.41	0.38	0.36
7	0.57	0.46	0.40	0.56	0.46	0.39	0.44	0.38	0.34	0.43	0.38	0.34	0.42	0.37	0.33	0.32
8	0.53	0.42	0.35	0.52	0.42	0.35	0.40	0.34	0.30	0.39	0.34	0.30	0.38	0.33	0.30	0.28
9	0.49	0.39	0.32	0.48	0.38	0.32	0.37	0.31	0.27	0.36	0.31	0.27	0.35	0.30	0.27	0.25
10	0.46	0.35	0.29	0.45	0.35	0.29	0.34	0.28	0.24	0.33	0.28	0.24	0.32	0.27	0.24	0.22

15 — High bay, open prismatic reflector, narrow
EFF = 61.4% % DN = 80.6 % UP = 19.4 Lamp = M400/C/U SC (along, across, 45°) = 1.1, 1.1, 1.1

RCR	70	50	30	70	50	30	50	30	10	50	30	10	50	30	10	0
0	0.70	0.70	0.70	0.67	0.67	0.67	0.62	0.62	0.62	0.56	0.56	0.56	0.52	0.52	0.52	0.49
1	0.65	0.62	0.60	0.62	0.59	0.57	0.55	0.53	0.52	0.50	0.49	0.48	0.46	0.45	0.44	0.42
2	0.60	0.55	0.52	0.57	0.53	0.50	0.49	0.47	0.45	0.46	0.44	0.42	0.42	0.41	0.39	0.37
3	0.56	0.50	0.46	0.53	0.48	0.44	0.45	0.42	0.39	0.42	0.39	0.37	0.39	0.37	0.35	0.33
4	0.52	0.45	0.41	0.49	0.44	0.40	0.41	0.38	0.35	0.38	0.35	0.33	0.36	0.33	0.32	0.30
5	0.48	0.41	0.37	0.46	0.40	0.36	0.38	0.34	0.31	0.35	0.32	0.30	0.33	0.31	0.29	0.27
6	0.45	0.38	0.33	0.43	0.37	0.33	0.35	0.31	0.28	0.33	0.30	0.27	0.31	0.28	0.26	0.25
7	0.42	0.35	0.30	0.40	0.34	0.30	0.32	0.28	0.26	0.30	0.27	0.25	0.29	0.26	0.24	0.23
8	0.40	0.32	0.28	0.38	0.31	0.27	0.30	0.26	0.24	0.28	0.25	0.23	0.27	0.24	0.22	0.21
9	0.37	0.30	0.26	0.36	0.29	0.25	0.28	0.24	0.22	0.26	0.23	0.21	0.25	0.22	0.20	0.19
10	0.35	0.28	0.24	0.34	0.27	0.23	0.26	0.22	0.20	0.25	0.22	0.19	0.23	0.21	0.19	0.18

Reproduced from the *IESNA Lighting Handbook, 9th Edition*,. 2000, courtesy of the Illuminating Engineering Society of North America.

Header (applies to all tables):

RCR	ρcc → 80			70			50			30			10			0
ρw →	70	50	30	70	50	30	50	30	10	50	30	10	50	30	10	0

16 — High bay, open prismatic reflector, medium

EFF = 59% %DN = 77.9% %UP = 22.1% Lamp = M400/C/U SC (along, across, 45°) = 1.3, 1.3, 1.2

RCR	70	50	30	70	50	30	50	30	10	50	30	10	50	30	10	0
0	0.67	0.67	0.67	0.64	0.64	0.64	0.58	0.58	0.58	0.53	0.53	0.53	0.48	0.48	0.48	0.46
1	0.62	0.59	0.57	0.59	0.57	0.55	0.52	0.50	0.49	0.48	0.46	0.45	0.43	0.43	0.42	0.40
2	0.57	0.53	0.50	0.55	0.51	0.48	0.47	0.45	0.43	0.43	0.41	0.40	0.40	0.38	0.37	0.35
3	0.53	0.48	0.44	0.51	0.46	0.43	0.43	0.40	0.37	0.40	0.37	0.35	0.36	0.35	0.33	0.31
4	0.50	0.44	0.39	0.47	0.42	0.38	0.39	0.36	0.33	0.36	0.34	0.32	0.34	0.32	0.30	0.28
5	0.46	0.40	0.35	0.44	0.38	0.34	0.36	0.33	0.30	0.33	0.31	0.28	0.31	0.29	0.27	0.26
6	0.43	0.37	0.32	0.41	0.35	0.31	0.33	0.30	0.27	0.31	0.28	0.26	0.29	0.27	0.25	0.23
7	0.40	0.34	0.29	0.39	0.33	0.28	0.31	0.27	0.25	0.29	0.26	0.24	0.27	0.24	0.23	0.21
8	0.38	0.31	0.27	0.36	0.30	0.26	0.28	0.25	0.22	0.27	0.24	0.22	0.25	0.23	0.21	0.20
9	0.36	0.29	0.24	0.34	0.28	0.24	0.26	0.23	0.21	0.25	0.22	0.20	0.23	0.21	0.19	0.18
10	0.33	0.27	0.23	0.32	0.26	0.22	0.24	0.21	0.19	0.23	0.20	0.18	0.22	0.19	0.18	0.17

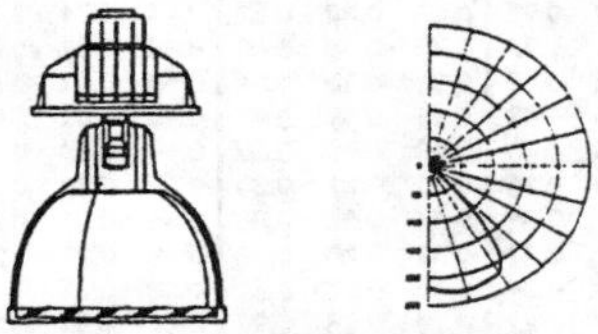

17 — High bay, open prismatic reflector, wide

EFF = 61.5% %DN = 83.7% %UP = 16.3% Lamp = M400/C/U SC (along, across, 45°) = 2.2, 2.2, 1.8

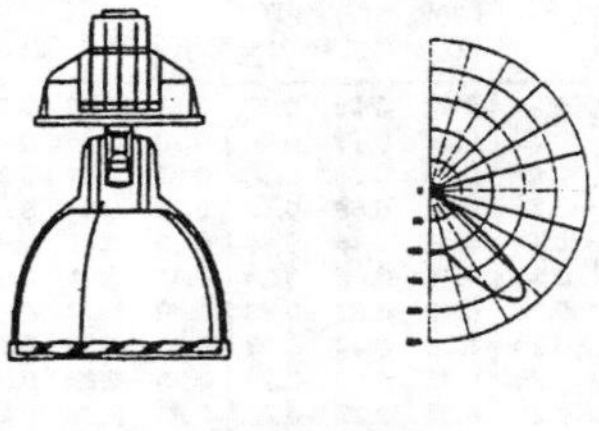

RCR	70	50	30	70	50	30	50	30	10	50	30	10	50	30	10	0
0	0.71	0.71	0.71	0.68	0.68	0.68	0.63	0.63	0.63	0.58	0.58	0.58	0.54	0.54	0.54	0.51
1	0.65	0.62	0.59	0.62	0.59	0.57	0.55	0.53	0.52	0.51	0.50	0.48	0.47	0.46	0.45	0.43
2	0.59	0.55	0.51	0.57	0.53	0.49	0.49	0.46	0.43	0.45	0.43	0.41	0.42	0.40	0.39	0.37
3	0.54	0.48	0.44	0.52	0.47	0.42	0.43	0.40	0.37	0.40	0.38	0.35	0.38	0.35	0.33	0.32
4	0.50	0.43	0.38	0.48	0.42	0.37	0.39	0.35	0.32	0.36	0.33	0.30	0.34	0.31	0.29	0.27
5	0.46	0.38	0.33	0.44	0.37	0.32	0.35	0.31	0.27	0.32	0.29	0.26	0.30	0.27	0.25	0.24
6	0.42	0.34	0.29	0.40	0.33	0.28	0.31	0.27	0.24	0.29	0.26	0.23	0.27	0.24	0.22	0.21
7	0.39	0.31	0.26	0.37	0.30	0.25	0.28	0.24	0.21	0.27	0.23	0.20	0.25	0.22	0.19	0.18
8	0.36	0.28	0.23	0.35	0.27	0.22	0.26	0.21	0.18	0.24	0.20	0.18	0.23	0.19	0.17	0.16
9	0.34	0.25	0.21	0.32	0.25	0.20	0.23	0.19	0.16	0.22	0.18	0.16	0.21	0.18	0.15	0.14
10	0.31	0.23	0.19	0.30	0.23	0.18	0.21	0.17	0.15	0.20	0.17	0.14	0.19	0.16	0.14	0.12

18 — Low bay with drop lens, narrow

EFF = 72.5% %DN = 97.8% %UP = 2.2% Lamp = M400/C/U SC (along, across, 45°) = 1.7, 1.7, 1.7

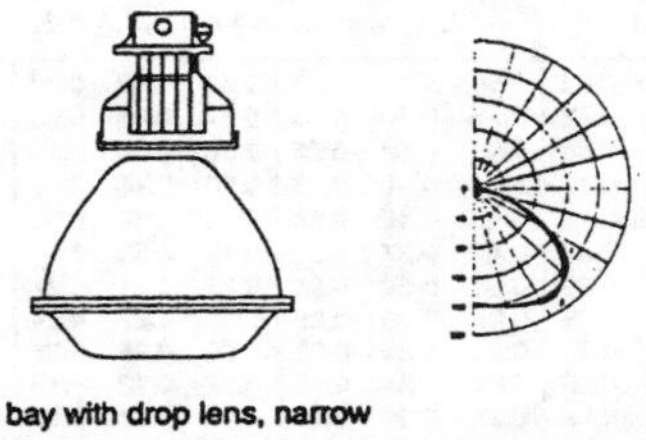

RCR	70	50	30	70	50	30	50	30	10	50	30	10	50	30	10	0
0	0.86	0.86	0.86	0.84	0.84	0.84	0.80	0.80	0.80	0.76	0.76	0.76	0.73	0.73	0.73	0.71
1	0.78	0.75	0.71	0.76	0.73	0.70	0.69	0.67	0.65	0.66	0.64	0.63	0.63	0.62	0.60	0.59
2	0.71	0.65	0.60	0.69	0.63	0.59	0.60	0.56	0.53	0.58	0.54	0.52	0.55	0.53	0.50	0.49
3	0.64	0.56	0.50	0.62	0.55	0.50	0.53	0.48	0.44	0.51	0.46	0.43	0.48	0.45	0.42	0.40
4	0.59	0.50	0.43	0.57	0.49	0.42	0.47	0.41	0.37	0.45	0.40	0.36	0.43	0.39	0.36	0.34
5	0.54	0.44	0.37	0.52	0.43	0.37	0.41	0.36	0.32	0.40	0.35	0.31	0.38	0.34	0.31	0.29
6	0.49	0.39	0.33	0.48	0.39	0.32	0.37	0.31	0.27	0.36	0.31	0.27	0.34	0.30	0.26	0.25
7	0.46	0.35	0.29	0.44	0.35	0.28	0.33	0.28	0.24	0.32	0.27	0.23	0.31	0.27	0.23	0.22
8	0.42	0.32	0.26	0.41	0.31	0.25	0.30	0.25	0.21	0.29	0.24	0.21	0.28	0.24	0.21	0.19
9	0.39	0.29	0.23	0.38	0.29	0.23	0.28	0.22	0.19	0.27	0.22	0.18	0.26	0.22	0.18	0.17
10	0.37	0.27	0.21	0.36	0.26	0.21	0.26	0.20	0.17	0.25	0.20	0.17	0.24	0.20	0.16	0.15

19 — Glowing suspended bowl, MH

EFF = 75.3% %DN = 8.1 %UP = 91.9 Lamp = M175/C° SC (along, across, 45°) = 1.3, 1.3, 1.5

RCR	70	50	30	70	50	30	50	30	10	50	30	10	50	30	10	0
0	0.73	0.73	0.73	0.63	0.63	0.63	0.45	0.45	0.45	0.29	0.29	0.29	0.13	0.13	0.13	0.06
1	0.66	0.63	0.60	0.57	0.55	0.52	0.39	0.38	0.36	0.25	0.24	0.23	0.11	0.11	0.11	0.05
2	0.60	0.55	0.51	0.52	0.48	0.44	0.34	0.32	0.30	0.21	0.20	0.19	0.10	0.09	0.09	0.04
3	0.55	0.48	0.43	0.47	0.42	0.37	0.30	0.27	0.25	0.19	0.17	0.16	0.09	0.08	0.07	0.03
4	0.50	0.42	0.37	0.43	0.37	0.32	0.26	0.23	0.21	0.17	0.15	0.13	0.08	0.07	0.06	0.02
5	0.46	0.37	0.32	0.39	0.33	0.28	0.23	0.20	0.18	0.15	0.13	0.11	0.07	0.06	0.05	0.02
6	0.42	0.33	0.28	0.36	0.29	0.24	0.21	0.18	0.15	0.13	0.11	0.10	0.06	0.05	0.05	0.02
7	0.39	0.30	0.24	0.33	0.26	0.21	0.19	0.16	0.13	0.12	0.10	0.09	0.05	0.05	0.04	0.02
8	0.36	0.27	0.21	0.31	0.23	0.19	0.17	0.14	0.12	0.11	0.09	0.07	0.05	0.04	0.03	0.01
9	0.33	0.24	0.19	0.28	0.21	0.17	0.15	0.12	0.10	0.10	0.08	0.07	0.05	0.04	0.03	0.01
10	0.31	0.22	0.17	0.26	0.19	0.15	0.14	0.11	0.09	0.09	0.07	0.06	0.04	0.03	0.03	0.01

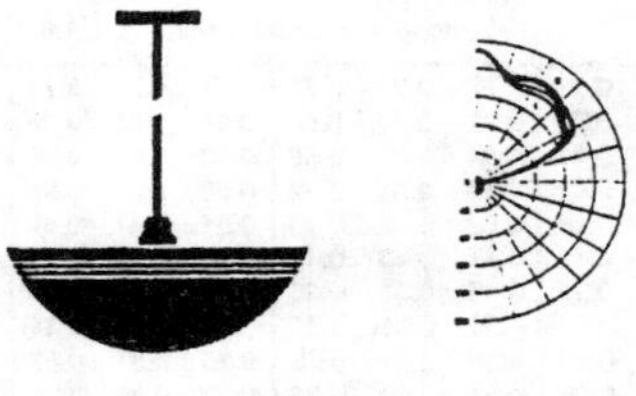

20 — Glowing suspended bowl, CFL

EFF = 81.5% %DN = 15.3 %UP = 84.7 Lamp = (4) FT39° SC (along, across, 45°) = 1.3, 1.3, 1.5

RCR	70	50	30	70	50	30	50	30	10	50	30	10	50	30	10	0
0	0.81	0.81	0.81	0.71	0.71	0.71	0.52	0.52	0.52	0.35	0.35	0.35	0.20	0.20	0.20	0.13
1	0.73	0.69	0.66	0.64	0.61	0.58	0.45	0.43	0.42	0.30	0.29	0.28	0.17	0.16	0.16	0.10
2	0.66	0.60	0.55	0.58	0.53	0.49	0.39	0.36	0.34	0.26	0.25	0.23	0.15	0.14	0.13	0.08
3	0.60	0.53	0.47	0.53	0.46	0.42	0.34	0.31	0.29	0.23	0.21	0.20	0.13	0.12	0.11	0.06
4	0.55	0.47	0.40	0.48	0.41	0.36	0.30	0.27	0.24	0.20	0.18	0.17	0.11	0.10	0.09	0.05
5	0.50	0.41	0.35	0.44	0.36	0.31	0.27	0.23	0.20	0.18	0.16	0.14	0.10	0.09	0.08	0.05
6	0.46	0.37	0.30	0.40	0.32	0.27	0.24	0.20	0.18	0.16	0.14	0.12	0.09	0.08	0.07	0.04
7	0.42	0.33	0.27	0.37	0.29	0.24	0.22	0.18	0.15	0.15	0.12	0.11	0.08	0.07	0.06	0.04
8	0.39	0.30	0.24	0.34	0.26	0.21	0.20	0.16	0.13	0.13	0.11	0.09	0.08	0.06	0.05	0.03
9	0.36	0.27	0.21	0.32	0.24	0.19	0.18	0.14	0.12	0.12	0.10	0.08	0.07	0.06	0.05	0.03
10	0.34	0.25	0.19	0.30	0.22	0.17	0.16	0.13	0.11	0.11	0.09	0.07	0.06	0.05	0.04	0.03

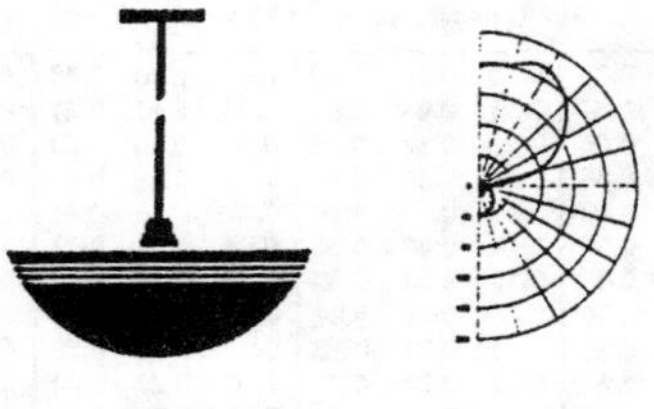

	ρcc →	80			70			50			30			10			0
Typical Luminaire	ρw →	70	50	30	70	50	30	50	30	10	50	30	10	50	30	10	0

21 — Industrial, white enamel reflector, 20% up
Lamp = (2) F40T12 — EFF = 90.5% — % DN = 78.2% — % UP = 21.8% — SC (along, across, 45°) = 1.3, 1.5, 1.5

RCR	70	50	30	70	50	30	50	30	10	50	30	10	50	30	10	0
0	1.03	1.03	1.03	0.98	0.98	0.98	0.90	0.90	0.90	0.82	0.82	0.82	0.74	0.74	0.74	0.71
1	0.93	0.89	0.85	0.89	0.85	0.81	0.77	0.74	0.72	0.70	0.68	0.66	0.64	0.62	0.61	0.58
2	0.84	0.77	0.71	0.80	0.74	0.68	0.67	0.63	0.59	0.61	0.58	0.54	0.56	0.53	0.50	0.47
3	0.77	0.67	0.60	0.73	0.64	0.58	0.59	0.53	0.49	0.54	0.49	0.45	0.49	0.45	0.42	0.40
4	0.70	0.59	0.51	0.66	0.57	0.50	0.52	0.46	0.41	0.48	0.43	0.39	0.44	0.39	0.36	0.33
5	0.64	0.53	0.45	0.61	0.51	0.43	0.46	0.40	0.35	0.43	0.37	0.33	0.39	0.35	0.31	0.29
6	0.59	0.47	0.39	0.56	0.45	0.38	0.42	0.35	0.31	0.38	0.33	0.29	0.35	0.31	0.27	0.25
7	0.55	0.43	0.35	0.52	0.41	0.34	0.38	0.32	0.27	0.35	0.30	0.26	0.32	0.28	0.24	0.22
8	0.51	0.39	0.31	0.48	0.37	0.30	0.34	0.28	0.24	0.32	0.27	0.23	0.29	0.25	0.21	0.19
9	0.47	0.35	0.28	0.45	0.34	0.27	0.32	0.26	0.21	0.29	0.24	0.20	0.27	0.23	0.19	0.17
10	0.44	0.33	0.26	0.42	0.31	0.25	0.29	0.23	0.19	0.27	0.22	0.18	0.25	0.21	0.17	0.16

22 — Industrial, white enamel reflector, down only
Lamp = (2) F40T12 — EFF = 86.9% — % DN = 100% — % UP = 0% — SC (along, across, 45°) = 1.3, 1.5, 1.5

RCR	70	50	30	70	50	30	50	30	10	50	30	10	50	30	10	0
0	1.03	1.03	1.03	1.01	1.01	1.01	0.97	0.97	0.97	0.92	0.92	0.92	0.89	0.89	0.89	0.87
1	0.94	0.90	0.86	0.92	0.88	0.84	0.84	0.81	0.79	0.81	0.79	0.76	0.78	0.76	0.74	0.72
2	0.85	0.78	0.72	0.83	0.76	0.70	0.73	0.68	0.64	0.70	0.66	0.63	0.67	0.64	0.61	0.59
3	0.77	0.68	0.60	0.75	0.66	0.59	0.64	0.58	0.53	0.61	0.56	0.52	0.59	0.55	0.51	0.49
4	0.70	0.60	0.52	0.68	0.58	0.51	0.56	0.50	0.45	0.54	0.48	0.44	0.52	0.47	0.43	0.41
5	0.65	0.53	0.45	0.63	0.52	0.44	0.50	0.43	0.38	0.48	0.42	0.38	0.47	0.41	0.37	0.35
6	0.59	0.47	0.39	0.58	0.47	0.39	0.45	0.38	0.33	0.43	0.37	0.33	0.42	0.37	0.32	0.31
7	0.55	0.43	0.35	0.53	0.42	0.35	0.41	0.34	0.29	0.39	0.33	0.29	0.38	0.33	0.29	0.27
8	0.51	0.39	0.31	0.50	0.38	0.31	0.37	0.30	0.26	0.36	0.30	0.26	0.35	0.29	0.25	0.24
9	0.48	0.36	0.28	0.46	0.35	0.28	0.34	0.28	0.23	0.33	0.27	0.23	0.32	0.27	0.23	0.21
10	0.45	0.33	0.26	0.43	0.32	0.25	0.31	0.25	0.21	0.31	0.25	0.21	0.30	0.24	0.21	0.19

23 — 2-Lamp bare strip
Lamp = (2) F40T12 — EFF = 89.3% — % DN = 86.4% — % UP = 13.6% — SC (along, across, 45°) = 1.3, 1.5, 1.6

RCR	70	50	30	70	50	30	50	30	10	50	30	10	50	30	10	0
0	1.03	1.03	1.03	1.00	1.00	1.00	0.92	0.92	0.92	0.86	0.86	0.86	0.80	0.80	0.80	0.77
1	0.93	0.88	0.83	0.89	0.84	0.80	0.78	0.75	0.72	0.73	0.70	0.68	0.67	0.65	0.63	0.61
2	0.83	0.75	0.68	0.80	0.72	0.66	0.67	0.62	0.58	0.62	0.58	0.55	0.58	0.55	0.52	0.49
3	0.75	0.65	0.57	0.72	0.63	0.56	0.58	0.52	0.47	0.54	0.49	0.45	0.50	0.46	0.43	0.40
4	0.69	0.57	0.49	0.65	0.55	0.47	0.51	0.45	0.40	0.48	0.42	0.38	0.44	0.40	0.36	0.34
5	0.63	0.51	0.42	0.60	0.49	0.41	0.46	0.39	0.34	0.43	0.37	0.32	0.40	0.35	0.31	0.29
6	0.58	0.45	0.37	0.55	0.44	0.36	0.41	0.34	0.29	0.38	0.32	0.28	0.36	0.31	0.27	0.25
7	0.53	0.41	0.33	0.51	0.40	0.32	0.37	0.30	0.26	0.35	0.29	0.25	0.33	0.27	0.24	0.22
8	0.50	0.37	0.29	0.47	0.36	0.29	0.34	0.27	0.23	0.32	0.26	0.22	0.30	0.25	0.21	0.19
9	0.46	0.34	0.26	0.44	0.33	0.26	0.31	0.25	0.20	0.29	0.24	0.19	0.27	0.22	0.19	0.17
10	0.43	0.31	0.24	0.41	0.30	0.23	0.29	0.22	0.18	0.27	0.21	0.18	0.25	0.20	0.17	0.15

24 — 2 × 4, 3-Lamp parabolic troffer with 3″ semi-spec. louvers, 18 cells
Lamp = (3) F32T8 — EFF = 72.7% — % DN = 100 — % UP = 0 — SC (along, across, 45°) = 1.3, 1.6, 1.6

RCR	70	50	30	70	50	30	50	30	10	50	30	10	50	30	10	0
0	0.87	0.87	0.87	0.85	0.85	0.85	0.81	0.81	0.81	0.77	0.77	0.77	0.74	0.74	0.74	0.73
1	0.81	0.78	0.76	0.79	0.77	0.74	0.74	0.72	0.70	0.71	0.69	0.68	0.68	0.67	0.66	0.65
2	0.75	0.70	0.66	0.73	0.69	0.65	0.68	0.63	0.61	0.64	0.61	0.59	0.62	0.60	0.58	0.57
3	0.69	0.63	0.58	0.68	0.62	0.57	0.60	0.56	0.52	0.58	0.54	0.52	0.56	0.53	0.51	0.49
4	0.64	0.56	0.51	0.62	0.55	0.50	0.54	0.49	0.46	0.52	0.48	0.45	0.51	0.47	0.44	0.43
5	0.59	0.51	0.45	0.58	0.50	0.44	0.48	0.44	0.40	0.47	0.43	0.40	0.46	0.42	0.39	0.38
6	0.55	0.46	0.40	0.53	0.45	0.40	0.44	0.39	0.35	0.43	0.38	0.35	0.42	0.38	0.35	0.33
7	0.51	0.42	0.36	0.50	0.41	0.36	0.40	0.35	0.31	0.39	0.35	0.31	0.38	0.34	0.31	0.30
8	0.47	0.38	0.32	0.46	0.38	0.32	0.37	0.32	0.28	0.36	0.31	0.28	0.35	0.31	0.28	0.27
9	0.44	0.35	0.29	0.43	0.35	0.29	0.34	0.29	0.25	0.33	0.29	0.25	0.32	0.28	0.25	0.24
10	0.41	0.32	0.27	0.40	0.32	0.27	0.31	0.26	0.23	0.31	0.26	0.23	0.30	0.26	0.23	0.22

25 — 2 × 4, 3-Lamp parabolic troffer with 4″ semi-spec. louvers, 18 cells
Lamp = (3) F40T12 — EFF = 66.2% — % DN = 100 — % UP = 0 — SC (along, across, 45°) = 1.3, 1.6, 1.5

RCR	70	50	30	70	50	30	50	30	10	50	30	10	50	30	10	0
0	0.79	0.79	0.79	0.77	0.77	0.77	0.74	0.74	0.74	0.70	0.70	0.70	0.68	0.68	0.68	0.66
1	0.74	0.72	0.69	0.72	0.70	0.68	0.67	0.66	0.64	0.65	0.64	0.62	0.62	0.61	0.61	0.59
2	0.69	0.64	0.61	0.67	0.63	0.60	0.61	0.58	0.56	0.59	0.57	0.55	0.57	0.55	0.53	0.52
3	0.63	0.58	0.53	0.62	0.57	0.53	0.55	0.51	0.48	0.53	0.50	0.48	0.52	0.49	0.47	0.46
4	0.59	0.52	0.47	0.57	0.51	0.47	0.50	0.46	0.42	0.48	0.45	0.42	0.47	0.44	0.41	0.40
5	0.54	0.47	0.42	0.53	0.46	0.41	0.45	0.41	0.37	0.44	0.40	0.37	0.43	0.39	0.37	0.35
6	0.50	0.43	0.37	0.49	0.42	0.37	0.41	0.36	0.33	0.40	0.36	0.33	0.39	0.35	0.33	0.31
7	0.47	0.39	0.33	0.46	0.38	0.33	0.37	0.33	0.30	0.36	0.32	0.29	0.35	0.32	0.29	0.28
8	0.44	0.35	0.30	0.43	0.35	0.30	0.34	0.30	0.26	0.33	0.29	0.26	0.33	0.29	0.26	0.25
9	0.41	0.33	0.27	0.40	0.32	0.27	0.31	0.27	0.24	0.31	0.27	0.24	0.30	0.26	0.24	0.23
10	0.38	0.30	0.25	0.37	0.30	0.25	0.29	0.25	0.22	0.28	0.24	0.22	0.28	0.24	0.22	0.20

Luminaire 26 — 2 × 2, 3-Lamp with 3″ semi-spec. louvers, 9 cells

EFF = 67.8%   % DN = 100%   % UP = 0%   Lamp = (3) FT40   SC (along, across, 45°) = 1.3, 1.6, 1.6

| ρcc → | 80 | | | 70 | | | 50 | | | 30 | | | 10 | | | 0 |
RCR ↓	70	50	30	70	50	30	50	30	10	50	30	10	50	30	10	0
0	0.81	0.81	0.81	0.79	0.79	0.79	0.75	0.75	0.75	0.72	0.72	0.72	0.69	0.69	0.69	0.68
1	0.75	0.73	0.71	0.74	0.71	0.69	0.69	0.67	0.66	0.66	0.65	0.64	0.64	0.63	0.62	0.60
2	0.70	0.65	0.62	0.68	0.64	0.61	0.62	0.59	0.57	0.60	0.57	0.55	0.58	0.56	0.54	0.53
3	0.65	0.59	0.54	0.63	0.58	0.53	0.56	0.52	0.49	0.54	0.51	0.48	0.52	0.50	0.47	0.46
4	0.60	0.53	0.48	0.58	0.52	0.47	0.50	0.46	0.43	0.49	0.45	0.42	0.47	0.44	0.42	0.40
5	0.55	0.47	0.42	0.54	0.47	0.42	0.45	0.41	0.38	0.44	0.40	0.37	0.43	0.40	0.37	0.35
6	0.51	0.43	0.38	0.50	0.42	0.37	0.41	0.37	0.33	0.40	0.36	0.33	0.39	0.36	0.33	0.31
7	0.47	0.39	0.34	0.46	0.39	0.33	0.38	0.33	0.30	0.37	0.32	0.29	0.36	0.32	0.29	0.28
8	0.44	0.36	0.30	0.43	0.35	0.30	0.34	0.30	0.27	0.34	0.29	0.26	0.33	0.29	0.26	0.25
9	0.41	0.33	0.28	0.40	0.32	0.27	0.32	0.27	0.24	0.31	0.27	0.24	0.30	0.26	0.24	0.23
10	0.39	0.30	0.25	0.38	0.30	0.25	0.29	0.25	0.22	0.29	0.25	0.22	0.28	0.24	0.22	0.20

Luminaire 27 — 2 × 2, 2-Lamp (U) parabolic troffer with 3″ semi-spec. louver, 16 cells

EFF = 50.8%   % DN = 100%   % UP = 0%   Lamp = (2) F31T8/U/6   SC (along, across, 45°) = 1.2, 1.5, 1.4

| ρcc → | 80 | | | 70 | | | 50 | | | 30 | | | 10 | | | 0 |
RCR ↓	70	50	30	70	50	30	50	30	10	50	30	10	50	30	10	0
0	0.61	0.61	0.61	0.59	0.59	0.59	0.56	0.56	0.56	0.56	0.56	0.56	0.54	0.54	0.54	0.52
1	0.57	0.55	0.53	0.55	0.54	0.52	0.52	0.50	0.49	0.52	0.50	0.49	0.50	0.49	0.48	0.47
2	0.53	0.49	0.47	0.51	0.48	0.46	0.47	0.45	0.43	0.47	0.45	0.43	0.45	0.43	0.42	0.42
3	0.49	0.44	0.41	0.48	0.43	0.40	0.42	0.39	0.37	0.42	0.39	0.37	0.41	0.38	0.37	0.38
4	0.45	0.40	0.36	0.44	0.39	0.36	0.38	0.35	0.32	0.38	0.35	0.32	0.37	0.34	0.32	0.34
5	0.42	0.36	0.32	0.41	0.35	0.32	0.34	0.31	0.29	0.34	0.31	0.29	0.33	0.31	0.28	0.30
6	0.39	0.33	0.29	0.38	0.32	0.28	0.31	0.28	0.25	0.31	0.28	0.25	0.30	0.27	0.25	0.27
7	0.36	0.30	0.26	0.35	0.29	0.25	0.29	0.25	0.23	0.29	0.25	0.23	0.28	0.25	0.22	0.24
8	0.33	0.27	0.23	0.33	0.27	0.23	0.26	0.23	0.20	0.26	0.23	0.20	0.26	0.22	0.20	0.22
9	0.31	0.25	0.21	0.31	0.25	0.21	0.24	0.21	0.18	0.24	0.21	0.18	0.24	0.20	0.18	0.20
10	0.29	0.23	0.19	0.29	0.23	0.19	0.22	0.19	0.17	0.22	0.19	0.17	0.22	0.19	0.17	0.19

Luminaire 28 — 1 × 4, 2-Lamp parabolic troffer with 3″ semi-spec. louver, 8 or 9 cells

EFF = 67.2%   % DN = 100%   % UP = 0%   Lamp = (2) F32T8   SC (along, across, 45°) = 1.3, 1.6, 1.5

| ρcc → | 80 | | | 70 | | | 50 | | | 30 | | | 10 | | | 0 |
RCR ↓	70	50	30	70	50	30	50	30	10	50	30	10	50	30	10	0
0	0.80	0.80	0.80	0.78	0.78	0.78	0.75	0.75	0.75	0.72	0.72	0.72	0.69	0.69	0.69	0.67
1	0.75	0.73	0.70	0.73	0.71	0.69	0.68	0.67	0.65	0.66	0.64	0.63	0.63	0.62	0.61	0.60
2	0.70	0.65	0.61	0.68	0.64	0.61	0.62	0.59	0.56	0.59	0.57	0.55	0.58	0.56	0.54	0.53
3	0.64	0.58	0.54	0.63	0.57	0.53	0.55	0.52	0.49	0.54	0.51	0.48	0.52	0.49	0.47	0.46
4	0.59	0.52	0.47	0.58	0.52	0.47	0.50	0.46	0.43	0.49	0.45	0.42	0.47	0.44	0.42	0.40
5	0.55	0.47	0.42	0.54	0.47	0.42	0.45	0.41	0.37	0.44	0.40	0.37	0.43	0.39	0.37	0.35
6	0.51	0.43	0.37	0.50	0.42	0.37	0.41	0.36	0.33	0.40	0.36	0.33	0.39	0.35	0.33	0.31
7	0.47	0.39	0.34	0.46	0.38	0.33	0.37	0.33	0.29	0.36	0.32	0.29	0.36	0.32	0.29	0.28
8	0.44	0.36	0.30	0.43	0.35	0.30	0.34	0.30	0.26	0.34	0.29	0.26	0.33	0.29	0.26	0.25
9	0.41	0.33	0.27	0.40	0.32	0.27	0.31	0.27	0.24	0.31	0.27	0.24	0.30	0.26	0.24	0.22
10	0.38	0.30	0.25	0.38	0.30	0.25	0.29	0.25	0.22	0.28	0.24	0.21	0.28	0.24	0.21	0.20

Luminaire 29 — 2 × 4, 3-Lamp parabolic troffer, spec. louvers, 18 cells, RP-1

EFF = 67.2%   % DN = 100   % UP = 0   Lamp = (3) F32T8   SC (along, across, 45°) = 1.3, 1.5, 1.5

| ρcc → | 80 | | | 70 | | | 50 | | | 30 | | | 10 | | | 0 |
RCR ↓	70	50	30	70	50	30	50	30	10	50	30	10	50	30	10	0
0	0.80	0.80	0.80	0.78	0.78	0.78	0.75	0.75	0.75	0.72	0.72	0.72	0.69	0.69	0.69	0.67
1	0.76	0.74	0.72	0.74	0.72	0.70	0.69	0.68	0.67	0.67	0.66	0.65	0.64	0.64	0.63	0.62
2	0.71	0.67	0.64	0.70	0.66	0.63	0.64	0.62	0.59	0.62	0.60	0.58	0.60	0.58	0.57	0.56
3	0.67	0.62	0.58	0.65	0.61	0.57	0.59	0.56	0.53	0.57	0.54	0.52	0.55	0.53	0.51	0.50
4	0.62	0.56	0.52	0.61	0.56	0.51	0.54	0.50	0.48	0.53	0.50	0.47	0.51	0.49	0.47	0.45
5	0.58	0.52	0.47	0.57	0.51	0.47	0.50	0.46	0.43	0.49	0.45	0.43	0.47	0.44	0.42	0.41
6	0.55	0.47	0.43	0.53	0.47	0.42	0.46	0.42	0.39	0.45	0.41	0.39	0.44	0.41	0.38	0.37
7	0.51	0.44	0.39	0.50	0.43	0.39	0.42	0.38	0.35	0.41	0.38	0.35	0.41	0.37	0.35	0.34
8	0.48	0.40	0.36	0.47	0.40	0.35	0.39	0.35	0.32	0.39	0.35	0.32	0.38	0.34	0.32	0.31
9	0.45	0.37	0.33	0.44	0.37	0.32	0.36	0.32	0.29	0.36	0.32	0.29	0.35	0.32	0.29	0.28
10	0.42	0.35	0.30	0.42	0.34	0.30	0.34	0.30	0.27	0.33	0.29	0.27	0.33	0.29	0.27	0.26

Luminaire 30 — 2 × 4, 3-Lamp parabolic troffer, 1.5 × 1.5 × 1.0″ silver louvers, RP-1

EFF = 51.4%   % DN = 100   % UP = 0   Lamp = (3) F32T8   SC (along, across, 45°) = 1.3, 1.3, 1.5

| ρcc → | 80 | | | 70 | | | 50 | | | 30 | | | 10 | | | 0 |
RCR ↓	70	50	30	70	50	30	50	30	10	50	30	10	50	30	10	0
0	0.68	0.61	0.55	0.67	0.60	0.55	0.58	0.53	0.50	0.56	0.52	0.49	0.54	0.51	0.48	0.46
1	0.63	0.55	0.49	0.61	0.54	0.48	0.52	0.47	0.43	0.50	0.46	0.42	0.49	0.45	0.42	0.40
2	0.58	0.49	0.43	0.57	0.48	0.42	0.47	0.42	0.38	0.45	0.41	0.37	0.44	0.40	0.37	0.35
3	0.54	0.44	0.38	0.52	0.44	0.38	0.42	0.37	0.33	0.41	0.37	0.33	0.40	0.36	0.33	0.31
4	0.50	0.40	0.34	0.49	0.40	0.34	0.39	0.34	0.30	0.38	0.33	0.29	0.37	0.33	0.29	0.28
5	0.47	0.37	0.31	0.46	0.37	0.31	0.36	0.31	0.27	0.35	0.30	0.26	0.34	0.30	0.26	0.25
6	0.44	0.34	0.28	0.43	0.34	0.28	0.33	0.28	0.24	0.32	0.27	0.24	0.31	0.27	0.24	0.23
7	0.41	0.32	0.26	0.40	0.31	0.26	0.30	0.25	0.22	0.30	0.25	0.22	0.29	0.25	0.22	0.21
8	0.00	0.00	0.00	0.00	0.00	0.00	0.00	0.00	0.00	0.00	0.00	0.00	0.00	0.00	0.00	0.00
9	0.00	0.00	0.00	0.00	0.00	0.00	0.00	0.00	0.00	0.00	0.00	0.00	0.00	0.00	0.00	0.00
10	0.77	0.77	0.77	0.76	0.76	0.76	0.72	0.72	0.72	0.69	0.69	0.69	0.66	0.66	0.66	0.65

Typical Luminaire	ρcc →	80			70			50			30			10			0
	ρw →	70	50	30	70	50	30	50	30	10	50	30	10	50	30	10	0
	RCR ↓																

31 — 2 × 2, 3-Lamp troffer, spec. louvers, 12 cells, RP-1
EFF = 64.6% | % DN = 100% | % UP = 0% | Lamp = (3) F31T8/U/6 | SC (along, across, 45°) = 1.3, 1.5, 1.3

RCR	70	50	30	70	50	30	50	30	10	50	30	10	50	30	10	0
0	0.77	0.77	0.77	0.75	0.75	0.75	0.72	0.72	0.72	0.69	0.69	0.69	0.66	0.66	0.66	0.65
1	0.73	0.71	0.69	0.71	0.69	0.68	0.67	0.65	0.64	0.64	0.63	0.62	0.62	0.61	0.60	0.59
2	0.68	0.65	0.62	0.67	0.64	0.61	0.61	0.59	0.57	0.59	0.58	0.56	0.58	0.56	0.55	0.54
3	0.64	0.59	0.55	0.63	0.58	0.55	0.56	0.54	0.51	0.55	0.52	0.50	0.53	0.51	0.50	0.48
4	0.60	0.54	0.50	0.59	0.53	0.50	0.52	0.49	0.46	0.51	0.48	0.45	0.49	0.47	0.45	0.44
5	0.56	0.50	0.45	0.55	0.49	0.45	0.48	0.44	0.41	0.47	0.43	0.41	0.46	0.43	0.41	0.39
6	0.53	0.46	0.41	0.51	0.45	0.41	0.44	0.40	0.37	0.43	0.40	0.37	0.42	0.39	0.37	0.36
7	0.49	0.42	0.38	0.48	0.42	0.37	0.41	0.37	0.34	0.40	0.36	0.34	0.39	0.36	0.34	0.33
8	0.46	0.39	0.34	0.45	0.39	0.34	0.38	0.34	0.31	0.37	0.33	0.31	0.36	0.33	0.31	0.30
9	0.43	0.36	0.32	0.43	0.36	0.31	0.35	0.31	0.28	0.34	0.31	0.28	0.34	0.31	0.28	0.27
10	0.41	0.33	0.29	0.40	0.33	0.29	0.33	0.29	0.26	0.32	0.28	0.26	0.32	0.28	0.26	0.25

32 — 2 × 4, 3-Lamp troffer with A12 lens
EFF = 75.6% | % DN = 100% | % UP = 0% | Lamp = (3) F32T8 | SC (along, across, 45°) = 1.3, 1.3, 1.4

RCR	70	50	30	70	50	30	50	30	10	50	30	10	50	30	10	0
0	0.90	0.90	0.90	0.88	0.88	0.88	0.84	0.84	0.84	0.80	0.80	0.80	0.77	0.77	0.77	0.76
1	0.83	0.79	0.76	0.81	0.78	0.75	0.75	0.72	0.70	0.72	0.70	0.68	0.69	0.67	0.66	0.65
2	0.76	0.70	0.65	0.74	0.69	0.64	0.66	0.62	0.59	0.64	0.61	0.58	0.61	0.59	0.57	0.55
3	0.70	0.62	0.57	0.68	0.61	0.56	0.57	0.53	0.50	0.57	0.53	0.50	0.55	0.52	0.49	0.47
4	0.64	0.56	0.49	0.63	0.55	0.49	0.53	0.48	0.44	0.51	0.47	0.43	0.49	0.46	0.43	0.41
5	0.59	0.50	0.44	0.58	0.49	0.43	0.48	0.42	0.38	0.46	0.41	0.38	0.45	0.41	0.37	0.36
6	0.55	0.45	0.39	0.53	0.45	0.38	0.43	0.38	0.34	0.42	0.37	0.33	0.41	0.37	0.33	0.32
7	0.51	0.41	0.35	0.50	0.41	0.35	0.39	0.34	0.30	0.38	0.33	0.30	0.37	0.33	0.30	0.28
8	0.48	0.38	0.31	0.46	0.37	0.31	0.36	0.31	0.27	0.35	0.30	0.27	0.34	0.30	0.27	0.25
9	0.44	0.35	0.29	0.43	0.34	0.28	0.33	0.28	0.24	0.33	0.28	0.24	0.32	0.27	0.24	0.23
10	0.42	0.32	0.26	0.41	0.32	0.26	0.31	0.26	0.22	0.30	0.25	0.22	0.29	0.25	0.22	0.21

33 — 2 × 4, 3-Lamp troffer with A19 lens
EFF = 72.4% | % DN = 100% | % UP = 0% | Lamp = (3) F32T8 | SC (along, across, 45°) = 1.3, 1.3, 1.3

RCR	70	50	30	70	50	30	50	30	10	50	30	10	50	30	10	0
0	0.86	0.86	0.86	0.84	0.84	0.84	0.80	0.80	0.80	0.77	0.77	0.77	0.74	0.74	0.74	0.72
1	0.80	0.77	0.75	0.78	0.76	0.73	0.73	0.71	0.69	0.70	0.68	0.67	0.67	0.66	0.65	0.63
2	0.74	0.69	0.65	0.72	0.68	0.64	0.65	0.62	0.59	0.63	0.60	0.58	0.61	0.59	0.57	0.55
3	0.69	0.62	0.57	0.67	0.61	0.56	0.59	0.55	0.52	0.57	0.54	0.51	0.55	0.52	0.50	0.48
4	0.64	0.56	0.50	0.62	0.55	0.50	0.53	0.49	0.45	0.52	0.48	0.45	0.50	0.47	0.44	0.43
5	0.59	0.51	0.45	0.58	0.50	0.45	0.48	0.44	0.40	0.47	0.43	0.40	0.46	0.42	0.39	0.38
6	0.55	0.46	0.40	0.54	0.45	0.40	0.44	0.39	0.36	0.43	0.39	0.35	0.42	0.38	0.35	0.34
7	0.51	0.42	0.36	0.50	0.42	0.36	0.41	0.36	0.32	0.40	0.35	0.32	0.39	0.35	0.32	0.30
8	0.48	0.39	0.33	0.47	0.38	0.33	0.37	0.32	0.29	0.36	0.32	0.29	0.36	0.32	0.29	0.27
9	0.45	0.36	0.30	0.44	0.35	0.30	0.35	0.30	0.26	0.34	0.29	0.26	0.33	0.29	0.26	0.25
10	0.42	0.33	0.28	0.41	0.33	0.28	0.32	0.27	0.24	0.31	0.27	0.24	0.31	0.27	0.24	0.23

34 — 2 × 2, 3-Lamp troffer with A12 lens
EFF = 68.4% | % DN = 100 | % UP = 0 | Lamp = (3) FT40 | SC (along, across, 45°) = 1.2, 1.3, 1.3

RCR	70	50	30	70	50	30	50	30	10	50	30	10	50	30	10	0
0	0.81	0.81	0.81	0.80	0.80	0.80	0.76	0.76	0.76	0.73	0.73	0.73	0.70	0.70	0.70	0.68
1	0.75	0.72	0.70	0.73	0.71	0.69	0.68	0.66	0.64	0.65	0.64	0.62	0.63	0.62	0.60	0.59
2	0.69	0.64	0.60	0.68	0.63	0.59	0.61	0.58	0.55	0.59	0.56	0.54	0.58	0.54	0.52	0.51
3	0.64	0.57	0.52	0.62	0.56	0.52	0.54	0.50	0.47	0.53	0.49	0.46	0.51	0.48	0.46	0.44
4	0.59	0.52	0.46	0.58	0.51	0.46	0.49	0.45	0.41	0.47	0.44	0.41	0.46	0.43	0.40	0.39
5	0.55	0.47	0.41	0.53	0.46	0.41	0.44	0.40	0.36	0.43	0.39	0.36	0.42	0.38	0.35	0.34
6	0.51	0.42	0.37	0.49	0.42	0.36	0.40	0.36	0.32	0.39	0.35	0.32	0.38	0.35	0.32	0.30
7	0.47	0.39	0.33	0.46	0.38	0.33	0.37	0.32	0.29	0.36	0.32	0.29	0.35	0.31	0.28	0.27
8	0.44	0.35	0.30	0.43	0.35	0.30	0.34	0.29	0.26	0.33	0.29	0.26	0.32	0.29	0.26	0.24
9	0.41	0.33	0.27	0.40	0.32	0.27	0.31	0.27	0.24	0.31	0.27	0.23	0.30	0.26	0.23	0.22
10	0.39	0.30	0.25	0.38	0.30	0.25	0.29	0.25	0.22	0.29	0.24	0.22	0.28	0.24	0.21	0.20

35 — 2 × 2, 2-Lamp troffer with A12 lens
EFF = 57.1% | % DN = 100 | % UP = 0 | Lamp = (2) F31T8/U/6 | SC (along, across, 45°) = 1.2, 1.3, 1.4

RCR	70	50	30	70	50	30	50	30	10	50	30	10	50	30	10	0
0	0.68	0.68	0.68	0.66	0.66	0.66	0.63	0.63	0.63	0.61	0.61	0.61	0.58	0.58	0.58	0.57
1	0.64	0.62	0.60	0.62	0.60	0.59	0.58	0.57	0.55	0.56	0.55	0.54	0.54	0.53	0.52	0.51
2	0.59	0.55	0.52	0.58	0.54	0.51	0.52	0.50	0.48	0.51	0.49	0.47	0.49	0.47	0.46	0.45
3	0.55	0.50	0.46	0.53	0.49	0.45	0.47	0.44	0.42	0.46	0.43	0.41	0.44	0.42	0.40	0.39
4	0.51	0.45	0.40	0.49	0.44	0.40	0.43	0.39	0.36	0.41	0.38	0.36	0.40	0.38	0.36	0.34
5	0.47	0.40	0.36	0.46	0.40	0.36	0.39	0.35	0.32	0.38	0.34	0.32	0.37	0.34	0.31	0.30
6	0.43	0.37	0.32	0.42	0.36	0.32	0.35	0.31	0.28	0.34	0.31	0.28	0.33	0.30	0.28	0.27
7	0.40	0.33	0.29	0.39	0.33	0.29	0.32	0.28	0.25	0.31	0.28	0.25	0.31	0.27	0.25	0.24
8	0.38	0.31	0.26	0.37	0.30	0.26	0.29	0.26	0.23	0.29	0.25	0.23	0.28	0.25	0.23	0.22
9	0.35	0.28	0.24	0.34	0.28	0.24	0.27	0.23	0.21	0.27	0.23	0.21	0.26	0.23	0.21	0.20
10	0.33	0.26	0.22	0.32	0.26	0.22	0.25	0.21	0.19	0.25	0.21	0.19	0.24	0.21	0.19	0.18

Reproduced from the *IESNA Lighting Handbook, 9th Edition,*. 2000, courtesy of the Illuminating Engineering Society of North America.

Luminaire 36 — 1 × 4, 2-Lamp troffer with A12 lens

EFF = 65.1% | % DN = 100% | % UP = 0% | Lamp = (2) F32T8 | SC (along, across, 45°) = 1.3, 1.3, 1.3

ρcc →	80			70			50			30			10			0
ρw →	70	50	30	70	50	30	50	30	10	50	30	10	50	30	10	0
RCR																
0	0.77	0.77	0.77	0.76	0.76	0.76	0.72	0.72	0.72	0.69	0.69	0.69	0.66	0.66	0.66	0.65
1	0.71	0.69	0.66	0.70	0.67	0.65	0.64	0.62	0.61	0.62	0.60	0.59	0.59	0.58	0.57	0.56
2	0.66	0.61	0.57	0.64	0.59	0.56	0.57	0.54	0.51	0.55	0.52	0.50	0.53	0.51	0.49	0.48
3	0.60	0.54	0.49	0.59	0.53	0.48	0.51	0.47	0.44	0.49	0.46	0.43	0.48	0.45	0.42	0.41
4	0.55	0.48	0.43	0.54	0.47	0.42	0.46	0.41	0.38	0.44	0.40	0.37	0.43	0.40	0.37	0.36
5	0.51	0.43	0.38	0.50	0.43	0.37	0.41	0.37	0.33	0.40	0.36	0.33	0.39	0.35	0.33	0.31
6	0.47	0.39	0.34	0.46	0.39	0.33	0.38	0.33	0.29	0.36	0.32	0.29	0.35	0.32	0.29	0.28
7	0.44	0.36	0.30	0.43	0.35	0.30	0.34	0.30	0.26	0.33	0.29	0.26	0.33	0.29	0.26	0.25
8	0.41	0.33	0.27	0.40	0.32	0.27	0.31	0.27	0.24	0.31	0.27	0.23	0.30	0.26	0.23	0.22
9	0.39	0.30	0.25	0.38	0.30	0.25	0.29	0.25	0.21	0.28	0.24	0.21	0.28	0.24	0.21	0.20
10	0.36	0.28	0.23	0.35	0.28	0.23	0.27	0.23	0.19	0.26	0.22	0.19	0.26	0.22	0.19	0.18

Luminaire 37 — 1 × 4, 2-Lamp lensed wrap-around, surface mounted

EFF = 68.9% | % DN = 91.4% | % UP = 8.6% | Lamp = (2) F32T8 | SC (along, across, 45°) = 1.3, 1.5, 1.5

ρcc →	80			70			50			30			10			0
ρw →	70	50	30	70	50	30	50	30	10	50	30	10	50	30	10	0
RCR																
0	0.81	0.81	0.81	0.78	0.78	0.78	0.73	0.73	0.73	0.69	0.69	0.69	0.65	0.65	0.65	0.63
1	0.74	0.71	0.68	0.71	0.68	0.66	0.64	0.62	0.60	0.61	0.59	0.58	0.57	0.56	0.55	0.53
2	0.68	0.62	0.58	0.65	0.61	0.56	0.57	0.54	0.51	0.54	0.51	0.49	0.51	0.49	0.47	0.45
3	0.62	0.55	0.50	0.60	0.54	0.49	0.51	0.47	0.44	0.48	0.45	0.42	0.46	0.43	0.41	0.39
4	0.57	0.50	0.44	0.55	0.48	0.43	0.46	0.41	0.38	0.44	0.40	0.37	0.41	0.38	0.36	0.34
5	0.53	0.45	0.39	0.51	0.43	0.38	0.41	0.37	0.33	0.39	0.35	0.32	0.38	0.34	0.31	0.30
6	0.49	0.40	0.35	0.47	0.39	0.34	0.38	0.33	0.29	0.36	0.32	0.29	0.34	0.31	0.28	0.26
7	0.46	0.37	0.31	0.44	0.36	0.31	0.34	0.30	0.26	0.33	0.29	0.26	0.31	0.28	0.25	0.24
8	0.42	0.34	0.28	0.41	0.33	0.28	0.32	0.27	0.23	0.30	0.26	0.23	0.29	0.25	0.23	0.21
9	0.40	0.31	0.26	0.38	0.30	0.25	0.29	0.24	0.21	0.20	0.24	0.21	0.27	0.23	0.20	0.19
10	0.37	0.29	0.23	0.36	0.28	0.23	0.27	0.22	0.19	0.26	0.22	0.19	0.25	0.21	0.19	0.17

Luminaire 38 — 2 × 2, semi-recessed troffer

EFF = 54.2% | % DN = 99.2% | % UP = 0.5% | Lamp = (2) FT36° | SC (along, across, 45°) = 1.3, 1.5, 1.5

ρcc →	80			70			50			30			10			0
ρw →	70	50	30	70	50	30	50	30	10	50	30	10	50	30	10	0
RCR																
0	0.65	0.65	0.65	0.63	0.63	0.63	0.60	0.60	0.60	0.58	0.58	0.58	0.55	0.55	0.55	0.54
1	0.58	0.55	0.53	0.57	0.54	0.52	0.52	0.50	0.48	0.49	0.48	0.46	0.47	0.46	0.45	0.44
2	0.53	0.48	0.44	0.51	0.47	0.43	0.45	0.41	0.39	0.43	0.40	0.38	0.41	0.39	0.37	0.36
3	0.48	0.42	0.37	0.46	0.41	0.36	0.39	0.35	0.32	0.37	0.34	0.31	0.36	0.33	0.31	0.30
4	0.43	0.37	0.32	0.42	0.36	0.31	0.34	0.30	0.27	0.33	0.29	0.27	0.32	0.29	0.26	0.25
5	0.40	0.33	0.27	0.39	0.32	0.27	0.31	0.26	0.23	0.30	0.26	0.23	0.28	0.25	0.23	0.21
6	0.37	0.29	0.24	0.36	0.29	0.24	0.28	0.23	0.20	0.27	0.23	0.20	0.26	0.22	0.20	0.19
7	0.34	0.26	0.21	0.33	0.26	0.21	0.25	0.21	0.18	0.24	0.20	0.18	0.23	0.20	0.17	0.16
8	0.32	0.24	0.19	0.31	0.24	0.19	0.23	0.19	0.16	0.22	0.18	0.16	0.21	0.18	0.15	0.14
9	0.29	0.22	0.17	0.29	0.22	0.17	0.21	0.17	0.14	0.20	0.17	0.14	0.20	0.16	0.14	0.13
10	0.28	0.20	0.16	0.27	0.20	0.16	0.19	0.15	0.13	0.19	0.15	0.13	0.18	0.15	0.13	0.12

Luminaire 39 — 9" Wide, thin profile, wide spread indirect

EFF = 84.1% | % DN = 0 | % UP = 100 | Lamp = (2) F32T8 | SC (along, across, 45°) = N/A

ρcc →	80			70			50			30			10			0
ρw →	70	50	30	70	50	30	50	30	10	50	30	10	50	30	10	0
RCR																
0	0.80	0.80	0.80	0.68	0.68	0.68	0.47	0.47	0.47	0.27	0.27	0.27	0.09	0.09	0.09	0.00
1	0.73	0.69	0.66	0.62	0.59	0.57	0.41	0.39	0.38	0.23	0.23	0.22	0.07	0.07	0.07	0.00
2	0.66	0.61	0.56	0.56	0.52	0.48	0.36	0.33	0.31	0.21	0.19	0.18	0.07	0.06	0.06	0.00
3	0.60	0.53	0.48	0.51	0.46	0.41	0.31	0.28	0.26	0.18	0.17	0.15	0.06	0.05	0.05	0.00
4	0.55	0.47	0.41	0.47	0.40	0.35	0.28	0.25	0.22	0.16	0.14	0.13	0.05	0.05	0.04	0.00
5	0.50	0.41	0.35	0.43	0.36	0.30	0.24	0.21	0.19	0.14	0.12	0.11	0.05	0.04	0.04	0.00
6	0.46	0.37	0.31	0.39	0.32	0.27	0.22	0.19	0.16	0.13	0.11	0.10	0.04	0.04	0.03	0.00
7	0.42	0.33	0.27	0.36	0.28	0.23	0.20	0.16	0.14	0.11	0.10	0.08	0.04	0.03	0.03	0.00
8	0.39	0.30	0.24	0.33	0.26	0.21	0.18	0.14	0.12	0.10	0.09	0.07	0.03	0.03	0.02	0.00
9	0.36	0.27	0.21	0.31	0.23	0.18	0.16	0.13	0.11	0.09	0.08	0.06	0.03	0.02	0.02	0.00
10	0.34	0.24	0.19	0.29	0.21	0.16	0.15	0.11	0.09	0.08	0.07	0.06	0.03	0.02	0.02	0.00

Luminaire 40 — V-shaped, completely indirect

EFF = 88.3% | % DN = 0 | % UP = 100 | Lamp = (2) F32T8 | SC (along, across, 45°) = N/A

ρcc →	80			70			50			30			10			0
ρw →	70	50	30	70	50	30	50	30	10	50	30	10	50	30	10	0
RCR																
0	0.84	0.84	0.84	0.72	0.72	0.72	0.49	0.49	0.49	0.28	0.28	0.28	0.09	0.09	0.09	0.00
1	0.76	0.73	0.70	0.65	0.62	0.60	0.43	0.41	0.40	0.25	0.24	0.23	0.08	0.08	0.07	0.00
2	0.70	0.64	0.59	0.59	0.54	0.50	0.37	0.35	0.33	0.22	0.20	0.19	0.07	0.07	0.06	0.00
3	0.63	0.56	0.50	0.54	0.48	0.43	0.33	0.30	0.27	0.19	0.17	0.16	0.06	0.06	0.05	0.00
4	0.58	0.49	0.43	0.49	0.42	0.37	0.29	0.26	0.23	0.17	0.15	0.14	0.05	0.05	0.05	0.00
5	0.53	0.43	0.37	0.45	0.37	0.32	0.26	0.22	0.20	0.15	0.13	0.12	0.05	0.04	0.04	0.00
6	0.48	0.39	0.32	0.41	0.33	0.28	0.23	0.19	0.17	0.13	0.11	0.10	0.04	0.04	0.03	0.00
7	0.45	0.35	0.28	0.38	0.30	0.24	0.21	0.17	0.15	0.12	0.10	0.09	0.04	0.03	0.03	0.00
8	0.41	0.31	0.25	0.35	0.27	0.22	0.19	0.15	0.13	0.11	0.09	0.08	0.03	0.03	0.03	0.00
9	0.38	0.28	0.22	0.32	0.24	0.19	0.17	0.13	0.11	0.10	0.08	0.07	0.03	0.03	0.02	0.00
10	0.35	0.26	0.20	0.30	0.22	0.17	0.15	0.12	0.10	0.09	0.07	0.06	0.03	0.02	0.02	0.00

Reproduced from the *IESNA Lighting Handbook, 9th Edition*,. 2000, courtesy of the Illuminating Engineering Society of North America.

	ρcc →	80			70			50			30			10			0
Typical Intensity Distribution / Typical Luminaire	ρw →	70	50	30	70	50	30	50	30	10	50	30	10	50	30	10	0

41 — Indirect with performated metal underside
EFF = 82.4% % DN = 5.4% % UP = 94.6% Lamp = (2) F32T8 SC (along, across, 45°) = N/A

RCR ↓	70	50	30	70	50	30	50	30	10	50	30	10	50	30	10	0
0	0.80	0.80	0.80	0.69	0.69	0.69	0.48	0.48	0.48	0.30	0.30	0.30	0.12	0.12	0.12	0.04
1	0.72	0.69	0.66	0.62	0.59	0.57	0.42	0.40	0.39	0.27	0.25	0.24	0.11	0.10	0.10	0.04
2	0.66	0.60	0.55	0.56	0.52	0.48	0.37	0.34	0.32	0.22	0.21	0.20	0.09	0.09	0.09	0.03
3	0.60	0.53	0.47	0.51	0.45	0.41	0.32	0.29	0.27	0.20	0.18	0.17	8.08	0.08	0.07	0.02
4	0.55	0.46	0.40	0.47	0.40	0.35	0.28	0.25	0.23	0.17	0.16	0.14	3 07	0.07	0.06	0.02
5	0.50	0.41	0.35	0.43	0.36	0.30	0.25	0.22	0.19	0.16	0.14	0.12	0.07	0.06	0.05	0.02
6	0.46	0.37	0.30	0.39	0.32	0.27	0.23	0.19	0.17	0.14	0.12	0.11	0.06	0.05	0.05	0.02
7	0.42	0.33	0.27	0.36	0.28	0.23	0.20	0.17	0.14	0.13	0.11	0.09	0.05	0.05	0.04	0.01
8	0.39	0.29	0.24	0.33	0.26	0.21	0.18	0.15	0.13	0.11	0.09	0.08	0.05	0.04	0.04	0.01
9	0.36	0.27	0.21	0.31	0.23	0.18	0.17	0.13	0.11	0.10	0.08	0.07	0.04	0.04	0.03	0.01
10	0.33	0.24	0.19	0.29	0.21	0.16	0.15	0.12	0.10	0.05	0.06	0.06	0.04	0.03	0.03	0.01

42 — Semi-indirect, 2-lamp, v-shape, parabolic baffles
EFF = 83.2% % DN = 21.6% % UP = 78.4% Lamp = (2) F32T8 SC (along, across, 45°) = N/A

RCR ↓	70	50	30	70	50	30	50	30	10	50	30	10	50	30	10	0
0	0.83	0.83	0.83	0.74	0.74	0.74	0.56	0.56	0.56	0.39	0.39	0.39	0.24	0.24	0.24	0.17
1	0.76	0.73	0.70	0.67	0.65	0.62	0.49	0.48	0.46	0.35	0.34	0.33	0.22	0.21	0.21	0.15
2	0.70	0.64	0.59	0.62	0.57	0.53	0.44	0.41	0.39	0.31	0.30	0.28	0.20	0.19	0.18	0.13
3	0.64	0.57	0.51	0.56	0.50	0.46	0.39	0.36	0.33	0.28	0.26	0.24	0.18	0.17	0.16	0.12
4	0.58	0.50	0.44	0.52	0.45	0.40	0.35	0.31	0.28	0.25	0.23	0.21	0.16	0.15	0.14	0.10
5	0.54	0.45	0.39	0.47	0.40	0.35	0.31	0.27	0.25	0.23	0.20	0.18	0.15	0.13	0.12	0.09
6	0.49	0.40	0.34	0.44	0.36	0.31	0.28	0.24	0.21	0.20	0.18	0.16	0.13	0.12	0.11	0.08
7	0.46	0.36	0.30	0.40	0.32	0.27	0.25	0.22	0.19	0.19	0.16	0.14	0.12	0.11	0.10	0.07
8	0.42	0.33	0.27	0.37	0.29	0.24	0.23	0.19	0.17	0.17	0.15	0.13	0.11	0.10	0.09	0.07
9	0.39	0.30	0.24	0.35	0.27	0.22	0.21	0.17	0.15	0.16	0.13	0.11	0.10	0.09	0.08	0.06
10	0.37	0.27	0.22	0.32	0.24	0.20	0.19	0.16	0.13	0.14	0.12	0.10	0.10	0.08	0.07	0.06

43 — Semi-indirect, 2-lamp, thin profile, parabolic baffles, 70% up
EFF = 85.2% % DN = 28.7% % UP = 71.3% Lamp = (2) F32T8 SC (along, across, 45°) = N/A

RCR ↓	70	50	30	70	50	30	50	30	10	50	30	10	50	30	10	0
0	3.87	0.87	0.87	0.78	0.78	0.78	0.61	0.61	0.61	0.45	0.45	0.45	0.31	0.31	0.31	0.24
1	3.80	0.77	0.74	0.72	0.69	0.66	0.54	0.53	0.51	0.41	0.40	0.39	0.29	0.28	0.28	0.22
2	0.73	0.68	0.63	0.66	0.61	0.57	0.48	0.46	0.44	0.37	0.35	0.34	0.26	0.25	0.24	0.20
3	0.67	0.60	0.55	0.60	0.54	0.50	0.43	0.40	0.37	0.33	0.31	0.29	0.24	0.23	0.22	0.18
4	0.62	0.54	0.48	0.55	0.49	0.43	0.39	3.35	0.32	0.30	0.28	0.26	0.22	0.20	0.19	0.16
5	0.57	0.48	0.42	0.51	0.44	0.38	0.35	0.31	0.28	0.27	0.25	0.23	0.23	0.18	0.17	0.14
6	0.53	0.43	0.37	0.47	0.39	0.34	0.32	0.28	0.25	0.25	0.22	0.20	0.18	0.17	0.15	0.13
7	0.49	0.39	0.33	0.44	0.36	0.30	0.29	0.25	0.22	0.23	0.20	0.18	0.17	0.15	0.14	0.11
8	0.45	0.35	0.29	0.41	0.32	0.27	0.26	0.22	0.20	0.21	0.18	0.16	0.15	0.14	0.12	0.10
9	0.42	0.32	0.26	0.38	0.30	0.24	0.24	0.20	0.18	0.19	0.16	0.14	0.14	0.13	0.11	0.09
10	0.39	0.30	0.24	0.35	0.27	0.22	0.22	0.18	0.16	0.18	0.15	0.13	0.13	0.12	0.10	0.09

44 — Direct/indirect, 2-lamp, thin profile, parabolic baffles, 60% up
EFF = 88.3% % DN = 40.3 % UP = 59.7 Lamp = (2) F32T8 SC (along, across, 45°) = N/A

RCR ↓	70	50	30	70	50	30	50	30	10	50	30	10	50	30	10	0
0	0.93	0.93	0.93	0.84	0.84	0.84	0.69	0.69	0.69	0.55	0.55	0.55	0.42	0.42	0.42	0.36
1	0.85	0.82	0.79	0.77	0.75	0.72	0.61	0.60	0.58	0.49	0.48	0.47	0.38	0.37	0.37	0.31
2	0.78	0.72	0.67	0.71	0.66	0.62	0.54	0.51	0.49	0.44	0.42	0.40	0.34	0.33	0.32	0.27
3	0.71	0.64	0.58	0.65	0.58	0.53	0.48	0.45	0.42	0.39	0.37	0.34	0.31	0.29	0.28	0.24
4	0.65	0.57	0.50	0.60	0.52	0.46	0.43	0.39	0.36	0.35	0.32	0.30	0.29	0.26	0.24	0.21
5	0.60	0.50	0.44	0.55	0.46	0.40	0.39	0.34	0.31	0.32	0.28	0.26	0.25	0.23	0.21	0.18
6	0.55	0.45	0.38	0.50	0.42	0.36	0.35	0.30	0.27	0.28	0.25	0.23	0.23	0.20	0.19	0.16
7	0.51	0.41	0.34	0.47	0.38	0.32	0.32	0.27	0.24	0.26	0.23	0.20	0.21	0.18	0.16	0.14
8	0.47	0.37	0.30	0.43	0.34	0.28	0.29	0.24	0.21	0.24	0.20	0.18	0.19	0.16	0.15	0.13
9	0.44	0.34	0.27	0.40	0.31	0.25	0.26	0.22	0.19	0.22	0.18	0.16	0.17	0.15	0.13	0.11
10	0.41	0.31	0.25	0.37	0.28	0.23	0.24	0.20	0.17	0.20	0.17	0.14	0.16	0.14	0.12	0.10

Reproduced from the *IESNA Lighting Handbook, 9th Edition,*. 2000, courtesy of the Illuminating Engineering Society of North America.

REVIEW QUESTIONS

ESTIMATE FOOTCANDLES AND DOLLARS

Lighting professionals estimate illuminance levels and costs for proposed designs. Answer the following questions to refine your estimating skills. Questions 1 through 8 are based on the example footcandle calculation summarized below, but refer back to pages 83 and 84 before answering.

Example Classroom Summary Pages 83–84

Dimensions: 20′ W, 30′ L, 9′ H, area = 600 sq. ft.

Fixture: #2, 24″ × 48″, 3 lamp fluorescent

Lamps: F40 CW (34 watt), 2,900 lumens each

Reflectance: pc = 80%, pw = 50%, pf = 20%

Cavity ratios: CCR = 0, RCR = 2.7, FCR = 1.0

Coefficient of utilization: CU = 60%

Light loss factor: LLF = 70%

No. of fixtures required = 8.2 for 50 footcandles

No. of fixtures provided = 9 for 55 footcandles

1. Classroom lighting operates 2,500 hours each year, and electricity costs $0.08 per kWh. Find the *annual* cost of lighting the classroom.
2. Classroom lamps are cleaned once a year with a labor cost of $1 per fixture. Fixture re-lamping is scheduled every six years (75% of lamp life) at a cost of $6 per fixture for labor and lamps. Find the *annual* cost of cleaning and re-lamping.
3. Estimate the annual *cost per square foot* for classroom lighting.
4. If classroom ceiling reflectance is changed to 50% (instead of 80%), less light will reach the work plane. Find the CU change and the footcandle change.
5. If classroom dimensions are changed to 10′ × 60′ (ugh!), both RCR and CU will change. Estimate the footcandle reduction.
6. Classroom illumination must be increased to 80 fc. Make two proposals for lighting changes, without doing a new calculation.
7. Increase the example classroom ceiling height to 10′ so that indirect fixtures can be mounted 20″ below the ceiling. Replace the example fixtures with nine #8 fluorescent fixtures. Calculate footcandles if each new fixture uses two F48 CW/HO (60 watt) lamps.
8. Nine #8 indirect fluorescent fixtures offer diffuse lighting, but they're a poor choice for uniform lighting in the example classroom. Why?
9. Make a preliminary comparison of 400 watt MH-HID and 120 watt fluorescent fixtures being considered for a new 10,000 sq. ft. grocery store. The 400 watt MH lamp produces 36,000 initial lumens. Two 60 watt T8 lamps in the 96″ fluorescent fixture produce 12,000 initial lumens.

   If both fixtures have a CU of 64% and an LLF of 70% in the grocery store, estimate the number of fixtures required to provide 50 footcandles.
10. If electricity costs $0.08 per kWh and grocery store lighting operates for 8,670 hours each year, estimate annual lighting costs for ninety-three 120 watt fluorescent fixtures.

ANSWERS

1. **$184 per year.**
   Each of nine fixtures has three 34 watt lamps.

   (34 watts)(9 fixtures)(3 lamps) = 918 watts

   (0.918 kW)(2,500 hr.)($0.08) = $183.60

   Electrical costs vary, and the average cost per kWh must include demand charges.
2. **$18 per year.**

   Clean ($1 per fixture)(6 yr.)(9 fixtures) = $54

   re-lamp ($6 per fixture)(9 fixtures) = $54

   annual cost, $108 ÷ 6 yr. = $18 per year

   Actual costs will be higher because some ballasts and lamps will fail during six years of operation.
3. **$0.34 per square foot per year.**

   $184 + $18 ÷ 600 sq. ft. = $0.34
4. **CU and fc drop about 5%.**

   Example CU was 60%, new CU = 57%.

   Example fc was 55, new fc = 52.
5. **fc drop 10% from 55 to 50.**

   Example RCR was 2.7, new RCR = 3.8.

   Example CU was 60%, new CU = 55%.

   Example fc was 55, new fc = 50.

6. **Two ways to get 80 fc.**

   - Replace 34 watt, 2,900 lumen lamps with 60 watt, 4,300 lumen lamps (and ballasts) to yield 80 fc.
   - Nine #2 fixtures produce 55 fc, so a new layout using 14 #2 fixtures will yield 80 fc. Which requires the most watts?

7. **37 footcandles** (with fixture #8). Review the footcandle calculation on pages 83–84 and use the following blank Footcandle Calculation form to check this answer. If you don't get 37 footcandles, recheck each entry on your calculation form as follows. New ceiling cavity height is 1.67¢. New room cavity height is 5.83¢.

   F 48 CW/HO (60 watt) lamps yield 4,300 lumens, and with two lamps FL = 8,600.

   The new CCR (ceiling cavity ratio) is 0.7 because the ceiling cavity height is 1.67'.

   $(5)(1.67)(20 + 30) \div 600 = 0.69$, rounded to 0.7

   The new RCR (room cavity ratio) is 2.4 because the new room cavity height is 5.83'.

   $(5)(5.83)(20 + 30) \div 600 = 2.42$, rounded to 2.4

   The new pcc (effective ceiling cavity reflectance) is 70% (CCR = 1.67, interpolate).
   The new CU = 41% (fixture #8, RCR 2.4, pcc = 70%, pw = 50%, interpolate).
   LLF = 70% no change, assumes frequent lamp cleaning in these open fixtures.
   Calculate footcandles with 9 fixtures = 37.

   $9 = (fc)(600 \text{ sq. ft.}) \div (8,600)(41\%)(70\%)$

8. **Nine indirect fixtures a poor choice.**

   - More watts, less footcandles.
   - Less uniform, fixture #8 works better in continuous rows because light distribution at fixture ends is poor.
   - Less efficient light distribution because #8 has a recommended 24″ drop, not 20″.

9. **31 MH or 93 fluorescent fixtures.** Compare 400 watt MH-HID fixtures and 120 watt fluorescent fixtures at 50 footcandles in a 10,000 sq. ft. grocery store.

   MH lamp = 36,000 initial lumens.

   Fluorescent lamps = 12,000 initial lumens.

   Both fixtures have a CU of 64% and a LLF of 70%.
   Solution:
   Both fixtures will deliver 45% of initial lamp lumens as usable footcandles.

   $(64\% \text{ CU})(70\% \text{ LLF}) = 45\%$

   50 lumens per square foot = 50 footcandles, so this 10,000 sq. ft. store requires 500,000 lumens.

   $(10,000 \text{ sq. ft})(50 \text{ lumens/sq. ft.}) = 500,000$

   31 MH-HID fixtures are required.

   $500,000 \text{ lm} \div (36,000 \text{ lm})(45\%) = 30.86$

   total watts = 14,260 (460 incl. ballast)(31)

   93 fluorescent fixtures are required.

   $500,000 \text{ lm} \div (12,000 \text{ lm})(45\%) = 92.59$

   total watts = 11,160 (120)(93)

   Danger! This comparison is incomplete; initial costs, lamp life, mean footcandles, operating costs, maintenance costs, and many other considerations are omitted.

10. **$7,741 per year.**

    $(93)(120)(8,760 \text{ hr})(\$0.08) \div 1,000 \text{ W/kW}$

    Perception begins with light. Color, contrast, form, and pattern are names that describe reflected light, and design.

    Lighting numbers can be interesting, but they are not metrics for design or perception.

Footcandle Calculation   project _____

Fixture Information

Manufacturer - model #
_____

option(s) _____

lamp(s) _____

_____

Fixture Lumens ↙

FL └____

ballast _____

Room View

1, 2, 3

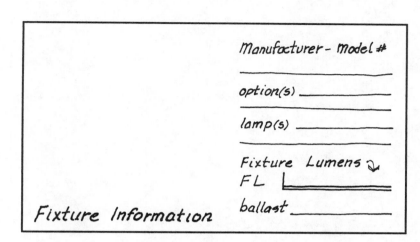

ceiling reflectance ↙ └__ pc

ceiling cavity height

room cavity height

floor cavity height

average wall reflectance
└__ pw

floor reflectance └__ pf

Room Plan & Fixture Layout

---

by _____ date _____

Room _____
dimensions _____ x ____
area _____ sqft.

Selected Fixture
lamp lumens _____
# of lamps _____
Fixture lumens _____

FL

1
cavity height & reflectance
ceiling ____' pc _____%
wall ____' pw _____%
floor ____' pf _____%

2
Cavity Ratios
CR = (5 hc) (L+W) ÷ (LxW)
ceiling ~ CCR _____
room ~ RCR _____
floor ~ FCR _____

3
effective ceiling cavity
reflectance ~ pcc _____

4
Coefficient of Utilization
_____ CU

5
Light Loss Factor _____
LLF

6
Footcandles required _____
FC

7
Number of Fixtures =

(FC)(area)÷(FL)(CU)(LLF)

_____

PART II
ELECTRICAL

☐

CHAPTER 8
Electrical Basics

Electricity seems destined to play a most important part in the arts and industries. The question of its economical application to some purposes is still unsettled, but experiment has already proved that it will propel a street car better than a gas jet and give more light than a horse.

Ambrose Bierce, 1906

Electrical distribution lines and equipment are essential building subsystems. Design and construction professionals study these electrical components so they can make competent judgments about:

- Lighting and power requirements
- Space required for switchgear and transformers
- Optimum location for major components
- Space required for horizontal and vertical distribution
- Construction and operating costs
- Occupant safety

Build your understanding of this short chapter by completing the ending sample problems and reviewing the preceding list of terms.

8.0 ELECTRICAL TERMS

To work with the people who electrify your projects you should speak their language. Skim this list of terms before you begin Chapter 8 and review it after each chapter to build a language foundation that supports effective communication.

Ampere   Unit of electrical current (flow of electrons). A 1 volt force, acting in a circuit with a resistance of 1 ohm, produces a current of 1 ampere.

AC   Alternating current. Electrons flow forward, then backward to complete a cycle. In 60 cycle AC, electron flow direction reverses 120 times each second.

Alternator   A device that transforms mechanical energy into AC electrical energy.

Bus   A rectangular metal conductor used to carry large currents.

Capacitor   An assembly of layered conductors and insulators. After passing through a capacitor, AC amperes will lead volts.

Circuit   A closed loop that accommodates electron flow.

Circuit breaker   A switch that opens when current exceeds the breaker trip rating.

Coil   A conducting spiral. Coils or windings are found in motors, transformers, and magnetic ballasts.

Conductors   Materials like copper and aluminum that accommodate electron flow.

Conduit   A tube used to protect conductors.

Current   Number of electrons passing a point in one second, measured in amperes. One amp = 6.28×10^{18} electrons/second.

DC   Direct current. Electrons flow in one direction.

Delta   A three-phase service geometry.

Demand   Volt-amperes or kVA.

EMT   Electrical metallic tubing.

Feeders   Conductors that serve large electrical loads.

Generator   A device that transforms mechanical energy into DC electrical energy.

GFI   Ground fault interrupter. A circuit breaker that opens if circuit current varies.

Ground   Electrical connection to earth.

Hz   Frequency. AC cycles per second.

Inductive reactance   A characteristic of conducting coils or windings. After passing through a coil, AC amperes will lag volts.

Insulators   Materials like rubber and glass that resist electron flow.

kVA   Volt-amperes ÷ 1,000.

kW   Kilowatt (1,000 watts). Unit of electrical power.

kWh   Kilowatt hour. Unit of electrical work or energy.

Megawatt   1,000,000 watts.

NEC   National Electric Code of the National Fire Protection Agency (NFPA).

NEMA   National Electrical Manufacturers Association.

Ohm   Unit of electrical resistance. Tendency to resist current flow.

Ohm's law   Relation between ohms, amps, and volts. $E = IR$. Volts = (amps) (ohms).

Panel   An enclosure where electric power is allocated among individual circuits.

Parallel circuit   A circuit where equal voltage is applied to each electrical load.

Power   Watt or kW.

Resistance   Tendency to resist current flow.

Series circuit   A circuit where equal current flows through all electrical loads.

Service   The conductors and equipment that bring electrical power to a building.

Strip heat   Heat caused by current flow in nichrome wire.

Switchgear   Large electrical load centers that allocate power.

Transformer   A device used to exchange AC volts and amperes.

VA   Volt-ampere. Product of amps and volts.

Volt   Unit of electrical pressure or force. A force of 1 volt, acting in a circuit with a resistance of 1ohm, causes a current of 1 ampere.

Voltage drop   Voltage lost to resistance, a problem when circuits serve distant loads.

Watt   Unit of electrical power. The rate at which work is done or energy is expended. One watt = 1/746 horsepower.

Winding   A series of conducting loops. Windings or coils are found in motors, transformers, and magnetic ballasts.

Wye   A balanced three-phase service geometry.

8.1 ELECTRICITY AND MAGNETISM

ELECTRICITY

Electricity is the flow of electrons in a given direction. Previous courses described matter as atoms consisting of a nucleus surrounded by orbiting electrons. Electrons orbit in a series of "shells" at increasing distances from the nucleus. Outermost shell electrons flow in electrical circuits.

Atoms of many metals hold their outermost shell electrons "loosely." These metals are called *conductors* because a small energy input will cause their electrons to flow from one atom to another. Silver, copper, gold, and aluminum are all good electrical conductors. Other materials such as rubber, glass, wood, air, and most plastics resist electron flow and are called *insulators*.

Circuits

An electrical circuit is a loop designed to accommodate electron flow. The loop consists of a conductor that extends from an electron source, to an electrical device, and back to the electron source. Electrons will not flow without a complete circuit that begins and ends at the source of electrical energy. *Parallel* circuits are used in most building applications while *series* circuits are used extensively in electronics (see Figure 8.1).

MAGNETISM

Magnetism and electricity are related phenomena. A magnetic field is an area of electrical influence, and a flow of electrons can create a magnetic field.

Magnets are surrounded by force fields called *magnetic flux*. The strength or intensity of a magnetic field is called *flux density*. When a conductor moves through a magnetic field, an electrical force is generated. If the conductor is looped to form a circuit, electrons will flow around the loop (see Figure 8.2).

A moving electric field creates a magnetic field, and a moving magnetic field creates an electric field. Because of the relationship between electricity and magnetism, a magnetic field forms around conductors carrying electrical energy. The *left-hand rule* describes this magnetic field based on the direction of electron flow. If a conductor is grasped by the left

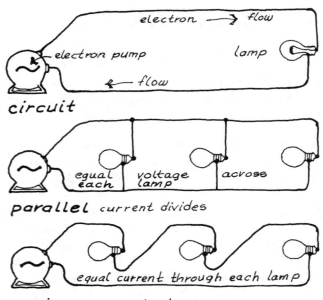

FIGURE 8.1

circuit + motion in field = current

FIGURE 8.2

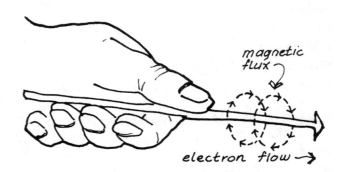

FIGURE 8.3

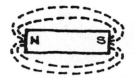 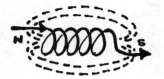

FIGURE 8.4

hand with the thumb pointing in the direction of electron flow, the fingers describe the magnetic field's flux from north to south (see Figure 8.3).

Electromagnets can be made by forming conductors in a coil shape. When electric current flows, each coil loop adds to the magnetic field, producing the same effect as a natural magnet. An iron bar centered in the coil will increase magnetic flux (see Figure 8.4).

8.2 **ELECTRICAL UNITS**

The electromotive force or pressure that causes electron flow is measured in **volts**, and the quantity of electrons flowing is measured in **amperes**. Even good conductors like copper wire offer some resistance to electron flow. Electrical resistance is measured in **ohms**. One ampere is defined as the quantity of electrons that will flow through a resistance of 1 ohm when pushed with a pressure of 1 volt. Electrical power is measured in **watts**. An electrical current of 1 ampere pushed by a force of 1 volt is defined as 1 watt.

The analogy of water flow in a pipe may help you visualize electrical terms (see Figure 8.5). The force or pressure pushing water through a pipe is measured in pounds per square inch and is analogous to electrical volts. The quantity of water flowing in a pipe is measured in gallons per minute and is analogous to amperes. Rough pipe offers more resistance to water flow than smooth pipe. In a water supply system, this resistance is called *pipe friction*, and it is measured as pressure drop when water flows. Similarly, the voltage in an electrical circuit drops when electrons flow. The resistance of electrical conductors causes this voltage drop.

Power required to pump water can be quantified in horsepower. Electrical power is measured in watts—actually thousands of watts or kilowatts.

8.3 **CALCULATIONS AND EXAMPLES**

RELATIONSHIPS

Electrical calculations use **E** (electromotive force) for voltage, **I** (intensity) for current in amperes, and **R** for resistance in ohms. Three formulas relate these electrical terms.

$$E = IR \qquad \text{volts} = (\text{amps})(\text{ohms})$$

defines the relative magnitude of electrical pressure in volts, current in amperes, and resistance in ohms. This

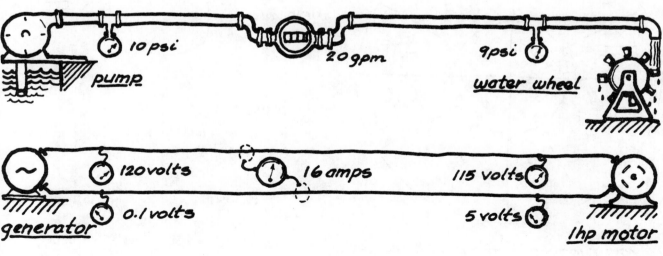

FIGURE 8.5

formula is called Ohm's law (for Georg Simon Ohm, 1787–1854) and is useful for calculating voltage drop in long electrical circuits.

Electrical power is measured in **watts** (honoring James Watt, 1736–1819). Increasing flow or pressure in a water system requires an increase in pump power, and increasing current or voltage in an electrical circuit requires an increase in electrical power.

The power formulas:

$$EI = W \qquad \text{(volts)(amps) = watts}$$

or

$$I^2R = W \qquad \text{(amps)}^2\text{(ohms) = watts}$$

relate volts, amperes, ohms, and power (see Figure 8.6).

CALCULATION EXAMPLES

Most building electrical calculations solve for *current* or *voltage drop*. Current sets conductor size, and excess voltage drop may require an increase in conductor size. The following examples are typical of the electrical math that will be applied later in the text to size building electrical circuits and services.

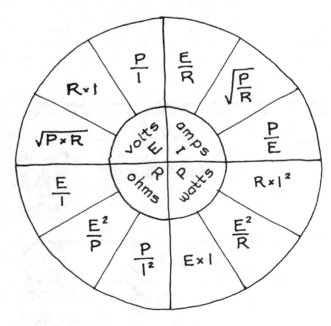

12 variations of 3 formulas

FIGURE 8.6

Example 1

A light bulb is rated at 100 watts. Find the current flow if the bulb operates at 120 volts.

$$EI = W$$
$$(120)(I) = 100 \qquad I = 0.83 \text{ ampere}$$

Example 2

Find the resistance of the light bulb.

$$E = IR$$
$$(120) = (0.83)(R) \qquad R = 144 \text{ ohms}$$

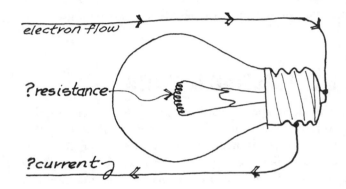

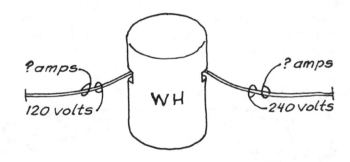

Example 3

A water heater is rated at 4,500 watts. Find the current required at 120 volts and at 240 volts.

$$EI = W$$
$$(120)(I) = 4,500 \qquad I = 37.5 \text{ amps}$$
$$EI = W$$
$$(240)(I) = 4,500 \qquad I = 18.8 \text{ amps}$$

Increased voltage means reduced current and smaller conductors.

Example 4

The total connected electrical load for all the lights and kilns in a small pottery studio is 125 kW. Find the current required for the studio's electrical service at 240 volts.

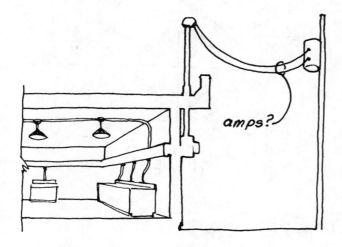

(1 kW = 1,000 watts)

$$EI = W$$
$$(240)(I) = 125,000 \qquad I = 521 \text{ amps}$$

Example 5

A current of 4 amperes is flowing in a conductor that has a resistance of 3 ohms per 1,000 feet. Find the voltage drop if a circuit run totals 2,000 feet from the source to the load and back to the source.

total ohms for 2,000 feet = 6

$$E = IR$$
$$E = (4)(6) \qquad E = 24 \text{ volts drop}$$

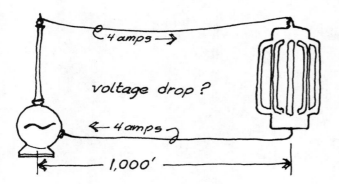

Low voltage can cause motor overloads and light dimming. Voltage drop is a design concern when electrical devices are located more than 100 feet distant from the electrical service entry.

Example 6

A conductor has a resistance of 20 ohms per 1,000 feet of length. The conductor serves a lamp 200 feet from the power source. If the lamp has a resistance of 72 ohms and source voltage is 120, find voltage drop across the light bulb.

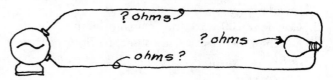

First find total circuit ohms:
The 200-foot circuit loop includes 4 ohms for supply run, 4 more ohms for return run, and 72 ohms for the lamp.

$$4 + 72 + 4 \qquad R = 80 \text{ ohms}$$

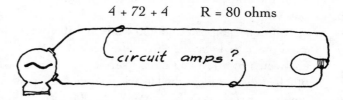

Next find circuit current.

$$E = IR$$
$$120 = (I)(80) \qquad I = 1.5 \text{ amps}$$

Current is constant throughout the circuit; voltage drops in proportion to resistance.

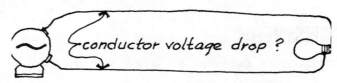

Next find the voltage drop through the circuit conductors.

$$E = IR$$
$$E = (1.5)(414) \qquad E = 12 \text{ volts drop}$$

These 12 volts were spent overcoming conductor resistance.

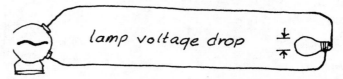

Finally, find the actual voltage drop across the lamp.

$$E = IR$$
$$E = (1.5)(72) \qquad E = 108 \text{ volts}$$

Light output will be reduced because conductor resistance cuts lamp voltage.

REVIEW QUESTIONS

1. A 115 volt fluorescent lamp is rated at 0.43 amps (430 milliamperes). Find the watts that will be consumed by the lamp and its ballast. (*115 volts is specified instead of 120 volts to allow for circuit voltage drop.*)

2. Consider 100 watt and 300 watt lamps. Which has the most resistance (ohms)? (*Try to pick the correct answer without referring to the formulas in Figure 8.7.*)

3. Find total current (amperes) for a home with a 230 volt electrical service, if the estimated maximum electrical load from all lights and appliances is 21,000 watts.

4. Find the amperes drawn by a 12 kW range at 230 volts.

5. An outdoor lamp is rated 250 watts at 130 volts. Find the actual voltage across the lamp if it is connected to a 115 volt circuit serving the lamp 500 feet distant from the electrical service. Circuit conductor resistance is 3 ohms per 1,000 feet.

6. Calculate the actual operating watts for the lamp in question 5. How will actual watts affect light output and lamp life?

7. One value is a constant for the lamp in question 5. Which one?

ANSWERS

1. 49 watts (*actually a bit less because of ballast characteristics*)

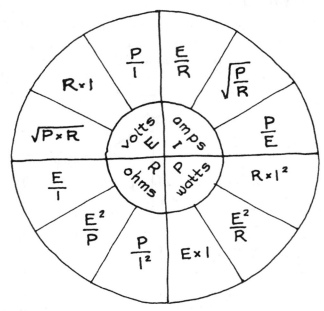

FIGURE 8.7

2. 100 watt
3. 91 amps (*nearest standard size is 100 amps*)
4. 52 amps (*1 kW = 1,000 watts*)
5. 110.1 volts
6. 178 watts, less light, longer life. *Conductor voltage drop is 4.9, so actual lamp watts is (110.1) (1.62) = 179 circuit amps @ 115 V = 1.62.*
7. Is the value amps, ohms, volts, or watts?

CHAPTER 9
Generation and Distribution

Work is of two kinds: first altering the position of matter at or near the earth's surface relatively to other such matter; second telling other people to do so. The first kind is unpleasant and ill paid; the second is pleasant and highly paid.

Bertrand Russell, 1935

This chapter describes the production and distribution of electrical energy. It introduces terms and concepts that influence the costs of electrical systems and the cost of electricity. These costs are substantial components of building construction and operating budgets.

Read and reread until you're comfortable explaining:

* Why demand and consumption set energy prices for commercial buildings
* Why transformers reduce AC distribution costs
* Why most electrical installations are grounded

Review the terms introduced earlier in the text to support your new vocabulary and work the ending review problems.

9.0 DEMAND AND CONSUMPTION

POWER, WORK, AND ENERGY

Power is the rate at which work is done. It may be measured in kilowatts (kW) or horsepower (hp). A kilowatt is a rate of 1,000 joules per second, and a horsepower is a rate of 746 joules per second, or 550 footpounds per second.

Work is defined as power multiplied by time, or force multiplied by distance. Work is measured in units of power and time; two common measurement units are the kilowatt-hour (kWh) and the horsepower-hour (hph).

Energy can be defined as the ability to do work. "Energy" and "work" are often used interchangeably. However, the word *energy* is also used to describe the heat content of fuels.

DEMAND AND CONSUMPTION

Electric utilities charge commercial customers for both power demand and energy consumption (see Figure 9.1). Power *demand* charges are based on the largest kW load connected during any 15-minute time period. Demand costs can be most of the utility bill for churches or theaters that require large quantities of power for short periods of time. Control systems that limit demand by sequencing electrical loads can reduce demand costs for commercial buildings.

Homes are not metered for demand. Utilities bill energy *consumption* in kWh, using different rate schedules for residential and commercial customers. Utility rate schedules vary remarkably; *don't* attempt to estimate electrical energy costs without an accurate rate schedule.

ELECTRICAL CURRENT, DC OR AC

Electron flow can be caused by batteries or generators. Batteries convert chemical energy to electrical energy and cause external electron flow in one direction—from the negative battery terminal to the positive battery terminal. This flow is called *direct current* (DC). Batteries are used in buildings for emergency lighting and in some cases for emergency power, computer power, and phones.

Generators use conductors, magnetic flux, and motion to convert mechanical energy into electrical energy. Generators may be designed to deliver DC like a battery, or *alternating current* (AC) where electron flow reverses direction at regular intervals (see Figure 9.2).*

Electric utilities produce AC power because AC can be distributed at high voltage, minimizing line losses, and then easily transformed to safer low voltages suitable for building use.

9.1 GENERATING EFFICIENCY

ELECTRICAL GENERATION

Utilities use falling water or burning fuel to turn generators. About 10% of U.S. energy is obtained from hydroelectric sources; fossil fuels or nuclear fission reactors account for the remaining 90%. Heat from fuel or nuclear sources is used to turn turbines that drive generators.

The bar graph in Figure 9.3 estimates 2,000 U.S. electrical energy sources. The fuel energy used to

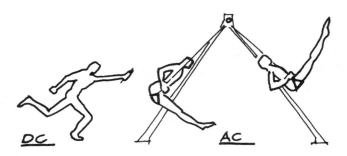

FIGURE 9.2

*The correct name for a device that produces AC current is "alternator," but "generator" is widely misused to describe a mechanical device that causes electron flow.

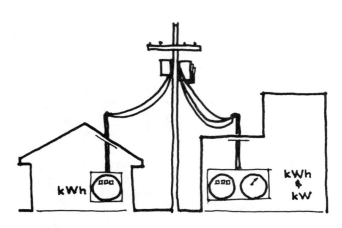

FIGURE 9.1

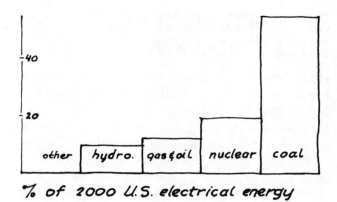

% of 2000 U.S. electrical energy

FIGURE 9.3

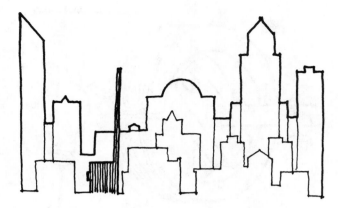

FIGURE 9.5

produce electricity is nearly equal to the fuel energy used for transportation.

GENERATING EFFICIENCY

The second law of thermodynamics limits the theoretical efficiency of steam turbine-driven generators to about 60%, and actual industry experience is near 35%.* A typical electric utility converts 10 units of fuel energy into 3.5 units of electrical energy and 6.5 units of waste heat (see Figure 9.4). Nuclear plants are less efficient than fossil fuel plants because they operate with reduced steam temperature.

Although the average thermal efficiency of an electrical generating plant is not nearly as good as an

industrial or residential furnace, it is much better than the typical new car.

COGENERATION

Cogeneration and *total energy* are terms used to describe private power producers that conserve fuel by improved utilization of excess heat or generating capacity. Cogeneration describes a power producer that sells surplus heat or electricity to a utility. Total energy usually refers to an on-site generating plant that uses waste heat for building heating, water heating, and refrigeration (see Figure 9.5).

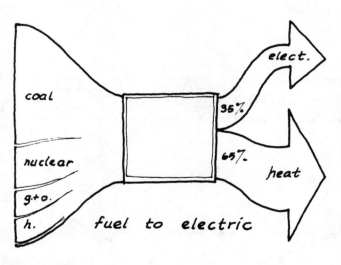

fuel to electric

FIGURE 9.4

9.2 ALTERNATING CURRENT

AC GENERATORS (*ALTERNATORS*)

Generators convert mechanical energy to electrical energy by rotating a magnetic field in a circle of conductors (see Figure 9.6). Voltage is generated when magnetic flux moves through a conductor, and if the conductor is connected to a circuit, current will flow.

Utilities in North America use steam-, gas-, or water-driven turbines to generate electric power. Generator drives rotate 60 times each second, causing electron flow to change direction or "alternate" 120 times each second. The resulting electrical power is called 60 cycle or 60 hertz alternating current (60 Hz AC).*

---

*Dual cycle generating plants can approach 50% efficiency. They drive one generator with a gas turbine, and a second with steam produced by turbine exhaust.

---

*50 Hz AC is used in Europe, and 400 Hz AC is used for some aviation applications.

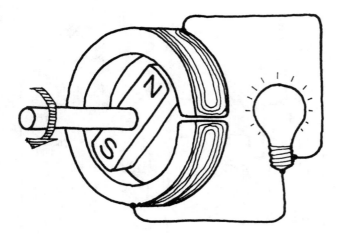

FIGURE 9.6

When reviewing Figure 9.7, assume a generator uses rotating conductors* in a stationary magnetic field to cause electron flow. As conductors rotate, their motion relative to the magnetic field is constantly changing. As the conductors cut through more magnetic flux, more voltage is generated and more current flows. The direction of current flow will reverse when the direction of conductor motion in the magnetic field reverses during each revolution.

---

*As shown in Figure 9.6, the magnetic field rotates and conductors are stationary.

9.3 SINGLE-PHASE AND THREE-PHASE POWER

High-voltage transmission lines use three conductors (or multiples of three). Utilities generate three-phase power. Their generators cause simultaneous current flow between three separate conductors. At one instant, electrons may be flowing from the power plant in one conductor while they flow back toward the plant in the other two. An instant later the direction and quantity of electron flow will change. Three-phase current and voltage peaks are offset 120 degrees between the three phases, and utilities connect their distribution systems so that nearly equal electrical loads are connected between all three phases. Equalizing connected loads means the energy produced by generators will be used in a complete or *balanced* way.

Residential neighborhoods usually have single-phase power that reduces the number of utility distribution lines needed to serve homes (see Figure 9.8). Utilities designate individual phases A, B, and C. If one neighborhood is served by A and B, the next will be connected across B and C, and the next to C and A so that the total load is balanced for each phase.

Commercial and industrial electrical customers use three-phase power to carry motor loads (see Figure 9.9). Motors are generators operating in reverse. Three-phase motors are more efficient than

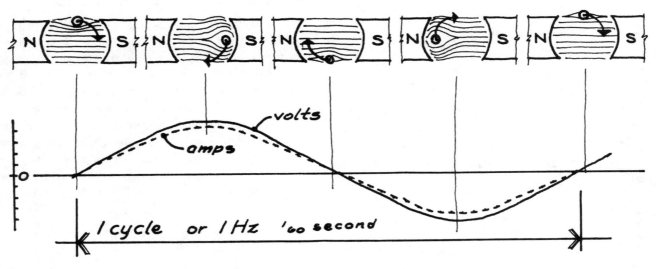

FIGURE 9.7

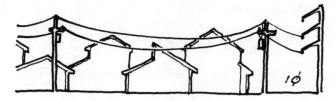

FIGURE 9.8

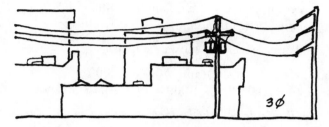

FIGURE 9.9

motor h.p. & phase

FIGURE 9.10

single-phase motors. Motors larger than 5 horsepower are usually three-phase, and where three-phase power is available, motors larger than 1/2 horsepower should be three-phase (see Figure 9.10).

Power in a three-phase electrical system is carried by three conductors. The basic power equation must be modified to account for current flow between these conductors where current peaks are staggered by 120 degrees. With three-phase power:

$$EI\sqrt{3} = W \qquad (volts)(amps)(\sqrt{3}) = watts$$

replaces the single-phase equation EI = W. *√3 or 1.73 is the cosine of 120° and accounts for the power vectors between three-phase conductors.*

Example Calculations

1. A single-phase, 120 volt electric device requires 20 kW. Find full load current.

$$EI = W$$
$$(120)(I) = 20,000 \qquad I = 167 \text{ amps}$$

167 amps flow in *both* the hot and the neutral conductors.

2. A single-phase, 240 volt electric device requires 20 kW. Find full load current.

$$EI = \backslash W$$
$$(240)(I) = 20,000 \qquad I = 83.3 \text{ amps}$$

83.3 amps will flow in each of *two* hot conductors. This is *not* two-phase.

3. A three-phase, 240 volt electric device requires 20 kW. Find full load current.

$$EI\sqrt{3} = W$$
$$(240)(I)(1.73) = 20,000 \qquad I = 48.2 \text{ amps}$$

48.2 amps flow in each of three hot wires. Three conductors carrying 48 amps use less copper than two conductors carrying 83 amps.

4. A three-phase, 480 volt electric device requires 20 kW. Find full load current.

$$EI\sqrt{3} = W$$
$$(480)(I)(1.73) = 20,000 \qquad I = 24.1 \text{ amps}$$

24.1 amps flow in each of three hot wires—much less copper.

9.4 **TRANSFORMERS**

POWER DISTRIBUTION

AC is the power of choice for utilities because it can be transmitted over fairly long distances without excessive distribution losses and then transformed to lower voltage at the customer's service entry. Very high voltages (up to 500,000 volts) are used on major distribution lines because for a given electrical load more voltage means less current. Reducing current flow reduces transmission losses. The insulators on electrical distribution lines indicate line voltage. Larger insulators are used for higher voltages.

TRANSFORMERS

Transformers permit power distribution at high voltage (minimizing line losses) and deliver a variety of lower voltages for building services. Residential customers may take 240 volts, while a commercial building nearby gets 480 volts. A large industrial plant may be served at 13,200 volts. Transformers let utilities serve many customers with differing electrical requirements using a single distribution network.

A transformer is a unique and efficient device without moving parts that trades AC voltage for current. Transformers include an iron core surrounded by two circuit loops called *windings*. One winding is connected to a power supply circuit, and the second is connected to building service circuits. *Primary* is the term for the high-voltage or input transformer winding. *Secondary* refers to the low-voltage or output winding.

AC current flow in the transformer primary windings creates a magnetic field that expands and contracts with changing current intensity and direction. This pulsating magnetic field induces current flow in the transformer secondary. Transformers change high-voltage primary power into low-voltage secondary power. Transformer capacity is rated in **kVA**, and the reduction of secondary voltage is proportional to the ratio of winding turns from the primary to the secondary side of the transformer.

Example Calculations

1. Find the transformer capacity, and the secondary voltage and current for a residential single-phase (1ø) transformer. Primary power is 13,200 volts at 1.9 amperes, and the ratio of winding turns is 55 to 1 (see Figure 9.11). Transformer capacity in kVA is:

$$(volts)(amps) \div 1,000$$
$$(13,200)(1.9) = 25,080 \quad \text{or } 25 \text{ kVA}$$

Transformer capacity and building service capacity are typically stated in thousands of volt-amperes (kVA) instead of kilowatts (kW) because of power factor considerations.

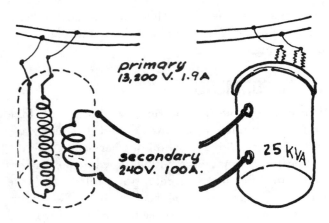

FIGURE 9.11

Secondary voltage will be 1/55 of the primary voltage because the winding ratio is 55 to 1.

$$13,200 \div 55 = 240 \quad 240 \text{ volts}$$

Secondary current will be 55 times the primary current of 1.9 amperes.

$$(55)(1.9) = 104.5 \quad \text{or } 100+ \text{ amperes}$$

2. Find the three-phase (3ø) transformer capacity required to serve a 60 ampere, 480 volt commercial building load.

Transformer capacity in kVA is:

$$(volts)(amps)(\sqrt{3}) \div 1,000$$
$$(480)(60)(\sqrt{3}) \div 1,000 = 49.8 \quad \text{or } 50 \text{ kVA}$$

Utilities frequently use three single-phase transformers instead of a three-phase transformer.
Large industrial customers can reduce utility costs by building their own electrical distribution network and buying power at 5,000 to 15,000 volts.

9.5 POWER FACTOR
POWER FACTOR

Transformers, motors, and ballasts include windings, and all windings have a property called *inductance*, which is a tendency to oppose changes in circuit current. *Inductive reactance** is a term describing this tendency to resist changes in current flow. Inductive reactance is a characteristic of windings in AC circuits. Its result is delayed current flow, so that current peaks occur after voltage peaks. Lagging current means lower metered kWh and less income for utilities.

Power factor is a number index used to quantify lagging current (see Figure 9.12). A building with no motors or coils will have a power factor of 1.0 (no lag), but buildings with substantial motor loads

---

*Inductive reactance is a characteristic of coils or windings in AC circuits. Capacitive reactance is a characteristic of capacitors in AC circuits.
The general term *inductance* is used to describe the combined effect of resistance, inductive reactance, and capacitive reactance in a circuit.

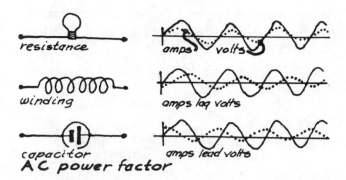

resistance

amps volts

winding

amps lag volts

capacitor

amps lead volts

AC power factor

FIGURE 9.12

and/or ballasted lighting loads will have a power factor of less than 1.

The calculation of real power for a building is given by the equations:

$$(amps)(volts)(PF) = Watts$$
for a single-phase service or
$$(amps)(volts)(\sqrt{3})(PF) = Watts$$
for a three-phase service

Buildings with low power factors use electrical energy less efficiently than similar buildings with power factors near 1, and utilities require capacitors or apply a rate surcharge to low PF buildings. Capacitors improve power factor by causing current peaks to lead voltage peaks.

Example Calculations

1. Calculate the power factor for a building if the connected electrical loads are:

   ♦ 40% resistive loads, including water heaters, incandescent lights, and electric heaters with a PF of 1.0
   ♦ 40% fluorescent lighting with a PF of 0.6*
   ♦ 20% electric motors with a PF of 0.8*

   Power factor is the weighted average of all electrical loads.

   $$(0.4)(1) + (0.4)(0.6) + (0.2)(0.8) = PF \quad PF \; 0.8$$

---

*Manufacturers will provide power factors for their motors or ballasts.

2. If the building in example 1 has a 480 volt three-phase service rated at 600 amperes, find the maximum value for *apparent power* (kVA) and *real power* (kW).

   Apparent power is $(A)(V)(\sqrt{3}) \div 1,000$
   $(600)(480)(\sqrt{3}) \div 1,000 = 498$ kVA
   Real power is $(A)(V)(\sqrt{3})(PF) \div 1,000$
   $(600)(480)(\sqrt{3})(0.8) \div 1,000 = 399$ kW

GROUNDING

Grounded means connected to earth. Electrical systems are grounded to dissipate fault currents caused by lightning or equipment failure. A grounded *neutral* conductor is used in most building electrical installations to complete single-pole circuits and to set a zero potential for safety and reference. The ground connection can be made to underground metal piping or a copper-coated steel rod driven into the earth.

Circuits may be described as loops from a generator to an electrical device and *back to the generator*. Utilities save money by using the earth instead of a conductor to provide a current return path for unbalanced loads (see Figure 9.13).

A high resistance in the neutral conductor supplying buildings can lead to elevated levels of neutral-to-earth voltage. A method of dealing with this problem is separation of equipment grounding from the neutral. All non-current carrying enclosures for electrical equipment or wiring are grounded using copper conductor connected between grounding electrode connection, at the source transformer, and at each enclosure and equipment frame.

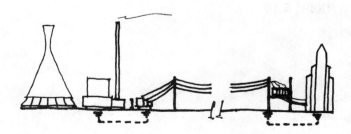

FIGURE 9.13

REVIEW QUESTIONS

1. A home has a single-phase, 240 volt, 150 ampere service. Find the transformer size needed to serve this home.
2. A commercial building has a three-phase, 480 volt, 600 ampere service. Find the required transformer size.
3. A commercial building has a total connected load of 500 kVA. Find the building power factor if the ballasted lighting load of 200 kVA has a power factor of 0.8 and the 150 kVA motor load has a power factor of 0.9. The remaining 150 kVA is resistance load with a power factor of 1.
4. Find the apparent power (kVA) and the real power (kW) for the building in question 3.
5. Find the current required for a 1,000 kVA load at 240 volts, 1ø.
6. Find the current required for a 1,000 kVA load at 240 volts, 3ø.

7. Find the current required for a 1,000 kVA load at 13,200 volts, 3ø.

Add these formulas (see Figure 9.14):

Apparent Power

$$(amps)(volts) = VA \text{ (volt-amperes)}$$
$$(amps)(volts) \div 1,000 = kVA$$
$$(amps)(volts)(\sqrt{3}) \div 1,000 = kVA \text{ (three-phase)}$$

Real Power

$$(amps)(volts) = watts \text{ (resistive loads)}$$
$$(amps)(volts)(PF) = watts \text{ (inductive loads)}$$
$$(amps)(volts)(\sqrt{3})(PF) = watts \text{ (three-phase)}$$

ANSWERS

1. 36 kVA, 1ø
   *Utilities don't size residential transformers for peak demand. They use a diversity factor and allow 5 to 8 kVA per home in residential subdivisions**
2. 500 kVA, 3ø (actually 498.8)
   Utilities size commercial customer transformers to suit building peak demand.
3. 0.89
   $[(200)(0.8) + (150)(0.9) + 150] \div 500 = 0.89$
4. apparent power = 500 kVA
   real power = 445 kW
   $(200)(0.8) + (150)(0.9) + 150 = 445$
5. 4,167 amperes
6. 2,408 amperes
7. 44 amperes

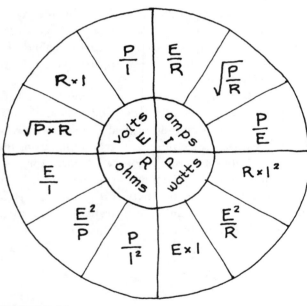

FIGURE 9.14

---

*Some utilities use high-voltage DC distribution lines because DC line losses are less than AC.

CHAPTER 10
Residential Electrical

A prophet is not without honor except in his own country, and among his own people, and in his own house.

Mark 6:4

This chapter describes residential electrical systems. It begins with a wall outlet and traces the wires and devices that connect the outlet to the home's electrical service. Then lighting circuits, dedicated circuits, and 240 volt circuits are described. Finally, circuit protection, the electrical panel, and a service calculation method are explained.

A competent electrician can wire a new home in two long days, one before and one after wallboard. Allow 6 to 8% of the total construction budget for electrical work, and plan to spend more if the owner selects light fixtures. A service calculation form and an electrical plan for the example home follow the text.

After reading this chapter you should be able to name, describe, and select the components of a residential electrical system. Develop an electrical plan, create a panel schedule, and then size the service for a house of your choice using the Residential Service Calculation form. Then review the terms introduced earlier in the text to support new vocabulary and complete the ending review questions to verify your abilities.

10.0 DUPLEX OUTLETS

RESIDENTIAL ELECTRICAL

The starting point for this description is a wall outlet that can serve one or two appliances. It is called a *duplex outlet* or a *duplex receptacle*. Follow the electrical components and materials connecting this outlet to the home's electrical service and then add lights, motors, and other electrical devices. A residential service is covered here, but many components and details are also used in commercial buildings.

In North America. 120 volt AC power is provided at duplex outlets. Because of voltage drop in home wiring, appliances are usually rated 115 volts (see Figure 10.1). Alternating current (AC) means the direction of electron flow cycles 60 times every second, and the power supply is described as 60 cycle AC current or 60 Hz AC.

The three plug blades shown in Figure 10.2 function as follows:

Hot

The hot blade is connected to a *black* conductor carrying 115 to 120 volt power. Electrons flow from the hot wire through appliances, and voltage drops as electrical energy does work or creates music. The hot con-ductor insulation is black to identify it as a wire carrying significant voltage.*

Neutral

The neutral blade is connected to a *white* conductor that provides a complete circuit for electron flow. Current is equal in the hot and neutral wires, but voltage in the neutral is minimal—just enough to overcome the resistance of the wire and complete the electrical circuit.

The hot and neutral openings in a duplex outlet differ in size to maintain *polarity* and minimize shock hazard. Hot is the smaller opening (see Figure 10.3). Polarity ensures that only the insulated components of an appliance receive higher voltage. The metal frame of an appliance is connected to the neutral conductor. Before approving an electrical installation, inspectors should check each outlet to verify polarity.

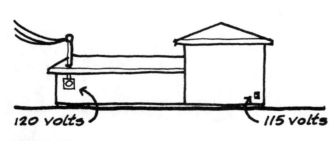

FIGURE 10.1

FIGURE 10.2

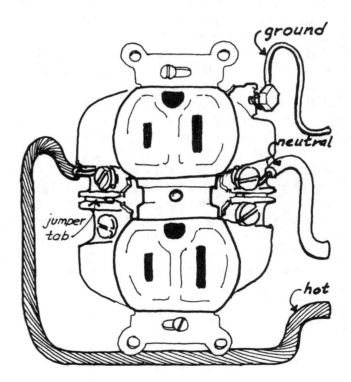

FIGURE 10.3

---

*In three-phase commercial installations, additional hot conductors will have red or blue insulation.

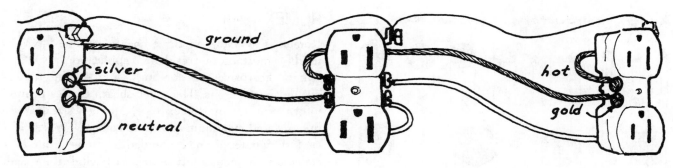

FIGURE 10.4

Ground

The ground opening in a duplex outlet is connected to a ground conductor. In residential applications the ground conductor is usually an uninsulated wire that provides an extra safe route for fault current. It duplicates the neutral wire throughout the residence, but while the neutral carries electric current to complete a circuit, the ground wire carries current *only* in the event of an electrical fault (see Figure 10.3).

Faults may result from worn insulation, loose connections, wet connections, a broken neutral, or inoperative circuit protectors. A very small current flow through the human body can cause death, and the ground conductor provides an alternate route for fault current. In commercial applications, ground wires may be bare or insulated and color coded green.

VARIATIONS

By breaking off a small jumper, duplex outlets can be wired so that separate conductors serve the top and bottom outlet positions. This feature allows two circuits to serve a kitchen appliance outlet, or a wall switch can control a table lamp. Some low-bid home builders provide switched duplex outlets for most rooms and charge extra for built-in wall or ceiling light fixtures.

INSTALLATION

Conductor connections to the terminals of a duplex receptacle are made by stripping insulation and then twisting the conductor around a screw or pushing the conductor into a spring-loaded slot.* Gold tint terminals are reserved for black hot wires, silver tint termi-

---

*Wiring to push in spring terminals is forbidden by some local codes.

nals for white neutral wires, and green terminals for ground wires (see Figure 10.4).

10.1 CONDUCTORS AND HARDWARE

Copper has less resistance than aluminum, but aluminum conductors are stronger and lighter. Copper conductors are used for *all* duplex and lighting circuits because aluminum wiring tends to fault when thermal expansion and contraction loosen terminal connections.

Aluminum conductors are used for service wiring between the weatherhead and the utility transformer. The National Electrical Code® permits aluminum wiring when terminals are designed to ensure fault-free connections, but many local codes prohibit aluminum wiring in buildings.

CONDUCTOR SIZES

Insulation melting temperature sets the current limit for building wiring. Heat is created when electrical current flows through a conductor. In building wiring, excess heat can cause insulation melting, short circuits, and fire. For a given current, larger conductors will produce less heat.

The smallest conductor permitted in building electrical installations is #14. Its copper core is about a sixteenth of an inch in diameter and will carry 15 amperes safely. It is used for switches and for circuit wiring in many older homes, but slightly larger #12 wire—which will carry 20 amperes—is the preferred wire size for home duplex and lighting circuits. Notice that American Wire Gauge (AWG) conductor

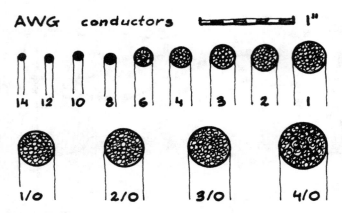

FIGURE 10.5

area increases as the gauge decreases (see Figure 10.5). A duplex circuit may use #12 wire, while a water heater uses #10, and a range may need #6. The largest AWG gauge designation is #4/0. A copper #4/0 conductor can carry a current of more than 200 amperes. Even larger conductors are used in commercial buildings, but their sizes are designated in MCM (thousands of circular mills) instead of AWG.

Copper conductors are made from solid or stranded wire in gauges #14 through #6. Gauge #4 and larger wires are stranded. Aluminum conductors are available in the same gauges. They have less current-carrying capacity than copper but are usually less costly.

RESIDENTIAL WIRING

The National Electrical Code® permits the use of type NM wiring *only* in residential occupancies, and *only* in concealed locations (inside a wall or ceiling) (see Figure 10.6). NM stands for the nonmetallic sheath that encloses and protects individual conductors. NM (or Romex) wiring offers low installed cost, but it's difficult to add or change circuits after Romex is in place.

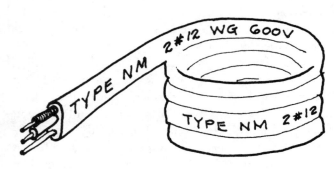

FIGURE 10.6

ROMEX

Romex is available in three sheath varieties. NM is used for most interior residential applications, NMC in wet or corrosive surroundings, and UF in underground installations. The outer sheath carries printed information about Romex type, number, size of conductors, and maximum allowable voltage. *UF 2#12 WG 600V* designates Romex intended for underground installation. It includes two size #12 wires (one with black insulation and one white) plus a bare #12 ground wire. The insulation is rated for use up to 600 volts.

BOXES

Wiring connections for switches and receptacles are made in protective boxes to prevent contact between conductors and construction materials (see Figure 10.7). Steel and plastic boxes are available. They are

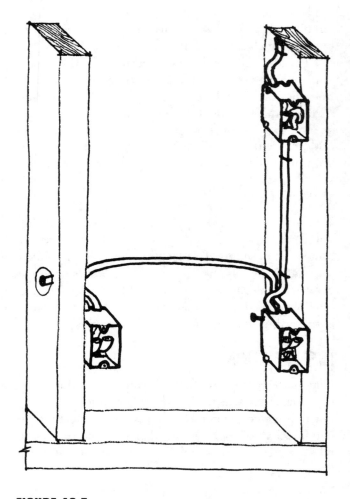

FIGURE 10.7

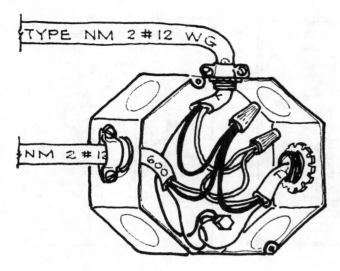

FIGURE 10.8

secured to wood studs or ceiling joists with nails. When using plastic boxes, the ground wire is connected to a green lug on the switch or receptacle frame. When steel boxes are used, the ground wire may be connected to a screw terminal inside the box and the green lug on each device.

Boxes are also used to enclose and protect wiring connections. Junction boxes may be round, octagonal, square, or rectangular. Codes require boxes to be accessible when construction is complete. As the number of wires connected in a box increases, larger boxes are required (see Figure 10.8).

Romex conductors serving residential circuits are secured in boxes with internal clamp fittings, and the Romex is also fastened to wood studs or joists supporting the outlet box with a large staple. After several duplex outlets are interconnected, Romex is extended to the power panel where a circuit breaker protects the wiring from excess current.

In wood stud walls, Romex is run through holes drilled in the center of each stud. Good builders are careful to fit insulation around Romex wiring in exterior walls to minimize building heat loss. In attics, Romex is stapled to ceiling joists. Where people or stored items are likely, Romex home runs to the panel must be protected with a plywood cover.

CONDUIT REQUIRED

Romex can only be installed in concealed locations. When an electrical power supply must run outside a wall, floor, or ceiling, the Romex is terminated in a box or switch, and conduit is installed to protect the exposed conductors.

Indoors, a flexible armored conduit called BX can be used. Outdoors, liquid-tight flexible conduit or rigid metal conduit will be used. The next chapter covers conduit materials and installation.

10.2 DUPLEX CIRCUITS AND THE PANEL

DUPLEX CIRCUITS

Many outlets can be connected on a single circuit by extending circuit wiring from one box to another. Each duplex outlet is counted as 180 watts (even though a single receptacle can supply a 1,500 watt hair dryer). In theory, a 20 ampere circuit can serve 13 outlets (see Figure 10.9), but electrical inspectors in most localities limit 20 ampere circuits to eight outlets. Duplex outlets are located so that a 6-foot-long cord can reach any point along every wall.

Dedicated Duplex Circuits

The National Electrical Code requires dedicated duplex circuits in residential construction. The kitchen must be served by at least two duplex circuits, and most localities require a third circuit serving only the refrigerator. These three circuits cannot serve a dishwasher, garbage disposal, or exhaust hood, so it is easy to dedicate at least six or more circuits in the kitchen alone. The laundry must have a separate circuit for ironing, and custom homes frequently add dedicated circuits for personal computers, security systems, pool pumps, landscape lighting, and shop tools.

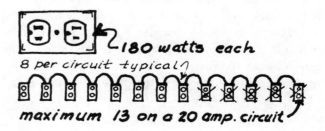

FIGURE 10.9

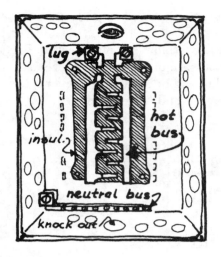
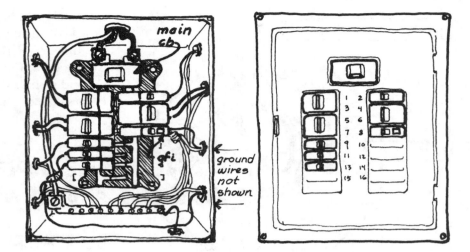

FIGURE 10.10

PANEL

Each residential circuit draws electrical current in a large metal box called a *panel* (see Figure 10.10). Panel components allocate current among individual circuits, and each circuit is protected by a *circuit breaker* that opens when current exceeds the breaker rating. The entire panel is protected by a larger main circuit breaker.

Inside the panel, *bus bars* facilitate connection of individual circuit breakers. A bus is a metal bar that carries current. Two hot wires from the utility transformer are connected through the main circuit breaker to two separate bus bars, and individual circuit breakers snap onto these busses. Circuit breakers serving a 120 volt circuit snap on one bus and protect one hot wire (see Figure 10.11); circuit breakers serving a 240 volt circuit snap on both busses and protect two hot wires.

Neutral conductors are **never** connected to breakers because they must provide an uninterrupted path from electrical loads to the system ground. All circuit neutral wires are connected to a grounded neutral bus. The utility neutral and the system ground are also connected to this neutral bus. The system ground is usually a buried metal water pipe or a copper-coated steel rod driven into the ground near the meter. A bare #6 copper wire connects the system ground to the panel neutral bus.*

---

*Modern residential panels have separate neutral and ground busses. They are bonded in the main panel but isolated in subpanels.

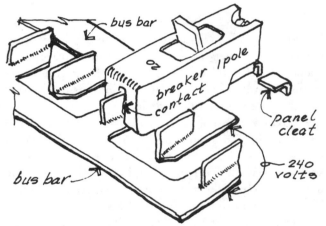

2 pole breaker contacts both bus bars

FIGURE 10.11

Circuit Breaker

A circuit breaker is a switch that opens to disconnect electrical power when it senses excess current (see Figure 10.12). Internal thermal *or* magnetic sensors open the breaker. Thermal sensors use a bimetal element that opens when excess current causes a temperature rise. Magnetic sensors use an electromagnet to open when excess current increases flux. After a circuit breaker opens, a latch prevents it from closing as it cools, or because of magnetic field collapse, breakers must be reset manually to resupply the protected circuit.

Residential circuit breakers are of one- or two-pole design. One-pole circuit breakers have one hot

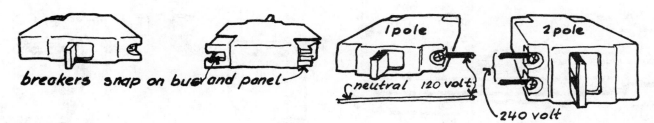

FIGURE 10.12

FIGURE 10.13

wire connected and serve 120 volt circuits. Two-pole circuit breakers protect two hot wires and serve 240 volt circuits.

GFI

A special circuit breaker may be used in residential panels serving circuits in wet locations. *Ground fault interrupter* (GFI) circuit breakers sense fault currents by comparing current in circuit hot and neutral wires (see Figure 10.13). If the currents are not equal, the GFI opens very quickly to minimize shock hazard.

GFI breakers in the panel protect an entire circuit. GFI protection in an individual outlet will protect downstream outlets. Codes require GFI protec-

tion for outlets in bathrooms, garages, and exterior walls; for outlets near kitchen sinks; and for circuits serving swimming pools or hot tubs.

MCB

The panel is protected by a two-pole *main circuit breaker* that will cut off all electrical power in the event of a high current short (see Figure 10.14). Panels should be easily accessible for firefighters and cannot be located in closets, bathrooms, or wet locations.

10.3 OTHER CIRCUITS

LIGHTING CIRCUITS

Commercial buildings separate lighting and duplex circuits, but in residential installations a single circuit can serve *both* duplex outlets and lighting fixtures. Lighting circuits add switches to control individual fixtures, and the NEC allows a maximum of 1,920* watts of lighting load on a 20 ampere 120 volt circuit.

Figures 10.15 and 10.16 show a typical lighting circuit switching connections for controlling lamps.

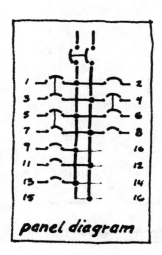
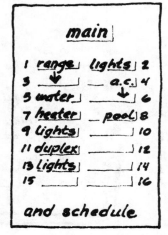

FIGURE 10.14

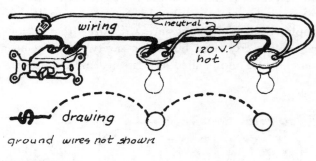

FIGURE 10.15

*1,920 watts is 80% of the theoretical circuit capacity. (120 volts)(20 amps) = 2,400 watts. 80% of 2,400 = 1,920.

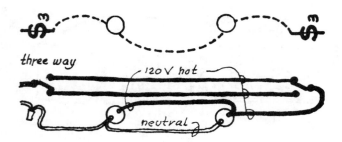

FIGURE 10.16

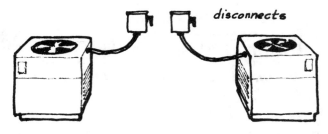

FIGURE 10.18

Three-way switches control lights from two locations and should be installed at top and bottom stair landings. Economical dimmer switches are available for incandescent lamps, but dimming fluorescent lamps is usually too expensive for residential work.

Some light fixtures include boxes for making electrical connections, and others are mounted on separate square, round, or octagon boxes.

240 VOLT CIRCUITS

Electric water heaters, ranges, ovens, dryers, air conditioners, heat pumps, large motors, and resistance heat strips are usually served by separate 240 volt circuits. Motors must have flexible connections—cord, BX, or Greenfield indoors and Liquidtight outdoors. Required circuit ampacity is usually listed on the manufacturer's nameplate.

The 240 volt circuits are protected by two-pole circuit breakers that draw power from both panel busses. Two hot wires serve all 240 volt loads, and a neutral or ground wire is usually run on 240 volt circuits for safety should a fault occur.

Current requirements vary for individual appliance circuits. Different outlet configurations are provided to ensure safe connection (see Figure 10.17). Appliances without plugs are *hard wired*—that is, connected directly in a box. For safety, a disconnect switch must be installed at strip heat units and cen-

tral air conditioners or heat pumps when the electrical panel supplying them is not visible to repair personnel (see Figure 10.18).

SERVICE

A residential *service* includes conductors, conduit, and accessories that connect the home to the utility transformer. The usual residential service is 120/240 volt, single-phase. The utility *service drop* includes two insulated (hot) wires and a bare neutral wire.

Before connecting electrical power to a new home, the installation must be inspected and approved for code compliance. Usually, the home builder purchases and installs a weatherhead,

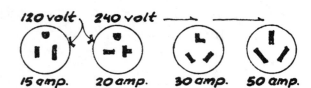

FIGURE 10.17

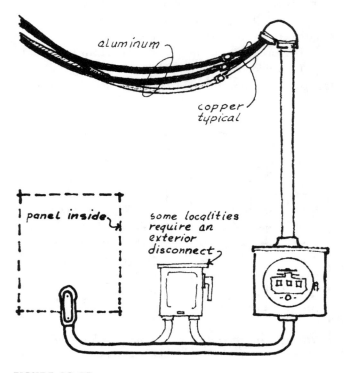

FIGURE 10.19

the meterbox, and all wire and conduit between the weatherhead and the building panel. For overhead services, the utility provides conductors from their transformer to the weatherhead and a meter (see Figure 10.19). If an underground service is used, the owner usually pays for trenching and underground cable. Removing a home meter disconnects power but exposes live sockets so a cover should be installed for safety whenever a meter is removed.

The service neutral can be smaller than the hot wires because the neutral carries only unbalanced current; 120 volt loads cause neutral current, but 240 volt loads do not. To calculate the maximum neutral current, total the watts *at 120 volts only* for each bus and divide the larger by 120. Service hot wires are sized to carry amperes equal to the main circuit breaker rating.

WEATHERHEAD

The weatherhead keeps rain and snow out of the service conduit. When an overhead service is used, check local codes and the National Electrical Code for required clearances above grade, roofs, driveways, and roads (see Figure 10.20).

Service Conductor Amperes*

Amperes	Aluminum	Copper
100	#2*	#4
125	#1/0*	#2
150	#2/0*	#1
175	#3/0*	#1/0
200	#4/0*	#2/0

*Verify ampere capacity with the local code authority. Many local codes prohibit aluminum conductors indoors.

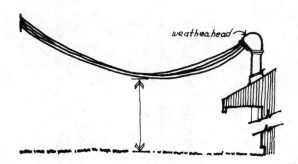

weatheahead

FIGURE 10.20

OTHER WIRING

Other home wiring serving TV or phones is run separately from AC power to minimize interference. TV uses a shielded coaxial cable, and phone uses a small four-conductor cable. Both are run unprotected in concealed locations like walls, floors, and ceilings.

Small transformers deliver the low-voltage power used for thermostats, doorbells, and security systems. Low-voltage control wiring can also be used to switch home lighting from many locations. Remote control lighting installations are expensive, but they allow you to turn many house lights on from the garage and then turn them all off from your bedroom. Control wiring is usually run in walls, floors, or ceilings, without protection.

Romex Amperes and Gauge

Verify ampacity with local code authority.

Amperes	Aluminum*	Copper
15	#12*	#14
20	#10*	#12
30	#8*	#10
40	#8*	#8
50	#6*	#8
60	#6*	#6

*Most local codes prohibit NM aluminum.

Type NM cable with THHN insulation @ 30°C ambient; for other conditions, see NEC table 310-16.

Service Conductor Amperes

Verify ampacity with local code authority.

Amperes	Aluminum*	Copper
100	#2*	#4
125	#1/0*	#2
150	#2/0*	#1
175	#3/0*	#1/0
200	#4/0*	#2/0

*Many local codes prohibit aluminum conductors in buildings.

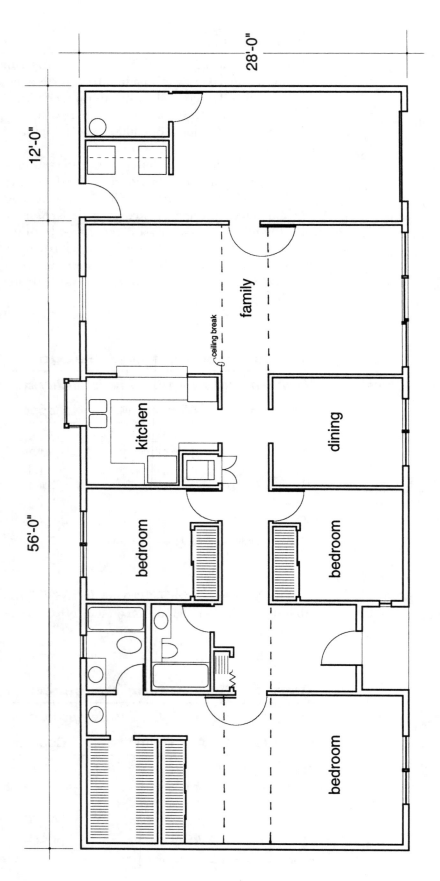

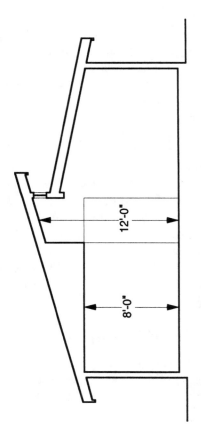

Example House

The 1,568-square-foot home shown here was used earlier in the text to illustrate lighting design. It's used again here as a template for the electrical plan on the next page. The electrical plan will satisfy many local code authorities, but it's schematic because the lighting and duplex circuit runs are not detailed. If your city requires detailed circuiting, examine the commercial electrical plans in Chapter 12.

Study the electrical plan and the panel schedule until you understand all the symbols and circuits and then check the service calculation that follows.

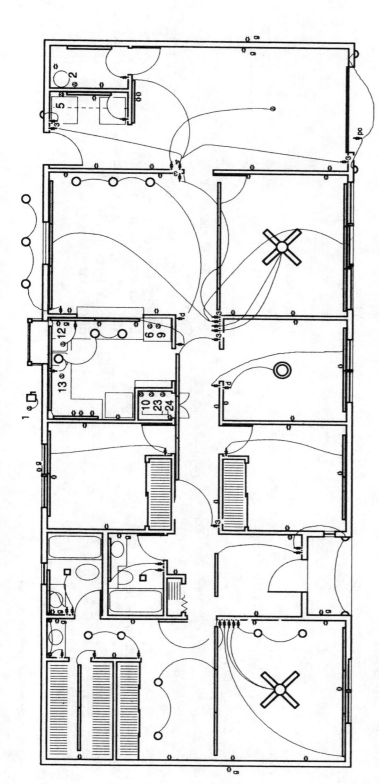

Electrical Plan

The plan shows all duplex outlets, light fixtures, and appliance circuits. Five duplex outlets marked "g" are GFI (ground fault interrupter) protected. The curved lines connecting switches and lights show which switch(es) control(s) which light(s). Circuits serving 240 volt and and/or hard wired appliances are numbered at their plan location. Each number corresponds to a panel pole, but only one number is used for 2 pole (240 volt) appliances.

Sizing the panel's 125 ampere main circuit breaker is explained on the following residential service calculation form.

Symbols

⊕ duplex outlet, count at 180 watts

⊘ 240 volt equipment OR hard wired equipment, # = circuit

$ switch, 3 = 3 way (2 locations), d = dimmer

Panel

125 Amp, Two-Pole Main Circuit Breaker	
1 air conditioner	2 water heater
3 d.o.	4 d.o.
5 dryer	6 oven
7 d.o.	8 d.o.
9 rangetop	10 furnace fan
11 d.o.	12 garbage disp.
13 dishwasher	14 lighting & duplex
15 lighting & duplex	16 d.o.
17 d.o.	18 d.o.
19 d.o.	20 d.o.
21 d.o.	22 refrigerator
23 elect. heat	24 elect. heat
25 d.o.	26 d.o.
27 landscape ltg	28 28–30 spares

Residential Service Calculation

Residential services are not sized for total connected load—an experience-based diversity factor keeps service size below 200 amperes for all but estate size dwellings. Utility services larger than 200 amperes require current transformers and are usually metered as commercial customers. The 1987 NEC optional calculation 220-30 is the basis of this form.*

NEC Calculation Requirements

• Lighting, allow 3 watts or VA per square foot 220-3b
• Provide a minimum of two kitchen appliance circuits—allow 1,500 watts or VA each—these cannot serve exhaust hoods, dishwashers, disposers, or other outlets 220-16
• Laundry circuit, allow a minimum of 1,500 watts or VA 220-4c
• Clothes dryer, allow 5,000 watts or VA or nameplate rating (use the larger value) 220-18

Appliance Loads estimated watt or VA values nameplate values are better

attic fan 1/4 hp	700	dishwasher	1,200
disposer 1/2 hp	1,200	dryer	5,000
furnace fan 1/3 hp	900	fan 1/2 hp	1,200
oven	4,500	range	12,000
cook top	6,000	wash. machine	1,200
water heater	4,500	each duplex outlet	180

Air conditioning—allow 1,000 to 1,200 watts per ton

NEC 220-30 Calculation

Select the largest of heating or air-conditioning load weighted as follows:

• AC equipment including heat pumps at 100% OR
• Central electric space heating, including supplemental heat in heat pumps at 65% OR
• Nameplate rating of electric space heating if less than four separately controlled units at 65% OR
• Nameplate rating of four or more separately controlled electric space heating units at 40%

Then ADD 100% of the first 10 kVA of other load and 40% of the remainder of all other load.

*NEC is a registered trademark of NFPA (National Fire Protection Association).

Worksheet

	Total Watts or VA
Heating OR Cooling (VA)[%]	
Appliance circuits @ 1,500 each	
Laundry circuits @ 1,500 each	
Lighting (floor area)[3]	
Clothes dryer (5,000 or nameplate)	
Cooktop	
Oven(s)	
Range (nameplate)	
Dishwasher	
Exterior lighting	
Freezer	
Garbage disposal	
Landscape lighting	
Pool pump(s)	
Shop	
air compressor	
welder	
saw	
other major tools	
Spa	
Trash compactor	
Water heater	
Other electrical items	

Total watts or VA

Calculate
first 10,000 VA @ 100% 10,000
remaining VA @ 40%

Service calculation total VA

Minimum required service amperes is the calculated total VA (above) ÷ 240 volts =

AMPERES

Use next larger standard size—100, 125, 150, 175, or 200A

Residential Service Calculation

Residential services are not sized for total connected load—an experience-based diversity factor keeps service size below 200 amperes for all but estate size dwellings. Utility services larger than 200 amperes require current transformers and are usually metered as commercial customers. The 1987 NEC optional calculation 220-30 is the basis of this form.*

NEC Calculation Requirements

- Lighting, allow 3 watts or VA per square foot 220-3b
- Provide a minimum of two kitchen appliance circuits—allow 1,500 watts or VA each—these cannot serve exhaust hoods, dishwashers, disposers, or other outlets 220-16
- Laundry circuit, allow a minimum of 1,500 watts or VA 220-4c
- Clothes dryer, allow 5,000 watts or VA or nameplate rating (use the larger value) 220-18

Appliance Loads estimated watt or VA values nameplate values are better

attic fan 1/4 hp	700	dishwasher	1,200
disposer 1/2 hp	1,200	dryer	5,000
furnace fan 1/3 hp	900	fan 1/2 hp	1,200
oven	4,500	range	12,000
cook top	6,000	wash. machine	1,200
water heater	4,500	each duplex outlet	180

Air conditioning—allow 1,000 to 1,200 watts per ton

NEC 220-30 Calculation

Select the largest of heating or air-conditioning load weighted as follows:

- AC equipment including heat pumps at 100% OR
- Central electric space heating, including supplemental heat in heat pumps at 65% OR
- Nameplate rating of electric space heating if less than four separately controlled units at 65% OR
- Nameplate rating of four or more separately controlled electric space heating units at 40%

Then ADD 100% of the first 10 kVA of other load and 40% of the remainder of all other load.

*NEC is a registered trademark of NFPA (National Fire Protection Association).

Worksheet

example house

1568 square feet say 3 w
Cooling = 3 tons say 3 w
heating = 22 kW allow
heating is larger

Worksheet 22,000 → ⟨65%⟩ **Total Watts or VA**

	Total Watts or VA
Heating OR Cooling (VA)(%)	14,300
Appliance circuits @ 1,500 each	3,000
Laundry circuits @ 1,500 each	1,500
Lighting (floor area)(3) (1568×3)	4,704
Clothes dryer (5,000 or nameplate)	5,000
Cooktop	6,000
Oven(s)	4,500
Range (nameplate)	
Dishwasher	1,200
Exterior lighting	900
Freezer	300
Garbage disposal	1,200
Landscape lighting	
Pool pump(s)	1,500 allow
Shop	
air compressor	
welder	
saw	
other major tools	
Spa	
Trash compactor	
Water heater	4,500
Other electrical items	
Total watts or VA	48,604

Calculate

first 10,000 VA @ 100%	10,000
remaining VA @ 40%	15,442
Service calculation total VA	25,442

AMPERES

Minimum required service amperes is the calculated total VA (above) ÷ 240 volts = 106

Use next larger standard size—100, (125) 150, 175, or 200A

REVIEW QUESTIONS

For questions 9 through 12, select circuit breakers and copper conductors to serve the circuits.

1. Which conductor should be connected to the silver terminal on a duplex outlet?
2. A residential duplex circuit is served by three conductors. Name them in order from highest to lowest voltage when the circuit is in use.
3. A residential duplex circuit is served by three conductors. Name them in order from highest to lowest current when the circuit is in use.
4. What type of Romex should be used for underground landscape lighting circuits?
5. The maximum number of duplex outlets that can be served by a single 20 ampere, 120 volt circuit is _____.
6. A two-pole circuit breaker is used to protect _____ volt circuits.
7. Where is GFI circuit protection required?
8. Why is the hot blade on a plug smaller than the neutral blade?
9. 115 volt lighting circuit with eighteen 100 watt lamps
10. 5,000 watt dryer (230 volts)
11. 11,000 watt heater (230 volts)
12. 1/2 hp fan (9.8 amps @ 115 volts)
13. Use a copy of the Residential Service Calculation form to work this problem:

A large home in Florida has gas heat and four 4-ton air conditioners rated at 5 kW each. Lighting circuits total 12 kW, and other dedicated circuits total 40 kW. Find service amperes and aluminum service conductor size @ 240 volts.

ANSWERS

1. the neutral (white) conductor
2. hot, neutral, ground
3. Hot and neutral carry equal current; ground carries none.
4. UF
5. 13, but check local codes. Canada uses 15 amp residential lighting and duplex circuits.
6. 240
7. wet locations
8. polarity
9. 20 amp, one-pole CB and 2 #12 wires
10. 30 amp, two-pole CB and 2 #10 wires
11. 60 amp, two-pole CB and 2 #6 wires (calculate 48 amps BUT add 25% for circuit heat buildup when in operation continuously—3 hours or more)
12. 20 amp, one-pole CB and 2 #12 wires (15 amp looks OK, but 20 amp needed to allow for the starting current surge)
13. 150 amps, #2/0 aluminum

CHAPTER 11
Commercial Electrical

When you believe in things you don't understand, then you suffer.

Stevie Wonder

Electrical energy claims a large fraction of operating costs, so commercial building owners pursue design and construction options that minimize costs. Electrical hardware is usually larger and a bit more complex in commercial buildings than in homes. Fortunately, however, there are more similarities than differences.

This chapter builds on the information developed for residential electrical systems and emphasizes differences between residential and commercial installations. Verify the new terms introduced here by checking each in the list of terms preceding Part 2. After reading, you should be able to name and describe the major components of a commercial building's electrical system and discuss the advantages of increased voltage and three-phase power.

Designing and drawing building electrical systems are covered in Chapter 12.

11.0 CONDUIT AND CONDUCTORS

CONDUIT

Conduit is a pathway for wiring that protects conductors, eases changes, and can serve as a ground (see Figure 11.1).

Duplex and lighting circuits in commercial buildings are nearly the same as their residential equivalents, but conduit is used instead of Romex. When conduit provides a continuous metal path to the building ground, it can replace the ground wire.

Steel Conduit

Many varieties of galvanized steel conduit are manufactured. The following five types are extensively used in commercial buildings. Conduit types are listed in order of increasing cost.

Type AC and MC. BX and Greenfield are armored steel conduits that can be installed quickly and economically. They flex to minimize vibration from motors and ballasts. Type AC (BX) includes conductors, and terminations must be made with care so that the steel armor does not cut conductor insulation. Conductors are pulled into Greenfield (type MC) after installation, and jacketed Greenfield (Liquidtight) is used in wet locations (see Figure 11.2).

EMT. Electrical metallic tubing is the lightest (least wall thickness) of the rigid steel conduits. Connections can be made with compression fittings or set screws, but compression fittings are preferred when EMT serves as a ground path (see Figure 11.3).

IMC and Rigid. Intermediate metal and rigid conduits are heavier conduits that assemble with threaded connections. Rigid conduit has the greatest wall thickness and offers maximum protection for conductors (see Figure 11.4).

The National Electric Code and local code authorities determine the type of conduit to be used in specific applications. Rigid conduit is commonly used on industrial projects and in areas like loading docks subject to forklift and truck traffic. EMT is used extensively in commercial projects, but flexible conduits are required for motor and fluorescent fixture connections to minimize vibration and noise transmission.

Some commercial lighting installations are prefabricated using Greenfield. This limits site labor to placing fixtures in the ceiling grid, placing switches, and connecting to a power supply.

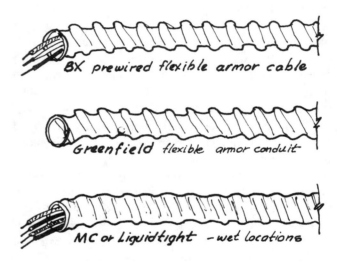

BX prewired flexible armor cable

Greenfield flexible armor conduit

MC or Liquidtight — wet locations

FIGURE 11.2

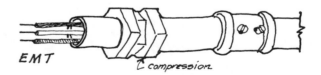

EMT   compression

FIGURE 11.3

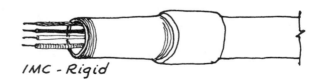

IMC - Rigid

FIGURE 11.4

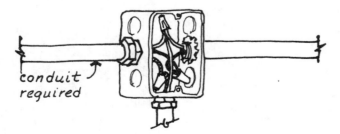

conduit required

FIGURE 11.1

Aluminum Conduit

Aluminum conduit can be substituted for galvanized steel in many applications, but it should *never* be installed in contact with concrete. Aluminum's lighter weight reduces installation labor, allowing competition with steel on a total cost basis.

Plastic Conduit

Plastic conduit (PVC and others) is the conduit of choice for underground installation where corrosion is likely. Plastic conduit is also permitted in some localities for exposed or concealed work. When using plastic, a ground conductor or individual equipment ground is required.

Many code authorities prohibit plastic conduit in buildings because of its possible performance in the event of a fire.

Fittings

Couplings, elbows, bends, pull-boxes, and assorted fittings and connectors are used to complete conduit installations. In hazardous locations—where electrical arcing could ignite flammable gasses, vapors, or dust—special *explosion-proof* fittings, switches, and devices are used. Gasketed, weather-tight fittings are used outdoors (see Figure 11.5).

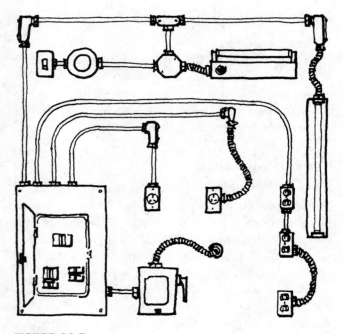

FIGURE 11.5

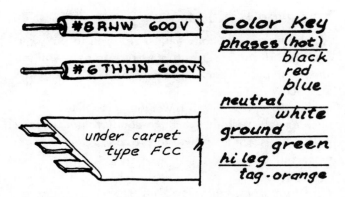

FIGURE 11.6

CONDUCTORS

Most commercial buildings are wired with copper conductors. Aluminum conductors are used for some high-current loads where special terminals are designed to ensure fault-free connections, but many local codes prohibit aluminum conductors in buildings.

Conductor Insulation

Thermoplastic or rubber insulation is used for most building wiring. **T** identifies *thermoplastic* and **R** *rubber*. **W** indicates insulation that can be used in *wet* locations, and **H** (*heat*) identifies insulation with an allowable operating temperature of 75°C (see Figure 11.6). A conductor can carry more current when its insulation has a higher temperature rating. For example

> #8 RHW is an 8 gauge conductor with rubber insulation, rated for both high heat (75°C) and wet location use.

> #6 THHN is a 6 gauge conductor with thermoplastic insulation rated for extra high heat (90°C) applications. (N *indicates nylon reinforcing*.)

Most building wiring is done with THHN, THWN, or XHHW (polyethylene). These insulations are thinner than equally rated rubber, and a thinner insulating jacket allows more conductors in a given conduit. Fewer wires are used in commercial power and lighting circuits because three-phase circuits can share a single neutral conductor.

Large buildings demand more amperes than residences, and large conductors are specified in MCM (thousand circular mil) instead of AWG (American Wire Gauge).

Installing Conductors

A fish tape is a flexible metal strip used to pull conductors into conduit. It is first pushed into the conduit and then pulled back with appropriate conductors attached.

Compressed air and vacuum are also used to push or draw fish line through conduit, and for long pulls a lubricant is applied to the conductors for easier installation.

Conductor Installation Practice

- A maximum of four 90° bends are permitted between pull-boxes.
- Do *not* mix high- and low-voltage conductors in the same conduit (high = over 600 volts).
- Do *not* mix control and power conductors in the same conduit.
- Do *not* mix phone and power conductors in the same conduit.
- *Do* place all three phases in the same conduit. (Single-phase conductors can induce current flow in the conduit.)

Special Conduits

Round conduit is not the only option for protecting and rewiring commercial installations. *Cable trays* and *busways* replace conduit where large currents must be distributed. Laboratory and manufacturing buildings substitute surface-mounted *raceways* for conduit, where many outlets are required in a small area. Office buildings use *under-floor ducts* to serve floor outlets (see Figure 11.7).

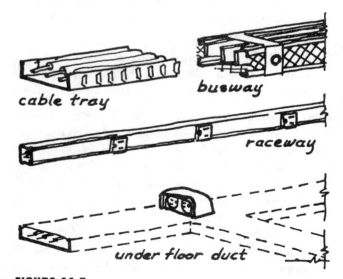

cable tray

busway

raceway

under floor duct

FIGURE 11.7

11.1 CIRCUITS

LIGHTING CIRCUITS

Commercial lighting circuits differ from their residential counterparts as follows:

Shared neutral

More amperes

Higher voltage

Also, duplex outlets are not connected on lighting circuits in commercial work.

Shared Neutral

Commercial lighting circuits often use a common neutral wire. With three-phase power, three hot wires can share a single neutral wire as they serve three circuits. A shared neutral reduces wiring costs.

More Amperes

When heavy-duty lamp holders are used, lighting circuits up to 50 amperes are allowed. However, most lighting circuits are protected with 20 ampere circuit breakers and are loaded to 80% of capacity (16 amperes). Circuit breakers are sometimes used to switch a group of light fixtures. When this is done, SWD (switch duty) breakers are specified.

Higher Voltage

In large buildings, fluorescent lighting ballasts are specified at 277 volts (347 volts in Canada) instead of 120 volts. This cuts electrical construction costs because each lighting circuit serves more fixtures. When 277 volt power is used for lighting, dry transformers are installed to carry the 120 volt loads.

POWER CIRCUITS

Commercial buildings use three-phase power for all but the smallest motors. Three hot wires carry motor loads, and grounding is provided by the conduit system, or by placing an individual ground at each motor. Large motors are equipped with *starters* to control starting current and protect from overloads.

Motors can be enclosed to minimize arcing danger in explosive environments. When enclosed, a ducted air supply is connected to each motor for cooling.

DEDICATED CIRCUITS AND UPS

In homes, separate circuits are used for appliances like ranges and air conditioners. In commercial buildings, similar separate circuits are *dedicated* for most motors, refrigeration equipment, and emergency power.

Building electrical equipment can produce harmonic currents that may overload common neutral conductors. Separate circuit neutrals or an oversized common neutral abate this problem.

Computers and computer-related equipment are sensitive to harmonic currents that distort power quality. Computers and stored data can be protected using dedicated UPS (uninterrupted power supply) circuits. UPS systems use batteries, isolating transformers, and control circuitry to ensure a clean, reliable power supply for computers and related accessories.

11.2 PANELS

LIGHTING AND POWER PANELS

Large commercial buildings use many panels to subdivide current and protect individual circuits. Separate panels are used for lighting, power, and emergency loads.

Lighting panels contain many single-pole circuit breakers, and designers *balance* lighting loads by allocating lighting watts equally among three phases. When power panels include single-pole loads, they are also balanced. Codes require separate panels, conductors, and conduit for all emergency circuits (see Figure 11.8).

PANEL CAPACITY

Panels of 100, 200, 225, or 400 ampere capacity are typical in commercial installations. Larger panels—up to 1,200 amperes—are called *switchboards*. When a building's total electrical load exceeds 1,200 amperes, electrical designers consider a high-voltage power supply. If a building service calculation totals 2,000 amperes at 240 volts, construction and utility costs can be reduced by taking 1,000 amperes at 480 volts or 125 amperes at 4,160 volts instead. Extra costs for transformers and switchgear may be offset by reduced power costs and smaller service components.

SWITCHGEAR

Switchgear is a term for large electrical load centers that allocate power to transformers, motors, and panels (see Figure 11.9). Switchgear usually includes overcurrent protection, so downstream panels are installed without main circuit breakers. The ampere capacity of a panel without main circuit breakers is designated MLO (main lugs only).

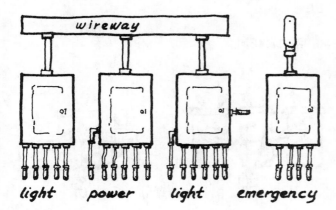

light   power   light   emergency

FIGURE 11.8

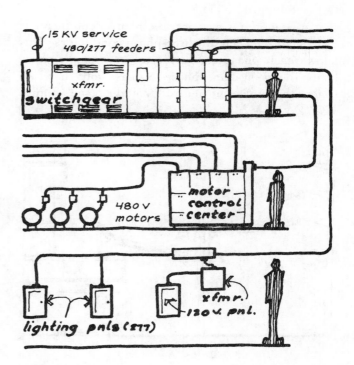

15 KV service
480/277 feeders
xfmr.
switchgear
480 V motors
motor control center
lighting pnls (277)
xfmr.
120 v. pnl.

FIGURE 11.9

11.3 SERVICES

THREE-PHASE SERVICES (3Ø)

Three-phase services usually include three hot phase conductors and a neutral. The neutral conductor is usually smaller than the phase conductors because the neutral only carries unbalanced current (see Figure 11.10).

When two or three hot wires serve a piece of equipment, they provide the circuit loop, and a neutral conductor is not required. Neutral current occurs only in single-phase (one-pole) circuits that serve loads like lighting or duplex outlets.

DELTA OR WYE

One of two electrical geometries can be used for building services. *Delta* geometry is usual in industrial applications, and *wye* geometry is typical for office buildings and shopping centers. The term *geometry* here refers to the relationship between the three-phase conductors and the neutral.

In delta systems, the neutral conductor is centered between two-phase conductors, and phase-to-phase voltage is double the phase-to-neutral voltage. Delta geometry's high leg conductor serves only three-phase loads and cannot be used with the neutral (see Figure 11.11).

In wye or star systems, the neutral conductor is connected between all three-phase conductors. This allows each phase conductor to serve single-phase loads. In wye systems, phase-to-phase voltage is 1.73 times the phase-to-neutral voltage (see Figure 11.12).

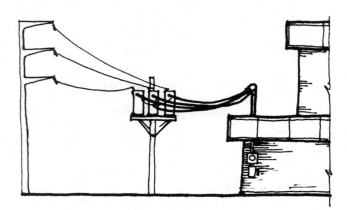

FIGURE 11.10

In delta installations, utilities balance loads by alternating the high leg in adjacent buildings. In wye systems, each building is designed for balanced electrical load on all three phases.

You can identify service geometry by observing service conductor size (see Figure 11.13). Each of the three-phase conductors is equal in wye systems, but in delta systems two conductors are large and one is small.

INCREASED SERVICE VOLTAGE

Lighting is often the largest consumer of electrical energy in commercial buildings. Typically, 1,920 watts is the maximum load connected to a 20 ampere, 120 volt circuit. When lighting is operated at higher voltage,

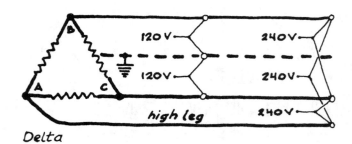

Delta

FIGURE 11.11

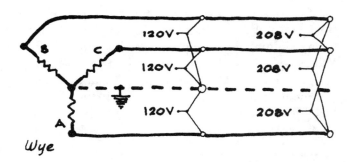

Wye

FIGURE 11.12

identify

FIGURE 11.13

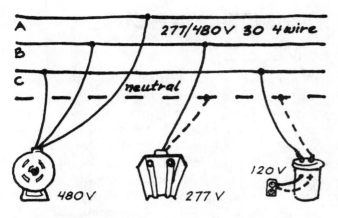

FIGURE 11.14

Typical Building Services

Service	Amps	Building Type
120/240 1ø	Up to 200	Home or
120/208 1ø		apartment
120/240 3ø	200 to 800	Small
120/208 3ø		commercial or
		manufacturer
240/480 3ø	400 to 2,000	Large
277/480 3ø		commercial or
347/600 3ø*		manufacturer
5–15 kV	Low	Very large

*Large commercial buildings in Canada.

each circuit carries more fixtures, reducing electrical construction costs. A single 277 volt lighting circuit easily serves more fixtures than two 120 volt circuits.

Motor loads for fans, pumps, elevators, refrigeration, and cooling towers may equal or exceed lighting energy in commercial buildings. Electric motors are designed for operation at a specific voltage, *but* higher voltage means less amperes and smaller conductors for a given motor horsepower. In larger commercial buildings, 480 or 600 volt motors are typical.

In the United States 277/480 volt, three-phase electrical systems are considered to serve building loads from 300 kVA to 1,600 kVA (see Figure 11.14). This service uses 480 volt, three-phase power for motors and 277 volt, single-phase power for most lighting. Small transformers throughout the building reduce 277 to 120 volts for other electrical loads. Canada uses 347/600 volt services in larger commercial buildings.

Very large buildings cut electrical costs by taking power at high voltage (5 to 15 kV). Utilities charge less for high-voltage power because the building owner installs and maintains transformers and switchgear.

TRANSFORMERS

Transformers are rated in kVA. If total building load is 600 amperes at 480 volts, find the transformer size required to serve the building.

$$(600)(480)(1.73) = +/- 500,000 \text{ or } 500 \text{ kVA}$$

Three-phase transformers are available, but for maintenance purposes, three single-phase transformers are usually specified.

EMERGENCY POWER

As building size and occupancy increase, emergency power requirements increase. A small, one-story building may need just a few exit lights and several battery-powered light fixtures to permit safe egress in a power outage.

Emergency lighting and night security lighting are frequently combined. One or two fluorescent fixtures in each room or space are connected to a circuit that is not switched. These lights stay on 24 hours a day, and they include a battery pack that will light one fluorescent lamp for about an hour in the event of a power outage.

In larger buildings, an emergency generator with automatic start is usually installed to ensure occupant safety. The generator is sized to carry items like fire pumps, emergency lighting, the fire alarm system, communications equipment, and smoke-control fans. Most codes also require emergency power sufficient to operate at least one elevator in multistory buildings.

Hospitals require more emergency power circuits because of the life-threatening implications of a power outage. Surgeries, intensive care units, and many medical instruments require a continuous power supply for safe operation. Emergency power for specific locations and devices is run in separate conduit and is isolated from the electrical system components used in normal building operations (see Figure 11.15).

Emergency circuits served by utility power are usually connected ahead of the building's main disconnect, so firefighters have the option of keeping emergency power on.

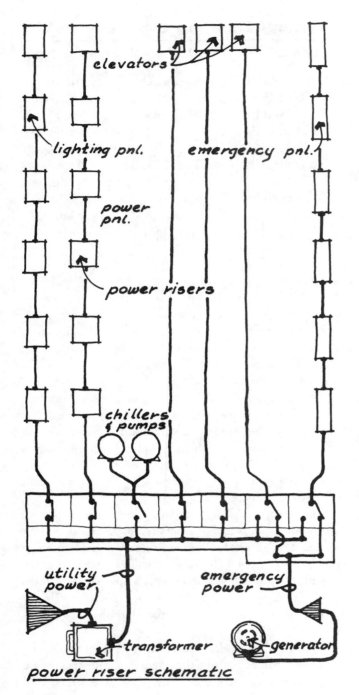

power riser schematic

FIGURE 11.15

11.4 SERVICE CAPACITY

SERVICE SIZE

The ampacity of an electrical service reflects the construction budget and anticipated future expansion plans. Sizing the service to meet the design load produces the lowest initial cost, but as time passes, building electrical loads tend to increase and it's difficult and expensive to increase the size of an existing service. Many building owners invest in added electrical capacity by oversizing the service and installing additional panels, conduit, and wiring to accommodate future growth without disrupting operations.

SERVICE DISCONNECTS

The power supply to a building can be turned off (disconnected) by a main circuit breaker or a fused switch (see Figure 11.16). Disconnecting building electrical power is an important first step in a fire emergency. Fuses have a better interrupting capacity than circuit breakers, so fused disconnects are often used to protect and disconnect the building service. The National Electric Code allows a maximum of six disconnects on a building service, but most local codes require *one* switch or breaker that will completely disconnect all building power.

The National Electrical Manufacturers Association (NEMA) sets standards for disconnects. For example, NEMA 1 is the indoor standard, and NEMA 3R is the rainproof standard.

LOW VOLTAGE

Most phone, sound, TV, thermostat, security, and control wiring is operated at low voltage supplied through small transformers. Low voltage (often 24 volts) can be run in open cable trays or in plastic conduit, but it *cannot* be run in power or lighting conduit.

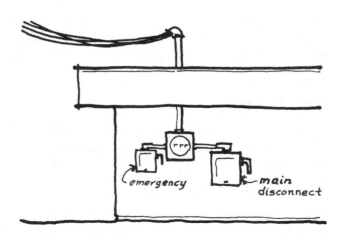

FIGURE 11.16

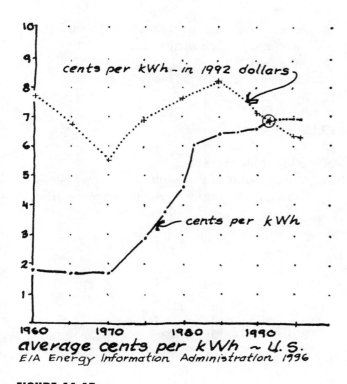

cents per kWh - in 1992 dollars

cents per kWh

average cents per kWh ~ U.S.
EIA Energy Information Administration 1996

FIGURE 11.17

ELECTRICAL COSTS

The kWh graph in Figure 11.17 shows decreasing real costs for electrical energy, but utilities bill commercial customers for energy (kWh) *and* peak demand (kW). A home might be billed 7¢ per kWh, while a nearby commercial building pays 2¢ per kWh *plus* $12 per kW of demand.

Because peak charges are substantial, commercial building owners are most interested in reducing demand, while home owners try to cut consumption. As a result, commercial building owners invest in peak limiting controls and thermal storage, while home owners can cut bills by improving the thermal performance of their walls, roof, and windows.

REVIEW QUESTIONS

1. Name three advantages of conduit vs. Romex.
2. Aluminum conduit must *not* be installed in contact with _____.
3. Select a steel conduit type to connect the starter to an electric motor.
4. What steel conduit type joins with compression fittings?
5. Select a conduit material for underground landscape lighting circuits.
6. Select a steel conduit type for surface mounting in a large warehouse.
7. Which of the following is the largest conductor?
   250 MCM RHW   #4/0 TW   #14 THHN
8. Which of the following should be used in wet locations?
   R   TW   THHN   THWN
9. Which of the following conductors has the highest safe operating temperature?
   250 MCM RHW   #4/0 TW   #14 THHN
10. Which of the following conductors will safely carry the largest current?
    #12 RHW   #12 THW   #12 THHN
11. Which of the following conductors has the thickest insulation?
    #12 RHW   #12 THW   #12 THHN
12. When very large currents are distributed, what is used to replace conduit?
13. How can three separate lighting circuits be served with only four wires?
14. Why do commercial lighting circuits use 277 or 347 volt power?
15. UPS is used for _____ power.
16. The designation 225 A - MLO means _____.
17. Is a 347/600 volt service delta or wye?
18. Which 3Ø geometry includes a "high leg" that can't serve single-pole loads?
19. Which 3Ø geometry is balanced by the utility?
20. Which 3Ø geometry is typical in large commercial buildings?
21. Why do very large buildings take power at high voltage?
22. A new office building owner asks designers to oversize the electrical service by 50%. Why?
23. Peak demand for a 20,000-square-foot commercial building is 140 kW. Monthly consumption is 40,000 kWh. Calculate the monthly utility bill if rates are $10 per kW plus 3¢ per kWh.

ANSWERS

1. protect, change, ground
2. concrete
3. BX or Greenfield
4. EMT
5. plastic
6. rigid or IMT
7. 250 MCM RHW
8. TW or THWN
9. #14 THHN
10. #12 THHN
11. #12 RHW
12. busways or cable trays
13. three individually protected phase conductors and a single common neutral
14. more fixtures per circuit = fewer circuits = lower electrical construction cost
15. computer
16. 225 amperes, main lugs only
17. wye
18. delta
19. delta
20. wye
21. reduced life cycle costs
22. Anticipated load growth. (50% was within the historical 10-year load growth range for office buildings during the 1960s.)
23. $2,600

CHAPTER 12
Example Office Building

Architecture is 90% business and 10% art.

Albert Kahn

Tables and example questions begin this chapter. They illustrate circuit breaker, conductor, conduit, panel, and transformer selection. Complete each example and check your answers so you will understand the following electrical plans and panel schedules.

The example office building* is used to illustrate electrical design and component selection. Drawings and panels are developed for the lobby and an office area. Then lighting, duplex, and power loads are estimated for the entire building, and an electrical service is designed. Study this chapter by reading the text, and then reinforce each text entry by identifying it on the accompanying drawing or diagram. Trace each circuit from plan to panel, and then follow the panel feeders to the building service.

*Lobby lighting design is covered in Chapter 6. More information about the building's HVAC systems is provided in Efficient Building Design Series, Volume 2: Heating, Ventilating, and Air Conditioning. Here, the example office is assumed to be an "all electric" building to illustrate as many electrical loads as possible.

12.0 TABLES

Use Tables 12.1 through 12.3 and the following instructions to select circuit breakers, conductors, and conduit for the example questions.

Select conductors with an ampere rating that equals or exceeds the circuit breaker trip amperes.

- Select copper conductors unless the local code authority specifically permits aluminum.
- Allow voltage drop for individual circuits. For example, 120 = 115, 208 = 200, 277 = 265.
- Use design voltage for transformers and feeders, that is, 120, 208, 240, 277, 480.
- Amps = watts ÷ volts (or volt-amperes ÷ volts) for single-phase loads.
- Amps = watts ÷ volts$\sqrt{3}$ (or volt-amperes ÷ volts $\sqrt{3}$) for three-phase loads.
- Add 25% to calculated amps for lighting loads or loads that operate continuously for three hours or more.
- Select circuit breaker amps next above calculated amps, and select circuit conductors with ampacity equal to or greater than their breakers.

Example Questions

1. Select a conductor to carry 45 amperes.
2. Select a copper conductor to carry 80 amperes. How many of these conductors are required for a 115 volt circuit?

3. A 115 volt lighting circuit totals 1,900 watts. Select circuit breaker, conductors, and conduit.
4. A 115 volt duplex circuit totals 1,080 watts. Select circuit breaker, conductors, and conduit.
5. Three 115 volt, 20 ampere lighting circuits will be run in a single conduit. How many #12 copper wires and what size conduit?
6. Find the maximum lighting watts allowable on a 20 ampere, 265 volt circuit.

TABLE 12.1

Circuit Breakers (standard trip amperes)

15–50 by 5s
   15   20   25*   30   35*   40 45* 50
(*25, 35, and 45 for motor circuits)
60–110 by 10s
   60   70   80   90   100 110
125–250 by 25s
   125 150 175 200   225 250
300–500 by 50s
   300 350 400 450   500
600–1,000 by 100s
   600 700 800   900 1,000
1,200–2,000 by 200s
1,200 1,400 1,600 1,800 2,000

TABLE 12.2

Conductors (sized by amperes)

Allowable amperes for three THWN (75°C) insulated conductors in a raceway based on an ambient temperature of 30°C.

Amperes	Aluminum *	Copper
15	#12*	#14
20	#10*	#12
25	#10*	#10
30	#8*	#10
35	#8*	#8
40	#8*	#8
45	#6*	#8
50	#6*	#8
60	#4*	#6
70	#3*	#4
80	#2*	#4
90	#2*	#3
100	#1*	#3
110	#1/0*	#2
125	#2/0*	#1
150	#3/0*	#1/0
175	#4/0*	#2/0
200	250 MCM*	#3/0
225	300 MCM*	#4/0
250	350 MCM*	250 MCM
300	500 MCM*	350 MCM
350	700 MCM*	400 MCM
400	900 MCM*	600 MCM

*Many local codes prohibit aluminum conductors inside buildings.
Conductors in air with insulation rated for higher temperature can carry more current.

TABLE 12.3

Conduit

Maximum number of type THWN or THHN conductors in conduit or tubing

Wire Size	1/2″	3/4″	1″	1 1/4″
14	13	24	39	69
12	10	18	29	51
10	6	11	18	32
8	3	5	9	16
6	1	4	6	11
4	1	2	4	7
3	1	1	3	6
2	1	1	3	5
1		1	1	3
0		1	1	3
00		1	1	2

Wire Size	1 1/2″	2″	2 1/2″	3″
8	22	36	51	79
6	15	26	37	57
4	9	16	22	35
3	8	13	19	29
2	7	11	16	25
1	5	8	12	18
0	4	7	10	15
00	3	6	8	13
000	3	5	7	11
0000	2	4	6	9
250	1	3	4	7
300	1	3	4	6
350	1	2	3	5

Wire Size	3 1/2″	4″	5″	6″
2	33	43	67	97
1	25	32	50	72
0	21	27	42	61
00	17	22	35	51
000	14	18	29	42
0000	12	15	24	35
250	10	12	20	28
300	8	11	17	24
350	7	9	15	21
400	6	8	13	19
500	5	7	11	16
600	4	5	9	13
700	4	5	8	11
750	3	4	7	11

7. A single-phase transformer rated at 15 kVA converts 277 volts to 120 volts. Find the aluminum conductor and conduit size for primary and secondary circuits. Allow 125% over current capacity for continuous operation.

8. A separate circuit will be run for a 230 volt water heater rated at 4,500 watts. Select circuit breaker, conductors, and conduit.

9. A separate circuit will be run for a 200 volt heater rated at 9,600 watts. Select a circuit breaker, conductors, and conduit.

10. Select circuit breaker, conductors, and conduit for a three-phase (3ø), 200 volt heater rated at 9,600 watts.

11. Select aluminum feeders and conduit for a single-phase 225 ampere panel.

12. Select aluminum feeders and conduit for a three-phase 400 ampere panel.

13. Select aluminum feeders and conduit for a three-phase 208 volt service if the total load is 287 kVA.

14. Select circuit breaker, conductors, and conduit for a three-phase, 575 volt, 25 horsepower motor.

Answers

1. #6 THWN aluminum or #8 THWN copper

2. #4 THWN, two (one hot and one neutral)

3. 20 ampere, one-pole CB, 2 #12 THWN copper wires, in 1/2″ conduit

4. 15 ampere, one-pole CB, 2 #14 THWN copper wires, in 1/2″ conduit, *but* 20 ampere circuits are recommended for all duplex and lighting loads.

5. With a three-phase Y system, four wires (three hot and a common neutral) in a 1/2″ conduit. The three hot wires *must* be connected to three separate panel busses.

6. 4,240 watts. For continuous loads like lighting, circuit loading is limited to 80% or $(265)(20)(80\%) = 4{,}240$ watts.

7. 2 #3 aluminum THWN, in 1″ conduit for the primary, 2 #4/0 aluminum THWN, in $1\frac{1}{2}″$ conduit for the secondary.

8. 30 ampere, two-pole CB, 2 #10 THWN copper wires, in 1/2″ conduit. Calculate 19.6 amps, but allow 125% (24.5 A) for circuits subject to continuous operation.

TABLE 12.4

Single-Phase Motors— 115, 200, and 230 Volts

Horsepower, full load amperes (FLA), dual element time delay fuse amperes, circuit breaker amperes, minimum THWN copper wire size.

Single-Phase 115-Volts (120)

HP	FLA	Fuse	CB	Wire
1/4	5.8	7	15	#14
1/3	7.2	9	20	#14
1/2	9.8	12	20	#14
3/4	13.8	15	25	#12
1	16	20	30	#12
1 1/2	20	25	40	#10
2	24	30	50	#10
3	34	45	70	#8
5	56	70	90	#4

Single-Phase 200-Volts

HP	FLA	Fuse	CB	Wire
1/4	3.3	5	15	#14
1/3	4.1	6.5	15	#14
1/2	5.6	10	15	#14
3/4	7.9	12	15	#14
1	9.2	15	15	#14
1 1/2	11.5	17.5	20	#14
2	13.8	20	25	#12
3	19.6	30	40	#10
5	32.2	50	60	#8
7 1/2	46	60	90	#6

Single-Phase 230-Volts (240)

HP	FLA	Fuse	CB	Wire
1/4	2.9	3.5	15	#14
1/3	3.6	4.5	15	#14
1/2	4.9	6	15	#14
3/4	6.9	8	15	#14
1	8	10	15	#14
1 1/2	10	12	20	#14
2	12	15	25	#12
3	17	20	35	#10
5	28	35	60	#8
7 1/2	40	50	80	#8

TABLE 12.5

Three-Phase Motors— 200 and 230 Volts

Horsepower, full load amperes (FLA), dual element time delay fuse amperes, circuit breaker amperes, minimum THWN copper wire size.

Three-Phase 200-Volts

HP	FLA	Fuse	CB	Wire
1/2	2.5	3	15	#14
3/4	3.7	5	15	#14
1	4.8	6	15	#14
1 1/2	6.9	10	15	#14
2	7.8	12	20	#14
3	11	17.5	20	#14
5	17.5	30	40	#12
7 1/2	25.3	40	50	#10
10	32.2	60	80	#8
15	48.3	80	90	#6
20	62.1	100	110	#4
25	78.2	125	125	#3
30	92	150	150	#2
40	120	200	200	#1/0
50	150	200	200	#3/0
100	285	400	400	500

Three-Phase 230-Volts

HP	FLA	Fuse	CB	Wire
1/2	2.2	3	15	#14
3/4	3.2	5	15	#14
1	4.2	6	15	#14
1 1/2	6	10	15	#14
2	6.8	15	20	#14
3	9.6	15	20	#14
5	15.2	25	30	#14
7 1/2	22	35	45	#10
10	28	45	60	#10
15	42	70	80	#6
20	54	90	100	#4
25	68	100	100	#4
30	80	150	150	#3
40	104	175	175	#1
50	130	200	200	#2/0
100	248	400	400	250

TABLE 12.6

Three-Phase Motors— 460 and 575 Volts

Horsepower, full load amperes (FLA), dual element time delay fuse amperes, circuit breaker amperes, minimum THWN copper wire size.

Three-Phase 460-Volts

HP	FLA	Fuse	CB	Wire
1/2	1.1	1.5	15	#14
3/4	1.6	2	15	#14
1	2.1	3	15	#14
1 1/2	3	5	15	#14
2	3.4	5	15	#14
3	4.8	8	15	#14
5	7.6	10	15	#14
7 1/2	11	20	25	#14
10	14	25	30	#14
15	21	35	40	#10
20	27	40	60	#10
25	34	60	70	#8
30	40	60	80	#6
40	52	90	100	#6
50	65	100	100	#4
100	124	200	200	#2/0

Three-Phase 575-Volts

HP	FLA	Fuse	CB	Wire
1/2	0.9	2	15	#14
3/4	1.3	2	15	#14
1	1.7	2	15	#14
1 1/2	2.4	3	15	#14
2	2.7	4	15	#14
3	3.9	6	15	#14
5	6.1	10	15	#14
7 1/2	9	10	15	#14
10	11	17.5	20	#14
15	17	25	35	#12
20	22	35	45	#10
25	27	40	60	#10
30	32	50	60	#8
40	41	70	80	#6
50	52	90	100	#6
100	99	175	175	#1

9. 60 ampere, two-pole CB, 2 #6 THWN copper wires, in 3/4″ conduit (allow 125%)

10. 40 ampere, three-pole CB, 3 #8 THWN copper wires, in 1/2″ conduit (allow 125%)

11. Three 300 MCM THWN aluminum wires, in 2″ conduit. This answer assumes neutral current is equal to the phase conductor current.

12. Seven 250 MCM THWN aluminum wires, in 3 1/2″ conduit. This answer assumes the neutral wire carries only half as much current as the phase conductors.

13. Fourteen 250 MCM THWN aluminum wires, in 5″ conduit. Each phase must carry almost 800 amperes, and this answer assumes a calculated maximum neutral current of 400 amperes. Twelve phase conductors and two neutral conductors.

14. 60 amperes, three-pole CB, 3 #8 THWN copper wires, in 1/2″ conduit. Check this answer in Table 12.6.

MOTORS

Use Tables 12.4 through 12.6 and the following notes to select motor circuit breakers and conductors. Check conduit sizes in Table 12.3.

- Amps = VA ÷ volts, for single-phase loads.
- Amps = VA ÷ volts$\sqrt{3}$, for three-phase loads.
- Use volt-amperes instead of watts for motor loads.
- Copper conductors are used in the tables, but aluminum conductors can be used for large motor circuits.
- Verify circuit breaker amps and wire size with motor manufacturer.
- To properly start and further protect electric motors, a starter must be installed nearby (see Figure 12.1).

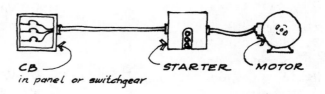

CB in panel or switchgear — STARTER — MOTOR

FIGURE 12.1

Example Questions

1. Find the circuit breaker, wire, conduit, and load in volt-amperes for a 3 horsepower, single-phase, 115 volt motor.
2. Find the circuit breaker, wire, conduit, and load in volt-amperes for a 3 horsepower, single-phase, 200 volt motor.
3. Find the circuit breaker, wire, conduit, and load in volt-amperes for a 3 horsepower, single-phase, 230 volt motor.
4. Find the circuit breaker, wire, conduit, and load in volt-amperes for a 3 horsepower, three-phase, 230 volt motor.
5. Find the circuit breaker, wire, conduit, and load in volt-amperes for a 3 horsepower, three-phase, 460 volt motor.
6. Find the circuit breaker, wire, conduit, and load in volt-amperes for a 30 horsepower, three-phase, 575 volt motor.
7. Find the circuit breaker, wire, conduit, and load in volt-amperes for a 30 horsepower, three-phase, 200 volt motor.
8. Find the circuit breaker, wire, conduit, and load in volt-amperes for a 30 horsepower, three-phase, 230 volt motor.

Answers

Breaker	Conductors	Conduit	Load
1. 70 A, one-pole	2 #8	1/2″	3,910 VA
2. 40 A, two-pole	2 #10	1/2″	3,920 VA

Load is carried by two hot wires, half on each.

3. 35 A, two-pole	2 #10	1/2″	3,910 VA
4. 20 A, three-pole	3 #14	1/2″	3,824 VA

Load is carried by three wires, 1,275 VA on each.

5. 15 A, three-pole	3 #14	1/2″	3,824 VA
6. 60 A, three-pole	3 #8	1/2″	31,870 VA
7. 150 A, three-pole	3 #2	1″	31,870 VA
8. 150 A, three-pole	3 #3	1″	31,870 VA

Time delay fuses are frequently used as motor circuit protectors. They are oversized by at least 125% of motor full load amperes to pass the starting current surge.

Circuit breakers protecting motors are oversized more than time delay fuses to pass the starting current surge. Motor circuits are an exception to the rule that circuit breaker trip amperes set circuit wire size. Some wires listed in the motor tables are rated for less amperes than the circuit breaker trip setting.

PANELS AND SWITCHBOARDS

Try to locate panels near the center of the electrical loads they serve to minimize voltage drop. Standard panel capacities are given in Table 12.7. Tabulate all electrical loads on panel form(s) and calculate amperes. Then select panel capacity to accommodate loads. Panels are ordered with or without main circuit breakers. Specify MCB if a main breaker is required, or MLO (main lugs only) when the panel has remote protection.

Panels are manufactured 14″ or 20″ wide, depths range from 3″ to 6″, and heights vary with capacity up to 60″.

TRANSFORMERS

Transformer size varies with capacity (see Figure 12.2). Because they hum and heat, location and ventilation are important considerations. Small transformers on each floor serve 120 volt loads in buildings with 277/480 volt services, but large transformers are located with switch gear in buildings with 5 or 15 kV services. Table 12.8 lists standard transformer capacities.

12.1 BUILDING SERVICE EXAMPLE

Select all the keyed components in the electrical service diagram in Figure 12.3 using the preceding tables; then check your selections with the answers that follow.

Select all panels, switches, fuses, lugs, conductors, conduit, and the transformer. Begin with panel D and work back to the main switch.

TABLE 12.7

Panels and Switchboards

Capacity in amperes for mains and mugs (MLO), and number of spaces or poles.

Single-Phase 120/208 or 120/240

40 or 70	2 or 4
100	8-12-16-20-24
125	2-4-6-8-12-16-20-24
150	12-16-20-24-30
200	4-8-12-16-20-24-30-40-42
225	24-30-42
400	24-30-42
600	24-30-42

Three-Phase 120/240 or 120/208

60	3
100	12-18-24-30
125	12-20-30
150	18-24-30-40-42
200	12-18-24-30-40-42
225	30-40-42
400	24-30-42
600	24-30-42

Three-Phase 277/480

125	12-18-30
225	30-42
400	30-42
600	42

Switchboard Amperes

480 volts	400	600	800		
600 volts	1,000	1,200	1,600	2,000	2,500

This is a partial listing of manufactured panels and switchboards. Verify availability and check capacity when multi-pole breakers are installed.

Compact *tandem* breakers are available. They pack two single-pole breakers in the space of a standard single-pole breaker.

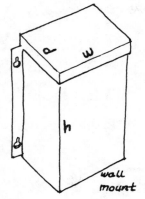

Dry Transformers

Single Phase
115°C temperature rise

KVA	h	w	d
1	8"	7"	6"
5	15"	10"	12"
10*	15"	10"	12"

wall mount   * 150°C temperature rise

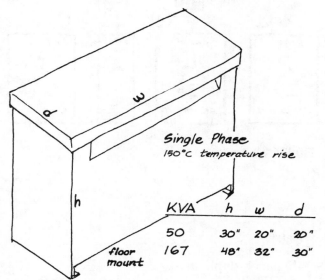

Single Phase
150°C temperature rise

KVA	h	w	d
50	30"	20"	20"
167	48"	32"	30"

floor mount

FIGURE 12.2

TABLE 12.8

Standard Transformer kVA

Single-Phase, kVA

1 - 3 - 5 - 7 1/2 - 10 - 15 - 25 - 37 1/2 - 50 - 67 1/2 - 100 - 125 - 167 1/2 - 200 - 250 - 333 - 500

Three-Phase, kVA

3 - 6 - 9 - 15 - 30 - 45 - 75 - 100 - 150 - 225 - 300 - 500 - 750 - 1,000 - 1,500 - 2,000 - 5,000 - 7,500 - 10,000

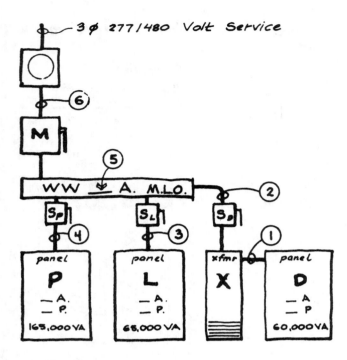

FIGURE 12.3

Given:

- **D** panel (duplex).
  3ø, 120/208 volts, serves duplex and other loads. Phase total 60,000 volt-amperes, neutral maximum 14,000 volt-amperes. Total poles = 30.
- **X** transformer.
  3ø, 277/480 to 120/208 volts.
- **L** panel (lighting).
  3ø, 277/480 volt lighting panel. Phase total 65,000 volt-amperes, neutral maximum 22,000 volt-amperes. 16 single-pole circuits.
- **P** panel (power).
  3ø, 277/480 volt power panel. Phase total 165,000 volt-amperes, no neutral load. 11 three-pole circuits.
- S_P S_L S_D switches.
  Fuse each for protected panel load.
- **WW** wireway.
  Size lugs for total load.
- **M** main switch.
  Fuse for service capacity.

Answers

- **D** Panel. Select:
  200 A, MLO with 40 pole spaces.
  $(60,000) \div (208)(\sqrt{3}) = 167$ A.
  Neutral current is $(14,000) \div (120) = 117$ A.
- **1.** Select:
  Three 250 MCM and one #3/0 aluminum THWN in $2\frac{1}{2}''$ conduit. Phase conductors sized for 200 A. Could have sized for 175 A, but limits expansion potential. Neutral could be sized for 117 A, but is permitted to be only two sizes smaller than the primary conductors.
- **X** Transformer. Select:
  75 kVA $(200)(208)(\sqrt{3}) = 72,000$
- S_D Switch. Select:
  100 A, three-pole fused switch
- **2.** Select:
  Three #1 and one #3 aluminum THWN in $1\frac{1}{2}''$ conduit. $(200)(208) \div (480) = 85$ A. Could size for 90A, but a 100 A fuse same price. Neutral could be sized for 117 A, but it is permitted to be only two sizes smaller than the primary conductors.
- **L** Panel. Select:
  100 A, MLO with 20 pole spaces.
  $(65,000) \div (480)(\sqrt{3}) = 78$ A. Could use 80 A.
  Neutral current is $(22,000) \div (277) = 79$ A.
- S_L Switch. Select:
  100 A, three-pole fused switch
- **3.** Select:
  Three #1 and one #2 aluminum THWN in 1 1/2" conduit.
- **P** Panel. Select:
  225 A, MLO with 42 pole spaces.
  $(165,000) \div (480)(H\sqrt{3}) = 198$ A. Could use 200 A to cut conductor cost.
- S_P Switch. Select:
  225 A, three-pole fused switch
- **4.** Select:
  Three 300 MCM aluminum THWN in 2" conduit.
- **5.** Wireway lugs. Select:
  400 A, MLO. Total connected load is 349 A. 400 allows almost 15% expansion. Many owners would specify a 600 A service.
- **M** Main disconnect. Select:
  400 A, three-pole fused switch
- **6.** Select:
  Six 250 MCM and one #3/0 neutral aluminum THWN in 30" conduit. Neutral current is 130 A.

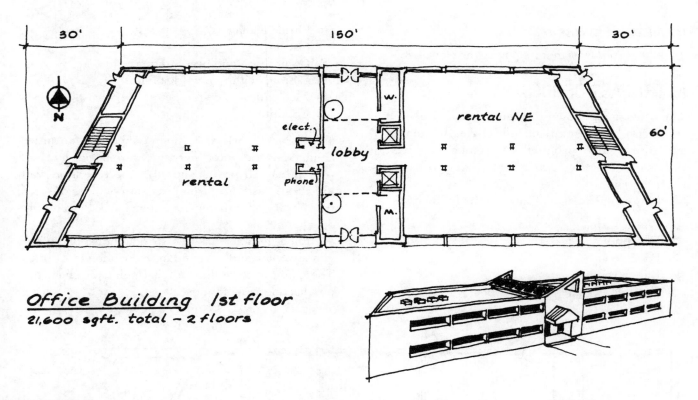

Office Building 1st floor
21,600 sqft. total – 2 floors

FIGURE 12.4

PLAN

This electrical installation example uses the building shell shown in Figure 12.4. Detailed electrical plans are illustrated for a rental office space on the northeast side of the ground floor and for the central two-story lobby.

The building's panels, feeders, and service are sized to serve the estimated electrical requirements for both floors, even though detailed drawings for the office and lobby account for only about 15% of the total building area. Utility service is three-phase, 120/208 volts.

ELECTRICAL LOADS

Lighting, duplex, and selected power outlets are shown on the detail plans and inferred for the remaining spaces. The following HVAC and elevator loads will be used to calculate the building service.

Cooling 66 kW

Nine air handlers serve eight rental office zones and the two-story lobby. Nine DX, split system, air-cooled condensing units are located on the building roof.

Calculated heat gain of 630,172 BTUH, or 52.5 tons, is met by five 5-ton and four 7.5-ton units. Larger units serve the four large office spaces. Second-floor air handlers circulate more air to carry roof heat gain. Electrical load is estimated at 1.2 kW per ton including the air handlers.

Heating 100 kW

Resistance heating elements are installed in each of nine air handlers.* Calculated heat loss for the example building is 406,850 BTUH, so 120 kW should be installed (406 ÷ 3.4 = 120). By reducing night ventilation rates and taking a partial credit for heat from lights, resistance heating capacity was reduced to 100 kW.

Ventilation

Two 1/2 horsepower exhaust fans serve the building rest rooms.

---

*Electric resistance heating is *not* energy-efficient, but it is used in the example to illustrate circuiting and load calculations.

Elevator Power

Two 30 horsepower motors power the elevators.

NORTHEAST OFFICES

The office area shown in Figure 12.5 is leased by a small graphic design firm specializing in computer-generated video product. Staff members include the owner, an office manager, two designers, and a receptionist.

Interior office spaces include the following:

1. Waiting area
2. Reception
3. Print room
4. Lunch room
5. Graphic designer workstations

6. Office manager
7. Owner
8. Conference and presentation
9. Library with two workstations
10. Mechanical and storage
11. Corridor

The firm's work is accomplished on an interconnected computer network supported by UPS (uninterrupted power supply). The reception-waiting area is used to display recent commissions, and the library maintains an extensive CD reference collection.

Each designer has two workstations, one with a window view and a second in the windowless library. The conference room is a focal point for client proposals and presentations.

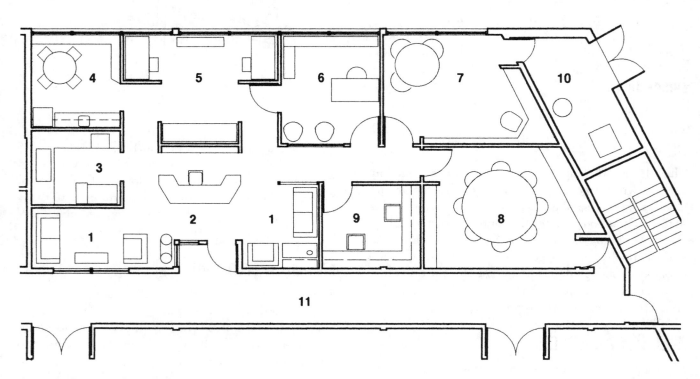

FIGURE 12.5 Northeast Offices

12.2 EXAMPLE OFFICE ELECTRICAL PLANS

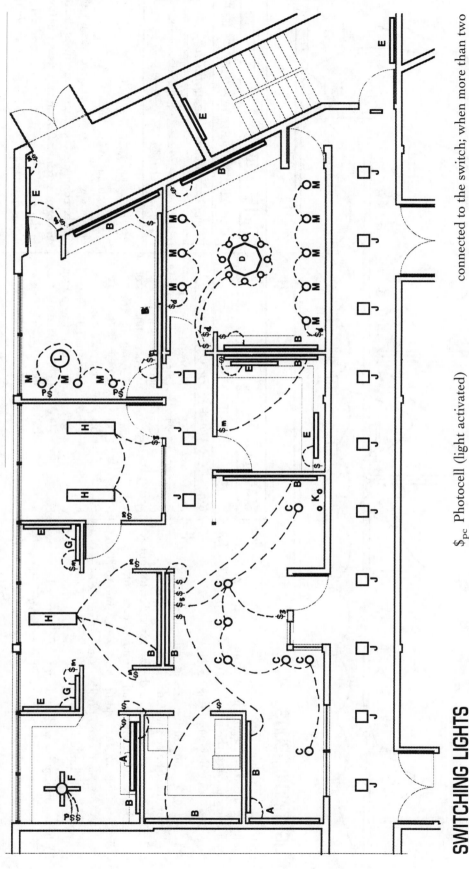

SWITCHING LIGHTS

Fixture design, selection, and layout were covered in previous chapters. This drawing shows fixtures and switching for the northeast office.

Switch Key

$ — Single-pole switch

$_3 — Three-way switch (two locations)

$_d — Dimmer

$_k — Key activated

$_m — Motion activated

$_{pc} — Photocell (light activated)

$_t — Timer activated

Study the drawing and notice the unswitched fixtures in halls and stairways. These fixtures operate 24 hours a day for safety and security reasons; they are controlled by a switch duty circuit breaker located in an emergency panel. Also notice there is no switch shown for fixture K. Two lamps designated K are installed in a wall-mounted display case which has a built-in switch.

Three-way switches are shown when fixtures must be controlled from two locations. Three-way designation indicates three wires connected to the switch; when more than two control locations are required, four-way switches must be used.

Dimmer switches are shown controlling some fixtures; dimmers are economical for incandescent lamps but expensive when used to control fluorescent lamps.

Motion-activated switches are used to save energy in unoccupied spaces. They are shown at switch locations, but they can be mounted in the fixture.

Key switches (rest rooms), photocell switches (exterior and perimeter), and timer switches are not shown on this drawing.

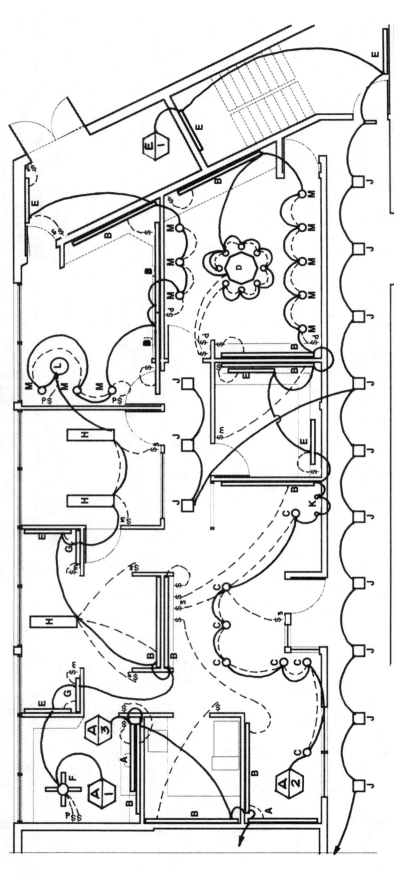

LIGHTING CIRCUITS

Curved solid lines are used to show circuits serving lighting fixtures. Begin circuit runs at a distant fixture and work toward the panel along a single path. The following rules guide circuiting:

1. When a group of fixtures is controlled by a switch, they must be connected to a single circuit.
2. Maximum circuit load is 1,920 watts. Try to connect at least 1,800 watts on all but the last circuit.
3. Use Table 12.9 to add lighting watts, then total and describe each circuit in the panel schedule as shown in Table 12.10.

A hexagon key is used to designate the panel and circuit; locate this key at either end of the circuit path.

Check circuit wattage tabulations for accuracy before going on to the following drawings.

Two additional circuits shown on the drawing are not tabulated in panel A. An emergency circuit shows 526 watts, but additional lighting will be connected before this circuit is run to the emergency panel. Circuit A-3 serves 610 watts in the reception, print, and lunch rooms. It continues into the rest room to serve additional lighting loads.

TABLE 12.9

Fixture Schedule

Key	Watts	Fixture
A	110	Fluorescent strip
B	130	Fluorescent strip
C	50	Incandescent can
D	500	Incandescent, custom
E	72	Fluorescent strip
F	200	Fan and light
G	55	Fluorescent strip
H	72	Recessed fluorescent
J	28	Recessed fluorescent
K	70	In display case
L	250	Incandescent pendant
M	50	Incandescent eyeball
=	18	Exit

TABLE 12.10

Circuit Tabulation for Panel A

Circuit	Watts	Description
1	1,942	Ltg. perimeter—NE
2	1,834	Ltg. recept. & conf.—NE

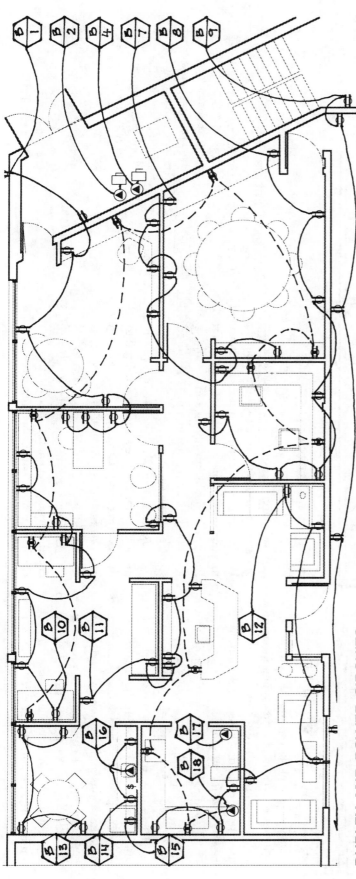

DUPLEX AND POWER CIRCUITS

Study the duplex circuits first. Codes allow a maximum of 13 duplex outlets on a single 20 ampere circuit, but usual commercial practice circuits 8. Each duplex is counted as 180 watts unless it serves a specific appliance.

Duplex and Power Key

⊕ Standard duplex outlet (15 A, 125 V)

⊕ Computer network duplex outlet

◄ Power outlet

This drawing is too small for a complete key. When preparing contract drawings, use a larger scale and add needed information at each duplex, for example:

  42″ Outlets above counters

  GFI Outlets in wet locations

Also add a Δ key for each phone outlet.

Curved solid lines show circuits, and hexagons indicate the panel and circuit number. Trace several duplex circuits and verify watts tabulated in the panel. One circuit is used for the refrigerator, and the lunch room has two circuits to accommodate appliances like microwaves and toasters.

UPS. Dashed lines are used to show computer network outlets served by an uninterrupted power supply.

Power outlets Include 120 volt, 208 volt, and 208 volt, three-phase loads. Usually, three-phase loads like strip heat are connected in large panels with elevator motors and condensing units. Here, strip heat is included in panel B for example purposes.

Table 12.11 tabulates each circuit shown on the plan, and Panel B (Figures 12.10 and 12.13) notes protection and conductors required for each circuit.

Review and verify these panel loads. They will be used to size panels and the building service in the following text.

TABLE 12.11

Circuit Tabulation for Panel B

Circuit	Watts	Description
1	1,440	Duplex—owner & manager
2	1,580	Air handler fan—3/4 hp
4	10,000	Strip heat—3 phase
7	1,440	Duplex—owner & conf.
8	1,440	Duplex—library & conf.
9	1,440	Duplex—hall
10	1,440	Duplex—manager & work
11	1,440	Duplex—manager & recept.
12	1,440	Duplex—waiting & print
13	720	Duplex—lunch
14	360	Duplex—lunch
15	300	Duplex—refrigerator
16	830	Duplex—garbage disposer
17	1,800	Duplex—copy machine
18	5,000	UPS (power supply)

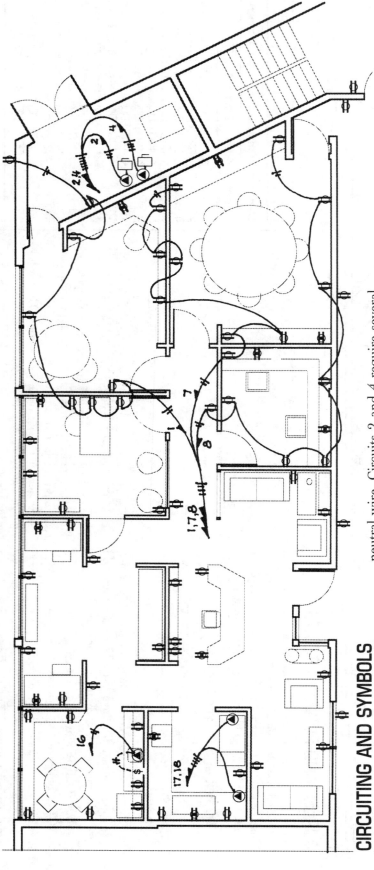

CIRCUITING AND SYMBOLS

Hexagon panel keys and dotted lines for switches or special circuits communicate effectively for a project like the example office. Another set of circuiting keys and symbols is illustrated above for information.

Drawings for government projects frequently detail each junction box and conduit run to encourage competitive bidding.

The plan shown here repeats selected circuits shown in the preceding figure to illustrate alternate circuit keys and symbols. Arrows replace hexagons and locate home runs for fully loaded circuits. Circuit numbers above the arrows designate the panel space assigned to each circuit. Slash marks show the number of conductors in a conduit. A small slash is used for a hot wire, and a larger slash designates a

neutral wire. Circuits 2 and 4 require several hot wires but share a common neutral. These 208 volt circuits don't cause neutral current, and the neutral is used as a ground conductor. When a grounded conduit system is connected to such loads, the neutral may be omitted. Even more detail, specifying and locating each junction box and conduit run, is required on some government work.

Lighting and duplex circuits in three-phase wiring systems can use a common neutral to reduce wiring costs. Circuits 1, 7, and 8 require separate hot wires, but they share a single neutral conductor.

A 1/2-inch conduit can hold ten 20 ampere conductors, so seven 120 volt circuits can be run in just one 1/2-inch conduit (seven hot and three neutral wires).

Selected Symbols

home run to panel LP-3, ckts. #1-2-3.

12 wires—1/2"C.

two hot wires, one neutral and one switch leg (S) in a 1/2" C.

phone outlet (never run with AC—use separate conduit)

duplex outlet in floor

ground fault duplex outlet 42" above finished floor

lighting panel LP-3

LOBBY

The lobby is a first experience for guests and tenants and sets the "building image." Planters and display cases flank the entry doors, and two large decorative vertical tapestries are suspended below a stained glass ceiling.

A bridge, accessed by elevators and a decorative circular stair, carries second-floor tenant and guest traffic between the east and west office spaces. A reception desk, building directory, and small waiting area share space beneath the bridge (see Figure 12.6).

Lobby Lighting

Scene controller dimmers set light intensity for seven lobby lighting circuits (see Table 12.12). Six scenes are preset, and a photocell override changes scene settings in response to changing daylight conditions. A-4 lights the stair, bridge, and building directory. A-5 and A-6 illuminate decorative tapestries and plants. A-7 and A-8 control backlights for the stained glass lobby ceiling 20 feet above the floor. The backlights are fluorescent fixtures equipped with electronic dimming bal-

lasts. A-9 and A-10 light the entrances, display cases, and reception spaces below the bridge.

Six lighting scenes are preset: morning, daylight winter, daylight summer, daylight overcast, evening, and midnight to sunrise.

Watts are noted for each fixture, so readers can verify panel load calculations that follow (see Table 12.13). Fixture selection and placement for lighting design are covered in Chapters 4 and 6.

TABLE 12.12

Scene Circuit Summary

Key	Watts	Fixtures Illuminate
A-4	1,138	Stair, bridge, and directory
A-5	1,200	Tapestry (two at 600 each)
A-6	1,750	Plants (seven at 250 each)
A-7	1,440	Stained glass backlights
A-8	1,440	Stained glass backlights
A-9	1,750	Display cases and under bridge
A-10	1,000	Exterior (ten at 100 each)

TABLE 12.13

Lobby Fixture Schedule

Key	Watts	Fixture
B	130	Fluorescent strip
C	50	Incandescent can
E	72	Fluorescent—wall mount
N	240	Industrial fluorescent (locate between the skylight and the stained glass ceiling)
O	600	Incandescent (a circle of twelve 50 watt MR-16 spots illuminates each tapestry)
P	460	Multi-vapor cylinder
R	100	Incandescent can
S	500	Incandescent—display case
T	250	Incandescent can
U	56	Fluorescent recessed
V	100	Incandescent wall mount
W	288	Fluorescent directory sign
	18	Exit lamp with battery

FIGURE 12.6

LOBBY FIXTURES

Light fixtures B, C, and E from the northeast offices, and N through W will be used to illuminate the lobby (see Figure 12.7).

LOBBY LIGHTING

Panel A serves building lighting loads. Seven lobby lighting circuits are regulated by a scene controller that dims or brightens selected circuits as sunlight intensity varies. Read the descriptions that follow, and trace each circuit on Figure 12.8 or swap for lighting plan.

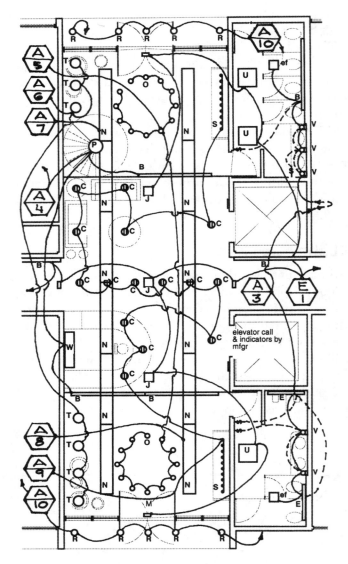

FIGURE 12.8 Lobby Lighting Plan

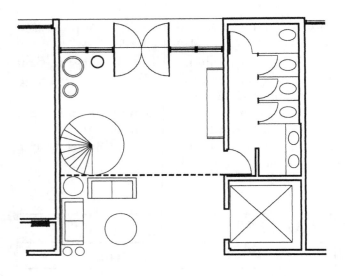

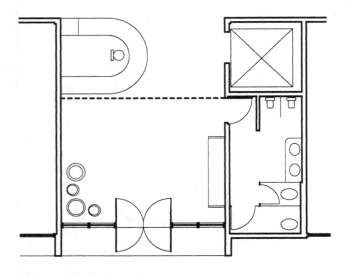

FIGURE 12.7

Emergency Lighting

Circuit E-1 serves exit lights and a few fixtures in the rest rooms and under the bridge. These lights will operate on emergency power in the event of a utility power failure. E-1 extends to ground-floor corridors and stairway lighting.

Rest Room Lighting

Circuit A-3 (1,884 watts) serves rest room lighting and also several fixtures in the northeast offices.

12.3 EXAMPLE OFFICE PANELS

OVERVIEW

Preparation of electrical plans is a four-step process:

1. Select and locate fixtures and switches.
2. Locate duplex and power outlets.
3. Circuit loads and size panels.
4. Size and detail the electrical service.

Plans illustrate steps 1 and 2 for the northeast offices and the lobby. Lighting and power loads for the other building tenants are inferred from the northeast offices.

Plans also show circuiting (step 3) for the northeast offices and the lobby. Look back and confirm the loads noted in the panel A schedule (see Figure 12.9).

LIGHTING PANEL A

Panels apportion electrical power among individual circuits. Single-pole, 20 ampere circuits are used for most building lighting, and connected loads are limited to 80% of circuit rating. This means a 20 ampere cir-

cuit can carry a maximum load of 16 amperes or 1,920 watts at 120 volts.

Each single-pole circuit breaker occupies one space in the panel, and #12 copper conductors are selected to safely carry 20 amperes.

PANEL SCHEDULE

The three-phase panel schedule shown in Figure 12.9 will be used for all examples. Start by filling in the panel key and voltage entries at the top of the form. Values for amperes, spaces, and feeders will be set after all panel circuits are tabulated.

Field entries on the form are divided into five vertical columns as follows:

Key is the number assigned to a given circuit on the plan drawing. Each key digit also represents a space in the panel, and each space can hold a single-pole circuit breaker. Usual lighting panel capacity choices are 20, 30, 40, or 42 spaces.

Description identifies the circuit location on the plan drawing.

PANEL A — Lighting

120/208 volts ___ amps MCB or MLO ___ spaces or poles

feeders __ # ___ and __ # ___ _____ in ___" C. or _____

Key	Description	CB A-P	wire	Load (watts or volt-amps)		
1	NE Offices	20-1	#12	1,942		
2	NE Offices	20-1	#12		1,834	
3	NE Offices & R.R.	20-1	#12			1,884
4	Lobby Scene stair bridge	20-1	#12	1,138		
5	Lobby Scene tapestry	20-1	#12		1,200	
6	Lobby Scene tapestry	20-1	#12			1,750
7	Lobby Scene ceiling	20-1	#12	1,440		
8	Lobby Scene ceiling	20-1	#12		1,440	
9	Lobby Scene display	20-1	#12			1,750
10	Lobby Scene exterior	20-1	#12	1,000		
11						
12						
13						

FIGURE 12.9

CB A-P is the column where circuit breaker amps and poles are entered for each circuit.

Wire notes conductor size for each circuit.

Load entries tabulate total connected watts or volt-amperes. Three entry columns alternate under the load heading. Each column represents one of the three busses in a three-phase panel.

RECEPTACLE (AND HVAC) PANEL B

Refer back to the northeast office Duplex and Power Circuits to check circuit loads. Then read the key summary in Figure 12.10 and verify each circuit in the panel schedule.

Duplex Circuits B-1 and B-7 through B-15

Commercial electrical installations connect up to eight duplex outlets on a 20 ampere circuit. Recall that each duplex outlet is counted as a 180 watt load, so a typical duplex circuit load is 1,440 watts.

Lunch room circuits B-13 and B-14 serve less than eight duplex outlets, so toasters, coffee pots, and

microwaves are less likely to overload them. Circuit B-15 serves just the refrigerator.

HVAC CIRCUITS

Air handlers and resistance heat strips are included in this panel to illustrate two- and three-pole loads. A typical office building would serve such loads in a separate power panel instead of mixing duplex and HVAC.

Circuit B-2

A single-phase, 3/4 horsepower, 208 volt air-handler motor is protected by a two-pole circuit breaker that takes two spaces in the panel. Motor loads are tabulated in volt-amperes, and this motor draws 7.9 amps. A 15 ampere circuit is specified to pass the motor's starting current surge.*

---

*The 3/4 hp motor load is shared equally by two panel busses so each carries 790 volt-amperes ($200 \times 7.9 \div 2 = 790$). Calculate with 200 volts instead of 208 volts to allow for voltage drop in the circuit conductors between the panel and the motor.

Key	Description	CB A-P	wire	Load (watts or volt-amps)		
	PANEL B *Power*					
1	duplex NE Offices	20 - 1	#12	1,440		
2	air handler NE (3/4 hp)	15 - 2	#14		790	
3	↓					790
4	strip heat NE (10 kW)	40 - 3	#8	3,333		
5	↓				3,333	
6	↓					3,333
7	duplex NE Offices	20 - 1	#12	1,440		
8	duplex NE Offices	↓	↓		1,440	
9	duplex Hall	↓	↓			1,440
10	duplex NE Offices	↓	↓	1,440		
11	duplex NE Offices	↓	↓		1,440	
12	duplex NE Offices	↓	↓			1,440
13	duplex - lunch room NE	↓	↓	720		
14	duplex - lunch room NE	↓	↓		360	
15	duplex - refrigerator	↓	↓			300
16	garbage disposer (1/3 hp)	15 - 1	#14	830		
17	copy machine	20 - 1	#12		1,800	
18	UPS (5 kVA)	30 - 2	#10			2,500
19				2,500		

FIGURE 12.10

PANEL [A] — *Lighting*

[120/208] volts [200] amps ~~MCB or~~ (MLO) [40] spaces or poles

feeders 4 # 250 and __ # ___ THWN in 2½" C. or _____
$\overline{MCM-AL}$

Key	Description		CB A-P	wire	Load (watts or volt-amps)		
1	NE Offices	1st floor	20-1	#12	1 942 *		
2	NE Offices	1st floor	20-1	#12		1 834	
3	NE Offices	1st floor	20-1	#12			1 884
4	Lobby scene	stair-bridge	20-1	#12	1 138		
5	Lobby scene	tapestry	20-1	#12		1 200	
6	Lobby scene	tapestry	20-1	#12			1 750
7	Lobby scene	ceiling	20-1	#12	1 440		
8	Lobby scene	ceiling	20-1	#12		1 440	
9	Lobby scene	display	20-1	#12			1 750
10	Lobby scene	exterior	20-1	#12	1 000		
11	SE Offices	1st floor	20-1	#12		1 800	
12							1 700
13					1 900		
14						1 700	
15	NW Offices	1st floor					1 600
16					1 800		
17						1 700	
18	SW Offices	1st floor					1 700
19					1 900		
20						1 800	
21							1 600
22	NE Offices	2nd floor			1 800		
23						1 700	
24							1 700
25	SE Offices	2nd floor			1 900		
26						1 800	
27							1 600
28					1 800		
29	NW Offices	2nd floor				1 700	
30							1 700
31					1 900		
32	SW Offices	2nd floor				1 800	
33							1 600
34					1 800		
35						1 700	
36							1 700
37							
38							
39							
40							
41							
42							

* cut ckt. 1 to 1920 — why?

$\dfrac{60778}{208\sqrt{3}}$ → connected load amps 169

$\dfrac{20,320}{120}$ → max. neutral amps 169

phase totals 20,320 | 20,774 | 20,284

panel total 60,778

by GJM date 9·1·99

FIGURE 12.11

169

Circuit B-4

Inefficient resistance heat strips are used to illustrate circuiting. The northeast offices need 34,000 BTUH or 10 kW for space heating. A three-phase strip heater has three internal resistance wire loops, and each loop is connected across two hot wires. Three two-pole circuit breakers could serve the heater, but using a single three-pole circuit breaker saves panel space and wiring.

Calculate circuit amperes = 28.9 (10,000 ÷ 2003 √3). Minimum circuit ampacity is 125% of the calculated value or 36.1 amperes, so a 40 ampere circuit breaker is selected. The 10,000 watt load is divided equally across the three panel busses.

Circuits B-16 and B-17

A dedicated circuit is usually provided for larger equipment or appliances.

Circuit B-18

The 5 kVA UPS (uninterrupted power supply) is served by a two-pole breaker. Calculated circuit amperes = 25, and the manufacturer specifies 30 ampere circuit protection. Conduit for UPS conductors is isolated from all other electrical conduit.

SCHEDULE PANEL A (LIGHTING)

The form in Figure 12.11 is a completed panel schedule. It describes each circuit and calculates loads to size the feeders. The northeast office's lighting load is about 2 watts per square foot of floor area, so two times floor area should provide a reasonable estimate of lighting load for all other office areas. The lobby and northeast office lighting totals about 15 kW. If the other offices add 45 kW, the lighting panel(s) should have a capacity of about 60 kW.

Lighting circuits for the lobby and the northeast offices fill 10 panel spaces. The 45 kW lighting load of the other offices will require 25* panel spaces or poles, so the panel should have at least 35 spaces.

Lighting circuits 11 through 36 are filled, allowing 2 watts per square foot. Designers work to balance the total watts connected to each panel bus so feeder loads are equal.

---

*45 kW at about 1,800 watts per circuit = 25.

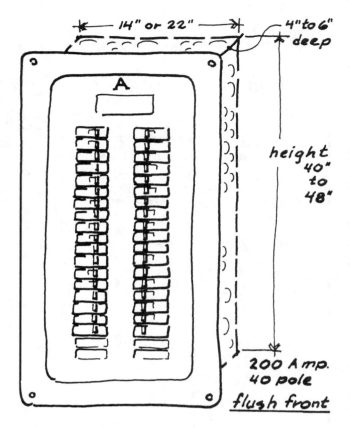

FIGURE 12.12

A 40 space lighting panel is specified. Typical installations include open spaces in panels for future expansion; this example reserves 10%. Consider including spare breakers in the open panel spaces and rough in wiring to an accessible junction box above the ceiling to ease future expansion (see Figure 12.12).

Calculate Panel Amperes

Add the watts connected to each panel bus, and then add three busses to find total panel load. This load is shared by three feeders. Each feeder carries about 169 amperes (60,778 ÷ (208)(√3) = 169).

The neutral feeder provides a return path for current serving 120 volt loads. Current flows in the neutral when panel loads are not balanced, and maximum unbalance occurs when all lights connected to one bus are operating, but all other lights are off.

To calculate maximum neutral current, total the 120 volt loads for each bus and then divide the largest by 120. The panel has a maximum neutral current of 169 amperes (20,320 ÷ 120 = 169).

PANEL B, Power

$120/208$ volts 200 amps ~~MCB or~~ (MLO) 40 spaces or poles

feeders 3 #250 MCM AL and 1 # 3 AL THWN in 2 " C. or _____

Key	Description	CB A-P	wire	Load (watts or volt-amps)		
1	duplex NE Offices	20 - 1	#12	1,440		
2	air handler NE (¾ hp)	15 - 2	#14		790	
3						790
4	strip heat NE (10 kW)	40 - 3	#8	3,333		
5					3,333	
6						3,333
7	duplex NE Offices	20 - 1	#12	1,440		
8	duplex NE Offices	↓	↓		1,440	
9	duplex Hall	↓	↓			1,440
10	duplex NE Offices	↓	↓	1,440		
11	duplex NE Offices	↓	↓		1,440	
12	duplex NE Offices	↓	↓			1,440
13	duplex - lunch room NE	↓	↓	720		
14	duplex - lunch room NE	↓	↓		360	
15	duplex - refrigerator	↓	↓			300
16	garbage disposer (⅓ hp)	15 - 1	#14	830		
17	copy machine	20 - 1	#12		1,800	
18	UPS (5 kVA)	30 - 2	#10			2,500
19				2,500		
20	duplex Lobby (not shown)	20 - 1	#12		1,440	
21	air handler SE (1 hp)	15 - 2	#14			920
22				920		
23	strip heat SE (12 kW)	50 - 3	#8		4,000	
24						4,000
25				4,000		
26	duplex SE Offices	20 - 1	#12		1,440	
27		↓	↓			1,440
28		↓	↓	1,440		
29		↓	↓		1,440	
30		↓	↓			1,440
31		↓	↓	1,440		
32		↓	↓		1,440	
33		↓	↓			1,440
34		↓	↓	1,440		
35		↓	↓		720	
36		↓	↓			720
37						
38						
39						
40						
41						
42		(neutral	totals)	(10,190)	(11,520)	(8,220)

phase totals: 20,943 | 19,643 | 19,763

$\dfrac{60,349}{208\sqrt{3}}$ → connected load amps 168

$\dfrac{11,520}{120}$ → max. neutral amps 96

panel total 60,349

by J O G date 9-9-99

FIGURE 12.13

Select Panel

Standard panel capacities are 100, 125, 150, 200, 225, 400, and 600 amperes. Select a 200 ampere panel to carry the 169 ampere load and accommodate future expansion. In a smaller building, the panel would include a three-pole main circuit breaker. For this example, panel protection will be located at the service entry. Specify 200 amperes MLO (main lugs only) for panel A. The panel lugs and busses are sized for 200 amperes but the panel does not have a main circuit breaker. Complete the second line in the panel schedule accordingly.

Size Panel Feeders and Conduit

Select four #3/0 copper panel feeders or four 250 MCM aluminum feeders to carry 200 amperes. Aluminum panel feeders are used in the example, but *many local codes prohibit aluminum* conductors inside buildings. Verify THWN feeder sizes on page 152. THWN is rated at 75°C. THHN rated at 90°C allows higher current for a given conductor size.

The neutral conductor could be #4/0 at 169 amps, but to allow for expansion, neutral ampacity serving a lighting panel is usually specified equal to the phase conductor ampacity. Use 2″ conduit for copper feeders or 2 1/2″ conduit for aluminum feeders.*

SCHEDULE PANEL B

The form in Figure 12.13 is a completed panel schedule. It describes each circuit and calculates loads to size the feeders. Duplex, air handler, and strip heat loads in the northeast offices use 18 panel spaces and total about 30 kW. Panel B will serve the northeast *and southeast* offices in 36 spaces with a total load of about 60 kW.

The northeast and southeast offices occupy about 25% of the building area, so when you calculate the building service allow for three more panels, each equal to panel B. Designated B_2, B_3, and B_4, they will serve duplex, air handler, and strip heat loads for the west offices and the second-floor offices.

A typical office building of this size would use separate panels for duplex loads and HVAC equipment. Panel B includes strip heat and air handlers

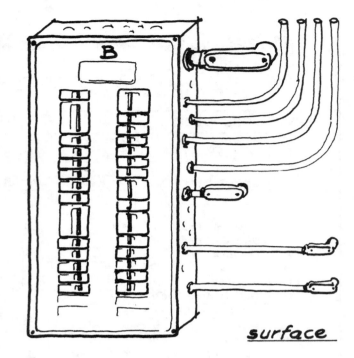

FIGURE 12.14

with the duplex circuits *only* to illustrate two- and three-pole loads and the neutral current calculation (see Figure 12.14).

Calculate Panel Amperes

Add the watts connected to each panel bus, and then add three busses to find total panel load. This load is shared by three feeders, and *each* feeder must carry 168 amperes. $(60,349 \div (208)(\sqrt{3}) = 168)$.**

Select Panel

Standard panel capacities are 100, 125, 150, 200, 225, 400, and 600 amperes. Select a 200 ampere panel to carry the 168 ampere load and accommodate future expansion. In a smaller building the panel would include a three-pole main circuit breaker, but for this example panel protection is located at the service entry. Specify 200 amperes MLO (main lugs only) for panel B. The panel lugs and busses are sized for 200 amperes, but the panel does not have a main circuit breaker. Complete the second line in the panel schedule accordingly.

---

*Verify conductor ampacity, conduit size, and panel poles (or spaces) on pages 152–157.

**The actual load on panels A and B totals *337 amperes*. A 350 ampere service would reduce conductor cost but limit future expansion.

Size Panel Feeders and Conduit

Select three #3/0 copper panel feeders or three 250 MCM aluminum feeders to carry 200 amperes. Aluminum panel feeders are used in the example because they're lighter and easier to install than copper, but *many local codes prohibit aluminum* conductors in buildings.

The neutral conductor provides a return path for current serving 120 volt loads, but 208 volt equipment does *not* cause neutral current. To calculate maximum neutral current, total *only* the 120 volt loads for each bus and then divide the largest by 120. The maximum neutral watts is 11,520 (line 42), so the maximum neutral current is 96 amperes (11,520 ÷ 120 = 96). Select a #3 copper or a #1 aluminum neutral to carry 100 amperes. Select 2" conduit for copper feeders or 2 1/2" conduit for aluminum feeders.*

SCHEDULE PANEL C (POWER)

The form in Figure 12.15 is a completed panel schedule. It describes each circuit and calculates loads to size the feeders. Panel C supplies condensing units and elevators. Five 5-ton condensers serve the lobby and smaller offices, while four 7.5-ton units carry the larger offices. Second-floor air handlers circulate a bit more air than first-floor air handlers, to carry the added cooling load caused by roof heat gain.

Condenser circuits (poles 1–27). Panel loads are estimated in volt-amperes based on each condenser's full load ampere (FLA) rating. The 7.5-ton units in panel C require 26 FLA or 9,000 volt-amperes at 200 volts (26 × 200 √3 = 9,000). This electrical load is shared equally by the panel feeders.

Circuit protection is specified by condenser manufacturers to accommodate the starting current surge. The rule of thumb for estimating motor circuit protection is 125% of FLA, but manufacturers frequently specify 150% or more.

Elevators (poles 28–35). The 30 horsepower elevator motors demand nearly 92 FLA, so 125 ampere circuit breakers are specified. Motors are an exception to the general rule that circuit wire ampacity must equal the circuit breaker

trip rating. The #2 copper wire used in these circuits is rated for only 115 amperes.

Spaces. A 40-space panel is selected. Three single-pole circuits fill spaces 34–36, and the remaining empty spaces allow for expansion or omissions (forgotten electrical loads).

Note: A typical office building of this size would include the HVAC strip heat and air-handler circuits from panel B in panel C. These loads were included in panel B only to illustrate two- and three-pole loads and the neutral current calculation.

Calculate Panel Amperes

Add the volt-amperes connected to each panel bus, and then add three busses to find total panel load. This load is shared by three feeders, and *each* feeder will carry 370 amperes (133,170 ÷ (208)(√3) = 370).

Select Panel

Standard panel capacities are 100, 125, 150, 200, 225, 400, and 600 amperes. Select a 400 ampere panel allowing 30 amperes for expansion. In a smaller building, the panel would include a three-pole main circuit breaker, but for this example panel protection is located at the service entry. Specify 400 amperes MLO (main lugs only) for panel C. The panel lugs and busses are sized for 400 amperes, but the panel does not have a main circuit breaker. Complete the second line in the panel schedule accordingly.

Size Panel Feeders

600 MCM copper or 900 MCM aluminum feeders will carry 400 amperes, but installation is difficult. Two #3/0 copper or two 250 MCM aluminum feeders for each phase allow easier installation. Aluminum panel feeders are used in the example, but *many local codes prohibit aluminum* conductors in buildings. Neutral current is less than 20 amperes, so a #10 aluminum neutral is specified.

Danger! Some code authorities require a full-size equipment ground conductor. Others allow the conduit system to serve as a ground for three-phase equipment.

Select 2 1/2-inch conduit for copper or 3-inch conduit for aluminum feeders.**

---

*Verify conductor ampacity, conduit size, and panel poles (or spaces) on pages 152–157.

**Verify conductor ampacity, conduit size, and panel poles (or spaces) on pages 152–157.

PANEL [C]   *Power (a.c. & elevator)*

[120/208] volts   [400] amps ~~MCB~~ or (MLO)   [40] spaces or poles

feeders 3-900 MCMand _1_ # _12_ AL _THWN_ in 3½" C. or 6-250 MCM & 1#12 in 3"C
          ↳ AL   *Aluminum ᵗ feeders*

Key	Description	CB A-P	wire	Load (watts or volt-amps)		
1	a.c. NE 1st floor↓ 5T	40 - 3	# 8	2,000		
2					2,000	
3						2,000
4	a.c. SE 7½T	50- 3	# 8	3,000		
5					3,000	
6						3,000
7	a.c. NW 5T	40 - 3	# 8	2,000		
8					2,000	
9						2,000
10	a.c. SW 7½ T	50 - 3	# 8	3,000		
11					3,000	
12						3,000
13	a.c. Lobby	40 - 3	# 8	2,000		
14					2,000	
15						2,000
16	a.c. NE 2nd floor↓ 5T	40-3	# 8	2,000		
17					2,000	
18						2,000
19	a.c. SE 7½T	50 - 3	# 8	3,000		
20					3,000	
21						3,000
22	a.c. NW 5T	40 - 3	# 8	2,000		
23					2,000	
24						2,000
25	a.c. SW 7½T	50 - 3	# 8	3,000		
26					3,000	
27						3,000
28	elevator N 30hp	125-3	#2	10,600		
29					10,600	
30						10,600
31	elevator S 30hp	125-3	#2	10,600		
32					10,600	
33						10,600
34	elevator controls	20 -1	#12	1,200		
35	elevator controls	20 - 1	#12		1,200	
36	security system	20 - 1	#12			1,130
37						
38						
39						
40						
41						
42						

phase totals [44,400 | 44,400 | 44,370]

133,170/208√3 ~

1200/120 allow↗ panel total 133,170

connected load amps __370__ max. neutral amps __20__ by __cb__   date 9·9·99

FIGURE 12.15

174

OTHER ELECTRICAL LOADS

Two tasks remain before sizing the building service and completing this long example:

* Select and size the emergency panel.
* Check for errors and omissions on all electrical plans and circuits.

Emergency and omissions loads are estimated, lest faithful readers grow weary studying panel forms.

Panel E (Emergency)

A 60 ampere, three-pole panel will be used for three emergency circuits that total 5,000 watts. The emergency panel carries the building fire alarm system, exit signs, stair lighting, hallway lighting, and some lighting in each office. Emergency lighting operates 24 hours a day, and selected light fixtures also include a battery pack to provide an hour of illumination during a power failure.

In small buildings with few emergency power requirements, battery packs can replace the emergency panel. However, in high-rise offices or hospitals, extensive emergency loads require panels, generators, and a separate distribution system for emergency circuits.

Panel O (Omissions)

What's missing? For this example, panel O is estimated at 28,000 watts, and a 100 ampere, 24-pole panel will be used. Good electrical designers don't need an omissions panel because they anticipate all electrical loads and allow expansion capacity in each panel. Panel O is used here, so careful readers can review the preceding pages and identify forgotten loads. Visualize the building *site* and think about electrical loads. Then, before you look at the following list, look back to panels B and C and see if you can identify two missing HVAC circuits.

Forgotten loads in panel O include:

* Air handlers and strip heat*
* Water heater
* Rest room exhaust fans
* Parking lot lighting
* Landscape lighting
* Christmas lights
* Site fountain (pump and lighting)
* Pumps for site drainage

---
*12 kW of strip heat and an air handler for the lobby.

SERVICE DIAGRAMS

An electrical service diagram is a schematic drawing identifying major components of a building's electrical service. Electrical subcontractors use this diagram and the electrical floor plans to price project labor and material.

Before sizing the service for the example building, consider a hypothetical *smaller building* where the total building electrical load is carried by two panels. Assume that these two panels are identical to example panels A and B, and that all electrical loads draw current simultaneously.

Smaller Building Service

The hypothetical smaller building has two 200 ampere panels. The building service will be sized for 400 amperes, and each panel will be protected by a 200 ampere circuit breaker located below a wireway. The wireway is an enclosure with 400 ampere lugs where the service is split (see Figure 12.16).

Single Main Disconnect

Many localities require a single switch to shut off all electrical power in the event of a fire. Fused disconnects are indicated in Figure 12.17 instead of circuit breakers; both work, and selection is usually based on cost.

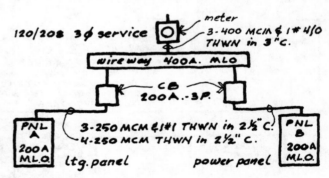

FIGURE 12.16

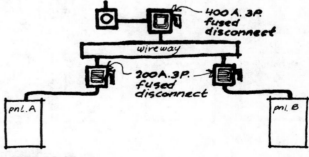

FIGURE 12.17

1,200 Amperes?

Some office buildings install separate electrical services for each tenant, but the example building will be detailed with a single service.

12.4 EXAMPLE OFFICE SERVICE

EXAMPLE BUILDING ELECTRICAL SERVICE

Using 120/208 volt power, the smallest possible standard service for the example building is 1,200 amperes. Review the following discussion and the illustrations until you understand and can explain this service ampacity.

Total capacity of eight panels in the example is 1,560 amperes. That's a lot of amperes, and the build-

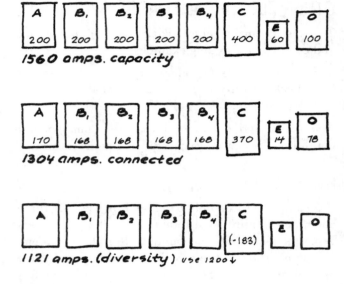

FIGURE 12.18

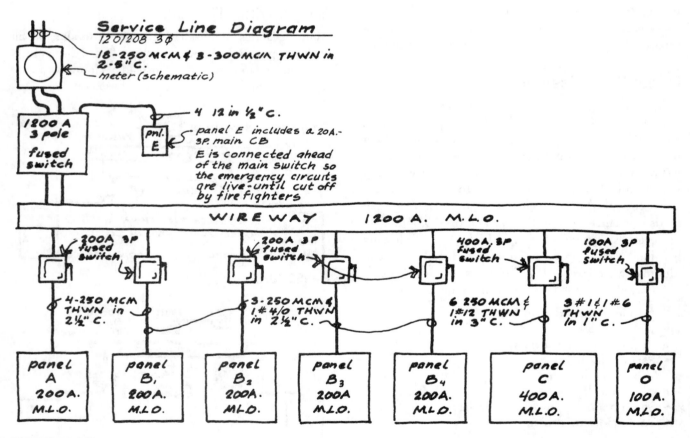

FIGURE 12.19

ing electrical service will be large and expensive (see Figure 12.18).

To reduce service size and cost, use the actual connected amperes instead of the panel capacity (this reduces the potential for future expansion). Actual connected amperes total 1,304.

When all building electrical loads operate simultaneously, the service must be sized to meet the total connected load. In the example building, heating and cooling are noncoincident loads, so the service size can be reduced to serve only the largest of these. Resistance heating totals 100 kW or 278 amps, and condensing units total 66 kVA or 183 amps. Subtract 183 amperes from the connected total because simultaneous heating and cooling is not anticipated, and the minimum service is 1,121 amperes.

Manufacturers' standard ratings for panels and lugs are 400, 600, 800, 1,200, 1,600, and 2,000 amperes. Select a 1,200 ampere service, allowing 79 amperes (28 kW) of expansion capacity (see Figure 12.19). Verify conductor and conduit sizes for the 1,200 ampere service and for the building neutral.

The meter shown is "schematic" because it is not actually connected across the main feeders. The emergency panel is located ahead of the main switch, so firefighters have the option of leaving emergency power on during a fire.

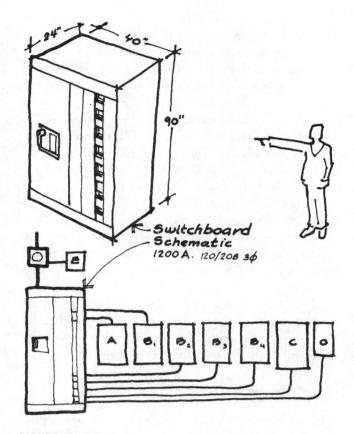

FIGURE 12.20

12.5 SERVICE OPTIONS

A BETTER WAY? SWITCHBOARD

In a building of this size with a single electrical service, a switchboard could replace the main switch, wireway, and seven fused switches. Wireways and disconnects are specified when each tenant is metered separately.

Switchboards act as large panels. They draw power from the service and apportion current to individual panels. The conductors, conduit, and panels shown in Figure 12.20 are identical to those in the preceding service diagram. Each fused switch has been replaced with a circuit breaker located in the switchboard.

277/480 VOLT SERVICE

Experienced electrical designers would consider increased service voltage for the example building. The diagram in Figure 12.21 illustrates a 277/480 volt service. Increasing service voltage reduces service size and cost.

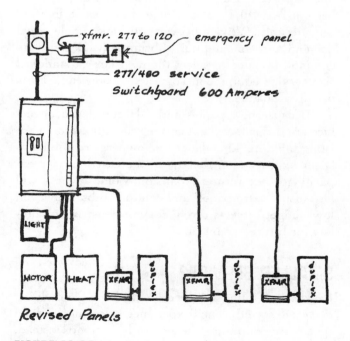

Revised Panels

FIGURE 12.21

The example building's service capacity is 432 kVA (1,200 × 208 √3 = 432,000).

Operating elevator motors, condensing units, and resistance heat strips at 480 volts cut current required by more than half. Ballasts designed for 277 volt operation reduce lighting current and the number of lighting circuits. Transformers that provide power for 120 volt loads are an added cost for the 277/480 volt system.

Calculations used for the 120/208 volt service are easily revised for a 277/480 volt installation (or for the 347/600 volt system of choice in Canada). The building's 432 kVA load requires 520 amperes at 480 volts. A 600 ampere service would be specified. Panels would be revised to serve motor, heating, and ballasted lighting at 277/480 volts. Duplex panels are served by 277 120 volt or 480 120/208 volt transformers.

Very large buildings cut electrical costs by taking power at high voltage (5 to 15 kV). Utilities charge less for high-voltage power because the building owner installs and maintains transformers and switchgear.

LOAD GROWTH

It's easy to calculate a building's electrical service capacity, but experience and judgment are necessary when oversizing to accommodate future loads. Historical 10-year load growth for office buildings has been in the 40 to 80% range, but recent lighting retrofits and HVAC improvements have cut electrical demand and consumption substantially. Because it's common practice to select the next largest standard size service equipment, some expansion capacity is usually available.

Because new panels and a larger electrical service are difficult and expensive renovations in an existing building, knowledgeable owners are willing to spend construction dollars on increased service capacity to ease future expansion. Some owners will also install extra panels and conduit stubs. However, low-budget projects rarely commit construction dollars for future expansion.

SMALLER SERVICE?

Resistance heating specified for the example building is *not* energy-efficient. Experienced designers would specify gas- or oil-fired heating or heat pumps instead. Efficient HVAC equipment can reduce annual heating costs and reduce the size of the electrical service.

The example also totals more than 5 watts per square foot for duplex loads. This allowance is at the very high end of the office building range. A minimum electrical service for the building using gas heat and allowing 2.5 watts per square foot for duplex loads could cut service size (cost and electric bills) by one-third (or 33%).

RIGHT SERVICE?

Two electrical services have been proposed for the example building and more follow. The right service is the one that best meets the construction budget and anticipated operating requirements.

Low-budget speculative rental office #1. 1,000 ampere, 120/208* volt service. Use gas heat and cut duplex load to 2.5 watts per square foot.

Low-budget speculative rental office #2. Five 200 ampere (or ten 100 ampere), 120/208* volt metered services. Use gas heat and cut duplex load to 2.5 watts per square foot. Individual meters pass utility costs to the tenants, but the number of services divides the building into 5 or 10 rental areas, which may not suit some potential tenants.

Medium-quality office for four tenants #1. 600 ampere, 277/480 volt service. With gas heat and 5 watts per square foot duplex load, this service allows for 25% load growth.

Medium-quality office for four tenants #2. Four metered 150 ampere, 277/480 volt services. Allows 25% load growth as above but limits building occupancy to four equal area tenants.

Medium-quality office for one owner. 800 ampere, 277/480 volt service. With gas heat and 5 watts per square foot duplex load, this service allows for almost 100% load growth.

High-quality office for one owner. 1,000 ampere, 277/480 volt service. With gas heat and 5 watts per square foot duplex load, this service would include electrical rough-in for future load growth.

---

*Or 120/240 delta depending on the utility.

12.6 TABLES AND PANEL FORMS

TABLE 12.14

Circuit Breakers (trip amperes)

15–50 by 5s

15	20	25*	30	35*	40	45*	50

(*25, 35, and 45 for motor circuits)

60–110 by 10s

60	70	80	90	100	110

125–250 by 25s

125	150	175	200	225	250

300–500 by 50s

300	350	400	450	500

600–1,000 by 100s

600	700	800	900	1,000

1,200–2,000 by 200s

1,200	1,400	1,600	1,800	2,000

Select conductors with an ampere rating that equals or exceeds the circuit breaker trip amperes.

TABLE 12.15

Conductors (sized by amperes)

Allowable amperes for three THWN (75°C) insulated conductors in a raceway based on an ambient temperature of 30°C.

Amperes	Aluminum	Copper
15	#12	#14
20	#10	#12
25	#10	#10
30	#8	#10
35	#8	#8
40	#8	#8
45	#6	#8
50	#6	#8
60	#4	#6
70	#3	#4
80	#2	#4
90	#2	#3
100	#1	#3
110	#1/0	#2
125	#2/0	#1
150	#3/0	#1/0
175	#4/0	#2/0
200	250 MCM	#3/0
225	300 MCM	#4/0
250	350 MCM	250 MCM
300	500 MCM	350 MCM
350	700 MCM	400 MCM
400	900 MCM	600 MCM

Conductors in air or with insulation rated for higher temperature operation can carry more current.

TABLE 12.16

Conduit

Maximum number of type THWN or THHN conductors in conduit or tubing

Wire Size	Conduit Size			
	1/2″	3/4″	1″	1 1/4″
14	13	24	39	69
12	10	18	29	51
10	6	11	18	32
8	3	5	9	16
6	1	4	6	11
4	1	2	4	7
3	1	1	3	6
2	1	1	3	5
1		1	1	3
0		1	1	3
00		1	1	2

Wire Size	Conduit Size			
	1 1/2″	2″	2 1/2″	3″
8	22	36	51	79
6	15	26	37	57
4	9	16	22	35
3	8	13	19	29
2	7	11	16	25
1	5	8	12	18
0	4	7	10	15
00	3	6	8	13
000	3	5	7	11
0000	2	4	6	9
250	1	3	4	7
300	1	3	4	6
350	1	2	3	5

Wire Size	Conduit Size			
	3 1/2″	4″	5″	6″
2	33	43	67	97
1	25	32	50	72
0	21	27	42	61
00	17	22	35	51
000	14	18	29	42
000	12	15	24	35
250	10	12	20	28
300	8	11	17	24
350	7	9	15	21
400	6	8	13	19
500	5	7	11	16
600	4	5	9	13
700	4	5	8	11
750	3	4	7	11

SINGLE PHASE

PANEL [ ] _____

[ ] volts [ ] amps <u>MCB or MLO</u> [ ] spaces or poles

feeders __ # ___ and __ # ___ _____ in ___" C. or _____

Key	Description	CB A-P	wire	Load (watts or volt-amps)
1				
2				
3				
4				
5				
6				
7				
8				
9				
10				
11				
12				
13				
14				
15				
16				
17				
18				
19				
20				
21				
22				
23				
24				
25				
26				
27				
28				
29				
30				
31				
32				
33				
34				
35				
36				
37				
38				
39				
40				
41				
42				

phase totals [ ] [ ] [ ]

panel total _____

connected load amps _____ max. neutral amps _____ by _____ date _____

PANEL ☐ _____

☐ volts ☐ amps <u>MCB or MLO</u> ☐ spaces or poles

feeders __ # ___ and __ # ___ _____ in ___" C. or _____

Key	Description	CB A-P	wire	Load (watts or volt-amps)
1				
2				
3				
4				
5				
6				
7				
8				
9				
10				
11				
12				
13				
14				
15				
16				
17				
18				
19				
20				
21				
22				
23				
24				
25				
26				
27				
28				
29				
30				
31				
32				
33				
34				
35				
36				
37				
38				
39				
40				
41				
42				

phase totals ☐ ☐ ☐

panel total _____

connected load amps _____ max. neutral amps _____ by _____ date _____

CHAPTER 13
Energy Profiles

. . . . how can the brute "causes" of finance be translated into the lasting stuff of profound aesthetic symbolism.

Louis Sullivan

Architects, engineers, and constructors aspire to create beautiful, functional, comfortable, economical, efficient buildings. Designers seek "profound aesthetic symbolism," and design development is a series of choices about orientation, fenestration, lighting, HVAC, and electrical systems that can support the symbolism. Responsible designers and constructors consider energy-conserving opportunities with leisurely paybacks because a building will consume energy for 50 years or more.

This short chapter begins with a discussion of electrical energy use in buildings and attempts to quantify lighting, duplex, and HVAC loads. Electric bill profiles for existing buildings are analyzed to explain probable energy consumption by new buildings.

After reading the chapter and answering the review questions, test your skills by obtaining, plotting, and analyzing a year's utility bills for a commercial building of your choice. Attempt to quantify electrical energy consumption by end use, and explain variations in annual consumption.

13.0 ESTIMATING ENERGY USE

Load density and operating hours set building energy consumption. Three end uses—lighting, duplex, and HVAC—can account for more than 80% of commercial building electrical energy costs. "Watts per square foot" is a convenient load index, and operating hours are easily estimated.

Watts per Square Foot*

Building Type	Lights	Duplex	HVAC
Office	1–2	3–5	4–7
Hospital	1–4	1–2	5–9
Hotel	1–2	1	3–7
Retail	3–5	1	5–8
School	1–2	1–2	3–5

*Very approximate ranges. Use actual watts ÷ floor area.

Electric bills can guide conservation strategies. Two bill profiles, climate driven or people driven, are typical in commercial buildings. Climate-driven buildings use significant amounts of energy for winter heating and summer cooling. People-driven buildings use nearly equal amounts of energy each month of the year regardless of outdoor conditions (see Figure 13.1).

ESTIMATING LIGHTING & HVAC

Estimating lighting energy is easy. Total the connected lighting watts, multiply by expected hours of operation, and divide by 1,000 to get kWh. Remember to count 24 hours daily for stair, hallway, and exit lighting. Allow extra lighting hours for cleaning and night or weekend overtime. Dollars are easier to understand than kWh, so determine an average cost per kWh based on past billing data and prepare estimates in dollars.

Begin by converting peak BTU of building heat gain or loss into kW. Then develop monthly heating and cooling hours based on local climate and proceed as with lighting.*

---

*Chapter 6 of *Efficient Buildings 2* provides more detailed HVAC cost estimates including heat pumps, boilers, and furnaces.

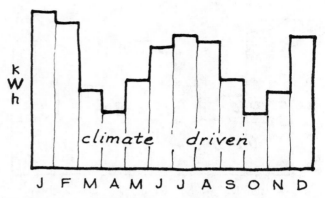

Monthly Electrical Consumption

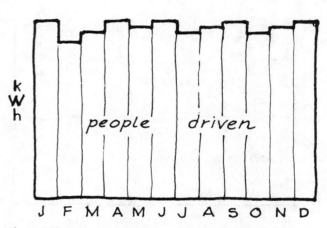

Monthly Electrical Consumption

FIGURE 13.1

Dollar Estimate Example (see Figure 13.2)

Lighting

Total connected lighting	40,000 watts
Monthly operating time	250 hours
Average $ per kWh*	$0.08
Monthly lighting cost	$800

Heating (HVAC)

Peak heat loss	136,000 BTU
Resistance heating capacity	40 kW
January heating hours	400 hours
Average $ per kWh*	$0.07
January heating cost	$1,120

Cooling (HVAC)

Peak heat gain	120,000 BTU
Cooling energy at 1 kW per ton	12 kW
July cooling hours	420 hours
Average $ per kWh*	$0.06
July cooling cost	$302

*$ per kWh arbitrary here. Use actual local utility costs.

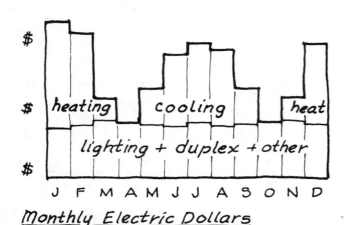

Monthly Electric Dollars

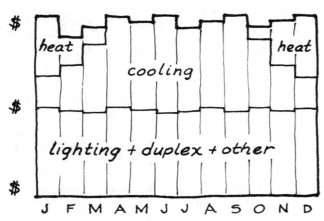

Monthly Electric Dollars

FIGURE 13.2

OTHER ENERGY USES

Lighting, duplex, and HVAC loads are large electrical energy end uses in many commercial buildings, but refrigeration, food preparation, and water heating can be substantial in selected occupancies.

Refrigeration

Supermarkets use more than half of their electrical energy for refrigeration. The good news is refrigeration replaces lots of HVAC load. Air lost from food coolers aids summer cooling, and heat rejected by refrigeration condensers can be recycled for winter heating.

Refrigeration equipment typically operates for at least 16 hours a day. Capacity is rated in horsepower, and 1 horsepower produces about 1 ton (12,000 BTUH) of cooling.

When preparing an energy estimate for a new building with lots of refrigeration, verify actual operating hours for similar existing buildings in the same climate.

Food Preparation

Restaurants use substantial amounts of energy for cooking and refrigeration. Moreover, kitchen ventilation requirements increase HVAC loads. Consult restaurant operators and kitchen equipment suppliers before developing energy estimates for these occupancies.

Water Heating

Hotels use lots of hot water between 6 and 8 A.M. for showers. Estimate water heating BTU for each shower at 1,200 BTU per minute of shower operation. Laundries also use a great deal of hot water. Washer manufacturers can provide water quantity and temperature requirements.

Electric water heating is usually expensive and inefficient. Use it *only* where gas, oil, or propane is not available.

Duplex Loads

Efficient lamps and ballasts have reduced building lighting loads, but increased duplex loads like computers, monitors, scanners, and printers eliminate lighting savings and increase total energy use.

13.1 BUILDING ENERGY PROFILES

Studying energy consumption in existing buildings will enhance your ability to develop valid estimates for new buildings. Electric bills for the past 12 months are easily available for most buildings, and historical bills are accessible without great difficulty. Bills tabulate energy consumption, and they reveal energy use patterns that suggest effective energy-conserving strategies.

A monthly graph of annual energy use is the ideal starting point for studying energy consumption. Two example "all electric" buildings are used to illustrate utility bill analysis.

BUILDING A

Building A is climate driven. The winter and summer graph mountains are heating and cooling energy (see Figure 13.3). Lighting-duplex energy is estimated using watts per square foot and operating hours. Other energy can include elevators, water heating, refrigeration, pumps, food preparation, landscape lighting, security systems, and so on.

Strategies for reducing energy consumption in building A should focus first on heating and cooling loads. Cost-effective energy-conserving renovations may include double glazing, added insulation, more efficient heating-cooling equipment, and reduced ventilation rates.

BUILDING B

Building B is occupant driven. The internal heat from people, lights, and equipment conceals heating and cooling requirements (see Figure 13.4). Actually, building B uses more cooling energy than building A, but it's hidden by the nearly uniform monthly totals. During cold weather, the heating system is warming the building perimeter, while the cooling system cools the "heat-rich" interior. Many large buildings meter heating and cooling loads separately. On the graph they're estimated as explained previously.

Strategies for reducing energy consumption in building B should focus first on internal loads. Cost-

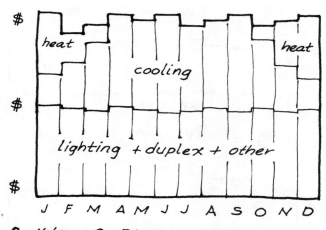

Building B Electric Bills

FIGURE 13.4

effective energy-conserving renovations will include new lighting, heating and cooling, and ventilation equipment. Renovations that upgrade the building envelope will not be cost-effective until the internal loads are minimized.

13.2 DEMAND CHARGES

The preceding examples are based on consumption, but don't despair as you read the following paragraphs about demand charges. An average cost per kWh for consumption is a valid analysis tool, even when demand charges are a large part of a customer's bill.

UTILITY DEMAND

Most electric utilities in the United States experience peak demand between 4 and 8 P.M. on a hot summer day (see Figure 13.5). Demand billing allocates the costs of generating capacity, and charges are usually based on the maximum kW demand metered in any 15-minute period during a billing month.

An ideal utility customer would impose a constant electrical demand 24 hours a day. With a constant load, utilities could operate plant and distribution lines at full capacity, maximizing operating efficiency. However, electrical demand varies on a daily, weekly, monthly, and annual basis. Utilities offer discount rates for customers who take power during "off peak" periods.

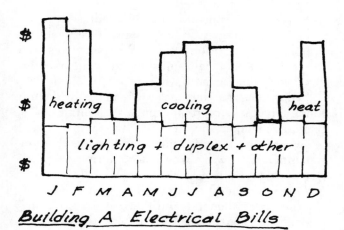

Building A Electrical Bills

FIGURE 13.3

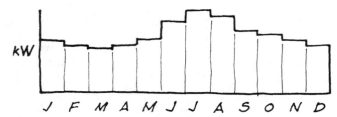

Demand ~ utility_year

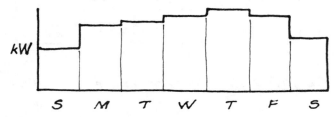

Demand ~ peak week

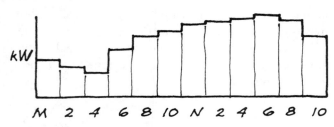

Demand ~ peak day

FIGURE 13.5

DEMAND COSTS

Commercial buildings are billed for demand plus consumption, while residential buildings are billed a higher rate for consumption only (see Figure 13.6). The following example illustrates commercial and residential billing.

Demand can be reduced by many of the same techniques that reduce energy consumption. However, if existing lighting, HVAC equipment, and controls are not upgraded, envelope improvements will probably *not* reduce electrical demand.

Billing Example

Commercial rate—3¢ per kWh plus $20 per kW demand. Store uses 15,000 kWh per month and has a peak demand of 35 kW.

monthly bill = $1,150

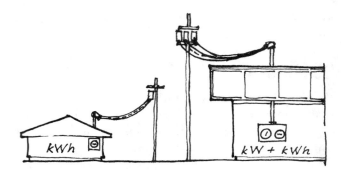

FIGURE 13.6

Residential rate—8¢ per kWh. Residence uses 1,500 kWh per month.

monthly bill = $120

Note: The average cost per kWh is nearly equal for these two customers. Envelope upgrades will be most effective in cutting utility costs for the home, while lighting and electrical upgrades will be most effective for the store.

13.3 REDUCE DEMAND kW

Utilities offer discounts to customers who reduce electrical demand during the peak load period. Programmed controls can cycle intermittent loads like refrigeration or water heating to prevent coincident operation. Thermal storage equipment makes ice at night and uses it to cool building air between noon and 8 P.M. the following day. Replacing inefficient equipment will also cut electrical demand.

- *Sequencing.* A control system can reduce demand in a cold storage warehouse with 10 refrigeration units serving 10 storerooms. When controls are set so that only eight units operate concurrently, refrigeration demand will be cut by 20%. Sequencing may increase the temperature swing in individual storerooms, but with a design storage temperature of −20°F the temperature swing caused by sequencing will be less than the temperature swing experienced during a typical storeroom loading cycle.
- *Delay.* Air conditioning is a large component of utility peak demand, and thermal storage can delay AC loads until off-peak night hours. Many utilities will pay for thermal storage equipment and discount electrical rates because thermal storage is cheaper than new generating

capacity. Bid proposals from sequencing control and thermal storage equipment suppliers frequently guarantee utility demand savings.

- *Equipment.* Replacing chillers, cooling towers, motors, lamps, and other equipment with more efficient units can reduce demand kW and consumption kWh.

DEMAND SAVINGS

With a current rate schedule and an accurate estimate of kW saved, estimating dollar savings resulting from reduced demand is easy. Review the following example and notice that demand savings can be used to re-calculate the *average* cost per kWh.

Billing Example	
Lighting	120 kW
Heating (gas)	0 kW
Chiller with pumps, and tower	80 kW
Air-handler fans	40 kW
Miscellaneous power	60 kW
Total demand	300 kW
Estimated consumption	100,000 kWh

Rate—3¢ per kWh plus $20 per kW

Monthly bill:

Demand	$6,000
Consumption	3,000
Monthly total	$9,000

Find monthly savings if peak demand is cut 50 kW by sequencing controls (consumption is unchanged).

50 kW at $20 saves $1,000

Update your energy analysis for this building by changing the average cost per kWh from 9¢ to 8¢.

13.4 REDUCE CONSUMPTION kWH

Reducing demand cuts utility costs, but sequencing and thermal storage don't cut energy consumption. Efforts in the following four opportunity areas can reduce energy demand AND consumption.

Efficient equipment. Motors and HVAC equipment, including fans, chillers, cooling towers, compressors, pumps, and controls.

Efficient lighting. Improved lamps and ballasts and reduced illuminance (the eye does not see illuminance; it adapts to field luminance).

Daylighting. Fenestration sized and located to maximize interior daylight with:

- Shading to exclude direct sun in summer
- Insulating covers or shades to minimize winter night heat losses

Efficient building envelope. Fenestration sized and located to minimize heat gains and losses, and maximize interior daylight. Sealing the envelope to minimize infiltration. Insulating the envelope to reduce heat gains and losses.

OTHER FORMS OF CONSERVATION
Cogeneration

Cogeneration is a term used to describe customers who sell power to their supplying utility. A law dating from the 1970s requires such purchases. The law's intent was increased efficiency of electrical generation through on-site use of waste heat. Its effect has been to encourage private investment in new generating capacity.

Cogeneration systems for institutional buildings have been used for many years. Building complexes such as colleges and prisons install generating equipment and use waste heat from the generation process for building heating and cooling. Efficient dual cycle gas-fired generating plants and utility deregulation make cogeneration less attractive when natural gas prices are low. Many schools installed similar generation and waste heat utilization equipment in the 1960s, calling their installations "total energy systems."

Photovoltaics

Converting sunlight to electrical energy with photo-voltaic cells is a viable way to provide small amounts of power in remote locations (railroad signals or sail-boat battery charging). Photovoltaics are not yet economically competitive with traditional building electrical power sources, but some future low-power applications are likely. Peak clear-day solar energy is about 1 kW per square meter, and peak efficiency for cells marketed in the 1990s was about 10%.

Motor Speed Control

Most electric motors operate at a constant speed called *synchronous speed*. With 60 Hz power the synchronous speed of an electric motor is determined by the number of pairs of motor poles as follows:

$$\text{SS (rpm)} = 3{,}600 \div \text{\# of pole pairs}$$

A motor with two poles will turn at 3,600 rpm, and one with six poles at 1,200 rpm.

When a motor is loaded, its speed decreases a bit, but motors try to run at synchronous speed. Two- and three-speed motors are made by using only part of the motor windings, but these motors are less efficient when operating at reduced speed.

Electronic motor speed control is expensive, but it will conserve energy when motor load varies.

REVIEW QUESTIONS

Use the following rate schedule to solve the following review problems.

Residential at 8¢ per kWh

Commercial at 3¢ per kWh plus $16 per kW

1. In March, a 1,500-square-foot home uses 1,000 kWh. Estimate the electric bill.
2. In July, the home in question 1 uses 2,000 kWh. Estimate the electric bill and the probable cause of the increase over May.
3. A home has a 240 volt, 100 ampere service. Find the home's potential peak demand (and the required transformer capacity).
4. A 15,000-square-foot commercial office building uses 35,000 kWh per month and has a peak demand of 150 kW. Estimate the monthly electric bill and the annual electric cost.
5. Find the average cost per kWh for the building in question 4.
6. The example office building used in the preceding chapter had a three-phase 120/208 volt, 1,200 ampere service. Estimate the monthly demand bill.
7. If the building service for the example office building in question 6 was changed to three-phase 277/480 volts, how many kW of demand would be saved?

8. The floor area of the example office building in question 6 is 21,600 sq. ft. Why is its demand charge so much higher than the building in question 4?
9. Get a 12-month utility bill summary for a building you are familiar with. Plot monthly dollars and then break out annual costs for lighting, HVAC, and other large electrical uses.

   ♦ If a lighting retrofit cuts the lighting load by 1 watt per square foot, calculate annual savings.
   ♦ If an HVAC retrofit cuts 20% (peak kW and kWh), calculate annual savings.

ANSWERS

1. $80
2. $160, probably air conditioning
3. Demand of 24 kW requires a 25 kVA transformer, but utilities allow 5 kW per home because of diversity so a 25 kVA transformer can serve five homes.
4. $3,450 per month, $41,400 per year
5. 9.9¢
6. $6,912 (1,200)(208)($\sqrt{3}$)($16)
7. None. Service amperes would be cut, but demand is unchanged.
8. Larger building, resistance heat, and high duplex load causes higher demand (432 kW instead of 150 kW).

PART III

HEATING, VENTILATING, AND AIR CONDITIONING

☐

CHAPTER 14
Terms, Comfort, and Psychrometrics

The best I have to offer you is the small size of the mosquitoes.

Basho

This is a small chapter, but it contains a surprisingly large quantity of information. The list of HVAC terms is the foundation of your ability to communicate effectively with professionals. The comfort section presents a brief overview of three ways that the human body maintains heat balance. Finally, the psychrometrics section introduces the interactions between heat, moisture, and air that affect comfort.

Start by reading through the terms twice. Read the definitions for R and HVAC three times because these terms are introduced in the psychrometrics discussion, but they are explained in greater detail later in the text. Plan to return to this list as you complete each section to review and reinforce the new terms covered in your readings.

As you study the topic of comfort, realize that it is quite subjective, and one person's ideal may be another's minimum. Fortunately, there is a general consensus concerning thermal comfort, and most building occupant needs can be met with a quality HVAC system.

The psychrometrics section introduces a variety of new terms and concepts. Psychrometric considerations impact efficient buildings in three significant ways:

1. They establish the air quantities and duct sizes required for a given building.
2. They determine the cooling capacity required to control moisture.
3. Psychrometric calculations are also used to predict condensation problems that may damage building components.

Most applications for psychrometric calculations are developed in later chapters, but make sure you understand and can use the psychrometric chart by solving the example questions that follow this chapter.

14.0 HVAC TERMS

Most professions use special terms (jargon) to describe the concepts and equipment they work with; this special language may also help ensure future income. To work and communicate effectively with building heating-cooling professionals, you must learn and speak their language. Reading this list before you begin the text and reviewing it after each chapter will establish a professional language foundation and permit you to communicate with the individuals who construct and equip your buildings.

Absorption   A refrigeration cycle where input energy is heat.

Adiabatic   Without gain or loss of heat.

Air conditioning   Controlling air temperature, humidity, and quality.

Air-cooled condenser   A heat exchanger that transfers building heat to outdoor air.

Air handler   Air-moving equipment that can change air temperature, humidity, and quality.

BTU   British thermal unit. Quantity of heat. One BTU will increase the temperature of 1 pound of water 1°F (1 Calorie = 4BTU).

BTU/H, BTUH   Rate of heat flow. BTU per hour.

C   The tendency to conduct heat for a standard thickness of material. *See* **U**.

CFM   Cubic feet per minute. Quantity of air.

Chiller   A heat exchanger where an evaporating refrigerant chills water.

Chill water   Water at about 45°F used in a heat exchanger to cool and dehumidify an air stream.

Coil   A heat exchanger (e.g., cooling coil).

Condensation   Change of state from vapor to liquid; heat is released.

Constant volume   HVAC equipment that controls room temperature by controlling the temperature of conditioned air.

Cooling tower   A heat exchanger that transfers building heat to outdoor air by evaporating water.

COP   Coefficient of performance. An efficiency rating for heating (cooling) equipment. BTU delivered (removed) divided by input BTU.

Deck   A heat exchanger (i.e., hot deck, cold deck).

Dehumidify   Remove moisture.

Dry-bulb temperature   Temperature indicated by a standard thermometer.

DX   Direct expansion. Equipment that uses an evaporating refrigerant to cool air.

EA   Exhaust air.

Economizer   Cooling with outdoor air.

Emissivity   Tendency of a surface to radiate heat.

Enthalpy   Total heat in an air/water vapor mixture.

Entropy   Tendency toward uniform inertness.

Evaporation   Change of state from liquid to vapor.

Evaporator   A heat exchanger where evaporating refrigerant cools an air stream.

FPM   Feet per minute. Velocity.

GPM   Gallons per minute. Quantity of fluid flow.

Grain   1/7,000 of a pound.

Head   Total mechanical energy in a fluid at a point in a piping system.

Heat exchanger   A device that transfers heat from one fluid to another.

Heat pump   A reversible refrigeration machine.

HSPF   Heating seasonal performance factor. The number of BTUs per watt delivered by a heat pump.

Humidify   Add moisture to air.

HVAC   Heating, ventilation, and air conditioning.

Hydronic   Equipment that circulates water to move heat.

Inches of water   Units for measuring pressure in fan and duct systems.

Infiltration   Air that leaks into a building.

k   Tendency to conduct heat for a 1-inch thickness of material. *See* **U**.

kW   Kilowatt. 1,000 watts of electrical energy. One kW = 3,400 BTU.

Latent heat   Hidden heat, absorbed or released when water changes state (liquid–vapor).

OA   Outside air.

Psychrometrics   Properties of air/water vapor mixtures.

Q   Rate of heat transfer.

R   Resistance. Tendency to resist heat flow. R = 1/C or 1/k. *See* **U**.

RA   Return air.

Refrigeration   Moving heat from a cool location to a warm location.

Relative humidity   A measure of water vapor held by air; 100% relative humidity is a saturated air/water vapor mixture.

SEER Seasonal energy efficiency ratio. A cooling efficiency rating for air conditioners and heat pumps. The number of BTU removed by each watt of energy input.

Sensible heat Dry heat, as measured by a dry-bulb thermometer.

Specific heat Number of BTU that will increase the temperature of 1 pound of a substance 1°F.

Sq. ft., sf Square feet area.

Strip heat Electric resistance heating.

Throw Horizontal distance an air stream travels.

Ton 12,000 BTUH cooling effect.

U The tendency of a construction to conduct heat. The number of BTU that will be conducted through one sq. ft. of a construction when the temperature difference across the construction is 1°F ($U = 1/R_t$).

Variable volume HVAC equipment that controls room temperature by controlling the quantity of conditioned air.

Ventilation Outside air brought into a building.

Watt A quantity of electrical energy. About 10 BTU are required to produce a watt at power plants. In a toaster or water heater, 1 watt yields 3.4 BTU. In an air conditioner or heat pump, 1 watt can move 8–16 BTU.

Wet-bulb temperature Temperature indicated by a thermometer when water is evaporating from a wet sleeve attached to the thermometer bulb. Wet-bulb temperature is an indicator of the total heat in an air/water vapor mixture.

Zone A group of similar spaces or rooms.

14.1 COMFORT

Comfort criteria for buildings have been developed for "Sedoc," an imaginary lightly clothed "sedentary occupant." (See Figure 14.1.) Surveys show most people prefer 70°F in winter and 75°F in summer, but authorities recommend 65°F in winter and 80°F in summer as comfortable and economical indoor temperatures. This text uses 70°F in winter and 75°F in summer to select heating and air-conditioning equipment. Occupants may then adjust indoor temperatures as desired for comfort or energy conservation.

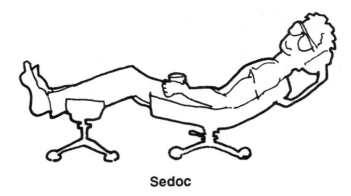

Sedoc

FIGURE 14.1

Physical comfort requires continuing dissipation of body heat by:

- Convection
- Radiation
- Evaporation

Convection is circulation of liquids or gasses caused by temperature difference. When air temperature is less than skin surface temperature, body heat can be lost by convection to the surrounding air. Increased air motion will increase convective heat losses.

Radiation is heat transfer by electromagnetic wave, from a warmer to a cooler surface. Body surfaces radiate heat to cooler surroundings and receive radiant heat from warmer surroundings. The magnitude of radiant heat flow is dependent on the temperature difference between source and receiver. Radiant heat flow situations in buildings often involve windows. Winter window surface temperatures can be 25°F below room air temperature, causing uncomfortable areas near windows.

Evaporation is a change of state from liquid to vapor. The human body continually dissipates heat by evaporation. Water vapor is expelled with each breath, and the evaporation rate can be increased by increased respiration or by perspiration. Humidity, the amount of water vapor in air, can affect comfort. However, the human body tolerates a wide range of humidity before becoming uncomfortable in very wet or very dry air.

Heat can also be lost by conduction (e.g., bare feet touching a cold floor), but most body heat losses are convective, radiant, or evaporative.

Some researchers spend a lifetime studying comfort, but exact results from such study are unlikely. Consider lightly clad skiers delighting in a brisk run

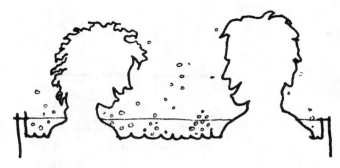

FIGURE 14.2

through cold air and later enjoying total immersion in a tub of hot water (see Figure 14.2). The temperature of surroundings affects comfort, but so do activity, attitude, and clothing.

14.2 PSYCHROMETRICS

The psychrometric chart in Figure 14.3 describes air/water vapor mixtures; it can be used to predict condensation problems and to calculate HVAC capacity. The chart shown here is applicable to standard atmospheric pressure at mean sea level. However, very little error is introduced for elevations up to about 2,000 feet above mean sea level. Seven characteristics of air/water vapor mixtures are noted on the chart. If any two unrelated characteristics are known, the other five can be read.

1. **Dry-bulb temperature** The air temperature (°F) indicated by a standard thermometer.
2. **Percent relative humidity** The amount of water vapor held in the air as a percent of the maximum amount of water vapor the air can hold *at a specific temperature* (warmer air holds more water vapor).
3. **Wet-bulb temperature** The temperature (°F) indicated by a thermometer with a wet wick attached to its bulb; the wick is located in a moving air stream to encourage evaporation.
4. **Humidity ratio** The weight of water vapor held in 1 pound of dry air (weight is measured in grains; 1 grain = 1/7,000 of a pound).
5. **Dew point** The air temperature at which condensation begins.
6. **Enthalpy** The total heat contained in an air/water vapor mixture (BTU per pound of dry air).
7. **Specific volume** The number of cubic feet that 1 pound of air occupies.

EXAMPLES

Use Figures 14.3 and 14.4 to:

1. Find the characteristics of an air/water vapor mixture at 75°F and 50% relative humidity.
2. Find the characteristics of an air/water vapor mixture at 95°F dry bulb, and 80°F wet bulb.

Characteristic	Example 1	Example 2
Dry-bulb temperature °F	**75**	**95**
Percent relative humidity	**50**	53
Wet-bulb temperature °F	63	**80**
Humidity ratio (grains)	65	132
Dew point °F	55	75
Enthalpy (BTU/lb)	28	44
Specific volume (cubic feet/lb)	13.7	14.4

Notice that the 95° air/water vapor mixture contains twice as much water as the 75° mixture, but its relative humidity is nearly the same because warmer air can hold more water.

SENSIBLE HEAT, LATENT HEAT, AND ENTHALPY

HVAC equipment must deal with two kinds of heat. Sensible heat is the dry heat in air as measured by a dry-bulb thermometer. Latent heat is the heat held in water vapor. Heating equipment may include humidifiers that add moisture to dry winter air, and a large part of summer cooling energy is used to remove moisture from air. Enthalpy, the total heat in an air/water vapor mixture, is the sum of sensible plus latent heat.

The heating or cooling capacity required to produce sensible and latent changes in an air/water vapor mixture can be accurately predicted using data from the psychrometric chart. The following formulas apply and will be developed in the next chapter.

$$Q_{sensible} = (CFM)(1.08)(TD) \qquad (1)$$
$$Q_{latent} = (CFM)(0.68)(GD) \qquad (2)$$
$$Q_{total} = (CFM)(4.5)(ED) \qquad (3)$$

where

Q = rate of heat transfer in *BTUH*

CFM = air quantity in cubic feet per minute

TD = temperature difference in °F

GD = water vapor difference in grains per pound of dry air

ED = enthalpy difference in *BTUH* per pound of dry air

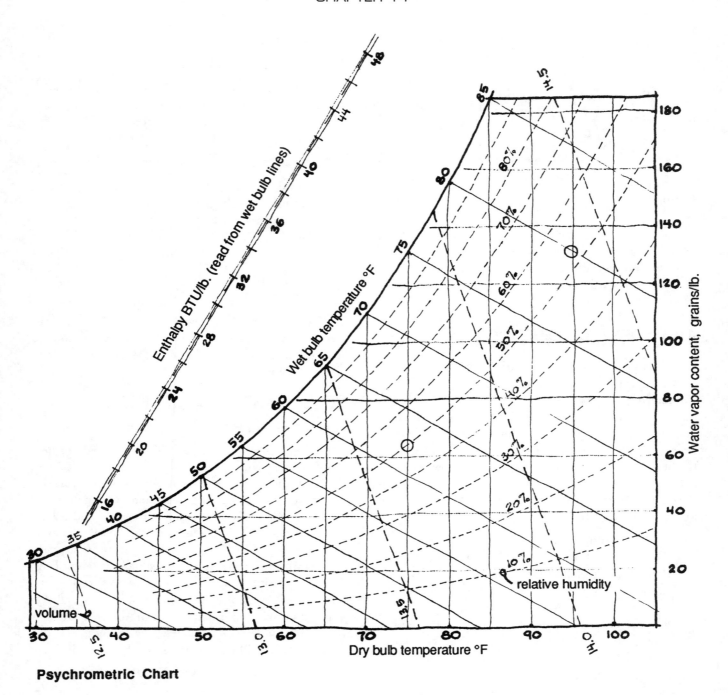

Psychrometric Chart

FIGURE 14.3

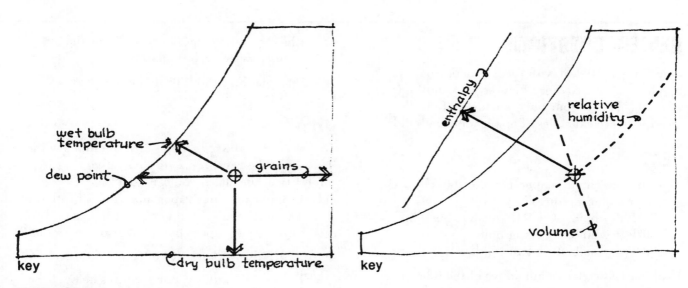

FIGURE 14.4

CONDENSATION

When a building component's surface temperature is below the dew point of the surrounding air/water vapor mixture, condensation can occur. Such condensation is a common winter problem on single-glass windows. In the next chapter you'll learn to calculate the surface temperature of windows and other building components so that condensation problems can be anticipated and corrected.

VAPOR BARRIERS

Many building materials are permeable; that is, water vapor can pass through them. Masonry, wood, sheet rock, and many insulations are permeable. Of course, water vapor will pass through any of the small cracks, joints, or holes that often penetrate a building's skin.

Cold air cannot hold as much water vapor as warm air. The driving force for water vapor movement is vapor pressure difference, and cold outdoor air usually has a lower vapor pressure than heated indoor air. When water vapor passes through a wall or roof, it will condense if it contacts a material whose temperature is below its dew point. Condensation inside a wall or roof construction can cause cosmetic and structural damage.

Vapor barriers include less permeable materials such as aluminum foil, plastic film, or waxed paper. They are installed on or near the **warm side** of a construction where they prevent vapor penetration and minimize internal condensation problems. Winter conditions usually govern building vapor barrier location, except in tropical climates or refrigeration work.

REVIEW QUESTIONS

Questions 1–10 deal with reading data from the psychrometric chart (see Figures 14.3 and Figure 14.5), 11–17 test your understanding of the chart, and 18–20 are examples of applications in buildings.

READ

Find the dew point temperature for the following air/water vapor mixtures.
1. 75°F dry bulb, 60% RH (relative humidity)
2. 80°F dry bulb, 70°F wet bulb
3. 90°F dry bulb, 30% RH

Find the RH (relative humidity) of the following air/water vapor mixtures.

4. 70°F dry bulb, 50°F wet bulb
5. 90°F dry bulb, 80°F wet bulb

Find the water vapor content in grains per pound for the following air/water vapor mixtures.

6. 95°F dry bulb, 40% RH
7. 70°F dry bulb, 70°F wet bulb

Find the enthalpy in BTU per pound for the following air/water vapor mixtures.

8. 85°F dry bulb, 65°F wet bulb
9. 100°F dry bulb, 10% RH

Find the dry-bulb temperature in °F for the following air/water vapor mixture.

10. 65°F wet bulb, 30 BTU/lb enthalpy

THINK

Try to answer the following seven questions without looking at the psychrometric chart; if you understand the chart, you can do it.
11. The two air/water vapor mixtures below have the same relative humidity. Which has the larger water vapor content in grains per pound?

   90°F db, 40% RH; or 80°F db, 40% RH

12. The two air/water vapor mixtures below have the same dry-bulb temperature. Which one has the higher relative humidity?

   80°F db, 60°F wb; or 80°F db, 50°F wb

13. The two air/water vapor mixtures below have the same relative humidity. Which one has the lower dew point temperature?

   70°F db, 50% RH; or 80°F db, 50% RH

14. The two air/water vapor mixtures below have the same dew point temperature. Which one has the higher enthalpy?

   80°F db, 50°F dp; or 90°F db, 50°F dp

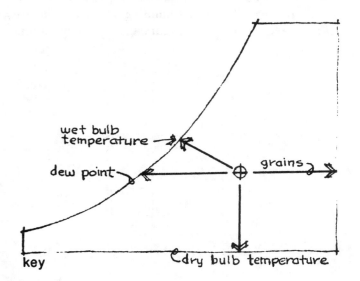

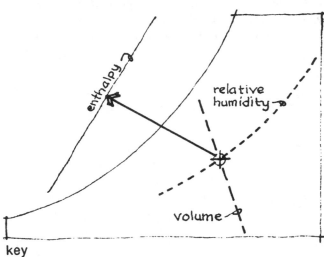

FIGURE 14.5

15. The two air/water vapor mixtures below have the same dew point temperature. Which one has the greater moisture content in grains?

    80°F db, 50°Fdp; or 90°F db, 50°F dp

16. The two air/water vapor mixtures below have the same wet-bulb temperature. Which one will have the lower dew point temperature?

    90°F db, 70°F wb; or 80°F db, 70°F wb

17. Which air/water vapor mixture will occupy the larger volume?

    90°F db, 70°F wb; or 80°F db, 70°F wb

APPLICATIONS

These questions are similar to the preceding 17, but they are stated as actual building applications.

18. The air supply for heating a room can be provided at 90°F db, 40% RH **or** 95°F db, 25% RH. Which air supply will heat the room with the lowest quantity of supply air (and the smaller duct system)?

19. Identical buildings will be constructed in Phoenix and New Orleans; both will have separate air-conditioning units to cool outdoor air for ventilation purposes. If summer air in Phoenix is 105°F db, 20% RH, and summer air in New Orleans is 90°F db, 60% RH, which building requires larger cooling capacity?

20. The surface temperature of an uninsulated steel duct carrying conditioned air is 55°F. If the duct passes through a space where the air surrounding it is 78°F db, 50% RH, will moisture condense on the surface of the duct?

ANSWERS

1. 60°F
2. 65°F
3. 55°F
4. 20%
5. 65%
6. 100 grains
7. 110 grains
8. 30 BTU/lb
9. 29 BTU/lb
10. No answer; both givens read on the same line.
11. 90°F dry bulb, 40% RH, because warm air holds more moisture.
12. 80°F db, 60°F wb
13. 70°F db, 50% RH
14. 90°F db, 50°F dew point
15. Grains of water vapor are the same for both.
16. 90°F db, 70°F wb
17. 90°F db, 70°F wb
18. 90°F db, 40% RH. This air supply has the higher enthalpy (total heat), and therefore less of it is required to heat the example room.
19. New Orleans. Higher enthalpy in 90°F db, 60% RH air means more heat to be removed.
20. Yes. The duct surface is below the dew point temperature of the surrounding air.

CHAPTER 15
Heat Loss and Gain

It is nice to read news that our spring rain also visited your town.

Onitsura

When you complete this chapter, you should be able to calculate heat loss and heat gain for residential or commercial buildings. This ability will enable you to size heating and cooling equipment. In Chapter 19 it will be used again as the basis of annual heating and cooling cost estimates.

Begin by learning about R values, U values, infiltration, and ventilation. These four factors are part of all heat loss and heat gain calculations, so read to understand what they are and where to use them.

Next learn the two simple formulas that quantify building heat loss. The first deals with heat flow through a building's skin, and the second concerns the outdoor air. A form is used to help organize heat loss calculations. Review it and then study the examples until you understand all the entries.

The heat gain analysis is done on the same form as heat loss because many aspects of both calculations repeat. However, summer heat gains differ from winter heat losses because they include added calculations covering solar, latent, and internal heat sources. Read to understand these differences and the five simple formulas that quantify them.

Finally, verify your new skills by doing the questions at the end of the chapter and by completing a new heat loss–heat gain summary form for a building of your choice.

The residence and the office building used for the example calculations will be reused in later chapters as discussion vehicles for equipment selection and annual operating cost estimates.

15.0 HEAT FLOW

Heat flows downhill like water. "Downhill" for heat means from high temperature to low temperature. Temperature difference is the driving force for heat flow; greater temperature difference between regions means greater heat flow between them. Three types of heat flow can cause building heat loss or gain: conduction in solids, convection in fluids, and radiation between warm and cool objects. Heat is measured in **BTUs** (British thermal units). A BTU is defined as the quantity of heat required to change the temperature of 1 pound of water 1°F. Heat flow in these units is expressed at an hourly rate in BTUH.

INSULATION

A Thermos bottle is a good example of a device that resists all three types of heat flow. The only path for conducted heat flow is across the neck of the bottle. A vacuum between the walls prevents convective flow, and the reflective wall surfaces reduce heat transfer by radiation.

Insulation slows heat flow but does not stop it (see Figure 15.1). Most building insulation uses many internal air spaces to limit conducted heat flow; polyurethane, fiberglass, and mineral wool are examples. Aluminum foil facing an air space can also be used as building insulation.

R IS RESISTANCE

Resistance values are an index of a material's tendency to resist heat flow. Larger R values mean more resistance.

R values are used as a starting point for calculating the rate of heat flow through a construction.* If

drink fast
or
add R

FIGURE 15.1

you are evaluating heat flow through a wall made of brick and block, you begin by adding R(brick) plus R(block) to get R(wall). Since the wall is not in a vacuum, you will add a bit more R for air films that naturally form on wall surfaces.

U IS CONDUCTANCE

The reciprocal of the sum of the R values in a construction is called U.

$$U = \frac{1}{R_1 + R_2 + R_3 + \cdots + R_n} \tag{1}$$

U is an index of a construction's tendency to conduct heat. U is defined as the number of BTUs that will be conducted through 1 square foot of a construction in one hour when there is a 1°F temperature difference driving heat flow.

$$U = (BTUH)(1 \text{ ft}^2)(1°F) \tag{2}$$

Calculate U for the wall shown:

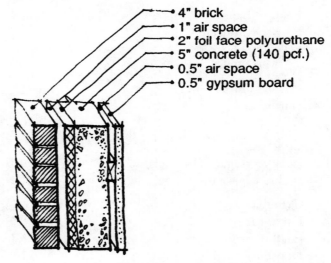

- 4" brick
- 1" air space
- 2" foil face polyurethane
- 5" concrete (140 pcf.)
- 0.5" air space
- 0.5" gypsum board

Rs		
	Air film outside	0.2
	4" brick	0.4
	Air space (facing foil)	3.0
	Polyurethane (2 × 6)	12.0
	Concrete (5 × 0.1)	0.5
	Air space	1.0
	Gypsum board (0.5 × 1)	0.5
	Air film inside	0.7
	R total	18.3
U value	U = 1/Rt = 1/18.3 = 0.054	

*"Construction" is used to describe walls, ceilings, floors, etc., which may include several materials, air spaces, or air films.

Do not confuse resistance to heat flow with heat storage capacity. Concrete is a poor insulator, but it has much more heat storage capacity than an equal volume of fiberglass. Heavy materials like concrete or masonry can be used to store heat and minimize indoor temperature changes from day to night.

15.1 R AND U VALUES

R VALUES FOR MATERIALS

The following are 1-inch thick unless noted otherwise.

Insulation		R per inch
Cellular glass		3
Cellulose		3–4
Mineral wool or fiberglass		3–4
Perlite		3
Polystyrene		4–5
Polyurethane		6
Polyisocyanurate		8
Vermiculite		2

Wood		
Plywood or softwoods		1.2
Hardwoods		0.9
OSB		0.9

Concrete		
Rock aggregate (140 pcf)		0.1
Perlite aggregate (40 pcf)		1.1

Other		
Earth (dry)		0.2
Earth (wet)		0.1
Gypsum board		1.0
Glass, steel, metals		Negligible

Masonry	R for thickness listed	
Brick	4"	0.4
Concrete block	4"	0.7
(sand and	8"	2.0
gravel aggregate)	12"	2.3
Stucco or plaster	1"	0.1

Roofing	R for usual thickness
Shingles (asphalt)	0.4
Shingles (wood)	0.8
Built-up roof (tar and gravel)	0.3
Metal	Negligible

Air films	R for usual thickness
Outside building	0.2
Inside building	0.7

Air spaces	R for nominal 1" space
Ordinary materials facing air space	1.0
Aluminum foil facing air space	3.0

U VALUES FOR CONSTRUCTIONS

U Values include air films

Walls	
3 1/2" wood studs, with exterior-type gypsum sheathing on both sides and no insulation	0.3
As above with R-11 insulation added	0.1
8" concrete block (sand and gravel aggregate)	0.5
8" concrete block with core insulation	0.3

Roofs	
No insulation	0.7
R-6 insulation	0.2
R-11 insulation	0.1

Windows	
Single glass	1.1
Single glass—low emissivity	0.9
Double glass	0.6
Double glass—low emissivity	0.3 to 0.4

Skylights	
Single glass or acrylic	1.2
Double glass or acrylic	0.7

Doors	
1 1/2" solid core wood	0.5
As above with storm door added	0.3
1 1/2" steel door with urethane core	0.2

Slab edge (heating ducts in slab)	
No insulation	1.0
1" insulation	0.4
2" insulation	0.2

Slab edge (no heating ducts in slab)	
No insulation	0.8
1" insulation	0.3
2" insulation	0.1

Note: Values given here are approximate (see Figure 15.2). They will change slightly with variations in material density, emissivity, temperature, and direction of heat flow. Try estimating a few R values for materials not listed here, and then check your esti-

FIGURE 15.2

mates on pages 312–315. In the next section, R values are used to predict internal temperatures for an example wall.

INTERNAL TEMPERATURE

The temperature inside a wall or roof is an important consideration for vapor barrier location. Because insulation delays heat flow, the internal temperature distribution is not uniform; instead, it varies in proportion to a construction's R value. The following example wall and R values are repeated from the previous page. They will be used to illustrate temperature distribution within the wall.

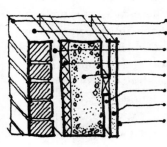

Material	R
Air film outside	0.2
4″ brick	0.4
Air space facing foil	3.0
2″ polyurethane	12.0
5″ concrete	0.5
Air space	1.0
1/2″ gypsum board	0.5
Air film inside	0.7
R total	18.3

Example

Assume indoor temperature is 70°F and outdoor temperature is −10°F. The total temperature difference from inside to outside is 80°F. Begin by finding the temperature of the inside surface of the gypsum board wall.

The only thermal resistance (R) between the 70°F indoor air and the gypsum board surface is the inside air film with an R value of 0.7. The tempera-

ture difference between the indoor air and the gypsum board surface can be calculated as follows:

1. Multiply the indoor to outdoor temperature difference by 0.7 and then divide the total by the wall's total R value.

$$(80)(0.7) / 18.3 = 3°F$$

2. Since 3°F is the difference between the indoor air and the gypsum surface, the gypsum surface temperature is 70°F − 3°F = 67°F.

Now calculate the two missing in-wall temperatures that are marked (?) below.

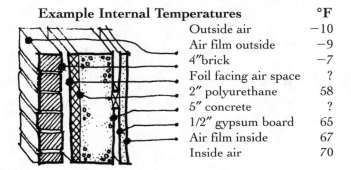

Example Internal Temperatures	°F
Outside air	−10
Air film outside	−9
4″ brick	−7
Foil facing air space	?
2″ polyurethane	58
5″ concrete	?
1/2″ gypsum board	65
Air film inside	67
Inside air	70

Surface temperatures can affect occupant comfort as well as vapor barrier placement. If the interior surfaces of a room are much cooler than room air, people will feel uncomfortable because of increased radiant heat loss from the skin to cooler surrounding surfaces. Cold windows frequently cause such comfort problems. The academically correct description for this phenomenon is low MRT (mean radiant temperature).

15.2 INFILTRATION AND VENTILATION

Building air quality is maintained by introducing a continuous supply of "fresh" outdoor air. Reducing the fresh air supply will save heating and cooling dollars, but reductions may cause health problems and encourage "sick building" lawsuits. Current "fresh air" recommendations are 15 CFM (cubic feet per minute) per person or more for most building occupancies. Most homes rely on infiltration for fresh air, while most commercial buildings are ventilated.

INFILTRATION

Infiltration describes air that leaks into a building because of construction quality, wind pressure, and temperature difference. Residences and small buildings often rely on infiltration to provide all necessary outdoor air. Special requirements such as cleaning, painting, or cooking odors are taken care of by opening doors and windows. Quantifying building infiltration is difficult because of the many variables involved. The estimates in Table 15.1 can be used in the absence of better information.

One air change per hour means that all the air in a building is replaced by outdoor air every hour. Since fan ratings and heating-cooling formulas are based on CFM, air change rates must be converted to CFM, as shown in the following example.

Example

A room measures 9'×12'×8', and 1.5 air changes per hour are expected. Find outdoor air CFM for the room. Use the following formula:

$$OA = \frac{(\text{Volume})(\text{ACH})}{60} \qquad (3)$$

where:

OA = Outdoor air in CFM
Volume = Volume of the room in cubic feet
ACH = Rate of air change per room

$$OA = \frac{(9 \times 12 \times 8)(1.5)}{60} = 21.6 \text{ CFM}$$

TABLE 15.1

Building Description	Air Changes per Hour	
	Winter	Summer
No insulation, many operable windows and doors	3.0	2.0
Insulated walls and roof, built between 1945 and 1972	1.5	1.0
Well insulated and sealed, a residence built after 1972 or a commercial building without operable windows	0.75	0.5

Winter air changes are 50% larger than summer changes because of wind and air temperature differences.

TABLE 15.2

Ventilation Recommendations

Allow 15 CFM per person for most spaces but increase or adjust as noted below.

Space Description	CFM/Person
Conference rooms	20
Dining rooms	20
Lobbies	20
Offices	20
Beauty shop	25
Hospital, patient room	25
Public rest rooms	50

Space Description	CFM/sq. ft.
Warehouse	0.05
Retail store	0.20–0.30
Locker rooms	0.50
Pet shops	1.00

Where smoking is permitted, allow 40 to 50 CFM/person. In hotels, allow 30 CFM for bedroom plus 35 CFM for bath.

VENTILATION

Ventilation refers to air brought into a building by fans or designed apertures. In large buildings with few operable windows, ventilation is necessary to ensure healthy and odor-free conditions. The quantity of ventilation air brought into a building is determined by the number of occupants and their activities. The recommendations in Table 15.2 can be used in the absence of better information to estimate required ventilation CFM. Use the maximum number of people expected to occupy the room or building as a multiplier (see Figure 15.3).

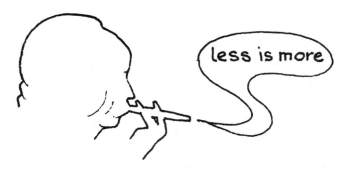

FIGURE 15.3

INFILTRATION OR VENTILATION?

Estimate infiltration and ventilation air quantities for a 10,000 square foot school classroom building built in 1954. The conditioned space is 12 feet high, and the total population is 320 students and teachers.

Infiltration CFM. Estimate 1.5 air changes in winter and 1 air change in summer.

$$\frac{(12)(10,000)(1.5)}{60} = 3,000 \text{ CFM in winter}$$

$$\frac{(12)(10,000)(1.0)}{60} = 2,000 \text{ CFM in summer}$$

Ventilation CFM. Estimate 15 CFM per person.

$$(15)(320) = 4,800 \text{ CFM all year}$$

Allow for heating and cooling capacity to satisfy the *larger* fresh air requirement—4,800 CFM of ventilation air for the classroom building.*

EFFICIENCY AND OUTDOOR AIR

Because it's expensive to heat or cool outdoor air, efficient buildings conserve energy by:

1. Minimizing outdoor air quantities.
2. Using heat exchangers to temper outdoor air.
3. Using an "economizer" instead of refrigeration equipment.

Minimize

Minimum outdoor air means minimum heating and cooling costs. Tight buildings with carefully controlled ventilation rates can save heating and cooling energy. Appropriate design strategies include quality construction and provision for reduced outdoor air on nights or weekends or whenever building occupancy drops.

Heat Exchangers

When outdoor air is brought into a building, it displaces conditioned air. Efficient buildings include heat exchangers to transfer heat between exhaust air and incoming outdoor air. Heat exchangers permit build-ing operators to use outdoor air for odor control without paying the full price of heating or cooling it.

Economizer

Many large buildings are heat rich. The heat from lights, people, and computers can exceed building heating requirements on moderate winter days. Such buildings operate air-conditioning equipment in the winter to remove internal heat.

Economizer describes air intakes and dampers that permit the use of outdoor air instead of refrigeration for building cooling when outdoor temperature conditions are right. Even better is an enthalpy-controlled economizer that evaluates both outdoor temperature and humidity, and then mixes appropriate quantities of outdoor and indoor air to achieve comfortable conditions without refrigeration.

15.3 ESTIMATING BUILDING HEAT LOSSES

A building's largest or "peak" heat loss is calculated to determine the size of heating equipment. Peak heat loss values will also be used in Chapter 19 to estimate annual operating costs and savings potential for energy-conserving alternatives. Heat leaves a building by only two routes: the building's skin conducts heat to colder outdoor surroundings, and cold outdoor air replaces heated building air.

CONDUCTED HEAT LOSSES

All parts of a building's skin lose heat to colder surroundings, and the rate of heat loss is determined by the U value of the skin. U is the number of BTUH conducted through 1 sq. ft. of a construction when a 1°F temperature difference drives heat flow. Conducted heat losses are given by the formula:

$$Q = (U)(A)(TD)$$

where:
Q = Quantity of heat flow in BTUH
U = Conductance value in BTUH per sq. ft. per °F
A = Area in sq. ft.
TD = Temperature difference in °F between indoor and outdoor

Each construction (wall, roof, etc.) with a unique area and U value is calculated separately, and then the

---

*Ventilation fans pressurize the building and thus overcome anticipated infiltration. More conservative designers add infiltration and ventilation CFM to allow for exhaust fans and door openings.

losses are summed to establish total building conducted heat losses. Since peak heat loss is used to size heating equipment, the outdoor temperature is selected for a typical cold winter night.

OUTDOOR AIR HEAT LOSSES

Cold winter air leaks into a building through doors, windows, and many other small openings found in the skin of a typical building. Well-sealed buildings constantly exhaust indoor air and replace it with outdoor air to maintain air quality. To ensure comfortable indoor conditions, heating equipment must be sized with adequate capacity to heat this air. The following formula establishes the magnitude of heat losses caused by infiltration or ventilation air:

$$Q_{sensible} = (CFM)(1.08)(TD)$$

where:

Q	= Quantity of heat flow in BTUH
CFM	= Quantity of outdoor air brought into a building by infiltration and/or ventilation in cubic feet per minute
TD	= Temperature difference in °F between indoor and outdoor
1.08	= Number of BTUH required to increase the temperature of 1 CFM of air by 1°F*

Heat losses calculated by this formula are added to conducted heat losses through the building skin to establish total building heat loss.

Conducted Examples

1. Find the heat loss through a 200 sq. ft. window if its U value is 1.1, the indoor temperature is 70°F, and the outdoor temperature is 10°F.

$$Q = (U)(A)(TD)$$
$$(1.1)(200)(70-10) = \textbf{13,200 BTUH}$$

---

*The constant 1.08 permits an answer in BTUH, although fresh air quantity is in CFM.

1 CFM of air flow totals 60 cubic feet per hour, or 4.5 pounds per hour. Air has a specific heat of 0.24 BTU, so 0.24 BTU will increase the temperature of 1 pound of air 1°F. Therefore, (4.5)(0.24) = 1.08 BTUH are required to change the temperature of 1 CFM (60 CF/hr.) 1°F.

2. Find the heat loss through a 1,400 sq. ft. wall if its U value is 0.1, the indoor temperature is 70°F, and the outdoor temperature is 0°F.

$$Q = (U)(A)(TD)$$
$$(0.1)(1,400)(70-0) = \textbf{9,800 BTUH}$$

3. Find the heat loss through a 12,000 sq. ft. roof if its U value is 0.15, the indoor temperature is 70°F, and the outdoor temperature is 10°F.

$$Q = (U)(A)(TD)$$
$$(0.15)(12,000)(70-10) = \textbf{108,000 BTUH}$$

Outdoor Air Examples

4. A building has an expected infiltration rate of 400 CFM. Find the BTUH heat loss when the indoor temperature is 70°F and the outdoor temperature is −10°F.

$$Q_{sensible} = (CFM)(1.08)(TD)$$
$$(400)(1.08)(70-\{-10\}) = \textbf{34,560 BTUH}$$

5. A building has an expected infiltration rate of 4,000 CFM *and* a ventilation rate of 6,000 CFM. Find the expected BTUH heat loss when the indoor temperature is 70°F and the outdoor temperature is 0°F.
   Use the largest value for infiltration *or* ventilation (i.e., 6,000 CFM here).

$$Q_{sensible} = (CFM)(1.08)(TD)$$
$$(6,000)(1.08)(70-0) = \textbf{453,600 BTUH}$$

FOUR SPECIAL CASES OF HEAT LOSS

Three cases don't fit the preceding heat loss formulas. They are ducts outside conditioned spaces, basement walls and floors, and insulated but unheated spaces adjacent to conditioned spaces. A fourth case, slab edge losses, will fit with an allowance.

Ducts Outside

Heating ducts outside the conditioned space—in a vented attic, for example—can lose a lot of heat (see Figure 15.4). Often they carry 130°F air in an attic where the ambient temperature can be below 0°F, yet duct insulation may be only 1 inch thick. Smart designers locate ducts inside the insulated zone; others add 10% to the sum of all other building heat losses to account for increased heat loss by the duct system.

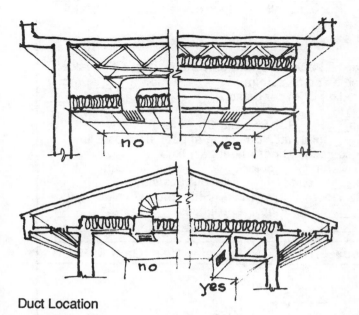

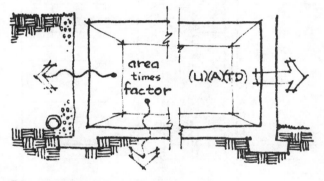

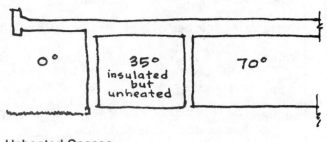

Unheated Spaces

FIGURE 15.6

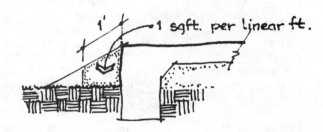

Duct Location

FIGURE 15.4

Slab Edge

FIGURE 15.7

Basement Walls and Floors

FIGURE 15.5

Basement Walls and Floors

Basement walls and floors in contact with the earth benefit from the earth's heat storage capacity and its resistance to heat flow (see Figure 15.5). To calculate heat loss through basement walls or floors (or slabs on grade), multiply the wall or floor area by the factor given in Table 15.3.

TABLE 15.3

Design Temperature	Factor BTUH per sq. ft.
Over 10°F	2
Under 10°F	4

Note: For basement walls not in contact with earth, use the equation (U)(A)(TD) 5 BTUH.

Unheated Spaces

An insulated but unheated space adjacent to a heated space will delay heat loss (see Figure 15.6). Spare rooms, closed attics, and enclosed crawl spaces are examples. In these cases, use the equation $Q = (U)(A)(TD)$, but reduce the TD by 50% to account for reduced heat flow.

Slab Edge

The formula $Q = (U)(A)(TD)$ can be used to calculate heat loss through the exposed edge of concrete floor construction (see Figure 15.7). Allow 1 square foot of area for each linear foot of slab edge.

15.4 CALCULATING BUILDING HEAT LOSS

The heat loss–heat gain summary form can be used to quantify peak heat loss (and later peak heat gain). Examine the form as you read this section; then work through the following examples. Heat loss calculations can be used to select heating equipment and to estimate annual heating costs.

_____   _____   _____   _____   _____
project name   location   floor area sqft.   calculated by   date

Design Conditions: **Project Conditions:**

winter   Outdoor Air; calculate 1&2 below but use only the **largest CFM** value:
____ °Fdb indoor   1. Infiltration based on air change rate_____winter _____summer
____ °Fdb outdoor   2. Ventilation based on CFM per person _____
____ TD
summer   Glass _____U value_____SC (Shading Coefficient)
____ °Fdb outdoor   Lighting _____ total watts operating at peak heat gain time
____ °Fdb indoor   People _____total occupants at peak heat gain time
____ TD   Ceiling-Roof _____U value _____color _____weight (lbs/sqft)
____ °Fwb outdoor   Walls _____U value _____color _____weight (lbs/sqft)
____ % RH indoor   Floor _____U value, Door _____U value, Slab Edge U value _____
____ GD   Equipment _____watts or hp, Appliances _____
____ time of peak gain   Other _____

Item Quantities	Winter Heat Loss = BTUH	Summer Heat Gain = BTUH
Outdoor Air		
winter _____CFM	(1.08)(___TD) = _____	
summer _____CFM		(1.08)(___TD) = _____
		(0.68)(___GD) = _____
Glass total _____sqft	(___U)(___TD) = _____	(___U)(___TD) = _____
N _____sqft		(___SF)(___SC) = _____
E _____sqft		(___SF)(___SC) = _____
S _____sqft		(___SF)(___SC) = _____
W _____sqft		(___SF)(___SC) = _____
HOR _____sqft		(___SF)(___SC) = _____
Lighting _____watts		(3.4) = _____
People _____#		(___sens) = _____
		(___latent) = _____
Ceiling-Roof_____sqft	(___U)(___TD) = _____	(___U)(___ETD) = _____
Walls _____sqft	(___U)(___TD) = _____	(___U)(___ETD) = _____
Floor bsmt _____sqft	(___factor) = _____	
slab _____sqft	(___factor) = _____	
crawl space _____sqft	(___U)(___TD) = _____	
above grade _____sqft	(___U)(___TD) = _____	
Slab Edge _____linft	(___U)(___TD) = _____	
Doors _____sqft	(___U)(___TD) = _____	(___U)(___TD) = _____
Equipment _____watts		(3.4) = _____
_____hp		(2500) = _____
Appliances _____		(___sensible) = _____
		(___latent) = _____
Other _____	_____	_____

Subtotals _____   _____

If ducts are outside the conditioned space add 10% _____   _____

TOTAL BTUH heat loss _____   heat gain _____

Check; heat loss = 20-60 BTUH persqft. (south to north); heat gain = 15-60 BTUH per sqft. (north to south). Allow 4% of area served for fan rooms, and 2% of gross building area for central plant equipment.

DESIGN CONDITIONS: PROJECT CONDITIONS

Select winter outdoor temperature from the map in Figure 15.8 or refer to the current *ASHRAE Handbook of Fundamentals* for more precise data. The following example uses a winter indoor temperature of 70°F, but 65°F will conserve energy.

Fill in the appropriate project conditions blanks for the building you are evaluating.

QUANTITIES

Calculate and enter the quantities indicated in the left-hand vertical column of the form.

Outdoor air: Use the *larger value* for infiltration or ventilation. Refer back to pages 202 and 203 to estimate infiltration and ventilation CFM.

Glass: For heat loss, you need only the total glass area (subdivide this area into appropriate orientations for later calculations of heat gain). Enter skylights as "HOR" (horizontal) glass.

Lighting: Calculate peak heat loss without lighting; doing so ensures adequate heating capacity to warm the building on a cold night.

People: Calculate peak heat loss without occupants because they may arrive late on a cold night.

Ceiling-roof, walls: Remember to deduct window and skylight areas. If your building has a variety of roof or wall constructions, develop an average U value.

Floor: Use only the appropriate construction type.

Slab edge: Use linear feet here.

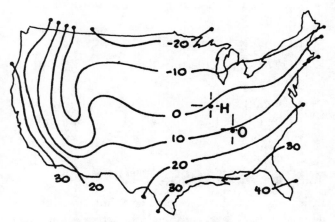

Winter design temperature °F (dry bulb)

FIGURE 15.8

CALCULATIONS

After entering quantities, insert the appropriate U, TD, or factor values in the winter heat loss column and complete the multiplications indicated. Your BTUH total is the minimum size needed for the heating plant, and it can be used to estimate annual heating costs. For typical buildings in the United States, the total heat loss should range between 20 and 60 BTUH per square foot of heated floor area (lower BTUH in the South and higher in the North).

15.5 HOUSE EXAMPLE, HEAT LOSS CALCULATION

DESIGN CONDITIONS

Winter temperature is 0°F (for location "H" in Figure 15.8). An indoor temperature of 70°F is assumed; temperature difference (TD) is 70.

PROJECT CONDITIONS

Outdoor air: The example house in Figure 15.9 was built in 1966 and has many windows. Estimate 1.5 air changes in winter; then estimate a minimum ventilation rate of 15 CFM per person (p. 202).

Glass: Windows are single glass. U = 1.1 (p. 200).

Ceiling-roof: Detail in Figure 15.10. U = 1/R total = 0.05.

$$[1/(0.7 + 19 + 0.6 + 0.7) = 1/21 = 0.05]$$

Note that R values for the plywood roof deck and the metal roofing were not included in the U calculation above because cold outside air will occupy the vented attic space between the roof deck and the top of the insulation. Also note that an air film R value of 0.7 was used for both air films. The R value of 0.2 for the outdoor air film is based on wind speed, and wind penetration of the vented attic space is not expected (p. 200).

Wall: See detail in Figure 15.10. U = 1/R total = 0.07.

$$[1/(0.2 + 0.1 + 1 + 11 + 0.5 + 0.7) = 1/13.5 = 0.07]$$

The R value of 1 is for the air space created by the 1 × 4 wood stripping (p. 200).

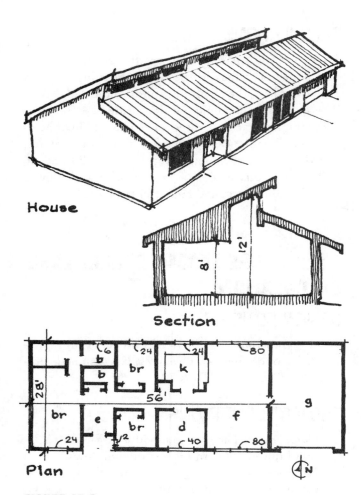

House

Section

Plan

FIGURE 15.9

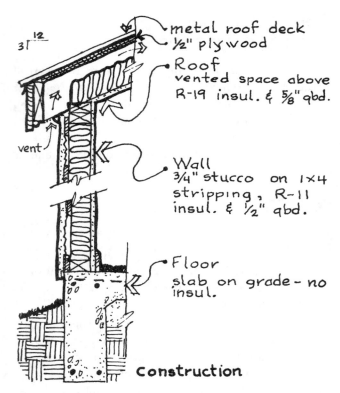

metal roof deck
½" plywood

Roof
vented space above
R-19 insul. & ⅝" gbd.

vent

Wall
¾" stucco on 1×4
stripping, R-11
insul. & ½" gbd.

Floor
slab on grade - no
insul.

Construction

FIGURE 15.10

Floor: Slab on grade. Factor is 4 BTUH per sq. ft. because the design temperature is below 10°F (p. 205).

Slab edge: U value for uninsulated slab edge is 0.8 (pp. 200 and 205)

Doors: 1 1/2″ solid core; U = 0.5 (p. 200).

QUANTITIES

See plan and section in Figure 15.11 for dimensions.

Outdoor air: The largest value comes from infiltration:

$$OA = \frac{(Volume)(ACH)}{(60)}$$

$$OA = \frac{(14,000)(1.5)}{60} = 350 \text{ CFM}$$

The estimated house volume of 14,000 CF includes the high ceiling areas of the house.

Glass: 320 sq. ft. including high windows at roof (small numbers on plan are low window areas).

Ceiling-roof: 1,620 sq. ft. including allowance for roof slope.

Walls: 1,090 sq. ft. including high end wall, wall at clerestory windows, wall at entry, and one-half of the garage wall (because the garage TD is half the indoor to outdoor TD) (p. 205).

Floor slab: 1,568 sq. ft.

Slab edge: 140 linear feet; excludes 28 ft. at garage.

Doors: 30 sq. ft. (actually 40 sq. ft., but the door to the garage has only half TD, so count half of this door area).

WINTER HEAT LOSS = BTUH

Insert appropriate U values, constants, factors, and temperature difference; then complete all multiplications and total the heat loss BTUH. Add 10% if ducts are located in an unconditioned attic.

House	_"H"(map p.18)_	_1,568_	_E J R_	_1·2·99_
project name	location	floor area sqft.	calculated by	date

Design Conditions:

winter _(p.16)_
- _70_ °Fdb indoor
- _0_ °Fdb outdoor
- _70_ TD

summer
- ____ °Fdb outdoor
- ____ °Fdb indoor
- ____ TD
- ____ °Fwb outdoor
- ____ % RH indoor
- ____ GD
- ____ time of peak gain

Project Conditions:

Outdoor Air; calculate 1&2 below but use only the <u>largest CFM</u> value:
1. Infiltration based on air change rate _350_ winter ____ summer
2. Ventilation based on CFM per person _75_ _(5@15)_

Glass _1.1_ U value ____ SC (Shading Coefficient)
Lighting _____ total watts operating at peak heat gain time
People ____ total occupants at peak heat gain time
Ceiling-Roof _0.05_ U value ____ color _6_ weight (lbs/sqft)
Walls _0.07_ U value ____ color _12_ weight (lbs/sqft)
factor Floor _4_ U value, Door _0.5_ U value, Slab Edge U value _0.8_
Equipment _____ watts or hp, Appliances _____
Other _____

Item	Quantities	Winter Heat Loss = BTUH	Summer Heat Gain = BTUH
Outdoor Air			
winter	_350_ CFM	(1.08)(_70_ TD) = _26,460_	
summer	____ CFM		(1.08)(____ TD) = _____
			(0.68)(____ GD) = _____
Glass total	_320_ sqft	(_1.1_ U)(_70_ TD) = _24,640_	(____ U)(____ TD) = _____
N	____ sqft		(____ SF)(____ SC) = _____
E	____ sqft		(____ SF)(____ SC) = _____
S	____ sqft		(____ SF)(____ SC) = _____
W	____ sqft		(____ SF)(____ SC) = _____
horiz	____ sqft		(____ SF)(____ SC) = _____
Lighting	____ watts		(3.4) = _____
People	_5_ #		(____ sens) = _____
			(____ latent) = _____
Ceiling-Roof	_1,620_ sqft	(_0.05_ U)(_70_ TD) = _5,670_	(____ U)(____ ETD) = _____
Walls	_1,090_ sqft	(_0.07_ U)(_70_ TD) = _5,341_	(____ U)(____ ETD) = _____
Floor bsmt	____ sqft	(____ factor) =	
slab	_1,568_ sqft	(_4_ factor) = _6,272_	
crawl space	____ sqft	(____ U)(____ TD) =	
above grade	____ sqft	(____ U)(____ TD) =	
Slab Edge	_140_ linft	(_0.8_ U)(_70_ TD) = _7,840_	
Doors	_30_ sqft	(_0.5_ U)(_70_ TD) = _1,050_	(____ U)(____ TD) = _____
Equipment	____ watts		(3.4) = _____
	____ hp		(2500) = _____
Appliances	____		(____ sensible) = _____
			(____ latent) = _____
Other	____		_____

Subtotals	_77,273_	_____
If ducts are outside the conditioned space add 10% _(inside)_		_____
TOTAL BTUH heat loss _77,273_ heat gain		_____

Check; heat loss = 20-60 BTUH persqft. (south to north); heat gain = 15-60 BTUH per sqft. (north to south). Allow 4% of area served for fan rooms, and 2% of gross building area for central plant equipment.

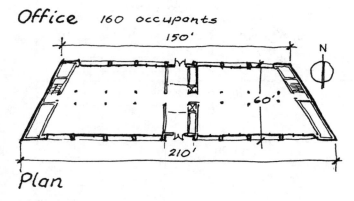

Office   160 occupants

150'

60'

210'

N

Plan

FIGURE 15.11

15.6 **OFFICE EXAMPLE, HEAT LOSS CALCULATION**

DESIGN CONDITIONS

Winter temperature is 10°F (for location "O" in Figure 15.8). An indoor temperature of 70°F is assumed; temperature difference (TD) is 60.

PROJECT CONDITIONS

Outdoor air: Estimate infiltration for the example office in Figure 15.11 at 0.75 air changes per hour in winter (0.5 air changes per hour in summer). Estimate ventilation rate at 15 CFM per person. Fresh air CFM due to infiltration is the largest value in winter (pp. 202–203).

Glass: Double glass. U = 0.6 (p. 200).

Lighting: Do not take heat gain credit because lights will be off on cold nights.

People: Estimate 160 occupants, but do not take heat gain credit because people are absent on cold nights.

Ceiling-roof: Detail is shown in Figure 15.12. U = 1/R total = 0.05.

$$[1/(0.2 + 18 + 0.7) = 1/18.9 = 0.05]\ (p.\ 200)$$

Wall: See detail in Figure 15.12. U = 1/R total = 0.07.

$$[1/(0.2 + 0.4 + 1 + 0.5 + 11 + 0.5 + 0.7) = 1/14.3 = 0.07]$$

Floor: Slab on grade. Factor is 2 BTUH per sq. ft. (p. 205).

Slab edge: Not shown. No insulation; no heating ducts. U = 0.8 (pp. 200 and 205).

Doors: Single glass. U = 1.1 (p. 200).

Equipment: Do not take a heat gain credit for equipment because it does not operate on cold nights.

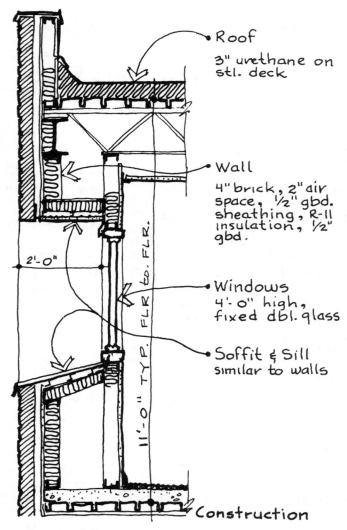

Roof
3" urethane on stl. deck

Wall
4" brick, 2" air space, 1/2" gbd. sheathing, R-11 insulation, 1/2" gbd.

2'-0"

11'-0" TYP. FLR. to FLR.

Windows
4'-0" high, fixed dbl. glass

Soffit & Sill
similar to walls

Construction

FIGURE 15.12

QUANTITIES

Refer again to Figures 15.11 and 15.12 for dimensions.

Outdoor air: Building volume is (21,600)(11) = 237,600 CF. At 0.75 air changes per hour, the expected infiltration is (237,600÷60)(0.75) = 2,970 CFM. Ventilation rate for 160 people at 15 CFM each is 2,400 CFM. Use the larger infiltration value for winter heat loss (pp. 202 and 203).

Glass: Windows are 4 ft. high. Total window area is 2,400 sq. ft.

Ceiling-roof: Area is half floor area (two floors), 10,800 sq. ft.

Walls: Net wall is 10,880 sq. ft. Gross wall area is 10,950 sq. ft. Add for soffits and sills 2,400 sq. ft.; deduct for windows and doors 2,470 sq. ft.

Floor: Slab on grade area = 10,800 sq. ft.

Slab edge: 494 linear feet.

Doors: Two at 5'×7' = 70 sq. ft. (excluding fire exits—they are counted as wall area).

WINTER HEAT LOSS = BTUH

Insert appropriate U values, constants, factors, and temperature difference; then complete calculations and total heat loss BTUH. Add 10% only if ducts are outside the conditioned space.

15.7 ESTIMATING BUILDING HEAT GAINS

Many of the concepts developed for calculating heat loss also apply to heat gain. However, three added factors—sun, internal loads, and latent heat—make calculation of heat gain a bit more complex.

SUN

The radiant energy delivered by the sun at the outer edge of the earth's atmosphere, known as the solar constant, is about 422 BTUH per square foot (see Figure 15.13). Radiant energy passes easily through glass, and it warms roofs and walls well above ambient air temperature. Since the sun is a moving heat source, its contribution to heat gain is continually changing, and a specific time must be used to estimate solar gain. Buildings will have different critical times for peak

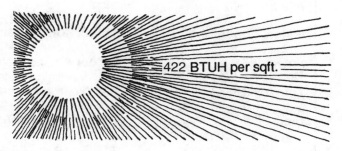

FIGURE 15.13

heat gain as a result of solar and occupancy factors. A church may experience peak heat gain at 11 AM, a gourmet restaurant at perhaps 9 PM, and an office building or residence at 4 PM. An office building with all windows facing southeast may peak at 10 AM.

Solar Heat Gain through Glass

Use the following equation to quantify solar gain through glass (see Figure 15.14):

$$Q_{solar} = (SF)(A)(TD)$$

where:

SF = Solar factor (sun's output) in BTUH per sq. ft.
A = Glass area in sq. ft.
SC = Shading coefficient (percent of solar energy passing through the glass).
Q_{solar} = Rate of heat flow in BTU per hour.

Note: Conducted heat gain through glass due to temperature difference occurs *in addition* to solar gain.

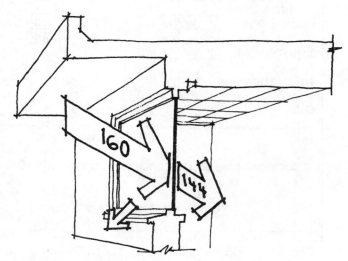

Solar BTU through glass

FIGURE 15.14

Office	_"O"_ (map p.18)	21,600	_GJIA_	1.3.99
project name	location	floor area sqft.	calculated by	date

Design Conditions:

winter (p.18)
70 °Fdb indoor
10 °Fdb outdoor
60 TD

summer
____ °Fdb outdoor
____ °Fdb indoor
____ TD
____ °Fwb outdoor
____ % RH indoor   _factor →_
____ GD
____ time of peak gain

Project Conditions:

3/4 A.C.

Outdoor Air; calculate 1&2 below but use only the **largest CFM** value:
1. Infiltration based on air change rate 2,970 winter ____ summer
2. Ventilation based on CFM per person 2,400 (160 @ 15)

Glass 0.6 U value _____ SC (Shading Coefficient)
Lighting _____ total watts operating at peak heat gain time
People 160 total occupants at peak heat gain time
Ceiling-Roof 0.05 U value ____ color ____ weight (lbs/sqft)
Walls 0.07 U value ____ color ____ weight (lbs/sqft) _glass doors_
Floor 2 U value, Door 1.1 U value, Slab Edge U value 0.8
Equipment _____ watts or hp, Appliances _____
Other _____

Item	Quantities	Winter Heat Loss = BTUH	Summer Heat Gain = BTUH
Outdoor Air			
winter	2,970 CFM	(1.08)(60 TD) = 192,450	
summer	_____ CFM		(1.08)(___ TD) = _____
			(0.68)(___ GD) = _____
Glass total	2,400 sqft	(0.6 U)(60 TD) = 86,400	(___ U)(___ TD) = _____
N	_____ sqft		(___ SF)(___ SC) = _____
E	_____ sqft		(___ SF)(___ SC) = _____
S	_____ sqft		(___ SF)(___ SC) = _____
W	_____ sqft		(___ SF)(___ SC) = _____
horiz	_____ sqft		(___ SF)(___ SC) = _____
Lighting	_____ watts		(3.4) = _____
People	_____ #		(___ sens) = _____
			(___ latent) = _____
Ceiling-Roof	10,800 sqft	(0.05 U)(60 TD) = 32,400	(___ U)(___ ETD) = _____
Walls	10,880 sqft	(0.07 U)(60 TD) = 45,700	(___ U)(___ ETD) = _____
Floor bsmt	_____ sqft	(___ factor) = _____	
slab	10,800 sqft	(2 factor) = 21,600	
crawl space	_____ sqft	(___ U)(___ TD) = _____	
above grade	_____ sqft	(___ U)(___ TD) = _____	
Slab Edge	494 linft	(0.8 U)(60 TD) = 23,700	(___ U)(___ TD) = _____
Doors	70 sqft	(1.1 U)(60 TD) = 4,600	(3.4) = _____
Equipment	_____ watts		(2500) = _____
	_____ hp		(___ sensible) = _____
Appliances	_____		(___ latent) = _____
Other	_____	_____	_____

	Subtotals	406,850	_____

If ducts are outside the conditioned space add 10% _inside_ _____

TOTAL BTUH   heat loss 406,850   heat gain _____

Check; heat loss = 20-60 BTUH persqft. (south to north); heat gain = 15-60 BTUH per sqft. (north to south). Allow 4% of area served for fan rooms, and 2% of gross building area for central plant equipment.

The following equation is used to quantify conducted heat gain:

$$Q_{conducted1} = (U)(A)(TD)$$

Heat Gain through Roofs and Walls

The sun can also increase the temperature of roofs and walls above ambient air temperature, and heat flow into the conditioned space is increased as a result. Heat gain through opaque roofs and walls may be calculated using the equation:

$$Q_{conducted2} = (U)(A)(ETD)$$

where:

U   = Conductance value of the roof or wall in BTUH per sq. ft. per °F.

A   = Area of roof or wall in sq. ft.

ETD = Equivalent temperature difference. An increased temperature difference that allows for heat gain caused by both air temperature difference and solar effect. (ETD values will vary, depending on the time of the day and the weight and color of the roof or wall.)

$Q_{conducted2}$ = Rate of heat flow in BTU per hour.

Example

Use the tables in Section 15.8 to find the total heat gain through a 50 sq. ft. single-glass window facing southwest at 4 PM. The window is not shaded; its U value is 1.1. Outdoor temperature is 100°F, and indoor temperature is 75°F.

$$Q_{solar} = (SF)(A)(TD)$$
$$(160)(50)(90\%) = 7,200 \text{ BTUH}$$

$$Q_{conducted1} = (U)(A)(TD)$$
$$(1.1)(50)(100-75) = 1,375 \text{ BTUH}$$
$$\text{Total heat gain } 8,575 \text{ BTUH}$$

Example

Use the tables in Section 15.8 to find the heat gain through a dark color 1,500 sq. ft. roof at 2 PM if the roof weight is 6 pounds per sq. ft. and the roof U value is 0.12.

$$Q_{conducted2} = (U)(A)(ETD)$$
$$(0.12)(1,500)(70) = 12,600 \text{ BTUH}$$

If the roof above is light color, find heat gain.

7,560 BTUH

If the roof is wet with water spray, find heat gain.

2,700 BTUH

INTERNAL LOADS

People, lights, appliances, motors, and food preparation add heat to building interiors. People and cooking can add both sensible (dry) and latent (wet) heat. Motors and lights add only sensible heat. HVAC equipment must remove internally generated heat to maintain comfortable interior conditions.

Example

Use the tables in Section 15.9 to find the total internal heat gain in a room caused by 10 people doing light work (standing), a 3 hp electric motor, and 12 light bulbs rated at 500 watts each.

(# people) (BTUH per person) = people gain	
(10)(750) =	7,500 BTUH
(hp)(2,500) = motor gain	
(3)(2,500) =	7,500 BTUH
(watts) (3.4) = lighting gain	
(12) (500) (3.4) =	20,400 BTUH
Total internal gain	35,400 BTUH

LATENT HEAT

Latent (phase change) heat exists in the water vapor carried by air (see Figure 15.15). As air temperature increases, its ability to carry water vapor increases.

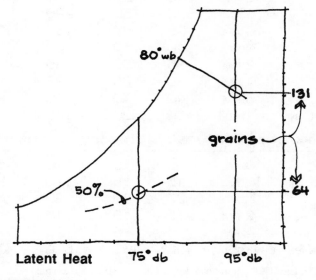

FIGURE 15.15

213

Significant air-conditioning capacity is required to dehumidify moist outdoor air because cool dry air is necessary for comfort. Calculate the air-conditioning capacity required to dehumidify outdoor air as follows:

$$Q_{latent} = (CFM)(0.68)(GD)$$

where:

CFM = Quantity of moist air to be dehumidified, measured in cubic feet per minute.

0.68* = A constant. It is the number of BTUH produced when 4.5 grains of water vapor condense (1 CFM = 60 CF per hour = 4.5 lbs per hour; 1 grain per lb = 0.68 BTUH).

GD = Moisture difference in grains per pound of air between moist outdoor air and dry indoor air (1 grain = 0.00014 lb).

Q_{latent} = Rate of heat flow in BTU per hour.

Example

Find the total heat gain caused by bringing 800 CFM of ventilation air into a building if the out-

---

*It takes 1,061 BTU to vaporize 1 pound of water at 212°F. This latent heat of vaporization is released when water condenses. Can you develop the latent heat formula constant −0.68?

door air is 95°F and 80°Fwb (wet-bulb temperature). Building air is 75°F and 50% RH (relative humidity).

$$Q_{sensible} = (CFM)(1.08)(TD)$$
$$(800)(0.68)(67) = 36,448 \text{ BTUH}$$

$$Q_{sensible} = (CFM)(1.08)(TD)$$
$$(800)(1.08)(95-75) = 17,280 \text{ BTUH}$$

$$\text{Total gain } 53,728 \text{ BTUH}$$

Note that the total heat gain due to outdoor air includes both latent heat and sensible heat. The same equation used to quantify winter heat loss:

$$Q_{sensible} = (CFM)(1.08)(TD)$$

is used to quantify sensible heat gain.

15.8 SOLAR, SHADING, AND ETD VALUES

SOLAR FACTORS (SF)

The following BTUs per hour per square foot are based on clear-day conditions.

Summer Values: July, Central United States

Solar Time	N	NE	E	SE	S	SW	W	NW	HOR
6 AM	40	120	140	70	10	10	10	10	30
8 AM	30	150	220	160	30	25	25	25	140
10 AM	35	60	150	160	80	35	35	35	230
Noon	40	40	40	80	110	80	40	40	260
2 PM	35	35	35	35	80	160	150	60	230
4 PM	30	25	25	25	30	160	220	150	140
6 PM	40	10	10	10	10	70	140	120	30
Clear day total	420	820	1,160	1,050	700	1,050	1,160	820	2,140

Winter Values: January, Central United States

Solar Time	N	NE	E	SE	S	SW	W	NW	HOR
8 AM	5	15	110	130	70	5	5	5	10
10 AM	15	15	120	240	210	50	15	15	100
Noon	20	20	20	180	250	180	20	20	130
2 PM	15	15	15	50	210	240	120	15	100
4 PM	5	5	5	5	70	130	110	15	10
Clear day total	120	130	500	1,170	1,600	1,170	500	130	700

Notes: Use *north* solar factors for glass with external shading; HOR = horizontal (e.g., a skylight).

SHADING COEFFICIENT (SC)

The following shows the percentages of solar BTUH that pass through glass into the conditioned space.

Single glass (no blind)	90%	Double glass (no blind)	85%
Single glass with dark color drape or blind	70%	Double glass with dark color drape or blind	65%
Single glass with light color drape or blind	50%	Double glass with light color drape or blind	50%

Heat-absorbing, reflective, or low e glass can be as low as 10%. Refer to the manufacturer's specifications.

EQUIVALENT TEMPERATURE DIFFERENCE (ETD)

ETD accounts for solar radiation, mass, color, and air temperature difference.

Roof ETD: Dark Color, July, Central United States

Roof Weight (pounds per sq. ft.)	8 AM	10 AM	Noon	2 PM	4 PM	6 PM	8 PM
0–10	15	40	60	70	60	30	10
11–50	5	25	45	60	60	45	20
51 plus	5	10	25	40	50	50	40
Wet roof (spray or pond)	0	0	10	15	20	20	15

Note: Use **60%** of the values above for a **light color roof** (e.g., dark value = 50, light = 30).

Wall ETD: July, Central United States

To find ETD, add the following values to air *TD* for *only* the orientations listed below.

Lightweight walls—0 to 50 pounds/sq. ft.   For light color, add 10 to the air TD value.
For dark color, add 20 to the air TD value.

Heavyweight walls—over 50 pounds/sq. ft.   For light color, use air TD value.
For dark color, add 5 to air TD value.

	8 AM	10 AM	Noon	2 PM	4 PM	6 PM	8 PM
Wall orientation	NE	NE E SE	E SE	S SW	SW W	SW W NW	NW

15.9 INTERNAL LOADS AND LATENT HEAT GAINS

INTERNAL LOADS

Internal loads include heat gains from lights, people, motors, kitchens, and other equipment.

Lighting

Lighting loads add heat to a building at the rate of 3.4 BTUH for each watt of lighting in operation. If lighting design is not complete, use the following data to estimate total lighting watts.

Building Type	Watts per sq. ft. of Floor Area		Building Type	Watts per sq. ft. of Floor Area	
	Range	Average		Range	Average
Apartments	1–4	2	Residence	1–4	2
Church	0–2	1	Schools	1–4	2
Factories	2–8	3	Shopping center	1–10	3
Hospitals	1–4	2	Supermarket	2–5	3
Hotels	1–3	2	Retail shops	1–10	5
Libraries	1–4	2	Restaurant, fast	1–6	3
Offices	1–4	2	Restaurant, gourmet	0–1	1

Miscellaneous

Electric appliances or heaters: 3.4 BTUH per watt hour; motors 2,500 BTUH per hp hour.
Residential kitchens: Allow 1,200 BTUH (average value for all kitchen appliances).
Other equipment or appliances: Refer to manufacturers' nameplate data.

People: BTUH for Selected Activities	Sensible	Latent	Total
Sleeping	200	100	300
Seated (awake)	250	150	400
Standing	250	200	450
Standing (light work)	275	475	750
Dancing (quietly)	300	550	850
Walking (or exercise)	375	625	1,000
Bowling	575	875	1,450
Heavy work or basketball	650	1,150	1,800

LATENT LOADS

Grain Difference (GD): Per CFM of Ventilation or Infiltration Air

(Values assume 95°F outdoor and 75°F and 50% RH indoor; for greater accuracy refer to the psychometric chart on Figure 1.3)

Outdoor °Fwb	GD	Outdoor °Fwb	GD
70°	5	76°	40
72°	15	78°	55
74°	27	80°	67

For example problems, see Chapter 14.

15.10 CALCULATING BUILDING HEAT GAIN

The heat loss–heat gain summary form can be used to quantify peak heat gain and peak heat loss. Examine the form as you read this section; then work through the following examples. Heat gain calculations can be used to select air-conditioning equipment and to estimate annual cooling costs.

DESIGN CONDITIONS: PROJECT CONDITIONS

Select summer outdoor dry-bulb temperature (°Fdb), and wet-bulb temperature (°Fwb) from the maps in Figures 15.16, 15.17, and 15.18, or refer to

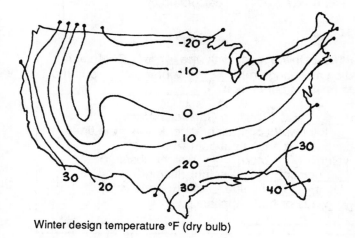

Winter design temperature °F (dry bulb)

FIGURE 15.16

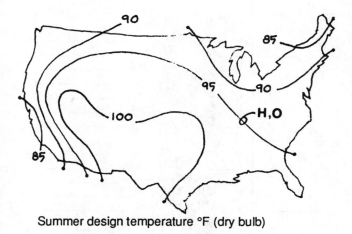

Summer design temperature °F (dry bulb)

FIGURE 15.17

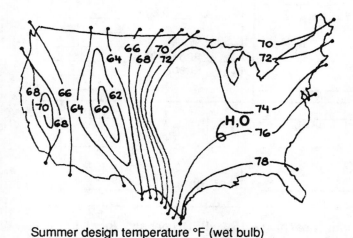

Summer design temperature °F (wet bulb)

FIGURE 15.18

the current *ASHRAE Handbook of Fundamentals* for more precise data. The following examples use 75°Fdb and 50% relative humidity as comfortable indoor design conditions. GD (grain difference) is most accurate from the psychrometric chart, but you can use the rough values when the outdoor design temperature is 95°F. The time of peak heat gain is 4 PM for most buildings, but churches may peak at 11 AM and dance halls at 11 PM, so think a bit before selecting the time of peak gain or test several times to be sure. Fill in the appropriate project conditions blanks for the building you are evaluating.

QUANTITIES

Most quantities indicated in the left-hand vertical column of the form were filled in for the heat loss calculations (pp. 208 and 211). Comments below emphasize changes for heat gain.

Fresh air: Use the larger value for infiltration or ventilation; expected summer infiltration is 67% of winter infiltration (pp. 202 and 203).

Glass: Use the total glass area to calculate the conducted heat gain; then subdivide total glass area into appropriate compass orientations for calculating solar gain. Enter skylights as "HOR" (horizontal) or reverse glass.

Lighting: Enter the total watts of lighting that will be operating at peak heat gain time.

People: Enter the total number of people that will occupy the building at peak heat gain time.

Ceiling-roof, walls: Remember to deduct window and skylight areas. If your building has a variety of roof or wall constructions, develop an average U value.

CALCULATIONS

After entering quantities, insert the appropriate U, TD, ETD, and factor values in the summer heat gain column and complete the multiplications indicated. Your BTUH total is the minimum size needed for air-conditioning equipment, and it can be used to estimate annual cooling costs.

For typical buildings in the United States, the total heat gain should range between 15 and 60 BTUH per square foot of air-conditioned floor area. Higher heat gains are for buildings with large internal heat loads.

project name _____ location _____ floor area sqft. _____ calculated by _____ date _____

Design Conditions:

winter
_____ °Fdb indoor
_____ °Fdb outdoor
_____ T D

summer
_____ °Fdb outdoor
_____ °Fdb indoor
_____ T D
_____ °Fwb outdoor
_____ % RH indoor
_____ GD
_____ time of peak gain

Project Conditions:

Outdoor Air; calculate 1&2 below but use only the <u>largest CFM</u> value:
  1. Infiltration based on air change rate_____winter _____summer
  2. Ventilation based on CFM per person _____

Glass _____U value_____SC (Shading Coefficient)
Lighting _____ total watts operating at peak heat gain time
People _____total occupants at peak heat gain time
Ceiling-Roof _____U value _____color _____weight (lbs/sqft)
Walls _____U value _____color _____ weight (lbs/sqft)
Floor _____U value, Door _____U value, Slab Edge U value _____
Equipment _____watts or hp, Appliances _____
Other _____

Item	Quantities	Winter Heat Loss = BTUH	Summer Heat Gain = BTUH
Outdoor Air			
winter	_____CFM	(1.08)(___TD) = _____	
summer	_____CFM		(1.08)(___TD) = _____
			(0.68)(___GD) = _____
Glass total	_____sqft	(___U)(___TD) = _____	(___U)(___TD) = _____
N	_____sqft		(___SF)(__SC) = _____
E	_____sqft		(___SF)(__SC) = _____
S	_____sqft		(___SF)(__SC) = _____
W	_____sqft		(___SF)(__SC) = _____
horiz	_____sqft		(___SF)(__SC) = _____
Lighting	_____watts		(3.4) = _____
People	_____#		(____sens) = _____
			(____latent) = _____
Ceiling-Roof	_____sqft	(___U)(___TD) = _____	(___U)(__ETD) = _____
Walls	_____sqft	(___U)(___TD) = _____	(___U)(__ETD) = _____
Floor bsmt	_____sqft	(___factor) = _____	
slab	_____sqft	(___factor) = _____	
crawl space	_____sqft	(___U)(___TD) = _____	
above grade	_____sqft	(___U)(___TD) = _____	
Slab Edge	_____linft	(___U)(___TD) = _____	
Doors	_____sqft	(___U)(___TD) = _____	(___U)(___TD) = _____
Equipment	_____watts		(3.4) = _____
	_____hp		(2500) = _____
Appliances	_____		(____sensible) = _____
			(____latent) = _____
Other	_____	_____	_____

	Subtotals	_____	_____

If ducts are outside the conditioned space add 10% _____ _____

TOTAL BTUH heat loss _____ heat gain _____

Check; heat loss = 20-60 BTUH persqft. (south to north); heat gain = 15-60 BTUH per sqft. (north to south). Allow 4% of area served for fan rooms, and 2% of gross building area for central plant equipment.

15.11 HOUSE EXAMPLE, HEAT GAIN CALCULATION

DESIGN CONDITIONS

Maps show a summer dry-bulb temperature of 95°F and a wet-bulb temperature of 76°F (Figures 15.17 and 15.18, location "H"). Desired indoor conditions are 75°F and 50% relative humidity. Time of peak heat gain is estimated at 4 PM. Temperature difference (TD) = 20. Grain difference (GD) = 40 (p. 216). See Figure 15.19 for the example house plan. Refer to Figure 15.20 for construction details.

PROJECT CONDITIONS

Outdoor air: Select largest of infiltration at 1 air change per hour or ventilation at 15 CFM per person (pp. 202 and 203).

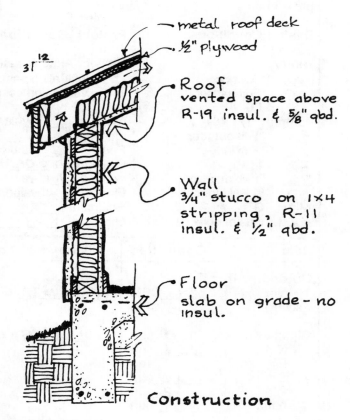

FIGURE 15.20

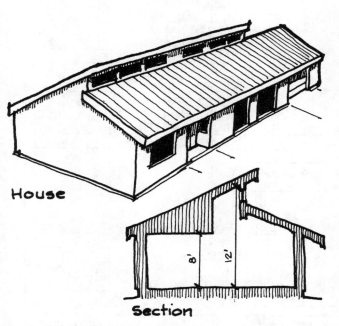

House

Section

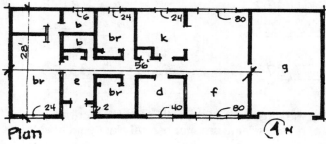

Plan

FIGURE 15.19

Glass: For single glass, U = 1.1. All windows have light-colored draperies to reduce solar gain. Shading coefficient (SC) = 0.5 (p. 215).

Lighting: Negligible for a residence. At 4 PM, most lights will be off.

People: Use seated value for five occupants unless the house is used for exercise classes at 4 PM.

Ceiling-roof: U = 0.05; weight is 6 pounds per sq. ft.; color is dark; ETD is 60 (p. 215).

Walls: U = 0.07; weight is 12 pounds per sq. ft.; color is light; weighted average ETD is 22 (ETD = 30 for west wall, 20 for other walls; p. 215).

Floor and slab edge: No significant heat gain.

Doors: U = 0.5 (p. 200).

QUANTITIES

Outdoor air: One air change per hour at 235 CFM is more than the 75 CFM recommended for ventilation.

House	"H" (map p. 28)	1,568	EJB	1.2.99
project name	location	floor area sqft.	calculated by	date

Design Conditions:

winter
- _70_ °Fdb indoor
- _0_ °Fdb outdoor
- _70_ TD

summer (p. 28)
- _95_ °Fdb outdoor
- _75_ °Fdb indoor
- _20_ TD
- _76_ °Fwb outdoor
- _50_ % RH indoor
- _40_ GD (p. 27)
- _4 P.M._ time of peak gain

Project Conditions:

Outdoor Air; calculate 1&2 below but use only the **largest CFM** value:
1. Infiltration based on air change rate _350_ winter _235_ summer
2. Ventilation based on CFM per person _75 (5@15)_ ← (⅔ of winter)

← light drapes

Glass _1.1_ U value _0.5_ SC (Shading Coefficient)
Lighting _—_ total watts operating at peak heat gain time
People _5_ total occupants at peak heat gain time
Ceiling-Roof _0.05_ U value _dark_ color _6_ weight (lbs/sqft)
Walls _0.07_ U value _light_ color _12_ weight (lbs/sqft)
Floor _4_ U value, Door _0.5_ U value, Slab Edge U value _0.8_
Equipment _____ watts or hp, Appliances _1,200_
Other _residential kitchen_ ↑

Item	Quantities	Winter Heat Loss = BTUH	Summer Heat Gain = BTUH
Outdoor Air			
winter	_350_ CFM	$(1.08)(70 \text{ TD}) =$ _26,460_	
summer	_235_ CFM		$(1.08)(20 \text{ TD}) =$ _5,076_
			$(0.68)(40 \text{ GD}) =$ _6,392_
Glass total	_320_ sqft	$(1.1 \text{ U})(70 \text{ TD}) =$ _24,640_	$(1.1 \text{ U})(20 \text{ TD}) =$ _7,040_
N	_134_ sqft		$(30 \text{ SF})(0.5 \text{ SC}) =$ _2,010_
E	_____ sqft		$(\text{ SF})(\text{ SC}) =$ _____
S	_184_ sqft		$(30 \text{ SF})(0.5 \text{ SC}) =$ _2,760_
W	_2_ sqft		$(220 \text{ SF})(0.5 \text{ SC}) =$ _220_
horiz	_____ sqft		$(\text{ SF})(\text{ SC}) =$ _____
Lighting	_____ watts		$(3.4) =$ _____
People	_5_ #		$(250 \text{ sens}) =$ _1,250_
			$(150 \text{ latent}) =$ _750_
Ceiling-Roof	_1,620_ sqft	$(0.05 \text{ U})(70 \text{ TD}) =$ _____	$(0.05 \text{ U})(60 \text{ ETD}) =$ _4,860_
Walls	_1,090_ sqft	$(0.07 \text{ U})(70 \text{ TD}) =$ _____	$(0.07 \text{ U})(22 \text{ ETD}) =$ _1,679_
Floor bsmt	_____ sqft	$(\text{ factor}) =$ _____	
slab	_1,568_ sqft	$(4 \text{ factor}) =$ _6,272_	
crawl space	_____ sqft	$(\text{ U})(\text{ TD}) =$ _____	
above grade	_____ sqft	$(\text{ U})(\text{ TD}) =$ _____	
Slab Edge	_140_ linft	$(0.8 \text{ U})(70 \text{ TD}) =$ _7,840_	
Doors	_30_ sqft	$(0.5 \text{ U})(70 \text{ TD}) =$ _1,050_	$(0.5 \text{ U})(20 \text{ TD}) =$ _300_
Equipment	_____ watts		$(3.4) =$ _____
	_____ hp		$(2500) =$ _____
Appliances	_____		$(\text{ sensible}) =$ → _1,200_
Other	_____		$(\text{ latent}) =$ _____
			residential kitchen ✓
Subtotals		_77,273_	_33,537_

If ducts are outside the conditioned space add 10% _(inside)_   _—_

TOTAL BTUH heat loss _77,273_   heat gain _33,537_

Check; heat loss = 20-60 BTUH per sqft. (south to north); heat gain = 15-60 BTUH per sqft. (north to south). Allow 4% of area served for fan rooms, and 2% of gross building area for central plant equipment.

Glass: Of the total 320 sq. ft., 134 sq. ft. face north, 184 sq. ft. face south (includes clerestory), and 2 sq. ft. face west.

Ceiling-roof: 1,620 sq. ft. total, including allowance for roof slope.

People: Each occupant produces 250 BTUH sensible plus 150 BTUH latent.

Walls: 1,090 sq. ft. (garage wall at 50%).

Doors: 30 sq. ft. (counting door to garage at 50%).

Appliances: 1,200 BTUH is an average allowance for residential heat gain due to appliances, cooking, showers, etc.

SUMMER HEAT GAIN = BTUH

Insert appropriate factors and temperature difference; then complete calculations and total BTUH. Add 10% if ducts are located in a hot attic.

Refer back to the heat loss example on page 208 if additional detail is needed.

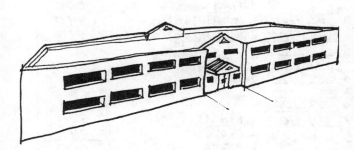

Office   160 occupants

Plan

FIGURE 15.21

15.12 OFFICE EXAMPLE, HEAT GAIN CALCULATION

DESIGN CONDITIONS

Maps show a summer dry-bulb temperature of 95°F and a wet-bulb temperature of 76°F (Figures 15.17 and 15.18, location "O"). Desired indoor conditions are 75°F and 50% relative humidity. Time of peak heat gain is estimated as 4 PM. Temperature difference (TD) = 20. Grain difference (GD) = 40 (p. 216). See Figure 15.21 for the example office building plan. Refer to Figure 15.22 for construction details.

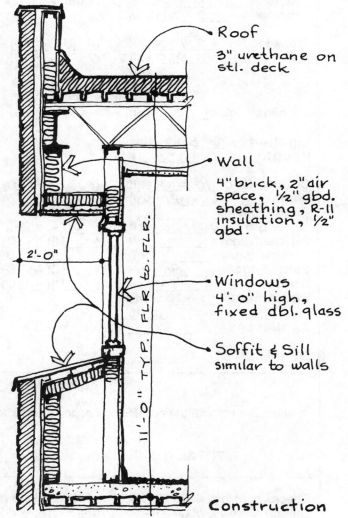

Roof
3" urethane on stl. deck

Wall
4" brick, 2" air space, ½" gbd. sheathing, R-11 insulation, ½" gbd.

Windows
4'-0" high, fixed dbl. glass

Soffit & Sill
similar to walls

Construction

FIGURE 15.22

Office	_"O"_ (map. p.28)	21,600	GJM	1·3·99
project name	location	floor area sqft.	calculated by	date

Design Conditions:

winter
70	°Fdb indoor
10	°Fdb outdoor
60	T D

summer (p. 28)
95	°Fdb outdoor
75	°Fdb indoor
20	T D
76	°Fwb outdoor
50	% RH indoor
40	GD (p.27)
4 P.M.	time of peak gain

Project Conditions:

½ A.C.

Outdoor Air; calculate 1&2 below but use only the **largest CFM** value:
1. Infiltration based on air change rate _2,970_ winter _1,980_ summer
2. Ventilation based on CFM per person _2,400_ (160 @ 15)

Glass _0.6_ U value _0.85_ SC (Shading Coefficient)
Lighting _64,800_ total watts operating at peak heat gain time
People _160_ total occupants at peak heat gain time
Ceiling-Roof _0.05_ U value _light_ color _7_ weight (lbs/sqft)
Walls _0.07_ U value _dark_ color _45_ weight (lbs/sqft)
Floor _2_ U value, Door ____ U value, Slab Edge U value _0.8_
Equipment _21,600 & 10_ watts or hp, Appliances —
Other _____

Item	Quantities	Winter Heat Loss = BTUH	Summer Heat Gain = BTUH
Outdoor Air			
winter	_2,970_ CFM	(1.08)(_60_ TD) = 192,450	
summer	_2,400_ CFM		(1.08)(_20_ TD) = 51,840
			(0.68)(_40_ GD) = 65,280
Glass total	_2,400_ sqft	(0.6U)(_60_TD) = 86,400	(_0.6_U)(_20_TD) = 28,800
N	_1,030_ sqft ← 960 + 70 doors		(_30_ SF)(_0.85_SC) = 26,265
E	____ sqft		(____SF)(____SC) =
S	_1,440_ sqft		(_30_ SF)(_0.85_SC) = 36,720
W	____ sqft		(____SF)(____SC) =
horiz	____ sqft		(____SF)(____SC) =
Lighting	_64,800_ watts		(3.4) = 220,320
People	_160_ #		(_250_ sens) = 40,000
			(_150_ latent) = 24,000
Ceiling-Roof	_10,800_ sqft	(0.05U)(_60_TD) = 32,400	(0.05U)(_36_ETD) = 19,440
Walls	_10,880_ sqft	(0.07U)(_60_TD) = 45,700	(0.07U)(_23_ETD) = 17,517
Floor bsmt	____ sqft	(____ factor) =	weighted average
slab	_10,800_ sqft	(_2_ factor) = 21,600	
crawl space	____ sqft	(____U)(____TD) =	
above grade	____ sqft	(____U)(____TD) =	
Slab Edge	_494_ linft	(_0.8_U)(_60_TD) = 23,700	
Doors	_70_ sqft	(_1.1_U)(_60_TD) = 4,600	(_1.1_U)(_20_TD) = 1,540
Equipment	_21,600_ watts		(3.4) = 73,440
	_10_ hp		(2500) = 25,000
Appliances	____		(____ sensible) =
Other	____		(____ latent) =
Subtotals		406,850	630,172

If ducts are outside the conditioned space add 10%   _____   _____

TOTAL BTUH   heat loss _406,850_   heat gain _630,172_

Check; heat loss = 20-60 BTUH persqft. (south to north); heat gain = 15-60 BTUH per sqft. (north to south). Allow 4% of area served for fan rooms, and 2% of gross building area for central plant equipment.

PROJECT CONDITIONS

Outdoor air: Select largest of infiltration at 0.5 air changes per hour or ventilation at 15 CFM per person (pp. 202 and 203).

Glass: Double glass U = 0.6. The SC (shading coefficient) for clear double glass is 0.85 (p. 215).

Lighting: Estimate high, 3 watts per sq. ft. of floor area for this office occupancy, and revise as necessary after completing lighting design. Office building lighting installations can range from 1 to 5 watts per sq. ft. depending on luminous intensity, lamp type, and fixture design (pp. 215–216).

People: 160 occupants seated doing light work.

Ceiling-roof: U = 0.05; weight is 7 pounds per sq. ft.; color is light; ETD = 36 (p. 215).

Walls: U = 0.07; weight is 45 pounds per sq. ft.; color is dark. The west wall is hot at 4 PM; its ETD is 40; the weighted average ETD for all walls at 4 PM is 23 (p. 215).

Doors: U = 1.1 (tempered single glass). Solar gain for doors is included in *Glass.*

Equipment: Allow 1 watt per sq. ft. for electrical equipment operating at the time of peak heat gain. Also allow 10 hp for air handler fans.

QUANTITIES

Outdoor air: Use largest value. Infiltration at 0.5 air changes per hour = 1,980 CFM.

$$(21,600)(11)(0.5) \div 60 = 1,980.$$

Ventilation at 15 CFM per person = 2,400 CFM.

$$(160)(15) = 2,400 \text{ (pp. 202 and 203)}$$

Glass: 1,030 sq. ft. face north (includes 70 sq. ft. of glass door area); 1,440 sq. ft. face south.

Lighting: Totals 64,800 watts at 3 watts per sq. ft.

People: Each occupant produces 250 BTUH sensible plus 150 BTUH latent (p. 218).

Ceiling-roof: Area is 10,800 sq. ft.

Wall: Net wall area is 10,880 sq. ft.

Doors: Area = 70 sq. ft.

Equipment: Allow 21,600 watts at 1 watt per sq. ft.; plus 10 hp for air handler fans.

SUMMER HEAT GAIN = BTUH

Insert appropriate factors and temperature difference; then complete calculations and total BTUH. Add 10% if ducts are located outside the conditioned zone. Refer back to the heat loss example on page 000 for more detail.

REVIEW QUESTIONS

Four kinds of example questions are presented for review purposes. Each question or group of questions begins with a type key and a page reference; solutions follow each question.

R and U Values, p. 200

1. Find the U value of Wall A shown below.

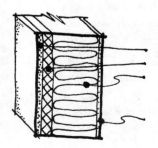

1" stucco
1 1/2" polystyrene
R-19 fiberglass
2 × 6 softwood studs @ 24" o.c. (not shown)
*1/2 " gypsum board

Wall A Section

R Values

Air film outside =	0.2
1" stucco =	0.1
1 1/2" polystyrene @ 4.0 per inch =	6.0
R-19 fiberglass =	19.0
1/2" gypsum board =	0.5
Air film inside =	0.7
R total =	26.5

U = 1/R total = 0.037 = **0.04**

2. If the outdoor temperature is −30°F and the indoor temperature is 70°F, find the temperature of the inside surface of the gypsum board wall. Remember that the temperatures in a construction are proportional to its R values.

♦ R from inside air to gypsum board surface is 0.7, and R total for Wall A is 26.5.

♦ Temperature difference is 100°F (70 − [−30]).

Gypsum surface temperature is **67.4°F** (70 − (100)(0.7)/(26.5) = 67.4

3. The actual cross section of Wall A includes 2 × 6 softwood studs at 24″ centers and 2 × 6 sill and plate members. Find the U value at a stud.

   R Values

Air film outside =	0.2
1″ stucco =	0.1
1 1/2″ polystyrene =	6.0
5 1/2″ softwood =	6.6
2 × 6 wd. is 1 1/2″ × 5 1/2″	
1/2′ gypsum board =	0.5
Air film inside =	0.7
R total =	14.1

   $$U = 1/R \text{ total } \mathbf{U = 0.07}$$

4. If 15% of Wall A is made of softwood (studs, sills, and plates), find the exact U value of the wall.

   U = 0.042 Why is the change so small?
   (15% of 0.07 + 85% of 0.037 = 0.042)

Heat Loss, pp. 203–207

5. Find the total heat loss through Wall A during one hour on a cold winter night if:
   - the wall area is 1,600 sq. ft.
   - the wall U value is 0.042
   - the outdoor temperature is −20°F
   - the indoor temperature is 68°F

   Heat loss = **5,914 BTUH**
   (1,600)(0.042)(68-{−20}) = 5,914

Heat Gain pp. 211–218

6. Find the total heat gain through Wall A during one hour on a hot July day if:
   - the wall area is 1,600 sq. ft.
   - the wall U value is 0.042
   - the outdoor temperature is 100°F
   - the indoor temperature is 78°F
   - the hour is 2 PM (1:30 to 2:30)
   - the wall weighs 20 lb/sq. ft.
   - the wall faces southwest
   - the stucco is white

   Heat gain = **2,150 BTUH**
   (1,600)(0.042)(32) = 2,150

7. Find an average ETD for Wall A if all the givens from question 6 are repeated except the wall area is equally divided among four compass orientations (i.e., 25% faces north, 25% faces east, 25% faces south, and 25% faces west).

   Average ETD = **24.5°F**
   (22 + 22 + 32 + 22)/4 = 24.5

R and U Values, p. 200

8. Find a U value for Roof B shown below.

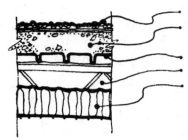

Built-up root 3″ lightweight concrete (30 pcf) steel deck-18 ga. 3″ air space R-11 fiberglass with foil vapor barrier facing air space?

DANGER: *see question* 9 wire mesh support for insulation

Roof B Section

R Values

Air film outside =	0.2
3/8″ built-up roof =	0.3
2″ lightweight concrete (30 pcf) =	2.0
Steel deck-18 ga. =	0.0
3″ air space* =	3.0
R-11 fiberglass with an aluminum foil vapor barrier facing the air space =	11.0
Wire mesh support =	0.0
Air film inside =	0.7
R total	17.2

$$U = 1/R \text{ total } = 0.058 = \mathbf{0.06}$$

9. If the outdoor temperature is 10°F and indoor conditions are 70°F dry bulb and 55°F wet bulb, will moisture condense on the aluminum foil vapor barrier in Roof B?
   *Begin by finding the temperature of the foil that is proportional to the R value:

   R from outside to foil = 5.5, R total = 17.2, temperature difference = 60°F, foil temperature = 29°F

   *Then find the dew point temperature of the indoor air on page 194; dew point = 43°F Since the foil is 29°F and the air dew point is 43°F, moisture **will condense** and ruin the ceiling insulation. Flip the insulation so that the vapor faces down and revise the U value calculation in question 8.

---

*An average air space R value has been used in this question. More precise references would give a lower R value for air spaces in roofs during winter and a higher R value for air spaces in roofs during summer. Why?

Heat Loss, pp. 203–207

10. Find the total heat loss through Roof B during one hour on a cold winter night if:

    • the roof area is 8,400 sq. ft.
    • the roof U value is 0.06
    • the outdoor temperature is 0°F
    • the indoor temperature is 70°F

    Heat loss = **35,280 BTUH**
    (8,400)(0.06)(70-0) = 35,280

Heat Gain, pp. 211–218

11. Find the total heat gain through Roof B during one hour on a hot July day if:

    • the roof area is 8,400 sq. ft.
    • the roof U value is 0.06
    • the hour is 4 PM (3:30 to 4:30)
    • the roof weighs 9.6 lb/sq. ft.
    • the roof color is black

    Heat gain = **30,240 BTUH**
    (8,400)(0.06)(60) = 30,240

    Find the heat gain if the roof color is changed from black to white:

    Heat gain = **18,144 BTUH**

12. Find the total heat gain through Roof B during one hour on a hot July day if all conditions are the same as question 10 except the roof is flooded with water.

    Heat gain = **10,080 BTUH**
    (8,400)(0.06)(20) = 10,080

13. A 4-foot square window faces southeast. Find the summer heat gain through the window at 8 AM and at noon if the window is single glass with dark blinds. Outdoor temperature is 95°F and indoor is 75°F. (U value, p. 200, SF and SC, p. 216)

    8 AM:

    Solar gain is (SF)(area)(SC)
    (160)(16)(70%) =   1,792 BTUH
    Conducted gain is (U)(A)(TD)
    (1.1)(16)(95-75) =   352 BTUH
                  Heat gain **2,144 BTUH**

    Noon:

    Solar gain is (SF)(A)(SC)
    (80)(16)(70%) =   896 BTUH
    Conducted gain is (U)(A)(TD)
    (1.1)(16)(95-75) =   352 BTUH
                  Heat gain **1,248 BTUH**

Heat Loss and Heat Gain

14. Find the winter heat loss or gain by a 4-foot-square window facing south on a clear January day at noon. Window is single glass with no blind. Outdoor temperature is 10°F, and indoor is 70°F.

    Conducted loss is (U)(A)(TD)
    (1.1)(16)(70-10) =   −1,056 BTUH
    Solar gain is (SF)(A)(SC)
    (250)(16)(90%) =   3,600 BTUH
    Net gain   **2,544 BTUH**

15. Same as question 14 but window faces west.

    Net loss   **−768 BTUH**

16. Estimate the total solar gain through 1 sq. ft. of single glass during three winter months if the glass faces south with no drape. Assume 50% clear days, 50% cloudy days, and a total of 90 days.

    Clear days
    (45)(1,600)(90%) =   64,800 BTU
    Cloudy days (use north value)
    (45)(120)(90%) =   4,860 BTU
    Heat gain   **69,660 BTU**

17. Consider the 1 sq. ft. glass area used in question 16. If the average outdoor temperature is 42°F and the indoor temperature is 70°F, will the glass gain or lose heat during three winter months? How much?

    Conducted loss is (U)(A)(TD)(90 days)(24 hr)
    (1.1)(1)(70-42)(90)(24) =   −66,528 BTU
    Solar gain from question 16 =   69,660 BTU
    Net gain   **3,132 BTU**

Outdoor Air

18. Find the expected infiltration CFM and ventilation CFM for a 20,000 sq. ft. department store.

    • 22′ high conditioned space
    • built in 1975, no operable windows
    • no smoking allowed

    Expected infiltration in winter = 0.75 air change
    (20,000)(22)(0.75)/60 =   **5,500 CFM winter**
                            3,666 CFM summer
    Design ventilation all year = 0.2 to 0.3 CFM per sq. ft.—estimate 0.25
    (20,000)(0.25) =   **5,000 CFM all year**

19. Find the expected winter peak hourly heat loss due to outdoor air for the department store in question 18.

- indoor temperature = 72°F
- outdoor temperature = 10°F
- use largest CFM from infiltration or ventilation

(CFM)(1.08)(TD) =   sensible loss
(5,500)(1.08)(72-10) =   **368,280 BTUH**

20. Find the expected summer heat gain due to outdoor air for the department store in question 18.

- indoor temperature 75°F
- indoor relative humidity 50%
- location: New York City (92°Fdb, 74°Fwb)
- use largest CFM from infiltration or ventilation
- use psychrometric chart

(CFM)(1.08)(GD) =   sensible gain
(5,000)(1.08)(92-75) =   91,800 BTUH
(CFM)(0.68)(TD) =   latent gain
(5,000)(0.68)(97-65) =   108,800 BTUH
Heat gain   **200,600 BTUH**

21. Find the heat gain due to lighting for a school with a total floor area of 6,000 sq. ft.

Estimate 2 watts per sq. ft.
Heat gain = (watts) (3.4) =   **40,800 BTUH**
(6,000)(2)(3.4) = 40,800

22. Find the heat gain caused by 100 people, standing, doing light work.

Heat gain is (# of people) (BTUH each)
Sensible gain (100)(275) =   27,500 BTUH
Latent gain (100)(475) =   47,500 BTUH
Heat gain   **75000 BTUH**

Latent gains are tabulated separately from sensible gains because cooling equipment capacity drops as the latent load decreases.

CHAPTER 16

Heating and Cooling Equipment

What a splendid day!
no one in all the village
doing anything!

Shiki

This chapter explains the equipment that produces or moves heat in buildings. Operating cycles are described so that readers can understand the factors that affect heating and cooling efficiency. Strive to accomplish the following four objectives as you read this chapter.

1. Build your knowledge and vocabulary by learning the differences between strip heat and furnaces; absorptive and compressive cooling cycles; and an "air conditioner" and a heat pump.
2. Explain the advantages of hydronic heating systems compared to forced-air, and argue in favor of water-air heat distribution in large buildings.
3. Learn that refrigeration and cooling cycles do not create "cool"; instead they remove heat from a cool place and discard heat in a warm place.
4. Be able to discuss the significant factors that affect heating and cooling equipment efficiency, and to calculate efficiency consequences of varying SEER and COP ratings.

On completion, check your new skills by completing the review questions at the end of the chapter.

16.0 HEATING EQUIPMENT

Heating equipment may use combustion or electrical energy. Selection is usually an economic decision based on initial and operating cost estimates. The quantity of heat produced by heating equipment is measured in BTU (British thermal units); 1 BTU is defined as the quantity of heat required to increase the temperature of 1 pound of water 1°F.

Boilers, furnaces, and heat pumps are rated in BTUH. They are selected to meet a building's calculated peak heat loss.

COMBUSTION HEATING

Fuels

Natural gas, oil, and coal are all potential fuels for building heating. See Table 16.1. When available, natural gas-fired equipment usually offers the lowest initial cost because fuel storage is not required. Oil-fired systems are more expensive to install because they include a storage tank and a more complicated burner. Coal is the most expensive installation because of storage, handling, and ash disposal requirements. Good operating cost projections should be developed to ensure selection of a heat source and equipment that will provide the lowest life cycle cost.

Furnace or Boiler

Furnaces heat air, while boilers heat water. Warm-air furnaces can provide effective heating in smaller buildings where the longest duct run extends less than 100 feet from the furnace. Boilers are preferred for larger

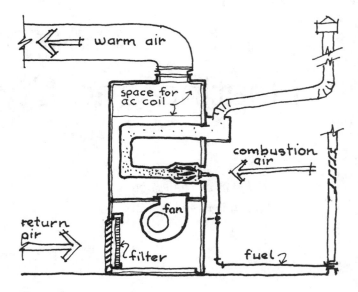

Furnace

FIGURE 16.1

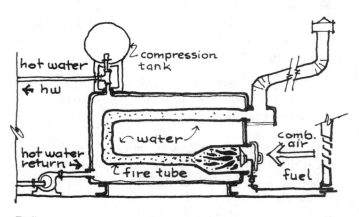

Boiler

FIGURE 16.2

projects because hot water carries heat over long distances more economically than warm air. In particular, water piping requires less building space than duct work, and pumps are more efficient than fans in moving large quantities of heat (see Figures 16.1 and 16.2).

Efficiency

The efficiency of combustion equipment depends on maintaining the correct air-fuel ratio and on minimizing flue heat losses. Thermal efficiency for well-maintained furnaces or boilers approaches 80%, and "pulse-type" burners that extract heat from flue gasses can reach 95%.

TABLE 16.1

Fuel Heat Content	BTU
Coal (anthracite) per lb	14,000
Electrical per kWh	3,400
Gasoline	125,000
Natural gas per MCF	1,000,000
Oil #2 per gallon	140,000
Propane per gallon	95,500
Wood per lb	5,000–7,000

Safety

Safe installations of combustion equipment require:

1. A continuing supply of combustion air to each furnace or boiler (*not* conditioned air from the building).
2. A fire-safe chimney or flue to discharge combustion gasses above the building roof. A test must be performed to verify that there is no carbon monoxide in the combustion gases.
3. Limit controls to shut off fuel supply if safe operating temperatures are exceeded.
4. Natural gas supply piping that rises above the ground before entering a building (to prevent underground gas leaks from following the pipe into a building).
5. Vented, fire-resistive chases for gas riser piping in multistory buildings.

ELECTRIC HEATING

Two different types of electric heating are used in buildings. Resistance heaters deliver 3,400 BTU for each kilowatt (kW) of electrical input. Heat pumps deliver more BTUs per kW because they take extra heat from the winter environment.

Resistance (Electric Furnace)

Resistance heating uses electric current to heat Nichrome wire (see Figure 16.3). Familiar small examples include hair dryers and coffee pots. In buildings, resistance heaters are called *strip heat* if they heat air, and *electric boilers* if they heat water. Resistance heaters also come in the form of baseboards that are equipped with built-in thermostatic controls. They are available from 500 watts up, in increments of about 250 watts. Standard-watt-density baseboards are normally 250 watts (an output of about 850 BTUH) per linear foot, while low-watt-density baseboards are 140 to 235 watts (an output of about 480 to 800 BTUH) per linear foot unit. Resistance heating is usually the least expensive equipment to install, and the most expensive to operate. Do not consider resistance heating unless building heating requirements are minimal or electricity costs are especially low.

Heat Pump

A heat pump is a reversible refrigeration machine that takes heat from a cool source and delivers it to a warmer location (see Figure 16.4). Initial cost for a heat pump is higher than strip heat and most combustion heating equipment, but a heat pump can cool a building during summer months.

Electric heat pumps can operate as economically as combustion equipment in mild climates because of the "free" heat they take from the outdoor environment. Like combustion furnaces or boilers, heat pumps can be designed to warm air or heat water depending on the size of the building served. Section 16.5 provides further information about heat pump operation and efficiency.

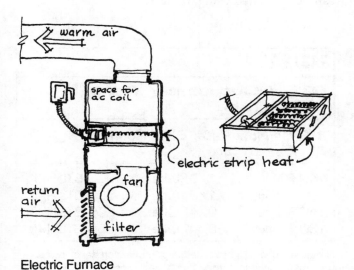

Electric Furnace

FIGURE 16.3

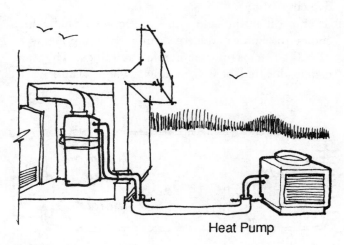

Heat Pump

FIGURE 16.4

Efficiency

The efficiency of electric heating is a complex topic. Almost all the electrical energy passing through a building meter is converted to heat by resistance heaters or heat pumps, and each kW of electrical energy yields 3,400 BTU. However, a generating plant that converts fuel energy into electrical energy will consume about 10,000 BTU in fuel for each kW produced. If the efficiency of the generating plant is considered, electric heating will be rated about 33% efficient. However, the extra heat that electric heat pumps take from the environment can make them competitive with combustion heating equipment in mild climates.

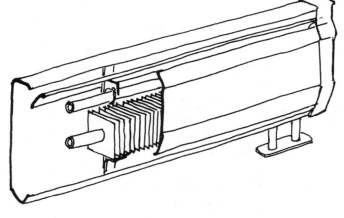

FIGURE 16.6

16.1 HYDRONIC HEATING

Four components shown in Figure 16.5 comprise a "hydronic" heating installation. Hydronic systems use water as a heat transfer fluid because water can carry BTUs more economically than air. Pumps use less energy than fans, and piping requires less building space than ducts.

Heating system designers choose one of three hydronic heat exchanger types:

* Convectors or radiators
* Radiant surface
* Forced-air

FIN-TUBE CONVECTORS

Convectors warm room air by contact. Warmed air then rises and is replaced by denser cool air (convective circulation).

Fin-tube convectors are an economical heating choice for homes and are frequently components of HVAC systems in commercial buildings. They are usually located below windows where rising warm air overcomes cold drafts and minimizes condensation. Fin-tube convectors are sometimes called "radiators," but most of their heat transfer is convective (see Figure 16.6).

Construction, Temperature, and Flow

Aluminum fins 2-1/2″ square are fitted on 1/2″ or 3/4″ copper tube. Two water supply temperatures, 140° or 180°F, are used. A protective cover 6 to 8 inches tall is usually attached, but fin-tubes can be built into counters or recessed in the floor below full-height windows. Design water flow through a fin-tube is usually about 1 GPM (gallon per minute).

Fin-tube length for a given BTUH heating load can be estimated using Table 16.2.

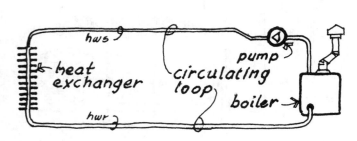

FIGURE 16.5

TABLE 16.2		
BTUH Output per Foot of Fin-Tube		
1 GPM Flow Rate, 65°F Room Air Temperature		
140°F input	1/2″ tube	300 BTUH/ft.
180°F input	1/2″ tube	550 BTUH/ft.
140°F input	3/4″ tube	300 BTUH/ft.
180°F input	3/4″ tube	550 BTUH/ft.

Note that tube size has little effect on heating capacity, but the smaller tube increases pump load. Increased flow (GPM) increases BTUH output slightly, but it's not worth the increased pumping cost.

Pipe Loops

Figure 16.7 shows *one-pipe* and *reverse return* piping loops used to serve fin-tube convectors. A one-pipe loop is economical, but water temperature drop in the loop means downstream fin-tubes must be longer for a given heat output. A reverse return loop is superior to a one-pipe loop because it provides the same water supply temperature at each fin-tube. Constant inlet temperature makes sizing easier and allows better regulation of heat output. It distributes proper quantities of water to each pipe because all loops are of the same length. The system is essentially self-balancing.

Sizing Examples

Given:

1 GPM flow through 1/2-inch fin-tube, 140°F water supply temperature, 65°F room air temperature (at floor).

1. Find the required fin-tube length for a room with a heat loss of 3,600 BTUH.
2. Find the required baseboard radiator length for a room with a heat loss of 1,800 BTUH.
3. Find the panel radiator area needed for a room with a heat loss of 1,800 BTUH.*

PANEL RADIATORS

Panel radiators can replace fin-tubes in rooms where available wall length is limited. They're made from sheet steel or aluminum and are available in a variety of sizes for horizontal or vertical installation. Panel radia-

tors project about 3 inches from the wall surface, but high-output panels have finned backs and may project 6 to 8 inches into the heated space. Typical supply piping is 1/2″ or 3/4″ inch, and heat output per square foot of panel is nearly equal to a fin-tube convector's output per linear foot under similar operating conditions.

BASEBOARD RADIATORS

Extruded aluminum baseboard radiators are a second type of panel radiator (see Figure 16.8). They are smaller than fin-tubes, and some find them better looking. Baseboard radiators snap onto wall-mounted clips. They project about 1 inch from the wall and stand about 5 inches high. Their heat output per foot of length is about half as much as a fin-tube convector, so required length is twice as long for a given heating load.

PUMPS AND ACCESSORIES

Air ejectors (also referred to as air separators) are installed to remove air in the piping loop. Air bubbles can damage the pump impeller and impede water circulation.

Manual air valves and automatic air vents should be used to purge air from the system. Even though hydronic systems are usually closed, fresh make-up water may contain air bubbles that must be released from the piping through air vents.

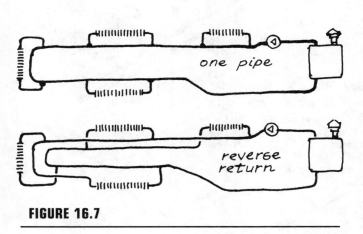

FIGURE 16.7

*Answers: 1. 12 feet 2. 12 feet 3. 6 sq. ft.

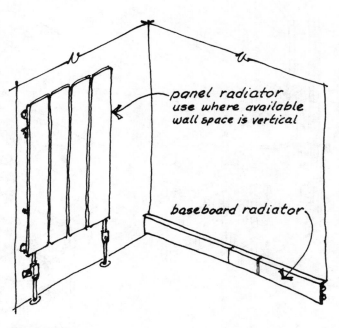

panel radiator
use where available
wall space is vertical

baseboard radiator

FIGURE 16.8

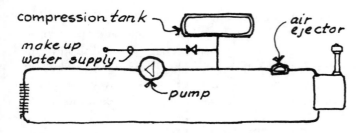

FIGURE 16.9

A compression tank must be installed in the piping loop (see Figure 16.9). It is located to maintain constant pressure at the pump outlet. An expansion tank may be installed as an alternative to the compression tank.

Flow control valves must be installed in the system. They are required to prevent the flow of water by gravity when the pump is not operating.

Most residential hydronic heating systems circulate less than 10 GPM, and the pump motor is usually less than 1/3 hp. Pumps and GPM for large buildings are discussed on the following pages.

16.2 **RADIANT HEATING**
RADIANT SURFACE

Radiant surface heating is efficient, quiet, invisible, draft-free, and comfortable. The radiating surfaces can be floors, walls, or ceilings, but only floors, warmed by water circulated through embedded piping, are covered here (see Figure 16.10).

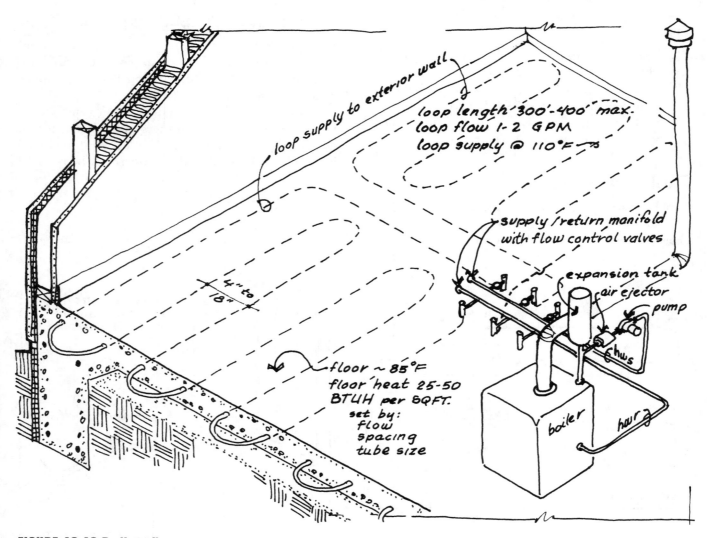

FIGURE 16.10 Radiant floor

Construction

Serpentine pipe loops are embedded in a concrete floor slab, or in lightweight concrete or gypsum poured on insulated floors. Pipe is usually PB (polybutylene) or PEX (polyethylene) plastic tubing in 3/8″ to 3/4″ sizes, but copper is also used.

Maximum tubing length is 300′ for 1/2″ tube and 400′ for 3/4″ tube. Tube spacing varies from 4″ to 18″, but 12″ center to center is fairly typical when heating requirements are not extreme.

Zones, Temperature, and Flow

Serpentine tubing loops form heating zones, and water flow in each zone is adjusted at a distribution manifold. The tube loop serving each zone is positioned so that the hottest water flows near exterior walls first.

A mixing valve limits water supply temperature to 110°F, and the design floor surface temperature is 80° to 85°F. Water flow in each zone ranges from 1 to 2 GPM, and heating capacity ranges from 25 to 50 BTUH per square foot of floor area.

A disadvantage of radiant floor heating is slow response time. It can take a long time to warm up a cold concrete slab, but once comfortable conditions are established they can be maintained economically.

16.3 FORCED-AIR HEATING

FORCED-AIR

When a fan is attached to a heat exchanger, the combination can deliver many more BTUH than the convective and radiant equipment illustrated on preceding pages. In residential applications, *central air*, *forced-air*, and *furnace* are appropriate descriptors.

In commercial buildings, a heat exchanger–fan combination is called an "air handler" or a "fan coil." Air handlers distribute air to many rooms through a large duct system, while smaller fan coils serve a single room or area without ducts (see Figure 16.11).

Fans increase heat transfer, but they also increase drafts and heating costs. However, air handlers and fan coils are used extensively in commercial buildings because they can be used for both heating *and* cooling.

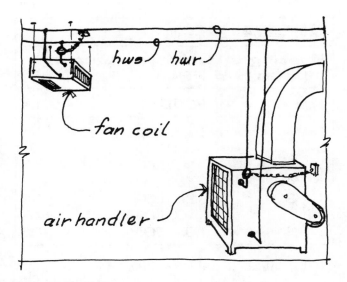

FIGURE 16.11

HYDRONIC GPM, BTUH, AND PIPING SIZES

Water flow and temperature drop across a heat exchanger determine heat output. Flow quantity and velocity set pipe and pump sizes.

Water flow is measured in GPM (gallons per minute), and heating requirements are calculated in BTUH (BTU per hour). The following formula is used to quantify heat output when water is used as a medium of heat transfer:

$$Q_{hydronic} = (GPM)(500)(TD)$$

where:

$Q_{hydronic}$ = Heat output in BTUH
GPM = Flow of water in gallons per minute
TD = Temperature difference in °F

In heating applications, a 20°F temperature drop across a heat exchanger is usual, so 1 GPM will yield about 10,000 BTUH (see Table 16.3).

TABLE 16.3

$Q_{hydronic}$ = (GPM)(500)(TD)

A 1 BTU change will alter the temperature of 1 pound of water 1°F.
1 gallon of water weighs 8.33 lb.
1 GPM = 60 gallons/hour = 500 lb/hour.
TD = temperature difference in °F.
At 20°F TD, 1 GPM yields (1)(500)(20) = 10,000 BTUH.

Note: In radiant floor installations, temperature drop is only 10°F, so 1 GPM yields only 5,000 BTUH.

TABLE 16.4

Heating BTUH (type L copper pipe at 3-5 FPS)

Size	20°TD	GPM
1/2"	30,000	3
3/4"	60,000	6
1"	100,000	10
1-1/4"	160,000	16
1-1/2"	240,000	24
2"	400,000	40
3"	1,000,000	100

To estimate forced-air heating GPM for a building or a room, divide the calculated heat loss in BTUH by 10,000. Water velocity in the piping loop is held below 5 feet per second (FPS) to limit pump load. Table 16.4 shows approximate loop piping sizes for various heating loads.

16.4 REFRIGERATION CYCLE

Cooling equipment moves heat from cool indoor spaces to warmer outdoor locations (see Figure 16.12). It moves heat by causing a refrigerant to evaporate and condense. Refrigerants capture a lot of heat when they evaporate, and the captured heat is released when refrigerant vapor condenses.

The evaporating or condensing temperature of a refrigerant fluid is dependent on the pressure acting on the fluid. Water evaporates (boils) at sea level pressure at 212°F. However, water on a mountaintop will boil at a lower temperature, and water in a pressure cooker must be raised to a higher temperature before boiling occurs. Water can be used as a refrigerant, but a deep vacuum is required to sufficiently lower its evaporating temperature. Most building cooling equipment uses halocarbon* refrigerant compounds instead of water.

---

*Chlorofluorocarbons (CFCs) were dominant refrigerants from 1931 until the mid-1990s when manufacture stopped to preserve upper atmosphere ozone. Hydrofluorocarbons (HFCs) are slowly replacing CFCs in existing refrigeration equipment. HCFC-22 used in most residential air conditioners and heat pumps is scheduled for phaseout in 2030. Water or ammonia is also used as a refrigerant, but negative circulating loop pressure (vacuum) is required. See Figure 16.14.

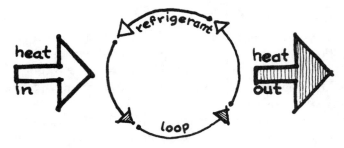

Cooling equipment moves heat

FIGURE 16.12

PRESSURE-TEMPERATURE

Cooling equipment maintains a low-pressure *evaporator* and a high-pressure *condenser* in a closed loop of circulating refrigerant. In the low-pressure evaporator, refrigerant boils at 45°F as it takes heat from indoor air. In the high-pressure condenser, hot refrigerant vapor releases heat to outdoor air when it condenses at 125°F.

A *compressor* circulates refrigerant through the loop, and an *expansion valve* maintains low pressure on the suction side of the compressor and high pressure on the discharge side (see Figure 16.13).

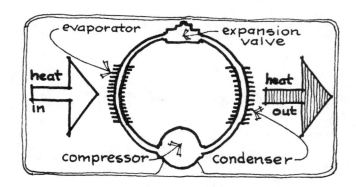

4 cooling cycle components

FIGURE 16.13

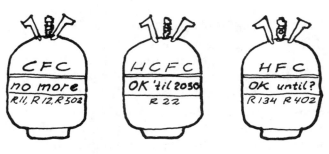

FIGURE 16.14

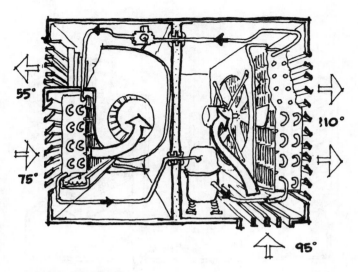

FIGURE 16.15

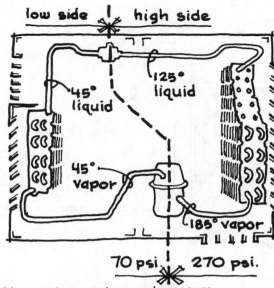

Refrigerant temperature and pressure

FIGURE 16.16

AIR CONDITIONER

The section view in Figure 16.15 shows a window-mounted air conditioner. As the indoor and outdoor temperatures indicate, it moves heat uphill from a cool location to a warm one. An air conditioner's efficiency is a function of indoor/outdoor temperature difference. Greater temperature difference means more input energy and less efficiency, just as driving a car uphill requires more gasoline input and yields fewer miles per gallon. Adding resistance heat strips or using a reversible refrigeration cycle will allow the air conditioner to heat or cool. Although just a small window unit is illustrated, the refrigeration cycle components and temperatures are similar in large equipment.

MOVING HEAT

The schematic drawing in Figure 16.16 shows temperatures and pressures within the refrigerant circulating loop. Low pressure on the suction side of the compressor permits the refrigerant to evaporate and capture heat from 75°F indoor air. High pressure on the discharge side of the compressor permits heat release (by refrigerant condensation) to 95°F outdoor air. The schematic shows a window unit, but the cycle is similar for large cooling equipment.

JARGON

A window-mounted air conditioner is called an air-cooled, DX, packaged unit. DX refers to the direct expansion of refrigerant as a means of cooling indoor air. In this system, the refrigerant is circulated in a coil and air moves around it. Larger models of this type of air conditioner are often located on building roofs and connected to duct systems serving several rooms. Rooftop units waste energy when they circulate conditioned air through a poorly insulated outdoor enclosure.

"Split" air-conditioning systems avoid energy waste by keeping conditioned air inside the conditioned space. The outdoor section shown in Figure 16.17 is called a condensing unit, and the indoor section is an *air handler*; the connecting refrigerant piping includes liquid and suction lines.

AIR-CONDITIONING EQUIPMENT

Miles per gallon (mpg) is an index of automobile efficiency. Air-conditioner efficiency is measured by a similar index called SEER (seasonal energy efficiency ratio). The SEER rating is an index of an air conditioner's miles per gallon. It is the number of BTUs removed by 1 watt of electrical energy input. Air conditioners are available with SEER ratings that range from 8 to more than 14.

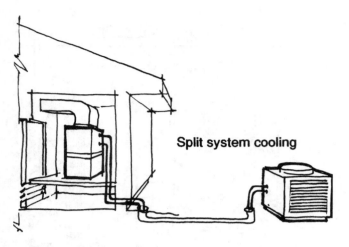

Split system cooling

FIGURE 16.17

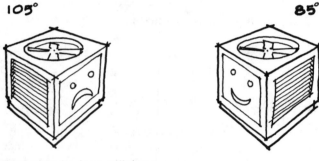

105°   85°

More load = less efficiency

FIGURE 16.18

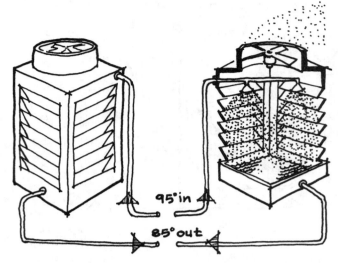

95° in

85° out

Return water is cooler than air

FIGURE 16.19

An automobile's mpg changes with load; a car will get less mileage going uphill and more mileage going downhill. An air conditioner's efficiency also depends on load (see Figure 16.18). Cooling load is the difference between heat source and heat sink temperature. Indoor air is the usual heat source, and outdoor air or cooling tower water is the usual heat sink. An air conditioner's efficiency can be increased by increasing indoor temperature or by lowering outdoor temperature. Either temperature change means less work for the equipment.

Manufacturers of efficient air-cooled equipment mate small compressors with big condensers. Increased condenser size reduces condensing temperature, cutting compressor load and input energy.

COOLING TOWERS

Water-cooled refrigeration equipment can achieve higher SEER ratings than air-cooled equipment because it is possible to cool water to near the wet-bulb

temperature of outdoor air. SEER ratings of chillers range from 15 to more than 24. A cooling tower is an outdoor shower that cools water by evaporation (see Figure 16.19). Cooling towers can cool water 10° to 15°F below ambient air temperature, and this cool water is used to lower condensing temperature and increase SEER. Each ton* of cooling load requires about 3 gallons per minute of cooling tower water, and about 5% of this water is lost to evaporation and drift. River, lake, or well water may offer lower temperatures and higher SEER potential.

ABSORPTION

Absorption-cycle cooling equipment uses heat instead of electricity as input energy (see Figure 16.20). A gas flame, hot water, steam, or other heat sources may be used to drive the absorption cycle.

Two components in absorption equipment perform the same jobs as their counterparts in the compressive refrigeration cycle (see Figure 16.21). Refrigerant picks up heat in the *evaporator* and releases heat in the *condenser*. Water can be used as the refrigerant in absorption equipment. An internal vacuum is created, allowing water to evaporate at low temperature.

Three absorption-cycle components replace the compressor: in the *absorber* a salt solution absorbs

---

*A ton is a heat-moving rate of 12,000 BTU per hour. Ice plants are rated in tons per day. A ton of water can be converted to a ton of ice by removing 288,000 BTU (12,000 BTU per hour for 24 hours).

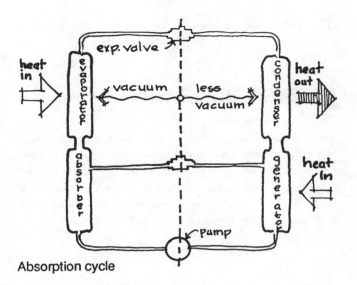

Absorption cycle

FIGURE 16.20

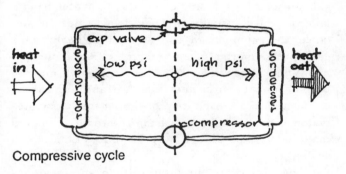

Compressive cycle

FIGURE 16.21

water vapor from the evaporator, a *pump* circulates the diluted salt solution and maintains system pressure difference, and a *generator* boils off water to reconstitute the salt solution.

Efficiency

Unfortunately, absorption-cycle equipment is less efficient than compressive refrigeration. The most efficient absorption equipment uses almost twice as much energy per ton as the best compressive competition. However, absorption equipment may be an excellent choice when waste heat is available from manufacturing processes, electrical generation, or environmental sources.

Absorption machines are large. They can occupy five times as much space as compressive machines of equal capacity (see Figure 16.22).

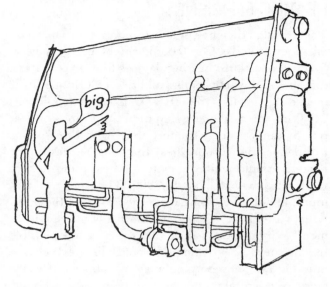

FIGURE 16.22

16.5 **HEAT PUMPS**

A heat pump is a reversible refrigeration cycle machine that can cool or heat. As a cooling unit, it takes heat from a building and rejects it to the warm summer environment. As a heater, it takes heat from a cool winter environment and delivers it to a building. A four-way reversing valve is used to reverse refrigerant flow, permitting the heat pump to heat or cool (see Figure 16.23).

Like an air conditioner, heat pump efficiency depends on the temperature difference between the environment and indoor air. As the environment gets cooler, a heat pump will deliver less heat.

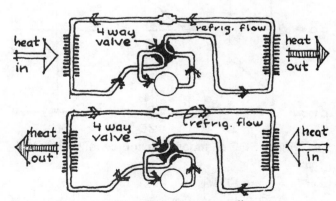

Reversing valve allows heating or cooling

FIGURE 16.23

HEATING EFFICIENCY

The heating efficiency of a heat pump is measured by an index called COP (coefficient of performance). Electric resistance heating devices like strip heaters, toasters, and water heaters all have a COP of 1. This means they deliver 3,400 BTU from each kilowatt (kW) of electrical energy input. COP ratings of heat pumps range from 1 to more than 4, so heat pumps can deliver much more heat than electric resistance heaters in mild winter climates. A heat pump with a COP of 3 will deliver 10,200 BTU from a 1 kW input; 3,400 BTU come from the kW; and the additional 6,800 BTU are taken from the winter environment. Winter COP values of 2 or more are possible for air-source heat pumps in cold climates because outdoor air temperatures may be above 20°F for many of the total winter heating hours.

As shown in Figure 16.24, heat pump performance improves with increasing environmental temperature and decreases with decreasing environmental temperature. Therefore, a heat pump produces less heat during cold weather, when it's needed most. To overcome the cold weather output problem, supplemental electric resistance heat is installed to augment heat pump output. Inefficient supplemental resistance heating is operated only when heat pump output is inadequate.

HSPF (heating seasonal performance factor) is a heating efficiency index used instead of COP for residential heat pumps. Like SEER, it is measured in BTU per watt, so an HSPF of 6.8 would equal a COP of 2.

AIR-SOURCE HEAT PUMPS

The air-source heat pump shown in Figure 16.25 uses outdoor air as a heat source or heat sink. During cold, damp weather, ice can form on the heat pump's outdoor coil, and a defrost cycle is required to permit continuous operation. Supplemental strip heaters must operate when defrosting because the heat pump is taking heat from indoor air to melt ice outdoors.

WATER-SOURCE HEAT PUMPS

Water-source heat pumps use water instead of outside air as a heat source or heat sink. Water from a lake or well is usually warmer than winter air and cooler than summer air, so water-source heat pumps can operate more efficiently than air-source units.

When water is not available, the earth may be used as a heat source and sink. A heat pump can be connected to a buried loop of piping that circulates water to give or take heat from the earth. The loop may be vertical like a well or horizontal like a sewer.

The illustration in Figure 16.26 shows an indoor DX air handler, but hydronic heat pumps that produce hot or chill water are available. This hot water could be circulated in a radiant floor-heating installation.

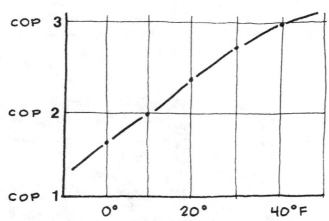

Heat Pump COP changes with source temperature
COP?
COP 1 = 3,400 BTU per kW
COP 2 = 6,800 BTU per kW
COP 3 = 10,200 BTU per kW

FIGURE 16.24

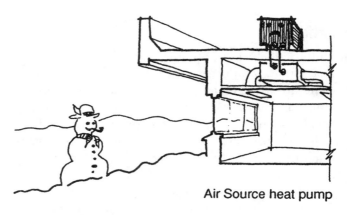

Air Source heat pump

FIGURE 16.25

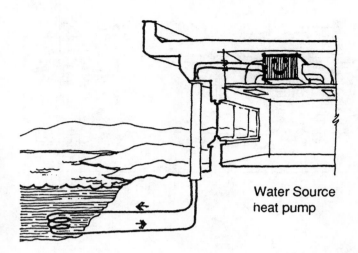

FIGURE 16.26

Water Source
heat pump

LOOP-CONNECTED HEAT PUMPS

Supermarkets often do all their winter heating with heat rejected by refrigeration equipment. In this ideal situation, refrigeration energy input is used twice. First it cools food, and then it heats the store. Like supermarkets, larger buildings often require heating and cooling at the same time. Heat required at the building perimeter during winter months may be equal to the cooling required at the building's center due to lights and people.

When building heating and cooling loads are balanced, heat pumps connected by a loop of water piping will deliver excellent heating and cooling efficiency (see Figure 16.27).

When heating and cooling loads are out of balance, a boiler and a cooling tower must be used to

keep loop water temperature in the range where the heat pumps operate efficiently.

16.6 **LARGE AIR-CONDITIONING EQUIPMENT**

The schematic in Figure 16.28 shows cooling components of an air-conditioning* installation for a large building.

Component 1 is an *air handler* that mixes fresh and return air, cools and dehumidifies the air mixture, and distributes conditioned air to building spaces. Many air handlers are used in large buildings to achieve comfort control for a variety of building spaces or zones.

Component 2 is called a *chiller*. It includes all parts of the refrigeration cycle described earlier (i.e., evaporator, condenser, compressor, and expansion valve), but instead of cooling air, it chills water that is pumped to individual air handlers. Using water to carry heat instead of air saves energy and space when transport distances exceed 200 feet.

Component 3 is a *cooling tower* that dumps building heat to the environment by evaporating water. Cooling towers can be more efficient than air-cooled condensers, and they require less space.

The sketch in Figure 16.29 shows physical examples of the components in Figure 16.28. Several considerations affect the placement of these components.

1. The **air handler** must have access to fresh and return air. Reserve 4% of the total building floor area to accommodate air handlers.
2. The **chiller** is usually located in the basement because of weight and noise. Allow 2% of the building's gross floor area for chillers, boilers, and electrical equipment. Also provide access for installing and servicing large components.
3. The **cooling tower** is noisy and wet. Select an appropriate outdoor location on the ground or roof. Allow 75 pounds per ton, and 1 square foot per ton for cooling tower weight and size. Creative designers may use fountains or pools to serve as cooling towers.

---

*Air conditioning is defined as controlling air temperature, humidity, and quality; so heating components should be added to the schematic before describing it as an air-conditioning installation. For the large equipment shown, heating components would typically include boilers, pumps, hot water piping, and heating coils in each air handler.

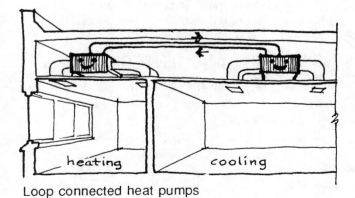

Loop connected heat pumps

heating   cooling

FIGURE 16.27

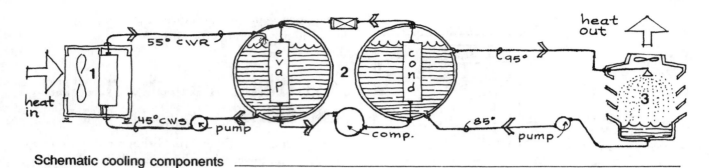

Schematic cooling components

FIGURE 16.28

Physical cooling components

FIGURE 16.29

16.7 HEATING AND COOLING

Air handlers and fan coils can be used for heating *and* cooling when hot and chilled water are circulated throughout a building. Boilers and chillers serve as heat sources or sinks, while air handlers and fan coils add or remove heat from occupied spaces. Moving heat in water instead of air cuts operating costs, so most large buildings use water as an intermediate heat carrier.

Three piping arrangements are utilized for heating-cooling applications (see Figure 16.30).

In *two-pipe* installations, the heat exchanger in each air handler or fan coil receives hot **or** chill water depending on the season. Two-pipe systems are economical, but changeover to and from heating or cooling is slow. The heat exchangers inside air handlers and fan coils are called "decks" or "coils."

In winter months, large office buildings heat the exterior zones and simultaneously cool the core zone to remove heat from lights and people. *Four-pipe* systems increase piping costs, but they allow coincident heating and cooling and easy and rapid change to and from heating or cooling. Thermostat-controlled valves select hot or chill water, and air handlers often include two heat exchangers appropriately named "hot deck" and "cold deck." (See Figure 16.31.)

Three-pipe installations save some piping dollars by using a common return. Comfort control is as good as a four-pipe system, but when some air handlers are heating and others are cooling, operating costs increase because the return water temperature is high for the chiller and low for the boiler.

AIR VELOCITY AND TEMPERATURE

Design air velocity at the face of the hot deck is usually about 700 FPM (feet per minute), and room air supply temperature varies from 80° to 130°F as controls regulate heat exchanger water flow. Design air velocity at the face of the cold deck is held near 500 FPM to encourage condensation. Room air supply temperature can be as low as 55°F. Refer to Chapter 18 for heating and cooling air quantity calculations and duct sizes.

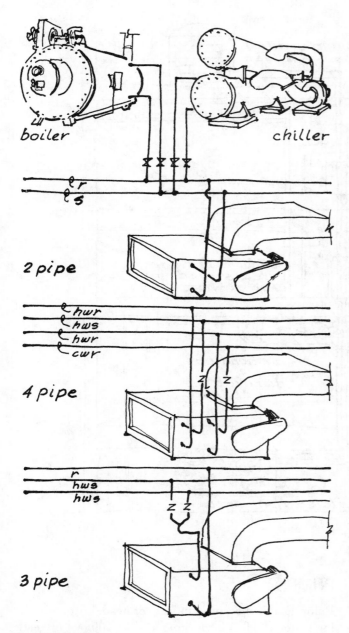

FIGURE 16.30

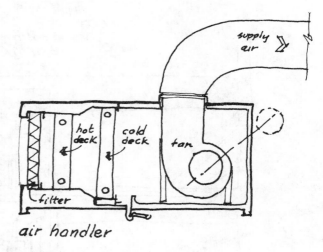

air handler

FIGURE 16.31

large heating systems circulate pressurized water at 300°F or more.

Fuel-fired boilers are 80 to 85% efficient at full load, so input for a 100,000 BTUH boiler is about 125,000 BTUH. Most heating boilers operate with a 20°F temperature difference between supply and return, so 1 GPM delivers about 10,000 BTUH.

CHILLERS

Calculated building heat gain sets chiller capacity. Many buildings can be served with a single chiller (or boiler), but designers frequently specify three or four smaller units so that a mechanical or electrical failure doesn't shut a building down. A typical chiller reduces loop water temperature by 10°F so each GPM circulated in the piping loop will carry about 5,000 BTUH.

Efficient modern chillers consume less than 0.8 kW per ton, but when the total energy required by cooling towers, circulating pumps, and air handlers is added, 3 to 4 kW per ton is not an unusual building HVAC load.

SPACE REQUIREMENTS

Boilers and chillers are large and heavy, so scheduling, delivery, and installation require careful planning. Combustion air intakes, fire-safe flues, and adequate space for retubing are design and construction considerations.

Remember to allow 4% of the floor area served for air handlers and 2% of total building floor area for boilers, chillers, pumps, and electrical switchgear.

BOILERS

The drawing in Figure 16.32 illustrates selected parts of a building heating-cooling installation using water as a heat transfer medium.

Calculated building heat loss sets total boiler capacity; oversizing boilers wastes fuel. Because water systems are safer and require less maintenance, most heating boilers deliver hot water instead of steam. Usual water supply temperature is 180°F, but very

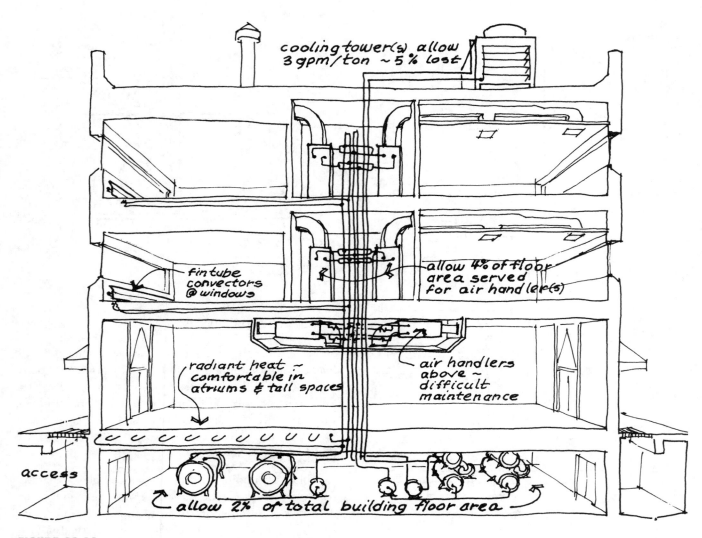

cooling tower(s) allow
3 gpm/ton ~ 5% lost

fin tube
convectors
@ windows

allow 4% of floor
area served
for air handler(s)

radiant heat ~
comfortable in
atriums & tall spaces

air handlers
above ~
difficult
maintenance

access

allow 2% of total building floor area

FIGURE 16.32

Large heavy equipment is usually located in basements or on mechanical floors in tall buildings.

BTUH, GPM, AND PIPE SIZE

The BTUH provided or removed by air-handler heat exchangers is determined by the water TD (temperature difference) from inlet to outlet and water flow in GPM. Many heating installations use a 180°F water supply temperature and design for a 20°F temperature drop across the hot deck. Cooling equipment provides chill water near 45°F and designs for a 10°F gain across the cold deck.

Designers use the chart in Table 16.5 and calculated heating and cooling loads to estimate GPM and pipe size for the entire building or individual air-handler zones.

PUMPS

Pump load is measured in "feet of head." A pump lifting water 100 feet is working against 100 feet of head. Head is caused by the weight of water and by friction as water flows through pipe, valves, and fittings. As head increases, pump output drops.

Two factors, GPM and head, must be established before selecting a pump for a particular application. GPM is set to satisfy the heating or cooling load, and head is the total friction in a closed circulating loop. In an open cooling tower, loop head is total friction plus lift.

Pump selection begins after detailing the entire pipe loop and calculating a system curve that illustrates the relationship between flow and head for a particular piping installation. An increase in loop GPM will increase pump head.

TABLE 16.5

Piping—Heating and Cooling Capacity (field numbers = BTUH)

Pipe Size	Heating 20°TD	Cooling 10°TD	GPM at 3–5 FPS*
1/2"	30,000	15,000	3
3/4"	60,000	30,000	6
1"	100,000	50,000	10
1-1/4"	160,000	80,000	16
1-1/2"	240,000	120,000	24
2"	400,000	200,000	40
3"	1,000,000	500,000	100
4"	2,000,000	1,000,000	200
6"	4,000,000	2,000,000	400
8"	7,000,000	3,500,000	700

*FPS = feet per second, water velocity. Remember to allow for at least 1" thick pipe insulation (more is better); also select hangers and supports that allow pipe movement without insulation damage.

Manufacturers provide pump curves that illustrate flow for various heads. Pumps are selected by superimposing pump curves on the system curve to identify a pump that will provide required GPM (see Figure 16.33).

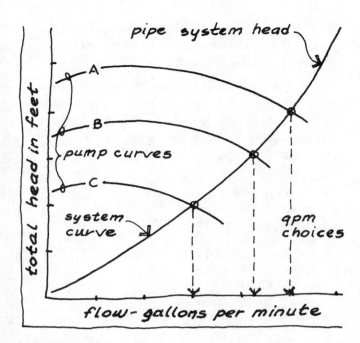

FIGURE 16.33

CONTROLS

A great variety of fittings, temperature sensors, valves, and control devices are not covered in this text. Owners seeking the best available HVAC equipment understand that an installation is only as good as its controls.

REVIEW QUESTIONS

1. Select oil or natural gas heating, choosing the most economical heating fuel based on the following:

   ◆ #2 fuel oil sells for $0.80 per gallon, and a 75% efficient oil furnace is available.
   ◆ Natural gas sells for $6.80 per MCF, and an 80% efficient furnace is available.

2. If electricity sells for $0.07 per kW, how much more expensive is electric resistance heating compared to oil heat from question 1?

3. A home has a peak heat loss of 25,000 BTUH. Estimate the total length of fin-tube convector needed to heat the home with 180°F water.

4. A 1,200 sq. ft. home with a peak heat loss of 24,000 BTUH will use hydronic radiant floor heating. Estimate total heating GPM.

5. Name the four refrigeration cycle components.

6. Which components absorb or reject heat?

7. A 4-ton, air-cooled, DX cooling unit has a SEER of 8. If the unit operates 1,200 hours during a typical summer, how many kWh (kilowatt hours) of electrical energy will it consume?

8. If a SEER 12 unit replaces the 4-ton cooling unit in question 7, what will be the annual % savings in electrical consumption?

9. The cooling tower for a 200-ton chiller will weigh about _____ pounds and circulate about _____ GPM.

10. If an electric heat pump operates with an average annual COP of 2.8, find the number of BTUs it will deliver for $1.00. Assume $0.07 per kWh.

11. How much space should be allowed for air handlers and heating-cooling equipment in a 50,000 sq. ft. building?

After completing questions 1 through 11, you should be able to set up and solve this last question before looking at the answer. Use the following heating-cooling loads, costs, annual operating hours, and efficiencies:

- oil cost is $0.78 per gallon
- electricity cost is $0.08 per kWh
- annual heating hours = 2,000
- annual cooling hours = 800
- building peak heat loss is 600,000 BTUH
- building peak heat gain is 400,000 BTUH
- boiler efficiency is 80%
- chiller SEER is 9
- heat pump SEER is 10
- heat pump COP is 2.3

12. A school board is considering two options for heating and cooling a new high school. Option A will use an oil-fired boiler for heating and an electric chiller for cooling. Option B will use electric heat pumps for heating and cooling. Find the annual heating-cooling cost for each option and select the option that will provide the lowest cost.

ANSWERS

1. Oil will be about 12% cheaper.
2. Electric resistance costs about 270% more.
3. 45 feet total
4. 5 GPM
5. compressor, condenser, evaporator, and expansion valve
6. Evaporator absorbs, condenser rejects.
7. 7,200 kWh
8. 33% savings (2,400 kWh less)

9. 15,000, 600
10. 136,000 BTU per $1
11. 2,000 sq. ft. for air handlers and 1,000 sq. ft. for heating-cooling equipment
12. Begin by finding the total annual BTU required for heating and cooling.

Heating:
$$(600,000)(2,000) = 1,200,000,000 \text{ BTU}$$

Cooling:
$$(400,000)(800) = 320,000,000 \text{ BTU}$$

Next find the heating cost with oil and heat pumps:

Oil heat cost per year = $8,357
$$(1,200,000,000)(\$0.78) \div (140,000)(80\%)$$

Heat pumps cost per year = $12,276
$$(1,200,000,000)(\$0.08) \div (3,400)(2.3)$$

Next find the cooling cost with chiller and heat pumps:

Chiller cost per year = $2,844
$$(320,000,000)(\$0.08) \div (9)(1,000^*)$$

Heat pumps cost per year = $2,560
$$(320,000,000)(\$0.08) \div (10)(1,000^*)$$
$$^*1,000 \text{ watts per kW}$$

Annual heating-cooling cost: Option A = $11,201
Annual heating-cooling cost: Option B = $14,836

Select Option A

Other factors that should be considered in the heating-cooling equipment selection process include initial costs, anticipated maintenance costs, and related energy costs for fans and pumps. Chillers are very efficient heat movers, but pumps and cooling tower fans cut their system SEER.

CHAPTER 17
Building Air Conditioning

My two plum trees are
so gracious . . . they flower
one now, one later.

Buson

This chapter builds on completed readings about heating-cooling equipment operation and efficiency. New information about air distribution is presented, and case studies are used to illustrate building applications. On completion you should be able to select air-conditioning equipment for a particular building project.

The chapter begins with a discussion of the components used to distribute conditioned air to individual rooms or zones in a building. Study four zoning alternatives that HVAC system designers consider for specific buildings to understand their comparative advantages and limitations. Also learn the difference between constant volume (CV) and variable air volume (VAV) air-conditioning systems. Some buildings are easily conditioned with one equipment type or configuration, but others require several.

Selecting air-conditioning equipment is an activity that seeks maximum quality at minimum cost. Unfortunately, increased quality usually increases *first* cost, but experienced system designers know that the benefits of a quality installation are found in improved comfort and reduced *operating* costs.

Case studies demonstrate the equipment selection process, and the residence and office examples developed in Chapter 15 are used to explore alternatives in selecting and zoning air-conditioning installations.

When you are comfortable with the material covered, complete the review questions at the end of the chapter.

17.0 **AIR CONDITIONING—ZONING**

COMPONENTS

Most air-conditioning installations use the following five components to control and distribute conditioned air in buildings (see Figure 17.1):

1. An *air handler* that controls air quantity, temperature, humidity, and quality as it filters and circulates air
2. A *supply duct* system that distributes conditioned air to each outlet in the building
3. A *return* path that carries air back to the air handler (may be ducted or an open plenum)
4. Provision for *exhaust* air to minimize odors and airborne contaminants (usually exhausted at kitchens, laboratories, and rest rooms)
5. A *fresh air* intake to maintain indoor air quality

Components 3, 4, and 5 are not shown on the following pages for simplicity, but they will be included in all quality air-conditioning installations.

ZONING ALTERNATIVES

At any given time, different rooms in a building may have different requirements for heating or cooling. On a winter day a lobby with large windows needs lots of heat, but an interior auditorium filled with people will need much less heat or perhaps cooling. HVAC system designers select from four zoning alternatives to distribute conditioned air in buildings.

1. **Individual units** separately heat or cool each room in a building.
2. **Single-zone** installations heat or cool all building rooms with a single conditioned air supply.
3. **Zoned** installations subdivide a building into zones and deliver discrete conditioned air supplies to each zone. A zone is a group of rooms with similar heating-cooling requirements.
4. **Multizone** installations heat or cool rooms (zones) by providing a discrete conditioned air supply to each room (zone).

The following discussion describes each zoning alternative. Very large buildings may use several equipment types and zoning schemes to satisfy a variety of heating and cooling requirements.

Type 1: Individual Units

Individual units are available for through-wall or in-room mounting (see Figure 17.2). They are an excellent choice for conditioning spaces like hotel guest rooms because they permit individual temperature control and prevent transfer of smoke or odors from

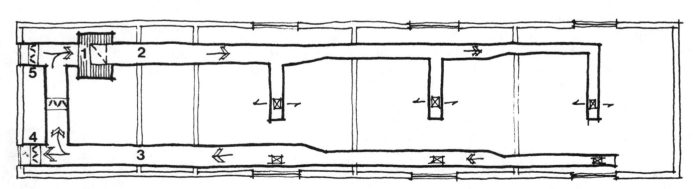

Plan - zoning components

FIGURE 17.1

Individual Unit

FIGURE 17.2

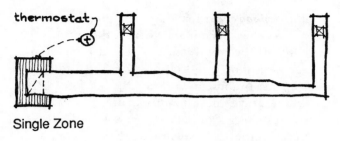

Single Zone

FIGURE 17.3

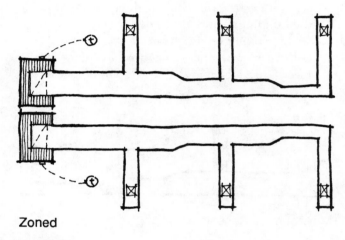

Zoned

FIGURE 17.4

one room to another. Two types of individual units are available:

1. *Fan coils* use hot or chilled water as a heat source or heat sink so boilers and chillers are required to serve them. Fan coils served with a two-pipe distribution system can heat or cool, but not at the same time. Four-pipe hot and chill water distribution is more adaptable because it permits some fan coils to deliver heat while others cool.

2. *Packaged units* contain heat-producing and heat-moving equipment. Examples include heat pumps and air conditioners with strip heat. Loop-connected heat pumps that use water circulating in a closed piping loop as a heat source or sink can be very efficient in buildings with simultaneous heating and cooling requirements.

Type 2: Single Zone

Single-zone installations are a good choice for conditioning a large room or a group of similar rooms (a zone). A single-zone system could be used for a building with a single row of identical adjoining rooms (see Figure 17.3). If each room has the same number of occupants, the same lighting, and the same window area and orientation, room air-conditioning requirements will be very similar, and a single-zone installation can provide and maintain comfortable conditions.

Type 3: Zoned

Consider a school where half the classrooms face south and the other half face north. On a sunny win-

ter day, the southern classrooms will enjoy substantial solar heat gain and therefore need less heat than the northern classrooms. A single-zone installation cannot provide optimum comfort for all classrooms because it delivers only a single air temperature.

Two single-zone air handlers, one serving southern classrooms and the other serving northern classrooms, can maintain comfortable conditions in all classrooms. When more than one single-zone air handler is used, a HVAC installation is described as "zoned" (see Figure 17.4).

Type 4: Multizone

Multizone equipment can provide excellent temperature control for buildings with a variety of air-conditioning requirements. A multizone installation can deliver warm air to one room or zone and, at the same time, provide cool air to an adjacent room or zone. Three types of multizone equipment are used in building applications: reheat, double duct, and individual duct.

Multizone reheat installations deliver cooled air to reheat terminals where a thermostat controls the supply air temperature for each room or zone (see Figures 17.5 and 17.8). Reheat equipment can provide excellent temperature and humidity control, but it wastes energy and is expensive to operate because the user pays twice (first to cool and then again to reheat).

Multizone double duct installations include separate warm and cool air ducts throughout the building. These ducts deliver conditioned air to "mixing boxes" where a thermostat adjusts inlet dampers to control air temperature (see Figures 17.6 and 17.8). Double duct

equipment can provide excellent temperature control, but double duct systems often operate in a mixing mode that wastes energy. (They mix cool and warm air to control zone temperature.)

Multizone individual duct installations include separate ducts from an air handler to each conditioned room or zone (see Figure 17.7). Thermostats control dampers in the air handler that proportion a mixture of warm and cool air for each zone served. Individual duct equipment is available in mixing or bypass configurations (see Figure 17.9).

The mixing type is essentially the same as a double duct installation except mixing occurs in a central air handler instead of remote mixing boxes. The bypass type does not mix warm and cool air; instead it blends warm or cool air with return air. Bypass installations offer good temperature control without the energy waste typical of mixing installations.

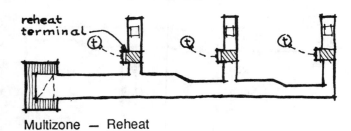

Multizone — Reheat

FIGURE 17.5

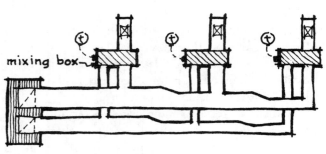

Multizone — Double Duct

FIGURE 17.6

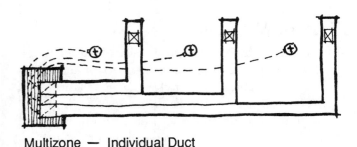

Multizone — Individual Duct

FIGURE 17.7

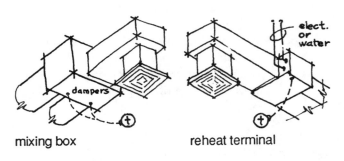

mixing box   reheat terminal

FIGURE 17.8

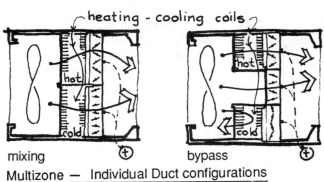

mixing   bypass

Multizone — Individual Duct configurations

FIGURE 17.9

CV OR VAV AIR SUPPLY?

Most of the preceding zoning alternatives can be equipped with CV (constant volume) or VAV (variable air volume) fans. In CV installations the temperature of conditioned air is changed to satisfy heating or cooling loads, and a constant quantity of air is provided as long as the fan operates. With VAV installations, fan output is controlled so that the quantity of conditioned air varies in response to changing heating or cooling loads.

VAV installations use less fan energy than constant-volume equipment, and fan energy can be a substantial part of total energy consumption in sealed buildings where fans may operate 8,760 hours each year. VAV equipment is specified to reduce annual energy costs. Reduced conditioned air quantity saves energy, but it can cause comfort problems including inadequate room air circulation and loss of humidity control.

A VAV installation looks similar to a constant-volume system except for a variable output fan and a VAV box or terminal that modulates the amount of air delivered to each space or zone (see Figure 17.10).

Two refinements have been developed to overcome the inadequate room air circulation problem typical of VAV installations operating with reduced air output. VAV induction terminals or VAV fan-powered terminals recirculate room air as the ducted air supply is cut back (see Figure 17.11). Unfortunately, both of these terminals increase system fan energy requirements.

SUMMARY

Individual units offer the lowest first cost for conditioning many small spaces. Single-zone equipment can offer low first cost for a large space or a group of similar spaces. Both individual units and single-zone installations can be operated economically when quality equipment is specified.

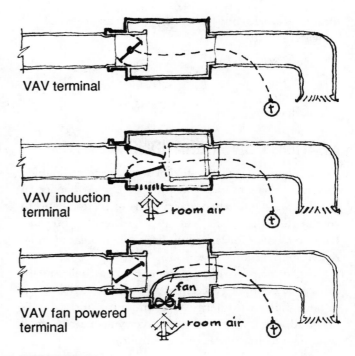

FIGURE 17.11

Zoned and multizone equipment can offer better temperature control than individual units or single-zone equipment. It is also more expensive to install and operate (except multizone individual duct bypass, which can be operated as economically as single-zone equipment).

VAV installations can cut fan energy compared to constant-volume equipment. However, reduced air movement and poor humidity control are potential VAV comfort problems that require careful zoning and a precise analysis of building heating-cooling load variations.

17.1 SELECTING HVAC EQUIPMENT

Performance requirements are the appropriate starting point for HVAC equipment selection. In a hospital, excellent temperature control is important. In a rare-book library, humidity control can be critical. In a recording studio, the allowable background noise level is a significant air-conditioning performance consideration. Designers selecting equipment for a specific building try to maximize performance and minimize cost. The selection process involves

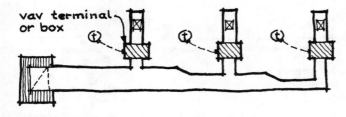

FIGURE 17.10

many variables, but four decision areas are essential considerations:

1. Heating
2. Cooling
3. Zoning
4. Efficiency

HEATING CONSIDERATIONS

Equipment alternatives for heating systems include furnaces, boilers, and heat pumps. Use a furnace in smaller buildings when warm air is to be delivered to rooms less than 100 feet away from the furnace. Choose a boiler for large buildings where heat must be delivered over long distances.

Select combustion equipment when fuel is available at a good price compared to electrical energy. Where electrical energy is economical and winters are mild, use heat pumps to deliver warm air in small buildings or hot water in large buildings.

COOLING CONSIDERATIONS

Equipment choices for cooling systems include air conditioners, heat pumps, or chillers. Use chillers in large buildings where cooling is provided by many air handlers. (Absorption chillers may be economical if waste heat is available.)

All cooling equipment must reject heat into the environment. Air-cooled condensing units or heat pumps reject heat from smaller DX equipment because they are easy to install and maintain. For larger air-conditioning installations (over 50 tons), consider cooling towers or evaporative condensers to reject heat because they require less space than air-cooled equipment and can offer improved efficiency. If lake or well water is available, it can replace a cooling tower and improve efficiency.

ZONING CONSIDERATIONS

Equipment choices for zoning include individual units, single-zone, or multizone equipment. Selection is based on operating requirements for each particular building.

EFFICIENCY CONSIDERATIONS

Designers who choose to save energy at every opportunity can produce efficient and excessively expensive buildings. It is not productive to invest in every possi-

ble energy-conserving opportunity because each added investment will return smaller energy savings.* Efficiency considerations must include specific building details. A new efficient air-conditioning system will save *less* energy in a well-insulated building than in an equal but poorly insulated structure.

The following checklist offers possible efficient equipment options.

1. Efficient furnace or boiler (up to 95%)
2. Efficient cooling equipment (SEER 12+)
3. Evaporative cooling in dry climates
4. Absorption equipment run on waste heat
5. Efficient heat pumps (COP 3 to 4 and a SEER of 12+)
6. Use of well water, lake water, or the earth as a heat source and heat sink
7. Heat recovery units to recycle heat from condensers, cooking, manufacturing, or from building exhaust air
8. Economizer equipment—dampers and controls that substitute outdoor air for cooling equipment when weather permits
9. On-site electrical generation with waste heat used for building heating and cooling
10. Use of variable-volume air distribution to reduce fan energy

17.2 CASE STUDY 1: RETAIL BUILDING

Discount Dan's" (see Figure 17.12) is a national chain retailer with a standard store plan (not a design award winner, but prices are low and profits are

FIGURE 17.12

---

*Chapter 19 of this text (Annual Costs) evaluates energy-conserving alternatives in terms of the first year's rate of return on investment.

good). The store interior is one large space. Dan's construction managers want a HVAC installation with these characteristics:

- Low initial cost
- Low operating cost
- Minimal equipment maintenance
- HVAC failure should not close the store

HEATING SELECTION

Select gas heat or heat pumps depending on the local climate and gas versus electric costs. Specify several small units instead of a single large one to permit continuing store operation should one unit fail. Contract for maintenance with a local HVAC firm that stocks replacement parts.

COOLING SELECTION

Select DX air-cooled condensing units when gas heating is chosen. (*Note:* building codes prohibit gas furnaces in plenums.)

Heat pumps and gas-electric units are available in single-package or split-system configurations (see Figures 17.13 and 17.14). Single-package units are easy to install, but split-system units are more efficient. Long refrigerant piping runs are not desirable, so locate split-system outdoor units near their indoor air handlers.

Outdoor equipment is noisy. Screen it from frequently used people areas or put it on the building roof. Rooftop equipment increases roof loads and

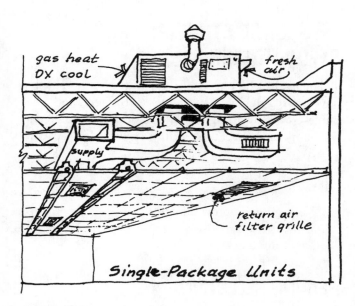

FIGURE 17.14

service access is difficult, but ground space is valued for parking and building access.

ZONING

Select single-zone, constant-volume equipment. A multizone installation is more expensive, and its ability to heat and cool at the same time is not required for the large retail space. Variable-volume equipment might save some energy, but temperature set back when the store is closed will cost less and save more.

The hung ceiling conceals equipment and ducts, and creates a plenum return path to the air handlers. Fresh air intakes are built into single-package units, but a duct is required for split-system equipment.

EFFICIENCY

Decisions about furnace efficiency, heat pump COP, and air-conditioning SEER will be based on a rate of return analysis for the more expensive high-efficiency equipment. Fan energy can be conserved by night temperature set back when the store is closed.

Comment: Single-package rooftop units are popular for stores like Dan's (a single-package unit includes heating, cooling, and air handling in one enclosure). Single-package equipment is a bit more expensive than split-system equipment, but installation is simpler. Split-systems are more efficient because they keep conditioned air inside the building instead of circulating it through a poorly insulated outdoor enclosure.

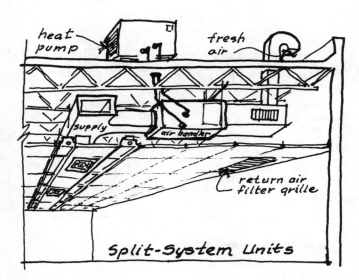

FIGURE 17.13

17.3 CASE STUDY 2: OFFICE TOWER

This elegant office tower is a tribute to a persuasive developer, a flush banker, a lucky architect, and a capable contractor. A glass salesman is a happier and richer person because of this tower (see Figure 17.15).

The developer's priorities for HVAC were as follows:

1. Very good comfort control
2. Low operating cost
3. Low initial cost

HEATING SELECTION

Select boilers because of building size. If gas or oil is not available, compare electric boilers with water-source heat pumps. Circulate hot water to air handlers throughout the building to meet heating requirements. Specify several boilers so that an equipment failure will not shut down the entire building in cold weather.

COOLING SELECTION

Select chillers and circulate chill water to air handlers throughout the building to satisfy cooling loads. Several chillers will be specified to provide partial capacity in the event of an equipment failure. Choose cooling towers to reject waste heat. Locate them at or near ground level if space is available. Otherwise, put them on the building roof.

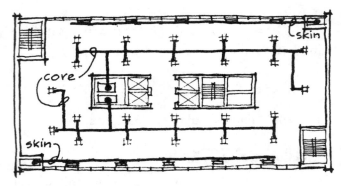

Typical Floor Zoning

FIGURE 17.16

ZONING

Divide each building floor into an interior *core zone* and exterior *skin zones* (see Figure 17.16).

Select a single-zone, VAV air handler for the core zone on each floor. The building's core will need constant cooling because of internal heat from people and lights, and cutting fan output on nights and weekends will save energy. Select zoned, constant-volume equipment to handle the building's skin loads. Provide a separate zone for each orientation to control skin heat gains and losses (see Figure 17.17). Deliver conditioned air below windows in cold climates or at the ceiling in warm climates.

FIGURE 17.15

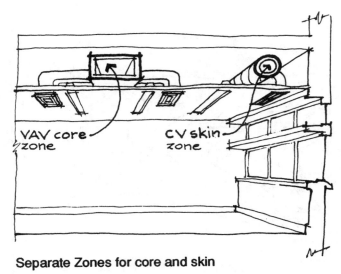

Separate Zones for core and skin

FIGURE 17.17

EFFICIENCY

VAV air handlers will save energy and operating dollars on nights and weekends when the core is unoccupied. Specify heat recovery equipment that recycles heat from exhaust air and economizer cycle controls that use outdoor air for core zone cooling during winter months.

17.4 CASE STUDY 3: CHURCH

The First Savior Church is planning a new sanctuary in a small southern town (see Figure 17.18). The church serves 600 members. Two services are conducted on Sunday morning, and a third is scheduled for Wednesday evening.

The congregation's HVAC requirements are:

♦ Low initial cost
♦ Low operating cost

HEATING SELECTION

Select a gas or oil furnace if reasonably priced fuel is available. The church will not be heated during unoccupied hours, so oversize the furnace to permit rapid building warm-up on a cold Sunday morning. If gas or oil is not available, a small heat pump can be used to store heat during the week.

FIGURE 17.18

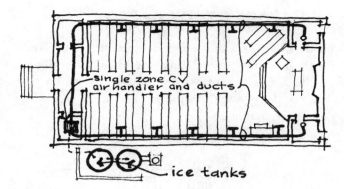

FIGURE 17.19

COOLING SELECTION

Select ice tank equipment for cooling (see Figure 17.19). It consists of a small refrigeration unit that produces ice in an insulated storage tank. When church cooling is required, the air handler uses chill water circulated through a heat exchanger in the ice storage tank.

Ice tank equipment is appropriate because the church requires a lot of cooling capacity for a short time. Electrical energy costs include a charge for peak demand, and large air-conditioning equipment demands a lot of electrical energy. Peak demand charges can account for most of a church's monthly electric bill. The ice tank lets a church have 20 tons of cooling capacity on Sunday morning without the peak demand (and electric bill) of a 20-ton air conditioner. Savings on peak charges can pay the extra cost of an ice tank installation in many utility service areas.

ZONING

A single-zone, constant-volume air handler is the appropriate choice for church heating and cooling. More expensive multizone equipment that can heat and cool at the same time is not needed in a single large interior space like a church.

EFFICIENCY

Efficiency considerations for a church like First Savior are very different from decisions for buildings with continuous occupancy. The church saves energy by limiting peak electrical demand and by turning HVAC equipment off during unoccupied hours instead of investing in added insulation.

17.5 **CASE STUDY 4: HOTEL**

Heart Hotel is a 500-room downtown facility with plush meeting and dining facilities (see Figure 17.20). Design features include a large atrium that delights the eye with twinkling elevators as well as pedestrian activity.

The hotel's HVAC requirements include:

- Individual temperature control for guest rooms
- Acoustical privacy for guest rooms
- Quick repairs for guest room equipment
- Individual temperature control for meeting and dining rooms with a variety of occupancy levels
- Low operating cost
- Heat for the atrium in winter

HEATING SELECTION

Select boilers for heating a building of this size and complexity. Hot water supply and return piping requires much less space than duct work and will help to minimize building height. Hot water will be piped to fan coils in each guest room and to single-zone air handlers serving each dining or meeting room. Heat the atrium using a radiant floor warmed by embedded hot water piping. Warm air heating will not work in the atrium because convection causes warm air to rise above the occupied area.

COOLING SELECTION

Select chillers to cool a building of this size. Circulating chilled (and hot) water throughout the building permits each fan coil or air handler to cool or heat as required by room occupants. Specify a four-pipe instead of three-pipe system to avoid mixing chill and hot water in a common return. Select cooling towers to reject waste heat because they offer less size and more efficiency than air-cooled condensing units.

ZONING

Specify a fan coil in each guest room to provide individual control and to ensure acoustical privacy. Detail fan coils for quick replacement in the event of failure and keep spares available in the hotel's maintenance department.

In meeting and dining rooms, there are two viable zoning options: select single-zone constant volume air handlers for each room, or choose single-zone VAV air handlers that can conserve fan energy when meeting or dining room occupancy is low. Remember two or more single-zone air handlers can be described as a "zoned" installation but not as multizone equipment.

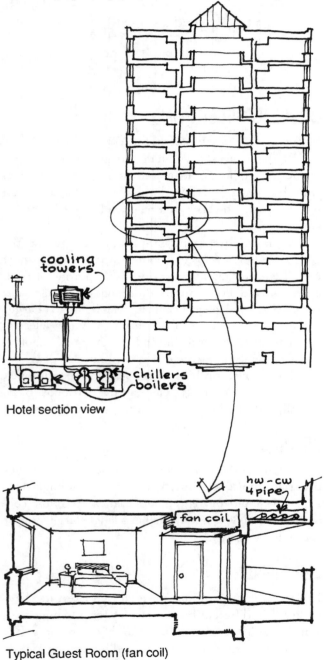

cooling towers

chillers boilers

Hotel section view

hw – cw 4 pipe

fan coil

Typical Guest Room (fan coil)

FIGURE 17.20

EFFICIENCY

Quality hotels select quality equipment and operate it only when conditioned spaces are occupied. They also invest in heat exchangers to recycle waste heat from laundries and kitchens to minimize winter heating costs.

Comment: Individual (through the wall) heat pumps or AC units with strip heat could be used instead of fan coils in guest rooms. These units are popular in motels where their noisy operation is used to mask outdoor sounds and the next room's TV, but quiet units are available for hotel applications.

HEART HOTEL: A MORE EFFICIENT WAY?

The four-pipe boiler/chiller installation with fan coils was typical for large hotels built before 1980. However, loop-connected water-source sink heat pumps are specified in many new hotels because they can reduce operating costs (see Figure 17.21).

Physically, installations of four-pipe fan coils and loop-heat pumps are quite similar. Both employ a boiler and cooling tower, and both use water to carry heat. The big advantage of loop-heat pumps is that they are very efficient when heating and cooling are required simultaneously (i.e., when guest rooms require heating at the same time, meeting rooms or dining areas require cooling because of people and lighting loads). A four-pipe system must operate boilers and chillers all year, but when heating and cooling loads are balanced, loop-heat pumps can

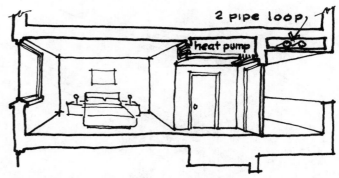

Typical Guest Room (heat pump)

FIGURE 17.21

heat and cool—without running boilers or cooling towers.

The initial cost of a loop-heat pump system is competitive with a four-pipe installation, and loop-connected heat pumps can provide substantial savings in operating costs. When building heating and cooling requirements are nearly equal, the loop's boiler and cooling tower need not operate. The heat pumps just shift heat between rooms. When heating is required, the boiler is used to maintain 50° to 60°F loop temperatures. This ensures a high COP for the heat pumps, good boiler efficiency, and minimal heat losses from system piping. When cooling is required, a cooling tower maintains loop temperatures in the 85°F range to ensure a high SEER for the heat pumps (see Figure 17.22).

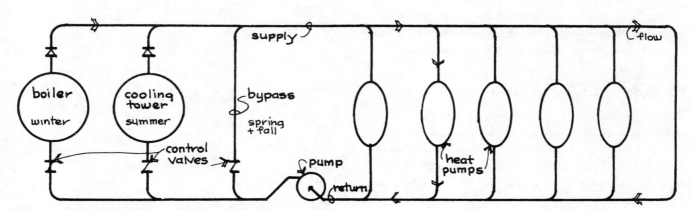

Schematic diagram - water loop and heat pumps

FIGURE 17.22

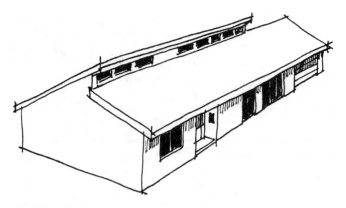

FIGURE 17.23

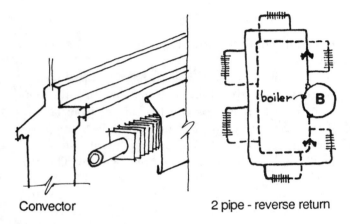

Convector   2 pipe - reverse return

FIGURE 17.24

17.6 **HOUSE EXAMPLE— HVAC OPTIONS**

Many different HVAC systems could be used to condition the example house from Chapter 15 (see Figure 17.23). This summary discusses four options with increasing initial cost and performance capabilities.

OPTION 1: CONVECTORS

Convectors can provide low-cost temperature control in climates where just heating and ventilation are sufficient for year-round comfort. The installation includes a boiler, a piping loop, and convectors located below windows. Convective air circulation distributes heat within individual rooms.

A single-pipe loop offers lowest initial cost, but it tends to deliver more heat to rooms near the boiler. A two-pipe reverse return installation will provide better heat distribution for a small increase in first cost (see

Figure 17.24). Boiler fuel is selected based on local availability. The boiler can also provide domestic hot water, so a separate water heater is not required.

OPTION 2: INDIVIDUAL UNITS

Individual heat pumps or air conditioners with strip heat can provide comfortable indoor conditions. They would be installed below windows of each room. Individual units can provide economical heating and cooling if they are operated only when the room they serve is occupied. (Bedroom units would run at night and living area units during the day.) Installing several HVAC units allows one unit to fail without making the whole house uncomfortable.

Disadvantages of this option are noisy equipment, wasted floor space in each room, and the appearance of outdoor heat exchangers (see Figure 17.25).

OPTION 3: SINGLE ZONE

A single-zone, constant-volume installation is the usual choice for a house like this example (see Figure 17.26). Heating equipment can be a furnace or a heat pump. If a furnace is selected, the cooling equipment will probably be a split system, DX unit, with an air-cooled condensing unit located outdoors.

Supply ducts for the installation will run from the furnace or heat pump to a register in each room. Ducts are usually run overhead in hot climates and under the floor in cold climates (with outlets below windows). Ducts are insulated as necessary to prevent condensation, and good designers keep ducts out of unconditioned crawl spaces or attics.

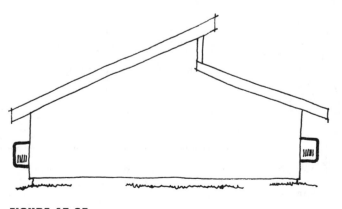

FIGURE 17.25

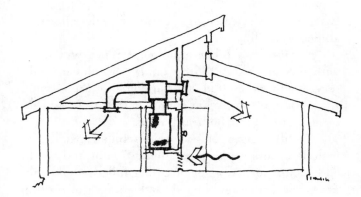

Option 3 schematic

FIGURE 17.26

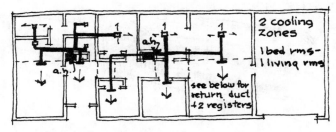

Cooling, 2 CV zones

FIGURE 17.28

Return air is drawn from rooms by undercutting doors, but this reduces acoustical privacy. Outdoor air is provided by infiltration or by opening windows.

OPTION 4: RADIANT HEAT PLUS ZONED COOLING

This is an expensive, high-performance installation. Heating is radiant. Embedded tubing maintains the floor slab at 85°F, providing quiet, efficient, draft-free heat. The radiant floor will keep a family comfortable through a very cold winter when good double- or triple-glass windows are specified to minimize the drafts single-glass windows create (see Figure 17.27).

Summer cooling is accomplished with two single-zone constant-volume air-conditioning units (see Figure 17.28). One serves the bedrooms, and the other serves the living areas. Two units permit different temperature settings and operating schedules for the rooms they serve. Supply ducts are oversized, and two registers serve each space to minimize air stream noise. Return air is carried in insulated return

ducts to ensure acoustical privacy. Outdoor air intakes with sensor-controlled dampers blend fresh air into the return air stream, and when weather conditions are ideal, 100% fresh air is used (see Figure 17.29). Electrostatic filters are installed on each air handler to minimize interior dust.

With separate heating and cooling equipment, ducted return, fresh air intakes, and electrostatic filters this installation is expensive. But it offers more flexibility, better control, greater comfort, and less noise than the three preceding options.

17.7 OFFICE EXAMPLE— ZONING OPTIONS

Air-conditioning equipment options for the example office building (see Figure 17.30) include heat pumps, furnaces with DX cooling, or boilers and chillers. This case study emphasizes zoning choices instead of heat-

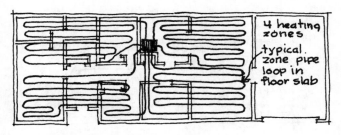

Radiant floor, 4 zones

FIGURE 17.27

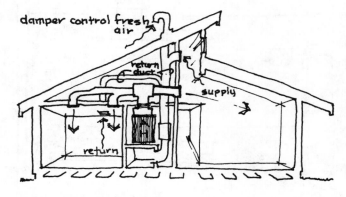

Option 4 schematic

FIGURE 17.29

FIGURE 17.30

ing-cooling equipment selection. Four possible zoning options are considered.

OPTION 1: ZONED, MANY UNITS

This option provides many single-zone units with individual air handlers located above the suspended ceiling (see Figure 17.31). The building owner sets a minimum rental area and specifies a heat pump to serve each tenant. Advantages of this option include minimum floor area for HVAC, individual temperature control for each tenant, and separate tenant billing for HVAC utilities. If a heat pump fails, only one tenant is disturbed. A disadvantage

is difficult service access to air handlers located above the ceiling.

At least two other equipment types will work with this option. A central boiler and chiller could serve ceiling-mounted air handlers in each rental space, or rooftop package units with furnaces could be substituted for the heat pumps.

Disadvantages of the boiler-chiller installation include excess capacity for limited night and weekend occupancy. A boiler or chiller failure could shut down the entire building. Disadvantages of rooftop package equipment include the loss of second-floor rental area because of ducts serving the first floor, and heating-cooling losses caused by circulating conditioned air through the poorly insulated rooftop units.

OPTION 2: ZONED, FOUR UNITS

This option assumes the building is occupied by one or two major tenants with open office plans. The building is divided into four zones (two on each floor) because of anticipated environmental heating and cooling loads. North and south sides of the building are separated to reflect different solar gains through windows (see Figure 17.32). First and second floors are separated because only the second floor experiences roof heat gain and loss. Each zone is served by a single, constant-volume air handler.

An advantage of this option is easy service access to air handlers. Disadvantages include the need to operate large equipment to serve reduced night or weekend occupancy, and also a large building area will be affected if an air handler fails. Heat pumps, furnaces with DX cooling units, or boilers and chillers can be used as heat sources and sinks for the four air handlers.

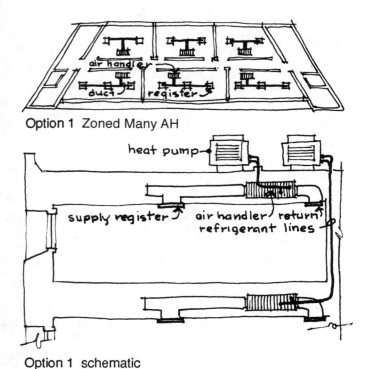

Option 1 Zoned Many AH

Option 1 schematic

FIGURE 17.31

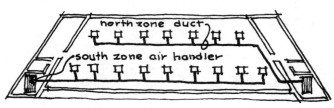

Option 2 Zoned 4 AH

FIGURE 17.32

OPTION 3: MULTIZONE

A single air handler can serve the entire building delivering warm or cool air to building spaces as dictated by individual thermostats (see Figure 17.33). The equipment delivers a constant volume of air to each outlet and varies air temperature to control space conditions. Many ducts are required to serve the building's four environmental zones and permit temperature choices for each tenant. *Individual duct bypass* multizone air handlers are specified to avoid mixing and to maximize operating efficiency. Duct work will require more space than the preceding options because a separate duct is required to serve each zone.

The advantage of this option is very good temperature control. Disadvantages are high first cost and loss of all heating and cooling should the air handler or the heating cooling equipment fail.

OPTION 4: ZONED, FOUR VAV UNITS

This option cuts the building into four environmental zones because concurrent heating and cooling may be required during spring and fall months. (A single VAV air handler cannot provide simultaneous

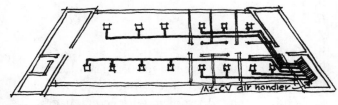

Option 3 MZ Individual Duct

FIGURE 17.33

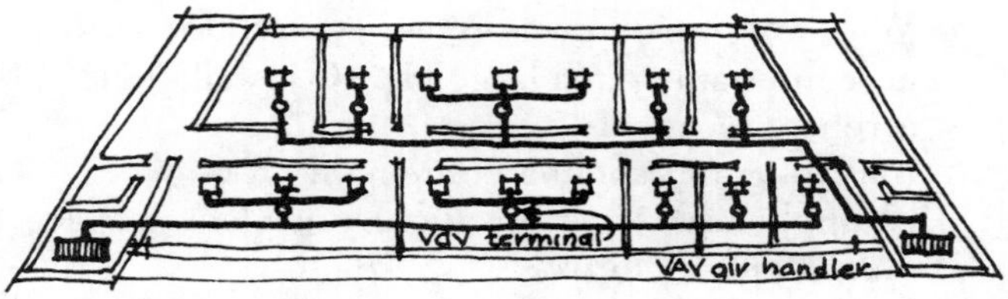

Option 4 Zoned VAV

FIGURE 17.34

heating and cooling.) Separate VAV air handlers are installed to serve the north and south sides of each floor. The air handlers vary air quantity to control space temperature. It is possible to heat with one air handler while another is cooling (see Figure 17.34).

Careful readers will note that this option is identical to Option 2 except for VAV. In fact, VAV could be used with all the preceding options. The potential advantage of a VAV installation is reduced fan energy, particularly when the building is partially occupied on nights or weekends. Disadvantages are the loss of conditioning for a large area should an air handler fail and possible inadequate air motion in conditioned spaces when VAV fans reduce air quantity during partial load hours. VAV installations with fan-powered terminals are a popular zoning technique, but real operating cost savings occur only with a comprehensive and well-designed control system.

BEST ZONING OPTION?

Good arguments can be made for each zoning option. For a given building the "best" option is the one that best meets tenant comfort requirements, operating cost estimates, and construction budget limits.

REVIEW QUESTIONS

1. Select the type of heating equipment likely to have the highest operating cost.
   a. oil furnace
   b. electric heat pump
   c. electric resistance (strip) heat
   d. natural gas-fired boiler
2. Select the type of heating equipment likely to provide the most comfortable indoor environment for occupants.
   a. convectors
   b. radiant
   c. warm air (or central air)
   d. wood stove
3. How much space should be allowed for air handlers on each floor of a multistory building?
   a. none (install above ceiling)
   b. 2%
   c. 4%
   d. 6%
   e. 10%
4. Boilers are used instead of furnaces in large buildings because:
   a. water holds heat longer than air
   b. boilers are cheaper than furnaces
   c. hot water will freeze faster than cold water
   d. fuel-fired boilers don't require flues
   e. water can carry heat more economically than air

5. Which of the following is *not* required for quality air conditioning in a large HVAC installation?
   a. reheat terminals
   b. heating and cooling equipment
   c. supply and return air ducting or plenum space
   d. exhaust air control
   e. outdoor air intake

6. The most efficient individual unit installation for dormitory heating and cooling in a temperate climate would be:
   a. fan coils
   b. packaged units
   c. duct furnaces
   d. loop-connected heat pumps

7. The most efficient type of multizone air-conditioning installation is:
   a. MZ reheat
   b. MZ double duct
   c. MZ individual duct mixing
   d. MZ individual duct bypass

8. The least efficient type of multizone air-conditioning installation is:
   a. MZ reheat
   b. MZ double duct
   c. MZ individual duct mixing
   d. MZ individual duct bypass

9. Reheat can be an effective way to control humidity. Try to think of a situation where reheat can be an efficient air-conditioning technique.

10. The best equipment type for air-conditioning small rooms where acoustical privacy is critical would be:
    a. individual units
    b. single zone
    c. zoned
    d. multizone

11. Where acoustical privacy is an important consideration in a building with large air handlers a ducted _____ should be specified.
    a. return
    b. exhaust
    c. fresh air intake
    d. boiler

12. An advantage of VAV (variable air volume) air handlers compared to constant volume is:
    a. improved temperature control
    b. improved humidity control
    c. improved air circulation
    d. reduced fan energy

Select answers for the next three questions from the following list (MZ = multizone; CV = constant volume; VAV = variable air volume). You may use answers more than once.
   a. single zone, CV
   b. zoned VAV units
   c. MZ double duct, CV
   d. MZ reheat, CV
   e. MZ individual duct, CV, mixing

13. Lowest operating cost for a large office building?

14. Lowest first cost and good performance for a church seating 200?

15. Low first cost and good temperature control for many different spaces, but high operating cost?

ANSWERS

1. c
2. b (correct for good radiant floor installations but incorrect for an individual radiant heater, which is about as comfortable as a wood stove)
3. c
4. e
5. a
6. d
7. d
8. a
9. *Economizer cycle* is a term for using outdoor air instead of refrigeration equipment to cool heat rich buildings. A problem with economizer operation is that sometimes outdoor air is too moist to do a good air-conditioning job. In this situation, reheat may be more economical than operating a chiller.

   A second efficient application of reheat is humidity control when the heat is "free." Waste heat rejected by the condenser of refrigeration equipment is a possible source.
10. a
11. a (a ducted return will minimize sound transmission typical in an open return plenum)
12. d
13. b
14. a
15. d

CHAPTER 18

Air Distribution

*When the autumn wind
scatters peonies, a few
petals fall in pairs.*

Buson

A somewhat different format is used in this chapter. Several tables are included in the text, and example problems are presented with each table. Work the examples as you read.

Following the tables is a presentation that develops a complete air distribution system for the house example used previously in Chapters 15 and 17. Refer back to the tables as you review illustrations and comments concerning air quantities, register selection, and duct sizing.

A discussion of fans and controls follows the illustrations on air distribution, and the chapter ends with review questions. When you complete this chapter, you should be able to size, select, and specify all the components of a low-velocity air distribution system. This ability will permit you to make proper space allowances for air conditioning most buildings.

Large buildings frequently use high-velocity air distribution systems to save space, but thoughtful designers allow adequate room for quieter low-velocity systems that use less fan energy. Carefully follow the air distribution example, and when you understand this material use it as a guide to size and select registers and ducts for a building of your choice.

FIGURE 18.2

18.0 **REGISTERS**

Registers are outlets that distribute and control conditioned air. They include two components: a *diffuser* that spreads an air stream and a *damper* to adjust air quantity (see Figure 18.1). Good registers have adjustable diffuser vanes and opposed blade dampers. Cheap registers have stamped diffuser vanes and multishutter dampers. Good registers are more expensive, but the small added cost buys quiet and complete distribution of conditioned air.

Registers are selected to distribute conditioned air without causing uncomfortable drafts. *Throw* is defined as the horizontal distance an air stream travels from a register before it slows to a velocity that is comfortable for people (usually 50 to 100 FPM). If a small register and a large register are delivering equal air quantities, the small register will provide a greater throw and make more noise (see Figure 18.2).

Register noise increases as throw increases. For quiet applications such as recording studios, designers increase the number of registers to minimize throw and noise.

WALL REGISTERS

Air distribution patterns for wall registers are straight, medium, and maximum (see Figure 18.3). Pick a pattern to distribute air throughout the conditioned space and select a throw of 75% of the distance to the far wall.

Selection Examples

1. Select a straight pattern wall register to throw 400 CFM a distance of 15 feet.
2. Select a maximum pattern wall register to throw 250 CFM a distance of 7 feet.
3. Select a medium pattern wall register to throw 700 CFM a distance of 15 feet.
4. Select a medium pattern wall register to throw 700 CFM a distance of 19 feet.
5–7. These three examples appear below Table 18.1 on page 263. Start by deciding the pattern. Figure the throw at 75% of the distance to the far wall. Select a register for each example.

Answers

1. Select 10×24.
2. Select 6×16 or 6×20.
3. Select 12×36.
4. Select 8×30 or 10×24 (the table indicates a throw of 20 feet—close enough).
5. Pattern = straight, throw = 12 feet, select 6×14.
6. Pattern = maximum, throw = 9 feet, select 8×20.
7. Pattern = medium, throw = 15 feet, select 8×24.

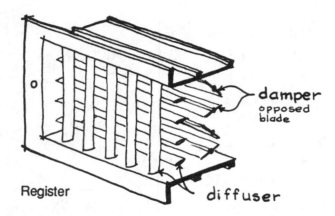

Register

FIGURE 18.1

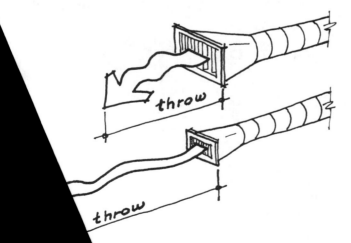

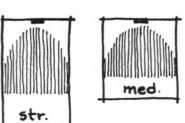

Wall Register Pattern

FIGURE 18.3

TABLE 18.1

Wall Register Size in Inches (size is opening dimension, actual register face is 2 to 3 inches larger)

CFM	4×8	4×10	4×12	6×10	6×12	6×14	6×16	6×20	6×24
50	3–6	3–5							
100	6–11	6–10	4–8	4–7					
125	7–13	6–12	5–10	5–9	4–8	4–7			
150			7–12	6–12	5–10	5–9			
175				7–13	6–11	6–10	5–9		
200				8–15	7–13	6–12	6–11	6–10	
225					8–15	8–14	7–12	6–11	
250					9–17	8–15	7–14	7–13	6–12
300						10–19	9–17	8–16	8–15
350							11–20	10–18	9–17
400								11–21	10–19

CFM	8×20	8×24	8×30	8×36	10×24	10×30	10×36	12×30	12×36
400	9–18	8–16			7–15				
500	12–22	10–20	9–18		9–18	8–16			
600	14–26	13–24	12–22	11–20	12–22	10–18	9–17		
700	16–31	15–29	14–26	13–23	14–26	12–23	11–21	11–21	10–20
800	19–35	17–32	16–29	14–26	16–29	14–26	13–24	13–24	12–22
900		19–36	17–32	16–30	17–32	15–29	14–27	14–27	13–25
1,000			19–36	18–34	19–36	18–32	17–30	17–30	15–26
1,200				21–40		21–38	19–35	19–35	17–33

Table field numbers are register throws (in feet). The *low number* is the throw for a *maximum* pattern register. The *high number* is the throw for a *straight* pattern register. *Interpolate for medium* pattern registers. Design throw dimension is 3/4 of the distance to the far wall.

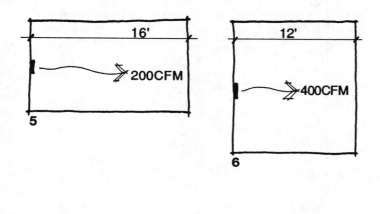

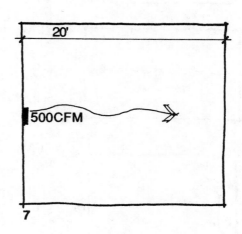

18.1 WALL REGISTER SELECTION

THROW *NOT* IN TABLE?

Throw information is omitted from Table 18.1 when air quantities fall above or below the working range of a given register. If no throw is given, try using two registers at half CFM for each, but remember to revise the register pattern.

18.2 CEILING REGISTERS

Tables 18.2, 18.3 and 18.4 provide throw distances for one-way, two-way, and four-way square ceiling registers. Pick a pattern to distribute air throughout the conditioned space and select a throw equal to the distance **to the far wall.**

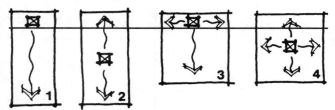

Examples

Cover the answers next to the examples and select the appropriate pattern, throw, and size for each.

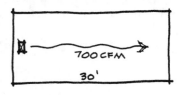

CFM: 700

Pattern: One Way

Throw: 30 feet

Select **15×15**

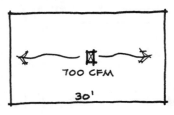

CFM: 700

Pattern: Two Way

Throw: 15 feet

Select **21×21**

CFM: 700

Pattern: Four Way

Throw: 15 feet

Select **15×15**

Danger! Use manufacturers' catalogs. CFMs and throws given here are generic. Register characteristics vary a lot from one manufacturer to another, so verify all CFM and throw data with your supplier before placing an order. Add a Hart & Cooley or Titus cata-

TABLE 18.2

One-Way Square Register Size (throw to far wall)

Size	6	9	12	15	18	21	24
CFM							
100	11	8					
125		9					
150		10					
175		12	9				
200		15	11				
250			13				
300			16				
400			22	18	14		
500				22	18		
600				26	21		
700				30	25	21	
800					29	25	21
900					32	28	24

TABLE 18.3

Two-Way Square Register Size (throw to far wall)

Size	6	9	12	15	18	21	24
CFM							
100	7	5					
125		6					
150		8	5				
175		9	6				
200		10	7				
250			9	7			
300			11	9	7		
400				12	10		
500				15	13		
600				19	15	13	
700				22	18	15	13
800					20	18	15
900					23	20	17

TABLE 18.4

Four-Way Square Register Size (throw to far wall)

Size	6	9	12	15	18	21	24
CFM							
75	4						
100	6	4					
125	7	5					
150		5					
175		6	4				
200		7	5				
225		8	6				
250			7	5			
300			8	7			
400			11	9	7		
500				11	9	7	
600				13	11	9	
700				15	13	11	9
800					14	13	11
900					16	14	13
1,000					18	16	14
1,200						19	17

Field numbers are throw in feet.

log to your library and use it instead of these pages. Both manufacture quality register lines, and they include the three-way patterns missing here.

18.3 FLOOR AND RETURN REGISTERS

FLOOR REGISTERS

Select floor registers so that their throw covers the vertical distance from the register to the ceiling (see Figure 18.4). Floor registers are often installed below windows to minimize cold winter drafts and to prevent condensation on glass.

RETURN AIR GRILLES

Sizes in Table 18.6 are for fairly quiet installations. Remember to provide ducts or a plenum for return air flow to the air handler. *Filter grilles* (return grilles with filters) can simplify filter maintenance when access to the air handler is difficult.

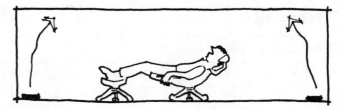

FIGURE 18.4

FLOOR AND RETURN REGISTER SELECTION

Select floor registers and return grilles for the following example problems using Tables 18.5 and 18.6.

1. Select a floor register for the room shown in Figure 18.5.

   Answer: Select a 3″ register 6′ long. It will deliver 600 CFM with a throw of 14′. You may add inactive register lengths if desired for aesthetic reasons. Best register location is below windows to minimize winter convective drafts.

2. Select a register for the room in the preceding example located in a 3′ high base cabinet.

   Answer: Select a 2″ register 10′ long. By interpolation, each foot will deliver 59 CFM with a throw of 11′ (590 CFM = nearly 600).

TABLE 18.5

Floor Registers: CFM for 1 Foot Length

Throw = ceiling height (above register)

Throw	6′	8′	10′	12′	14′	16′	20′
Size							
1 1/2″	27	36	42	54			
2″	34	44	54	64	78		
2 1/2″	42	50	60	73	86	100	
3″		60	71	86	100	112	
3 1/2″		67	76	89	105	118	
4″			88	98	114	132	160
5″			104	121	138	156	185
6″			129	150	170	190	230

TABLE 18.6

Return Grilles

Size	CFM	Size	CFM
6×12	250	24×30	2,000
12×12	500	30×30	2,500
12×18	700	30×36	3,000
12×24	900	24×48	3,200
20×24	1,400	36×48	4,800
24×24	1,700	48×48	6,000

Return Filter Grilles

Size	CFM	Size	CFM
16×20	800	20×25	1,200
20×20	1,000	20×30	1,400
16×25	1,000		

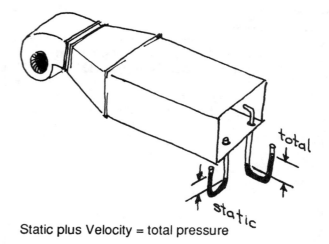

Static plus Velocity = total pressure

FIGURE 18.6

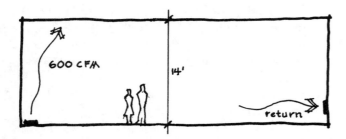

FIGURE 18.5

3. Select a return grille for the preceding example room.

   Answer: Select 12×18. Size to return all supply air. Oversize return grilles to minimize air stream noise.

4. Select a return air filter grille for a 2-ton air conditioner.

   Answer: Refer ahead to page 270.

   Two tons at 400 CFM per ton = 800 CFM. Select 16×20.

18.4 DUCT SIZING

STATIC REGAIN

A duct system is "balanced" when the design air quantity flows from each outlet. Duct system designers understand the tendency for outlets near a fan to deliver more air than outlets distant from a fan

because of duct friction. The *static regain method* is the most accurate way of sizing a well-balanced duct system. Static regain calculations adjust duct size to obtain equal static pressure and correct air quantity at each outlet.

The total pressure in a duct is the sum of static and velocity pressures (see Figure 18.6). Static pressure is the outward push of air against duct surfaces as a result of the fan's compressive action. Velocity pressure is the directional push of an air stream due to its speed. It is possible to trade static pressure and velocity pressure by changing duct size. A decrease

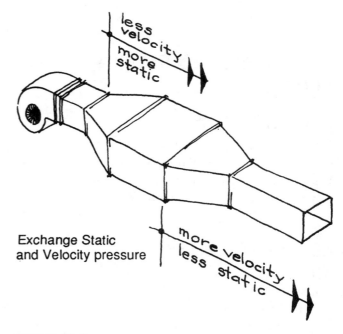

Exchange Static and Velocity pressure

FIGURE 18.7

in duct size forces air speed and velocity pressure to increase as static pressure decreases. Enlarging a duct will cause air speed and velocity pressure to decrease as static pressure increases (see Figure 18.7).

EQUAL FRICTION

Static regain calculations are complex because friction losses must be established for all elbows, transitions, and other components of the duct system.

The duct size chart in Table 18.7 is based on a simpler (and less accurate) method of duct sizing called *equal friction*.

The equal friction method of duct sizing can be used for low-velocity duct systems with a duct length of 100 feet or less. Ducts are selected by assuming a total friction loss of 1/10 inch of water. Duct system performance approximates static regain, but extractors and dampers are required to ensure balanced air distribution.

TABLE 18.7

Duct Size (CFM = cubic feet of air per minute)

CFM	Round	Rectangular Duct Size							
50	5	4×4	6×3						
100	6	4×7	6×5	8×4					
150	7	4×10	6×7	8×5	10×4				
200	8	4×12	6×8	8×6	10×5	12×4			
250	8		6×10	8×7	10×6	12×5			
300	9		6×11	8×8	10×7	12×6	14×5		
350	10		6×13	8×9	10×7	12×6	14×6		
400	10		6×14	8×10	10×8	12×7	14×6		
500	12		6×16	8×12	10×10	12×8	14×7	16×6	
600	12		6×20	8×14	10×12	12×10	14×8	16×7	
700	12			8×16	10×12	12×10	14×9	16×8	
800	14			8×18	10×14	12×12	14×10	16×9	18×8
900	14			8×20	10×16	12×12	14×10	16×9	18×9
1,000	14			8×22	10×16	12×14	14×12	16×10	18×9
1,200	16			8×24	10×18	12×16	14×14	16×12	18×10
1,400	16			10×20	12×18	14×16	16×14		18×12	20×10
1,600	18			10×24	12×20	14×18	16×14		18×14	20×12
1,800	18			10×26	12×20	14×18	16×16		18×14	20×12
2,000	18			10××0	12×24	14×20	16×18		18×16	20×14
2,500	20	20×16	24×14	28×12	32×12	36×10			
3,000	22	20×18	24×16	28×14	32×12	36×12			
4,000	24	20×24	24×20	28×16	32×16	36×14			
6,000	28	20×32	24×26	28×22	32×20	36×18	40×16		
8,000	32	20×40	24×34	28×28	32×26	36×22	40×20	48×18	
10,000	34	20×48	24×40	28×32	32×28	36×26	40×24	48×20	
20,000	44		24×70	28×60	32×50	36×46	40×40	48×34	60×20
40,000	60						40×80	48×66	60×52
80,000	90								60×100+

Notes:
• Dimensions are given in inches for the airway *inside* the duct. Most ducts are insulated to prevent condensation and minimize heat gain or loss. To determine outside duct dimensions, add two times the thickness of the duct insulation specified (standard duct insulation thicknesses are 1/2", 1", and 2"). Duct sizes assume a friction loss of 1/10" of water per 100' of length.
• Duct sizes given are only for low-velocity duct systems (less than 2,000 FPM air velocity). The maximum length of low-velocity supply duct runs is about 200'.
Size return air ducts 2" larger than supply ducts.

Examples

Use Table 18.7 to select airway dimensions for ducts. *Airway* means inside dimensions.

1. Select a round duct to carry 1,000 CFM.
2. Select a rectangular duct to carry 1,000 CFM.
3. A 16×12 duct carrying 1,000 CFM—duct must divide into two ducts, which will carry 600 CFM and 400 CFM, respectively. Select sizes for these two ducts.

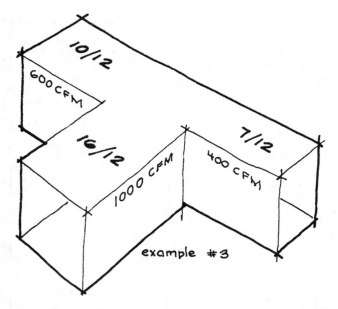

example #3

4. Select a rectangular duct to carry 6,000 CFM.
5. Select an insulated rectangular duct to carry 6,000 CFM through a 19″ high space.
6. The 40×16 duct from the preceding example must be run vertically to serve a floor above. Size the vertical duct section to fit in a 30″ square chase.

Answers

1. 14″ diameter
2. 12×14 (Always select rectangular duct sizes that are nearly square if you have enough space to accommodate them. Square ducts can be fabricated with less material.)
3. 7×12 and 10×12 (Where space permits, hold one dimension of the original duct. This permits easier fabrication of the transition.)
4. 2 ×26
5. 40×16 (Always give plan dimension of the duct first; 40×16 duct is 42×18 outside dimension with 1″ thick duct insulation.)
6. 24×26 (nearest square fit)

18.5 DUCT MATERIALS AND FABRICATION

Most rectangular ducts are fabricated from galvanized steel or foil-faced fiberglass. Fiberglass ducts are self-insulating. Insulation is applied to steel ducts to minimize heat loss, heat gain, and condensation. Insulating the inside surfaces of steel ducts cuts sound transmission and simplifies duct supports with a slight increase in duct friction.

STEEL

"Pittsburgh" (see Figure 18.8) snap lock, spiral lock, and flanged connections are used to connect steel duct sections. Pittsburgh joints permit ducts to be shipped flat and ease site assembly.

Metal thickness (gauge), support spacing, and connection types for various applications should meet SMACNA specifications (Sheet Metal & Air Conditioning Contractors National Association).

FIBERGLASS

Foil-faced fiberglass ducts are easier to fabricate than steel ducts (see Figure 18.9). Sheets of 1″ or 1 1/2″ duct board are cut with a razor knife and folded to size. End joints have an interlocking edge that is sealed with heat-sensitive tape.

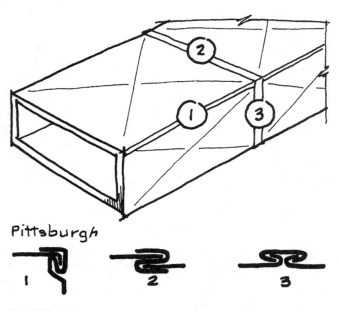

Pittsburgh

FIGURE 18.8

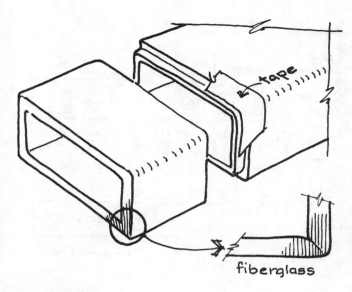

FIGURE 18.9

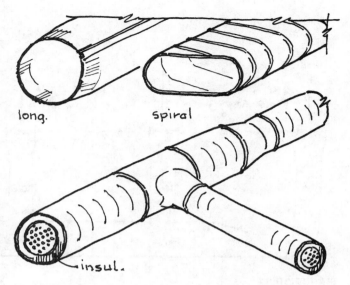

FIGURE 18.10

Round rigid fiberglass ducts are also available in diameters from 4″ through 30″, and fiberglass duct liner in thicknesses from 1/2″ to 2″ is used in steel ducts.

ROUND AND ELLIPTICAL

Round steel ducts are fabricated in light gauge sections with longitudinal joints or in heavier gauges using spiral lock joints. A variety of standard fittings are available for changing duct size and direction.

A circle is the most efficient duct shape in terms of material required and friction loss, but round ducts are sometimes rolled to elliptical shapes to reduce duct height where vertical clearance is limited. When round duct work is a visible design feature, insulated duct with a perforated internal steel liner to minimize air stream noise is often specified (see Figure 18.10).

FULL RADIUS

The drawings in Figures 18.11 and 18.12 illustrate full-radius and right angle duct configurations. Full-radius bends minimize duct friction and turbulent air flow. They require a lot of material and space, but they are the preferred bend type in quiet quality installations.

Right angle bends are cheaper than full-radius bends, but they are noisy and require more fan energy to overcome increased duct friction. Turning

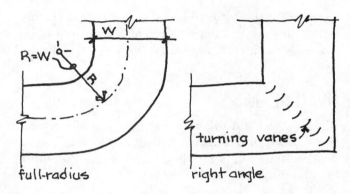

FIGURE 18.11

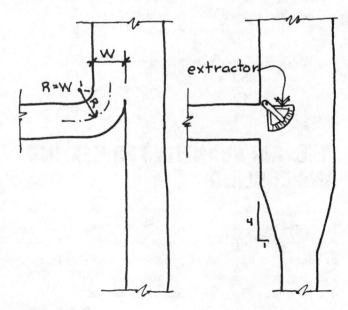

FIGURE 18.12

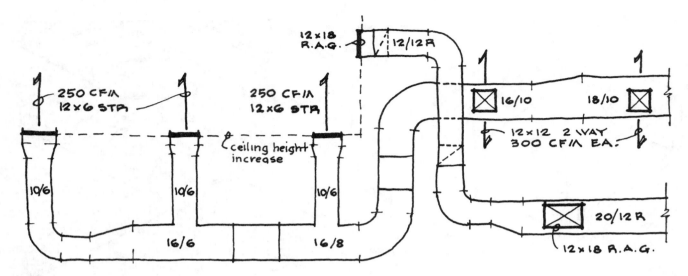

FIGURE 18.13

vanes are installed in right angle bends to limit turbulent air flow. Extractors are required at right angle branches to balance air flow.

HVAC PLAN

The view in Figure 18.13 is a partial plan of a typical supply and return duct installation. Dotted lines at wall registers locate a ceiling break. What do dotted lines on the return duct indicate?

- Duct sizes are noted with plan dimension first.
- Supply and return registers and grilles are emphasized with the heaviest line weight.
- CFM, size, and pattern are noted for each register.
- Extra clearance is required where the return duct crosses the supply duct.

18.6 AIR QUANTITY FOR HEATING AND COOLING

Heating-cooling systems use air to carry heat from one location to another. The quantity of air circulated in a duct system is determined by the quantity of heating or cooling required and by the temperature difference between conditioned air and room air. The following equation is used to determine air quantity:

$$RSH/(1.08)(TD) = CFM$$

where:

RSH (room sensible* heat) is the amount of sensible heat that must be added or removed to maintain room temperature (in BTUH).
1.08 is the sensible heat content of air (p. 204).
TD is the temperature difference between conditioned air and room air.
CFM is the quantity of air in cubic feet per minute.

EXAMPLE 1

A room has a peak winter sensible heat loss of 25,000 BTUH. How much 130°F air must be supplied to maintain room temperature at 70°F?

$$RSH/(1.08)(TD) \quad = CFM$$
$$25,000/1.08(130\text{-}70) = \textbf{386 CFM}$$

The same room has a summer peak sensible heat gain of 15,000 BTUH. How much 55°F air is required to maintain 75°F in the room?

$$RSH/(1.08)(TD) \quad = CFM$$
$$15,000/1.08(75\text{-}55) = \textbf{694 CFM}$$

---

*Sensible heat is measured with a dry-bulb thermometer; air quantities for heating or cooling are determined by sensible heat loads.

Latent heat, the phase change heat held in water vapor, is removed when water condenses on a cooling coil. When latent heat is a large part of the total cooling load, larger cooling coils are specified to reduce air velocity and increase the condensing surface area.

Note that summer cooling requires more conditioned air than winter heating even though heat loss exceeds heat gain. This is usually true due to the smaller TD typical in cooling operations. However, heat pumps deliver warm air in the 80° to 110°F range, and heating requirements may determine air quantity when heat pumps are used.

EXAMPLE 2

Calculate the quantity of heating and cooling air required for the following room:

Winter RSH = 35,000 BTUH; supply air is 100°F; room temperature is 65°F.
Summer RSH = 20,000 BTUH; supply air is 55°F; room temperature is 75°F.

Does heating or cooling set air quantity requirements for the room above?

Answer:

Heating air required

$$35,000/1.08(100-65) = \textbf{926 CFM}$$

Cooling air required

$$20,000/1.08(75-55) = \textbf{926 CFM}$$

Heating and cooling CFM are equal in this example.

400 CFM PER TON

One ton of air-conditioning capacity is defined as a heat removal rate of 12,000 BTUH, and each ton of air-conditioning load usually includes a sensible* component of 8,000 to 9,000 BTUH. Air handlers typically circulate about 400 CFM for each ton of capacity.

FOUR STEPS FOR AIR DISTRIBUTION

If you have not calculated sensible heat loads for each room in a building and you need a good preliminary estimate for conditioned air quantity, proceed as follows:

1. Multiply total tons by 400 to estimate total CFM.
2. Distribute CFM in proportion to floor area.
3. Adjust CFM to reflect anticipated variations in RSH due to specific building conditions (p. 202).
4. Round off air supply for each room to nearest 25 CFM.

DUCT EXAMPLES

Section 18.7 provides examples of air distribution, register selection, and duct sizing for the example house used in previous heat loss and gain calculations. Study the drawings and text until you understand the air distribution process and then prepare a schematic duct plan for the west half of the house. This air distribution method can also be used for commercial buildings like the example office and similar low-velocity air-conditioning installations.

18.7 HOUSE EXAMPLE, AIR DISTRIBUTION

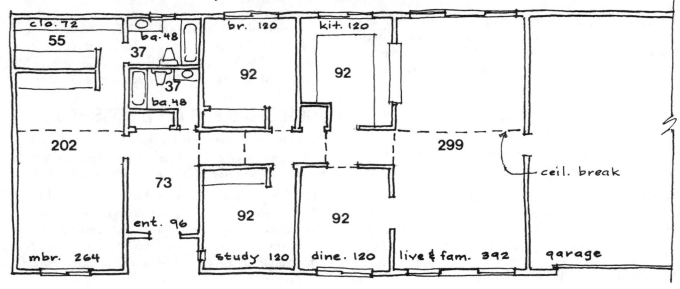

The plan above illustrates steps 1 and 2 of the air distribution method described in Section 18.6. The house example above has a floor area of 1,528 sq. ft. and a peak heat gain of 33,537 BTUH. A 36,000 BTUH (3-ton) cooling unit will be used because it is the standard equipment size closest to the peak cooling load. The 3-ton air conditioner will circulate about 1,200 CFM (400 CFM per ton).

Room areas in square feet are shown next to room names. CFM distribution, based on room area, is shown in the center of each room (1,200 CFM ÷ 1,568 sq. ft. = 0.76 CFM per sq. ft.).

On the plan below, room CFM has been adjusted in accordance with steps 3 and 4 to reflect variations in anticipated RSH (room sensible heat).

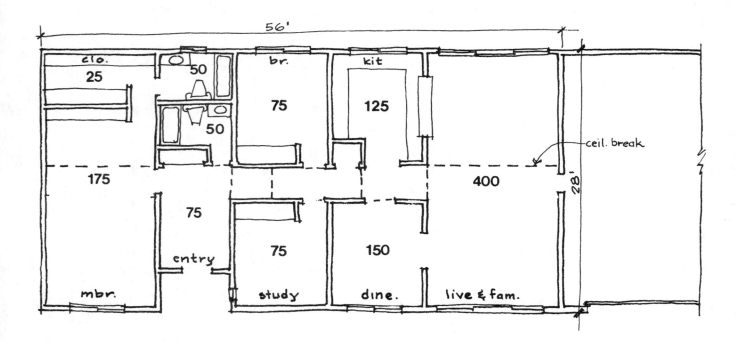

Adjustments were made as follows:

Bedrooms and study: Decrease CFM because of smaller glass area and cooler nights.

Dining: Increase CFM because of more potential occupants.

Kitchen: Increase CFM because of cooking.

Living and family: Increase CFM because of occupants and glass area.

Readers who desire more precise air quantity adjustments will calculate sensible heat gain for each room (RSH) and use the equation: CFM = RSH/1.08 TD.

Both ceiling and wall registers will be used to deliver conditioned air in the example house. Ceiling registers will serve the 8-foot-high spaces. They are sized to throw air to the surrounding walls. Wall registers will be used to serve the spaces with high walls and cathedral ceilings. They are sized to throw air 75% of the distance to the far wall. Register patterns are selected to distribute conditioned air throughout each room. The schematic section in Figure 18.14 shows register and duct location.

Registers shown on the plan in Figure 18.15 were selected using Tables 18.1 through 18.6. Selection was based on the CFM, pattern, and the throw necessary to cool each room. Designers may select alternate register shapes as long as the required throw distance is maintained. For example, a 4×18 can replace a 6×12. Wall registers are shown as a rectangle and ceiling registers as a square. The numbers near each register are CFM, throw, and pattern, and the underlined numbers are register size. Review the air quantities on page 272 and verify the register selections in Figure 18.15 by referring back to Tables 18.1 through 18.6.

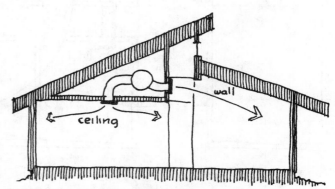

Register types for example house

FIGURE 18.14

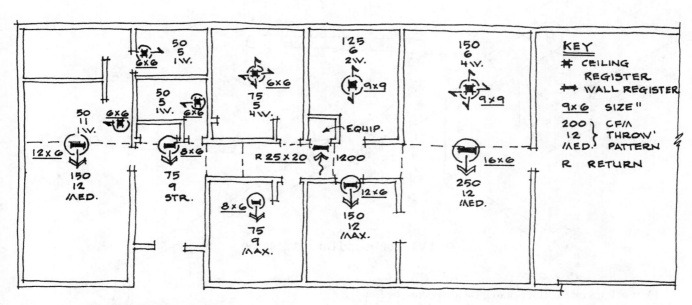

Register Size, CFM, Throw, and Pattern

FIGURE 18.15

The plan below shows a schematic round duct layout for the house example. Duct sizes were selected using Table 18.7. Check the results by working from the last outlet back to the air handler, increasing duct size as CFM increases.

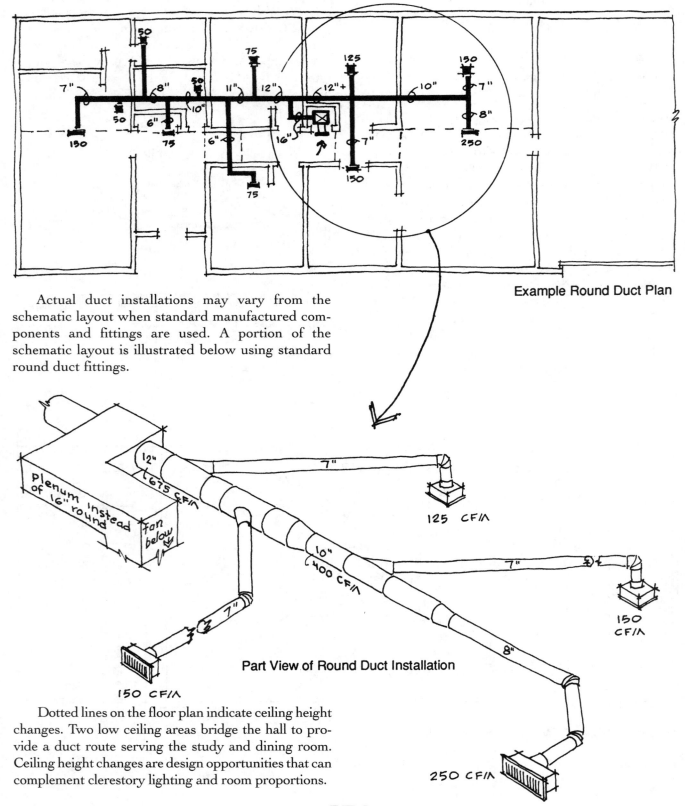

Example Round Duct Plan

Actual duct installations may vary from the schematic layout when standard manufactured components and fittings are used. A portion of the schematic layout is illustrated below using standard round duct fittings.

Part View of Round Duct Installation

Dotted lines on the floor plan indicate ceiling height changes. Two low ceiling areas bridge the hall to provide a duct route serving the study and dining room. Ceiling height changes are design opportunities that can complement clerestory lighting and room proportions.

Round duct provides the most efficient duct shape in terms of duct material and friction loss (see Figure 18.16). But because round duct requires more vertical space (and building height), rectangular duct is often substituted. *Aspect ratio* is the ratio of width to height for rectangular duct. Limiting rectangular duct aspect ratios to 3:1 or less will limit material waste and friction losses.

The schematics on page 276 illustrate conversion of round duct to rectangular duct for the house example. Rectangular ducts will be fabricated with 1″ thick insulation, and a maximum duct height of 10″ is assumed. Duct sizes shown are inside *airway* dimensions. Actual outside dimensions of the ducts will be 2″ greater than the airway dimensions because of insulation thickness.

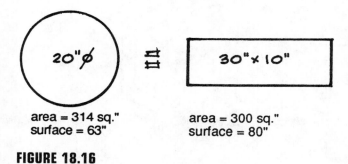

area = 314 sq."
surface = 63"

area = 300 sq."
surface = 80"

FIGURE 18.16

The following checklist explains considerations and details of the rectangular duct conversion:

1. Duct plan or width dimension is noted before duct height. The rule is, "note the dimension that is visible in a drawing first." Therefore, width is noted first in a plan view, but height is noted first in a section or elevation view.
2. Ducts larger than 12″ are fabricated in even inches, that is, 22×18 instead of 22×17.
3. Reductions in duct size are made with a gentle slope to avoid air turbulence—1:4 slope is a recommended maximum.
4. Ducts extend one duct width past ceiling registers to reduce air turbulence and noise.
5. Ducts are insulated to reduce heat loss or gain and to prevent condensation. When insulation is installed inside ducts, it will also reduce air stream noise.
6. Branch duct sizes are selected so that the width of the duct equals the width of the register they serve.
7. Duct work in commercial buildings would include extractors at the two right angle branches nearest the air handler and turning vanes in the end tee (so would a good residential installation).

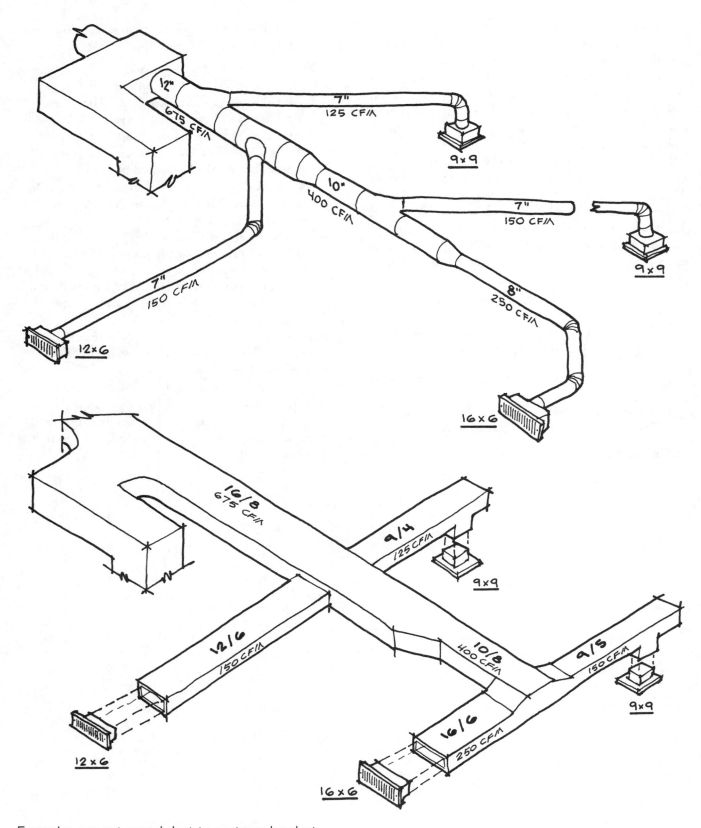

Example; convert round duct to rectangular duct

18.8 **FANS**

PRESSURE, VELOCITY, AND QUANTITY

In duct systems, pressure and fan load are measured in inches of water (see Figure 18.17). Air speed is measured in FPM (feet per minute) and air quantity is measured in CFM (cubic feet per minute).

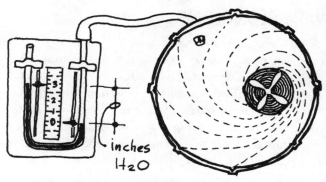

Pressure in fan and duct systems

FIGURE 18.17

Fans are similar to water pumps in that the amount of fluid they move varies with the load they overcome. Fan load is caused by friction as air moves through ducts, elbows, dampers, coils, registers, grilles, and filters. Duct system friction load increases with increasing air velocity.

FAN TYPES

Most air-conditioning installations use fans to distribute conditioned air. Centrifugal (rotating wheel) or axial (propeller) fan types are used (see Figure 18.18).

Centrifugal

Centrifugal fans offer quiet, efficient air distribution for ducted installations. Residential and small commercial installations use forward curved blades on the fan wheel. Larger buildings use backward curved fan blades because of their increased efficiency. The most efficient centrifugal fan blade design is the air-foil shape. It delivers the greatest air quantity per unit of energy input.

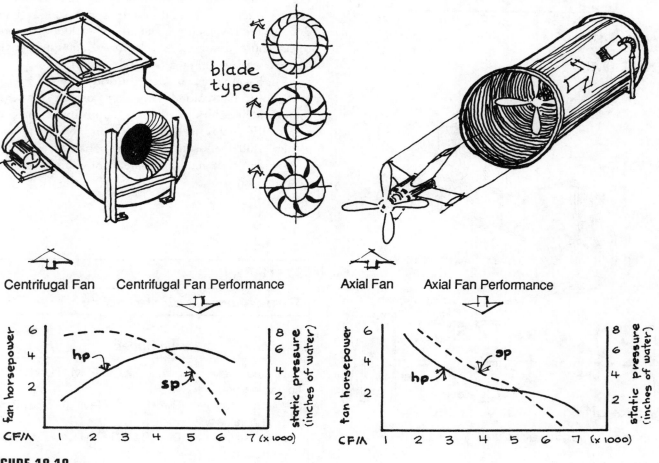

blade types

Centrifugal Fan   Centrifugal Fan Performance

Axial Fan   Axial Fan Performance

FIGURE 18.18

Axial

Axial fans are efficient, moving large quantities of air in open applications such as building exhaust. Recent design improvements also permit axial fans to be used in some ducted installations.

ACTUAL CFM

The actual quantity of air delivered by a fan/duct system is determined by superimposing the fan performance curve on the pressure profile for the duct system (see Figure 18.19). The duct system profile will show increasing pressure (due to friction) as

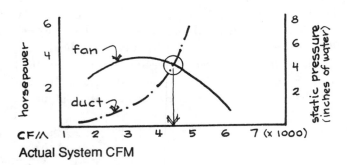

Actual System CFM

FIGURE 18.19

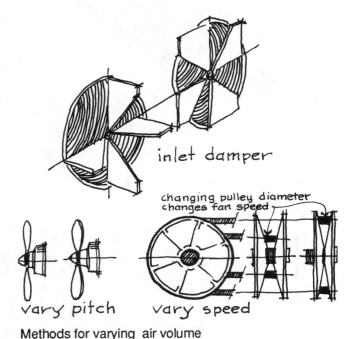

inlet damper

changing pulley diameter changes fan speed

vary pitch   vary speed

Methods for varying air volume

FIGURE 18.20

system CFM increases. Actual CFM will be found at the intersection of the duct pressure profile and the fan performance curve. As a first guess estimate of actual CFM in a low-velocity system, look at the 0.3-inch to 0.5-inch pressure range of the fan curve.

VARIABLE VOLUME

Variable-volume fans save energy by varying CFM (and fan power input) in response to variations in a building's heating or cooling load. Air quantity and fan energy can be decreased using inlet dampers, varying blade pitch, or controlling fan speed. Electronic motor speed control offers the greatest potential energy savings. Inlet dampers don't conserve energy (see Figure 18.20).

HIGH VELOCITY OR LOW VELOCITY

Large buildings use high-velocity air distribution to reduce duct size (and building space required for ducts). High-velocity installations include stronger ducts, sound attenuators, velocity-reducing terminals, and more powerful fans (see Table 18.8).

High-velocity air distribution wastes energy and should be avoided whenever possible. Low-velocity air handlers can deliver conditioned air economically as far as 200 feet from the fan room. Where 200 feet is inadequate, consider a second low-velocity air handler instead of a high-velocity system.

TABLE 18.8

Pressure and Velocity

Name	Air Speed	Inches H₂O
Low pressure (low velocity)	Under 2,000 FPM	Under 2"
Medium pressure (medium velocity)	Over 2,000 FPM	2" to 6"
High pressure (high velocity)	Over 2,000 FPM	6" to 10"

18.9 HVAC CONTROLS

First-time builders spend thousands of dollars for air-conditioning equipment and then try to reduce total building costs by purchasing the cheapest controls available. An air-conditioning installation is only as good as its controls, and control "cost cutting" always cuts occupant comfort and possibly occupant productivity.

Several fine companies operate divisions that specialize in building HVAC controls and devices. Designers, constructors, and owners are well advised when they employ competent control consultants and rely on their advice and recommendations concerning control type and function. "Smart" buildings rely on state-of-the-art control subsystems to maintain comfortable conditions and to conserve energy (see Figure 18.21).

TYPES OF CONTROL

On-off and proportional are two types of control typically used with building air-conditioning equipment (see Figures 18.22 and 18.23). The following paragraphs describe each type briefly, beginning with the simplest.

On-Off

On-off control is typical of most residential heating-cooling systems. The heating or cooling equipment starts and operates at full capacity until the require-

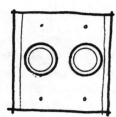

FIGURE 18.22

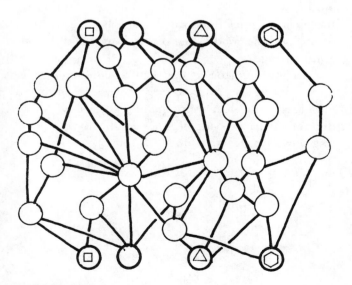

FIGURE 18.23

ment for heating or cooling is satisfied. Temperature in the controlled spaces is continually changing around a set point, and the magnitude of this change is called "swing" or "control differential."

On-off type control (and open-closed control for valves) is also used in large commercial buildings for specific applications. An air handler that is designed for immediate changeover from heating to cooling will use open-closed control for hot and chill water supply valves.

Proportional

Proportional control offers more precise control of temperature and humidity than on-off control. Proportional controls continually adjust the position of valves and dampers to maintain design temperature and humidity with minimum swing.

CONTROLS

HVAC control installations include three elements: sensors, signal switches and media, and controlled devices. Sometimes all three components are incorporated in a single device like a pressure relief valve that senses and operates without outside help. More often the control components are separate interconnected parts.

FIGURE 18.21

Sensors

Sensors respond to changes in temperature, pressure or velocity, and humidity by moving or by changing the resistance of an electric circuit. Mechanical temperature sensors move because of the expansion or contraction of a metal element or gas. Pressure or velocity sensors include bellows, diaphragms, and sail switches that move in response to changing pressure or air speed. Humidity sensors often rely on the movement of a human hair, although other materials and electrical circuits are also used. Electrical sensors change circuit resistance in response to changing temperature or humidity.

Switches and Signal Media

Switches and signal media convert small sensor movements or currents into larger and more powerful signals that can control large motors or valves. Building air-conditioning systems use compressed air and electrical signal media extensively. Compressed air signals position automatic valves and dampers, while electrical signals are used to control electric motors, compressors, fans, and pumps.

Controlled Devices

Controlled devices can include fan motors, compressors, pumps, valves, and dampers. These devices are essentially the same as their manually operated counterparts, but their operation or position can be changed by pneumatic or solenoid actuators in response to signal media commands (see Figure 18.24).

SMART BUILDINGS

Smart buildings use state-of-the-art controls in conjunction with computer programs for precise control of building temperature and humidity under varying internal and external load conditions. Numerous sensors and feedback loops, coupled with well-con-

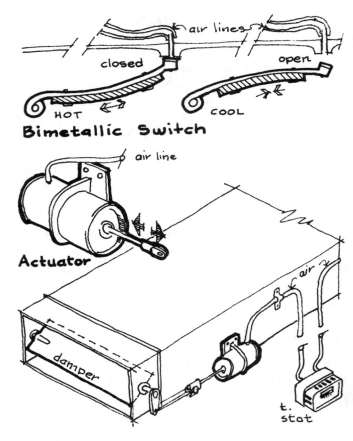

FIGURE 18.24

ceived operating programs, permit anticipatory control where the building can respond to an approaching cold front by starting boilers and warming circulating water temperatures prior to a building thermostat signal requesting heat. Building control operation is as good as the sensor inputs and the operating program.

An amusing story about the first computer-operated building control system is that the program to minimize peak electrical demand by selective shedding of electrical loads began load shedding by cutting power to the controlling computer!

REVIEW QUESTIONS

A number of problems have been included in the text near the appropriate tables. Solve them first.

1. A small commercial building has a peak heat loss of 300,000 BTUH and a peak heat gain of 240,000 BTUH. (Heat gain is 180,000 BTUH sensible and 60,000 BTUH latent.) Calculate the CFM that must be distributed by the air handler for heating and cooling. Use 75°F indoor design temperature, 115°F heating air, and 55°F cooling air.

2. Will heating or cooling set the air quantity used to size ducts in question 1?

3. Calculate cooling CFM again for question 1 using the rule of thumb value for CFM per ton.

4. Select a round duct to carry 8,000 CFM.

5. Select a rectangular duct to carry 2,500 CFM and fit in a space 14″ high. Allow for 1″ thick duct insulation.

6. A 24×12 rectangular duct is carrying 2,000 CFM. The duct must split into two branches. One will carry 600 CFM and the other 1,400 CFM. Select branch sizes.

7. Select a rectangular duct to supply 300 CFM to a 10×10 ceiling register.

8. Select a rectangular duct to supply 600 CFM to a 30×8 wall register.

9. Select a ceiling register to deliver 400 CFM with a two-way throw of 10 feet.

10. Select a wall register to deliver 300 CFM with a medium pattern throw of 13 feet.

11. Select a 4′ long floor register to deliver 200 CFM in a room with an 8′ ceiling height.

12. Select a return air filter grille for a 5-ton air handler.

13. Specify CFM for an exhaust fan that serves a laboratory fume hood. The fume hood inlet opening measures 3′×6′, and a minimum inlet air velocity of 50 FPM (feet per minute) is required to ensure safety for lab occupants.

ANSWERS

1. 7,500 CFM for heating, 8,333 CFM for cooling
   300,000/(1.08)(115–75) = 7,500 CFM
   180,000/(1.08)(75–55) = 8,333 CFM

2. Ducts should be sized to satisfy the larger CFM needed for *cooling*. CFM may be reduced in winter to minimize drafts.

3. 20 tons , 8,000 CFM
   240,000/12,000 = 20 tons, (20)(400) = 8,000
   Notice that the rule of thumb estimate uses total BTUH (sensible plus latent) instead of the more accurate CFM calculation, which uses only the sensible gain taken by conditioned air.

4. Select a 32″ diameter airway (plus insulation).

5. Select a 36×10 airway dimension. The outside dimension of the duct will be 38×12. A 32×12 airway duct would require less material, but there would be no clearance for duct support angles or stiffeners.

6. Split—10×12 = 600 CFM and 18×12 = 1,400 CFM. Hold a constant duct height for a simple transition.

7. Select a 10×7 duct to fit the register. Many ceiling registers are built with round flanges on the register boot. In such cases, a short section of flexible round duct is used between the register and the duct.

8. Use a 30×8 duct *if* there is a short distance (4 feet or less) from the main supply duct to the register. If a longer duct run is required, select a 14×8 duct and transition to 30×8 at the register.

9. 18×18

10. 6×16

11. 2 1/2″×48″

12. Select two at 20×20 because the 5-ton unit will circulate about 2,000 CFM.

13. This question is not specifically covered in the text, but a thinking reader will have no trouble working out an answer. Try before you read on. An air velocity of 1 FPM through an inlet opening with an area of 1 sq. ft. will result in an air flow of 1 CFM.
    Specify a 900 CFM exhaust fan.

(FPM)(area) = CFM   (50)(6×3) = 900 CFM

CHAPTER 19
Annual Costs

*In my small village
even the flies aren't afraid
to bite a big man.*

Issa

This short chapter lets you predict annual utility costs based on heat loss and heat gain calculations. It draws information from the preceding chapters and uses the house and office examples introduced in Chapter 15 to demonstrate annual cost estimates.

Study the example energy and operating cost estimates with patience. When you develop confidence in your ability to evaluate energy-conserving alternatives, you are well on the way to becoming a capable analyst of efficient buildings.

Building owners are particularly interested in costs for obvious reasons, and an important responsibility of designers and constructors is owner education concerning life cycle costs.

Lacking other information, building owners will always select the lowest first cost alternative, but informed owners will opt for high-quality equipment if it offers life cycle cost benefits. Competent building professionals select efficient equipment and communicate potential cost advantages to the owner. Proposing a more efficient (and more expensive) cooling tower with the explanation "it's better" will fall on deaf ears, but proposing the same cooling tower because "its added costs will be paid back in less than two years through utility savings" will earn the owner's respect.

Complete the problems at the end of this chapter and then apply these evaluation techniques to your specific building projects. Remember that your recommendations are only as good as the effort that produces them. They will be respected if they are backed with accurate estimates of initial cost and payback.

19.0 ESTIMATE ANNUAL ENERGY

Calculated values for heat loss and heat gain may be used to estimate annual energy requirements and costs. The heating and cooling hours maps in Figures 19.1 and 19.2 give estimated full-load operating hours for heating and cooling equipment. Maps are rough estimates; better values can be obtained from local utilities. Do *not* apply map values to buildings that are heated and cooled occasionally, like churches and dance halls, or to buildings that operate 24 hours a day, like hospitals and airline terminals.

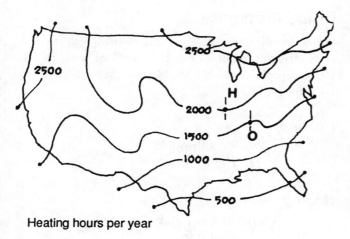

Heating hours per year

FIGURE 19.1

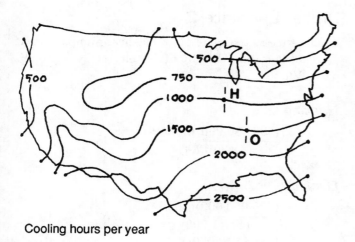

Cooling hours per year

FIGURE 19.2

ANNUAL ENERGY

To estimate annual energy requirements in BTU/yr (BTUs per year), multiply the peak heat loss and gain values for a building by the annual full-load heating and cooling hours shown on the maps. Heat loss and gain calculations for the house and office examples developed in Chapter 15 are used again here as the basis of energy and cost estimates.

House

The example house had a calculated peak heat loss of 77,273 BTUH and a peak heat gain of 33,537 BTUH. Estimate annual heating and cooling BTU at location H.

1. The maps show about 2,000 heating hours and 1,000 cooling hours per year at location H.
2. Heating BTU per year: 154,500,000 BTU/yr
   $$(2,000)(77,273) = 154,500,000$$
3. Cooling BTU per year: 33,500,000 BTU/yr
   $$(1,000)(33,537) = 33,500,000$$

Office

The example office building had a calculated peak heat loss of 406,850 BTUH and a peak heat gain of 630,172 BTUH. Estimate annual heating and cooling BTU at location O.

1. Maps show about 1,500 hours for heating and cooling at location O.
2. Heating BTU per year: 389,680,000 BTU/yr
   $$(1,500)(406,850) = 610,000,000 \text{ BTU/yr}$$

   However, heat contributed by lights* can be credited to total annual heating BTUs.

   Lighting heat is 220,320 BTUH. If the heating season lasts 20 weeks, and lights operate 10 hours per day, 5 days per week, they will provide $(220,320)(20)(10)(5) =$ 220,320,000 BTU during the heating season. Net office heating BTUs are:

   $$610,000,000 - 220,320,000 = 389,680,000$$

3. Cooling BTU per year: 945,000,000 BTU/yr
   $$(1,500)(630,172) = 945,000,000 \text{ BTU/yr}$$

---

*At 3.4 BTU per watt, lights are not efficient heaters, but they do reduce the annual heating load. They also increase the annual cooling load, but that increase is tabulated in the heat gain calculation.

19.1 ESTIMATE ANNUAL ENERGY COSTS

DOLLARS PER YEAR

BTU/yr can be converted to $/yr (dollars per year) if you know the cost of energy and the efficiency of the heating-cooling equipment. The following calculations use the preceding house and office examples with a variety of equipment efficiencies and energy costs. Indoor fan energy is *not* usually included in equipment efficiency ratings, so allow 10% extra for residential air handlers and 20% extra for commercial air handlers with longer duct runs.

Fuel Heat Content	BTU
Coal (anthracite) per lb	14,000
Electricity per kW	3,400
Natural gas per therm (100 CF)	100,000
Natural gas per MCF (1000 CF)	1,000,000
Oil #2 per gal	140,000
Propane per gal	95,500
Wood per lb	5,000–7,000

Change BTU/yr to $/yr with these equations:

Heating

$$\frac{(\text{BTU/yr})(\$/\text{energy unit})}{(\text{BTU/energy unit})(\text{efficiency})} = \$/\text{yr}$$

Cooling

$$\frac{(\text{BTU/yr})(\$/\text{energy unit})}{(\text{SEER} \times 1,000)} = \$/\text{yr}$$

House: Estimate 1

Heat with oil at $1.00 per gallon and a furnace that is 70% efficient. Cool using an air conditioner with a SEER of 12. Electricity costs $0.10 per kWh (1 gal. oil = 140,000 BTU; SEER 12 = 12,000 BTU/kW).

Heating

$$\frac{(154,500,000)(\$1)}{(140,000)(70\%)} = \$1,577/\text{yr}$$

Cooling

$$\frac{(33,500,000)(\$0.10)}{(12)(1,000)} = \$279/\text{yr}$$

Subtotal $1,856/yr
Add 10% for indoor fan = $2,042/yr

House: Estimate 2

Same example, but use a heat pump with a COP of 2.2 and a SEER of 10 (COP 1 = 3,400 BTU per kW).

Heating

$$\frac{(154,500,000)(\$0.10)}{(3,400)(2.2)} = \$2,066/\text{yr}$$

Cooling

$$\frac{(33,500,000)(\$0.10)}{(10)(1,000)} = \$335/\text{yr}$$

Subtotal $2,401/yr
Add 10% for indoor fan = $2,641/yr

Office: Estimate 1

Heat the office building with natural gas that is 70% efficient at $0.50 per therm. Cool with SEER 10 air conditioners at $0.08 per kWh (1 therm = 100,000 BTU).

Heating

$$\frac{(389,680,000)(\$0.50)}{(100,000)(70\%)} = \$2,783/\text{yr}$$

Cooling

$$\frac{(945,000,000)(\$0.08)}{(10)(1,000)} = \$7,560/\text{yr}$$

Subtotal $10,343/yr
Air handlers add 20% = $12,412/yr

Office: Estimate 2

Same office as Estimate 1 but use heat pumps with a COP of 2.4 and a SEER of 11, electricity at $0.09 per kWh.

Heating

$$\frac{(389,680,000)(\$0.09)}{(3,400)(2.4)} = \$4,298/\text{yr}$$

Cooling

$$\frac{(945,000,000)(\$0.09)}{(11)(1,000)} = \$7,732/\text{yr}$$

Subtotal $12,030/yr
Air handlers add 20% = $14,436/yr
(only a bit more than $1,600 a month)

Cautions! The preceding calculations can provide reasonable first estimates, but precision depends on accurate annual heating and cooling hours. Local electrical utilities and fuel suppliers are the best source of annual heating and cooling hour information for many building types.

The calculations can be applied with some confidence to residential and smaller commercial occupancies. However, they will not apply to 24-hour-a-day occupancies like hospitals and airline terminals, or intermittent occupancies like theaters and churches.

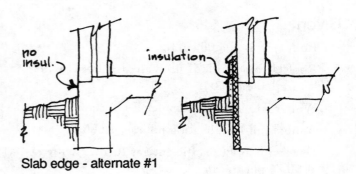

Slab edge - alternate #1

FIGURE 19.3

19.2 EVALUATE ENERGY-CONSERVING ALTERNATIVES

Section 19.1 proposed a method for estimating annual heating and cooling costs. The same method may be used to evaluate energy-conserving alternatives.

awesome savings !

Calculations in this section are based on the example office building used in Chapter 15. Alternates 1 through 3 assume 1,600 heating hours and 1,400 cooling hours. Alternates 4 and 5 use 1,500 hours for both heating and cooling.

Given

Slab edge length is 494 linear feet.

The slab edge U value without insulation is 0.8; with insulation U is 0.1.

Winter TD (temperature difference) is 60.

Annual full-load heating hours = 1,600.

Heat is electric resistance at $0.06/kWh (each kWh yields 3,400 BTUH).

Heating Savings

No insulation $\quad$ $(0.8)(494)(60) = 23{,}712$ BTUH
With insulation $\quad$ $(0.1)(494)(60) = \underline{2{,}964}$ BTUH
$\qquad\qquad$ Peak heat loss saving $20{,}748$ BTUH

First-Year Heating Savings

$$\frac{(20{,}748)(1{,}600)(0.06)(1.2)}{3{,}400} = \$703$$

Summary

Estimated simple payback: 2.8 years.

Payback = 2,000/703 = 2.8 years.

ALTERNATIVE 1: SLAB EDGE INSULATION

An insulation contractor proposes to reduce building heat loss by installing slab edge insulation for $2,000 (see Figure 19.3). Evaluate the proposal by estimating the payback period.

ALTERNATIVE 2: MORE ROOF INSULATION

The roofing contractor offers to increase the thickness of urethane roof insulation from 3 inches to 4 inches for $3,000. Evaluate this proposal by estimating the payback period.

Given

Roof area is 10,800 sq. ft.

Roof U value with 3″ urethane is 0.05; U with 4″ urethane is 0.04.

Winter TD is 60°F; summer ETD is 36°F.

Annual full-load heating hours = 1,600.

Heat with an 80% efficient gas furnace with gas at $0.75 per therm.

Annual full-load cooling hours = 1,400.

Cool with SEER-10 air conditioners; electricity costs $0.09 per kWh.

Heating Savings

3″ roof $(0.05)(10,800)(60)$ = 32,400 BTUH
4″ roof $(0.04)(10,800)(60)$ = 25,920 BTUH
Peak heat loss savings   6,480 BTUH

First-Year Heating Savings

$$\frac{(6,480)(1,600)(\$0.75)(1.2)}{(100,000)(80\%)} = \$116.64/\text{yr}$$

Cooling Savings

3″ roof $(0.05)(10,800)(36)$ = 19,440 BTUH
4″ roof $(0.04)(10,800)(36)$ = 15,552 BTUH
Peak heat gain savings   3,888 BTUH

First-Year Cooling Savings

$$\frac{(3,888)(1,400)(\$0.09)(1.2)}{(10)(1,000)} = \$58.79/\text{yr}$$

Summary

Estimated simple payback: 17.1 years.

Annual heating and cooling savings is $175.43 (calculations include 20% for air handlers).

Payback = 3,000/175 = 17.1 years.

ALTERNATIVE 3: LESS VENTILATION

The HVAC and insulation contractors propose to weatherstrip the building and cut the average ventilation rate to 2,000 CFM. Added costs for controls and weatherstripping will total $2,500.

Given

Original design was 2,970 CFM winter, 2,400 CFM summer. New design is 2,000 CFM for both winter and summer.

Winter TD is 60°F; summer TD is 20°F.

GD (grain difference) is 40.

Heat with a 75% efficient oil-fired boiler using $0.90 per gallon oil.

Annual heating hours = 1,600.

Annual cooling hours = 1,400.

Cool with a SEER-8 (net) electric chiller; electricity costs $0.07 per kWh.

HEATING SAVINGS

Original: $(2,970)(1.08)(60)$ = 192,456 BTUH
Proposal: $(2,000)(1.08)(60)$ = 129,600 BTUH
Peak heat-loss savings   62,856 BTUH

First-Year Heating Cost Savings

$$\frac{(62,856)(1,600)(\$0.90)(1.2)}{(140,000)(75\%)} = \$1,034/\text{yr}$$

Cooling Savings

Original sensible: $(2,400)(1.08)(20)$ = 51,840 BTUH
Original latent: $(2,400)(0.68)(40)$ = 65,280 BTUH
Original total 117,120 BTUH
New sensible: $(2,000)(1.08)(20)$ = 43,200 BTUH
New latent: $(2,000)(0.68)(40)$ = 54,400 BTUH
New total 97,600 BTUH
Peak cooling saving 19,520 BTUH

First-Year Cooling Cost Savings

$$\frac{(19,520)(1,400)(\$0.07)(1.2)}{(8)(1,000)} = \$287/\text{yr}$$

Total heating and cooling savings $1,321/yr

Summary

Estimated simple payback: 1.9 years.

Payback = 2,500/1,321 = 1.9 years.

Caution! Check the number of building occupants *before* reducing the ventilation rate. In many localities, 15 CFM/occupant is a minimum code requirement.

ALTERNATIVE 4: IMPROVED HEAT PUMPS

The HVAC subcontractor proposes to furnish more efficient heat pumps for $10,000 extra. Evaluate the proposal by estimating payback.

Given

The specified heat pumps have a COP of 2 and a SEER of 9. High-efficiency heat pumps have a COP of 2.8 and a SEER of 13.

Electricity costs $0.08/kWh.

Annual heat loss is 389,680,000 BTU per year.

Heat gain is 945,000,000 BTU per year.

Note: 1,500 full-load heating and cooling hours are used to calculate BTU per year for this example.

Heating Savings

Specified	$\dfrac{(389,680,000)(\$0.08)(1.2)}{(3,400)(2)}$	$5,501/yr
Efficient	$\dfrac{(389,680,000)(\$0.08)(1.2)}{(3,400)(2)}$	$3,929/yr
	First-year heating savings	$1,572/yr

Cooling Savings

Specified	$\dfrac{(945,000,000)(\$0.08)(1.2)}{(9)(1,000)}$	$10,080/yr
Efficient	$\dfrac{(945,000,000)(\$0.08)(1.2)}{(13)(1,000)}$	$6,978/yr
	First-year cooling savings	$3,102/yr

Total heating and cooling savings allowing 20% for air handlers $4,674/yr

Summary

Estimated simple payback: 2.1 years.
Payback = 10,000/4,674 = 2.1 years.

ALTERNATIVE 5: REFLECTIVE WINDOWS

The glazing contractor proposes to provide reflective windows* instead of clear windows for $2,470 extra. Evaluate the proposal by estimating payback.

Given

Glass area totals 2,470 sq. ft.

Shading coefficient (SC) for original glass was 85%; SC for proposed reflective glass is 45%.

Solar factor (SF) is 36 (weighted average for north and shaded south glass).

Annual cooling hours = 1,500.

Cooling SEER is 9; electricity costs $0.12 per kWh.

Heating Savings

None!   Reflective glass probably increases winter heating requirements because of reduced solar gain for the south-facing windows. However, the lower emittance of a reflective surface improves the U value of all window areas.

The example estimates *only* summer cooling savings.

Cooling Savings

Original glass:	(36)(2,470)(85%) =	75,582 BTUH
Reflective:	(36)(2,470)(45%) =	40,014 BTUH
	Peak cooling savings	35,568 BTUH

First-Year Cooling Cost Savings

$$\frac{(35,568)(1,500)(\$0.12)(1.2)}{(9)(1,000)} = \$854/yr$$

Summary

Estimated simple payback: 2.9 years.
Payback = 2,470/854 = 2.9 years.

---

*Alternative 5 may overestimate annual cooling savings credited to reflective glass because an average solar factor is used. More accurate estimates can be developed using actual solar data for the building site.

REVIEW PROBLEMS

Set up and solve each of the following problems *before* checking your solutions.

1. Evaluate Furnaces

Two furnaces are being considered for a proposed elementary school. Furnace **A** costs $1,200 and has a 75% operating efficiency. Furnace **B** is a condensing type that costs $2,000 and has a 95% operating efficiency. Find the simple payback period for furnace B.

Given:

Furnace capacity = 240,000 BTUH
School peak heat loss = 200,000
Natural gas at $0.60 per therm (100,000 BTU)
Annual full-load heating hours = 1,700

Solution

First find heating BTU per year; then $ per year for each furnace.

BTU per year = (peak heat loss)(full-load hr/yr)
  (200,000)(1,700) = 340,000,000 BTU
$ per yr = (BTU per yr)($ fuel)/
  (BTU-fuel)(efficiency)

Furnace A

$$\frac{(340,000,000)(\$0.60)}{(100,000)(75\%)} = \$2,720$$

Furnace B

$$\frac{(340,000,000)(\$0.60)}{(100,000)(95\%)} = \$2,147$$

Furnace B costs $800 extra and will save $573 in heating costs each year. The simple payback for furnace B is 1.4 years.

2. Evaluate Storm Windows

Storm windows for a residence will cost $500 ($2.50 per sq. ft. of window area). Find the simple payback period for storm windows.

Given:

U value existing windows = 1.1
U value with storm windows added = 0.6
Winter TD = 70°F
Electric heat pump COP = 2.4
Electricity at $0.07 per kWh
Annual full-load heating hours = 1,900
Summer TD = 20°F
Electric heat pump SEER = 10.0
Annual full-load cooling hours = 800

Solution *

Consider just 1 sq. ft. of window area. First calculate heat loss and gain; then convert BTU to $ per year.

Peak Heat Loss = (U)(area)(TD)
  No storm (1.1)(1)(70) = 77 BTUH
  Storm on (0.6)(1)(70) = 42 BTUH
Heat Loss/Year: (BTUH)(full-load hr)
  No storm (77)(1,900) = 146,300 BTU/yr
  Storm on (42)(1,900) = 79,800 BTU/yr
Heating $/Year:

$$\frac{(BTU)(\$ \text{ fuel})}{(BTU\text{-fuel})(\text{efficiency})}$$

No storm $\dfrac{(146,300)(0.07)}{(3,400)(2.4)}$ = $1.26

Storm on $\dfrac{(146,300)(0.07)}{(3,400)(2.4)}$ = $0.68

Storm windows save $0.58 per sq. ft.

Peak Heat Gain* = (U)(area)(TD)
  No storm (1.1)(1)(20) = 22 BTUH
  Storm on (0.6)(1)(20) = 12 BTUH
Heat Gain/Year: (BTUH)(full-load hr)
  No storm (22)(800) = 17,600 BTU
  Storm on (12)(800) = 9,600 BTU

---

*This solution is incomplete because it ignores solar gain reductions and the probability that infiltration rates will be reduced by storm windows. Use the tables on pages 214–215 to see if summer and winter solar gain changes cancel out. Problem 4 illustrates the procedure for estimating cost savings due to reduced infiltration or ventilation. The difficulty is estimating the infiltration CFM reduction caused by storm windows.

Cooling $/Year:

$$\frac{(BTU)(\$ \text{ per kW})}{(SEER)(1,000)}$$

No storm $\frac{(17,600)(0.07)}{(10)(1,000)}$ = $0.12

Storm on $\frac{(9,600)(0.07)}{(10)(1,000)}$ = $0.07

Storm windows save $0.05 per sq. ft.

Sum annual heating and cooling savings per sq. ft. of window area and add 10% for indoor fans:

$$(0.58 + 0.05)(110\%) = \$0.69$$

Storm windows cost $2.50 per sq. ft., and they should save $0.69 per sq. ft. in heating and cooling costs. The simple payback is 3.6 years.

3. Evaluate Cooling Equipment

Two 100-ton chillers are being considered for a department store (Figure 19.4). **Chiller A** costs $44,000 and has a net SEER of 7.1. **Chiller B** costs $50,000 and has a net SEER of 8.6. (Net SEER values include chill water pumps, cooling tower pumps and fans, and air-handler fans.) Find the simple payback period for chiller B.

Given:

Peak heat gain = 1,200,000 BTUH
Electricity at $0.08 per kWh
Annual full-load cooling hours = 1,800

FIGURE 19.4

Solution

First find cooling BTU per year, then $ per year for each chiller.

BTU per year = (peak heat gain)(full-load hr/yr)
(1,200,000)(1,800) = 2,160,000,000 BTU
$ per yr =(BTU per yr)($ per kW) (SEER)(1,000)

Chiller A

$$\frac{(2,160,000,000)(\$0.08)}{(7.1)(1,000)} = \$24,338$$

Chiller B

$$\frac{(2,160,000,000)(\$0.08)}{(8.6)(1,000)} = \$20,093$$

Chiller B savings/yr = $4,245

Chiller B costs $6,000 extra and will save $4,245/yr. The simple payback for chiller B is 1.4 years.

4. Evaluate Reduced Ventilation

A HVAC engineer recommends a 2,800 CFM reduction of an existing hospital's ventilation rate. The ventilation system maintains positive pressure in most areas, so infiltration is not a concern. Find the annual heating and cooling cost savings due to the reduced ventilation rate.

Given:

Winter TD = 66°F

Gas boiler heating at 80% efficiency

Natural gas at $0.60 per therm

Annual full-load heating hours = 1,100

Summer TD = 15°F

Summer GD = 40

Electric chiller, net SEER = 7.2

Electricity at $0.07 per kWh

Annual full-load cooling hours = 2,600

Solution

First find peak BTUH savings, then annual BTU saved, and finally annual $ saved.

Heat Loss Savings BTUH: (CFM)(1.08)(TD)

(2,800)(1.08)(66) = 199,584 BTUH

Heat Loss Savings BTU/Year: (BTUH)(full-load hours)

(199,584)(1,100) = 219,542,400 BTU

Heat $/Year Saved:

$$\frac{(BTU)(\$ \, fuel)}{(BTU\text{-}fuel)\,(efficiency)}$$

$$\frac{(219,542,400)(\$0.60)}{(100,000)(80\%)} = \$1,647$$

Heat Gain Savings BTUH: (CFM)(1.08)(TD) = sensible, plus (CFM)(0.68)(GD) = latent

Sensible BTUH

(2,800)(1.08)(15) = 45,360 BTUH

Latent BTUH

(2,800)(0.68)(40) = 76,160 BTUH
Total savings BTUH = 121,520 BTUH

Heat Gain Savings BTU/Year: (BTUH)(full-load hours)

(121,520)(2,600) = 315,952,000 BTU

Cooling $/Year Saved:

$$\frac{(BTU)(\$ \, per \, kW)}{(SEER)(1,000)}$$

$$\frac{(315,952,000)(\$0.07)}{(7.2)(1,000)} = \$3,072$$

Total annual savings due to reduced ventilation ($ heating + $ cooling)* is $4,719

5. Evaluate Wall Insulation

Wall A is made with 2×4 studs at 16″ centers and includes R-11 fiberglass insulation. It can be built at a cost of $1.00 per sq. ft. of wall area.

*No air handler % added—included in *net* SEER.

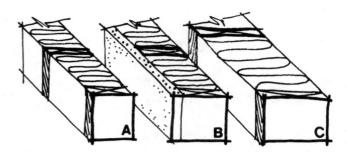

FIGURE 19.5

Wall B is the same as wall A except 1″ thick, R-5 polystyrene sheathing will be nailed to the studs at a total cost of $1.40 per sq. ft.

Wall C is made with 2×6 studs at 24″ centers and includes R-19 fiberglass insulation. It can be constructed for $1.32 per sq. ft. (see Figure 19.5).

You are to determine the heating and cooling costs associated with each wall and recommend one wall type for a new residence.

Given:

Winter TD = 70°F

Gas furnace heating at 90% efficiency

Natural gas at $0.70 per therm

Annual full-load heating hours = 2,000

Summer ETD = 25°F (weighted average for N, S, E, and W orientations)

Electric air conditioning SEER = 10.0

Electricity at $0.06 per kWh

Annual full-load cooling hours = 1,000

Solution

Consider just 1 sq. ft. of wall area to keep the numbers small. First calculate U values and heat loss-gain for each wall; then convert BTU to $ per year, and finally compare the walls.

U Values

Wall A 1/11 = 0.09

Wall B 1/16 = 0.06

Wall C 1/21 = 0.05

Peak Heat Loss: (U)(A)(TD)

Wall A (0.09)(1)(70) = 6.3 BTUH

Wall B (0.06)(1)(70) = 4.2 BTUH

Wall C (0.05)(1)(70) = 3.5 BTUH

Heat Loss/Year: (BTUH)(full-load hours)

Wall A (6.3)(2,000) = 12,600 BTU

Wall B (4.2)(2,000) = 8,400 BTU

Wall C (3.5)(2,000) = 7,000 BTU

Heating $ /Year: (BTU)($ fuel)/(BTU-fuel)(efficiency)

Wall A $\dfrac{(12,600)(\$0.70)}{(100,000)(90\%)}$ = \$0.098

Wall B $\dfrac{(8,400)(\$0.70)}{(100,000)(90\%)}$ = \$0.065

Wall C $\dfrac{(7,000)(\$0.70)}{(100,000)(90\%)}$ = \$0.054

Peak Heat Gain: (U)(A)(ETD)

Wall A (0.09)(1)(25) = 2.25 BTUH

Wall B (0.06)(1)(25) = 1.50 BTUH

Wall C (0.05)(1)(25) = 1.25 BTUH

Heat Gain/Year: (BTUH)(full-load hr)

Wall A (2.25)(1,000) = 2,250 BTU

Wall B (1.50)(1,000) = 1,500 BTU

Wall C (1.25)(1,000) = 1,250 BTU

Cooling $/Year: (BTU)
($ per kW)/(SEER)(1,000)

Wall A $\dfrac{(2,250)(\$0.06)}{(10)(1,000)}$ = \$0.014

Wall B $\dfrac{(1,500)(\$0.06)}{(10)(1,000)}$ = \$0.009

Wall C $\dfrac{(1,250)(\$0.06)}{(10)(1,000)}$ = \$0.0075

Sum annual heating and cooling costs per sq. ft. of wall area and then add 10% to each for indoor fan:

Wall A (0.098 + 0.014)(110%) = \$0.123

Wall B (0.065 + 0.09)(110%) = \$0.081

Wall C (0.054 + 0.0075)(110%) = \$0.068

The costs above are the annual heating-cooling costs per sq. ft. for each wall. Evaluate walls B and C compared to the least first cost wall A.

Wall B costs \$0.40 extra and will save \$0.042 per year (\$0.123 – \$0.081). This is a 9.5-year simple payback. Wall C costs \$0.32 extra and will save \$0.055 per year. This is a 5.8-year simple payback.

Recommend wall C, but explain that other alternatives may offer better payback.

CHAPTER 20
Efficient Designs

*One perfect moon
and the uncountable stars
drowned in a green sky.*

Shiki

Contemporary terms include *green* and *sustainable*, but whatever the wording, efficiency is an essential ingredient of quality design. Efficient buildings exploit solar, site, and climate opportunities to conserve heating and cooling energy. Talented designers include efficiency as a primary design priority from concept through construction. Efficient buildings don't just happen; they're the result of a continuing commitment by designers, constructors, owners, and operators.

OPPORTUNITIES

Design choices in four areas can reduce building heating and/or cooling costs.

1. **Solar designs.** The winter sun warms south-facing walls. South-facing windows can cut heating costs, and with proper shading they can welcome daylight but exclude direct summer sun.
2. **Thermal mass.** Heavy objects hold heat longer than light objects; earth temperature varies less than air temperature. The earth's thermal mass can be exploited by building underground, and designers can use building mass to delay heat flow.
3. **Evaporation.** In hot-dry climates, evaporative coolers and fountains are economical sources of cool (moist) air.
4. **Air movement.** Increased air velocity can make a warm room comfortable. In some locations, ocean breezes or convective air circulation can replace air conditioning. Where breezes are few, fans are a bargain compared to air conditioners.

In frigid climates where solar and site factors don't offer comfort opportunities, buildings must be designed to limit heat losses. Minimum surface area and maximum envelope insulation are appropriate design responses for such climates.

20.0 SOLAR DESIGNS

Solar control should be a primary design principle. Buildings that use south-facing windows can enjoy natural heat input during winter months, and summer heat can be excluded with minimal effort. Effective sun control and free winter heating are easy _only for south-facing windows!_ Windows that don't face south gain less heat in winter, and shading devices to exclude summer sun are more expensive and less effective. Architectural sun-control surfaces can make a plain building interesting and an interesting building exciting.

WHY SOUTH-FACING WINDOWS?

Tables 20.1 and 20.2 give the total BTU per square foot per day incident on windows with different orientations during clear winters and summers.

Example

Use Tables 20.1 and 20.2 to estimate the total solar gain through 1 square foot of south-facing single glass at 40°N during three winter months. Assume 60% clear days and 40% cloudy days, for 90 winter days. Use the Nov–Dec–Jan average for clear days and the average north value for cloudy days. Use 90% as the shading coefficient (SC) for clear glass.

Clear days   (54)(1,590)(90%) = _77,274 BTU_
Cloudy days   (36)(113)(90%) =   _3,661 BTU_
           Total solar gain = 80,935 BTU

If the _average_ winter temperature difference (TD) across the glass is 30°F, will the glass gain or lose heat during three winter months? How much?

Glass heat loss (U)(A)(TD)(total hr) = BTU
(1.1)(1)(30)(90)(24) = 71,280 BTU

One square foot of south-facing window will gain 9,655 BTU over three winter months.

SUN CONTROL

Effective shading devices can be designed to admit winter sun and exclude summer sun. For a given latitude (see Figure 20.1) you can easily find the sun's position at any time of the year. _Sun charts_ and the _section angle overlay_ (in the back of this text) indicate the sun's location and permit you to calculate actual sun lines on building drawings. The charts and overlay will help you set dimensions for shading devices. They are also useful in evaluating daylighting design proposals.

Examples

Use the 30°N sun chart in the following two examples. Sun charts for other latitudes can be found on pp. 297–299.

1. Using _only_ the 30°N sun chart, find the time and location of sunrise on 21 April.

Time = 5:30 AM
Location = 14° north of east

TABLE 20.1

32°N Latitude	N	S	E-W	Hor.
Dec	140	1,700	570	880
Nov or Jan	160	1,690	650	1,020
Oct or Feb	200	1,530	820	1,370
June	520	450	1,170	2,240
May or July	450	500	1,150	2,180
Apr or Aug	330	710	1,090	2,000

TABLE 20.2

40°N Latitude	N	S	E-W	Hor.
Dec	100	1,550	420	560
Nov or Jan	120	1,610	500	700
Oct or Feb	170	1,620	720	1,090
June	500	630	1,230	2,240
May or July	440	700	1,200	2,150
Apr or Aug	310	970	1,090	1,900

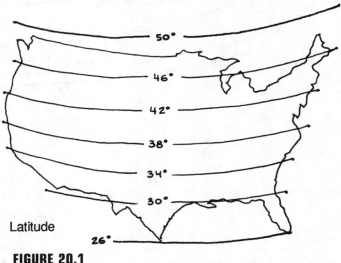

Latitude

FIGURE 20.1

2. Find the sun's bearing (compass direction from observer) at 2 PM on 21 September.

Bearing = 50° west of south

(Extend a line from chart center through the time-date to the bearing ring.)

The following two examples use the 30° sun chart and the **section angle overlay**. Use a pin through the center points of both, and set the "window faces" arrow by rotating the overlay. Read the section angle on the overlay directly above the time-date point on the sun chart.

3. Window faces south. Find the section angle at 3 PM on 21 August.

Section angle is 75°

4. Window faces 35° west of south. Find the section angle at 3 PM on 21 August.

Section angle is 50°

Notice that changing the window orientation reduced the section angle. The roof overhang shown in Figure 20.2 excluded direct sun from the south-facing window, but direct sun added to the air-conditioning load for the window facing 35° west of south.

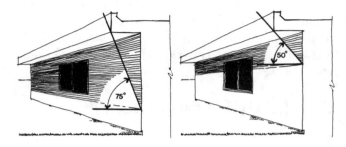

FIGURE 20.2

Sun Chart

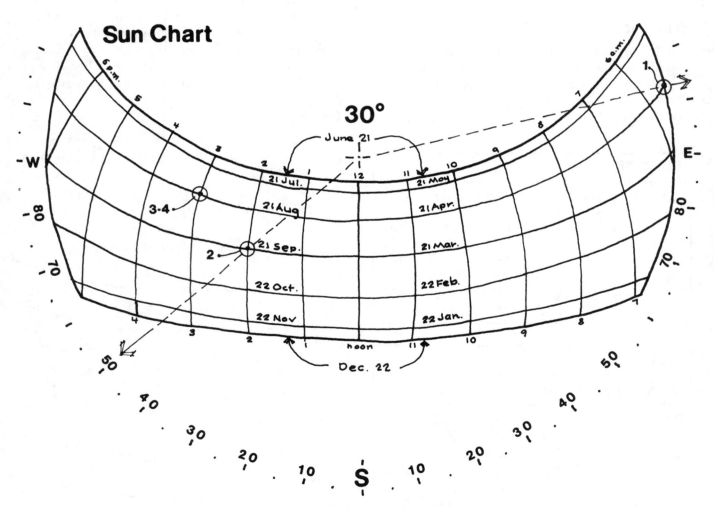

SOLAR DESIGNS

Skillful designers detail windows and shading devices to maximize heat gain during the winter and minimize heat gain during the summer.

Example

Optimize exterior shading for the following window:

1. Window faces south at 30°N latitude.
2. Overheated summer conditions are expected from 21 May through 21 August.
3. Winter heating is required from 22 November through 22 February.

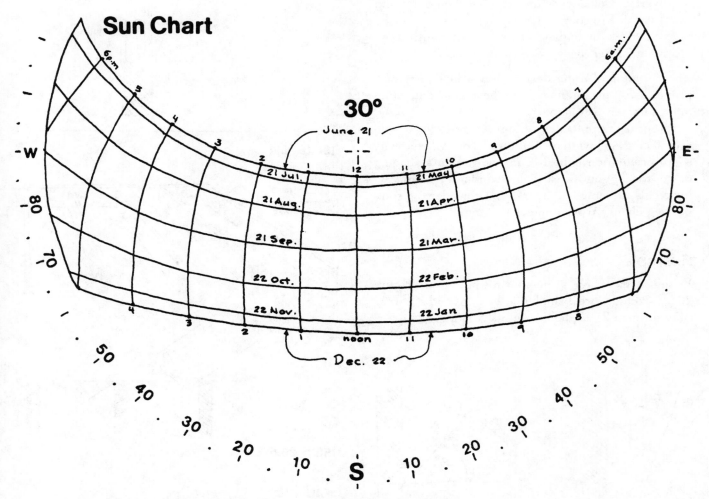

Developed with permission from material copyrighted by LIBBEY-OWENS-FORD CO.

Solution

1. Find the lowest summer section angle using the overlay. Consider summer sun between 7 AM and 5 PM. (Earlier or later times are not important for south-facing windows because of the sun's bearing.)

    70° is the lowest summer section angle (at noon 21 August).

2. Find the highest winter section angle using the overlay. Consider winter sun between 8 AM and 4 PM due to short solar days.

    48° is the highest winter section angle (at noon 22 February). See Figure 20.3.

The optimum shading design shown in Figure 20.4 provides for maximum winter heat gain and minimum summer heat gain.

Repeat the preceding design exercise for a window facing 45° west of south and you will see why south orientation is best. Can you design a shading device for the southwest-facing window? The following trellis example will help if you consider the sun's bearing change from summer to winter.

Example

Design a trellis for a west-facing patio (see Figure 20.5). Exclude direct sun between 4 PM and 7 PM, 21 May through 21 September at 30°N latitude.

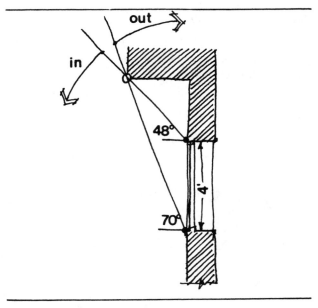

FIGURE 20.4

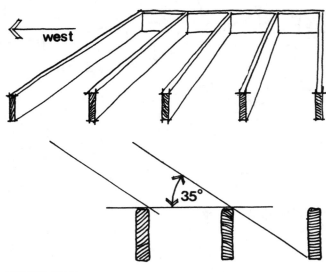

FIGURE 20.5

Solution

Set the section angle overlay facing due west on the 30° sun chart. Find the highest section angle for the given time-date period.

35° is the highest section angle (at 4 PM on 21 June). Use a 35° angle to space trellis joists.

Talented designers will choose south-facing windows and horizontal shading surfaces at every opportunity. Selected challenges and design details are briefly discussed here.

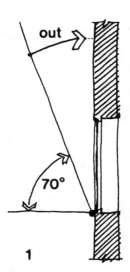
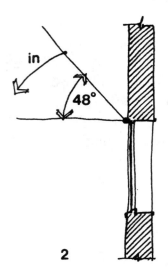

FIGURE 20.3

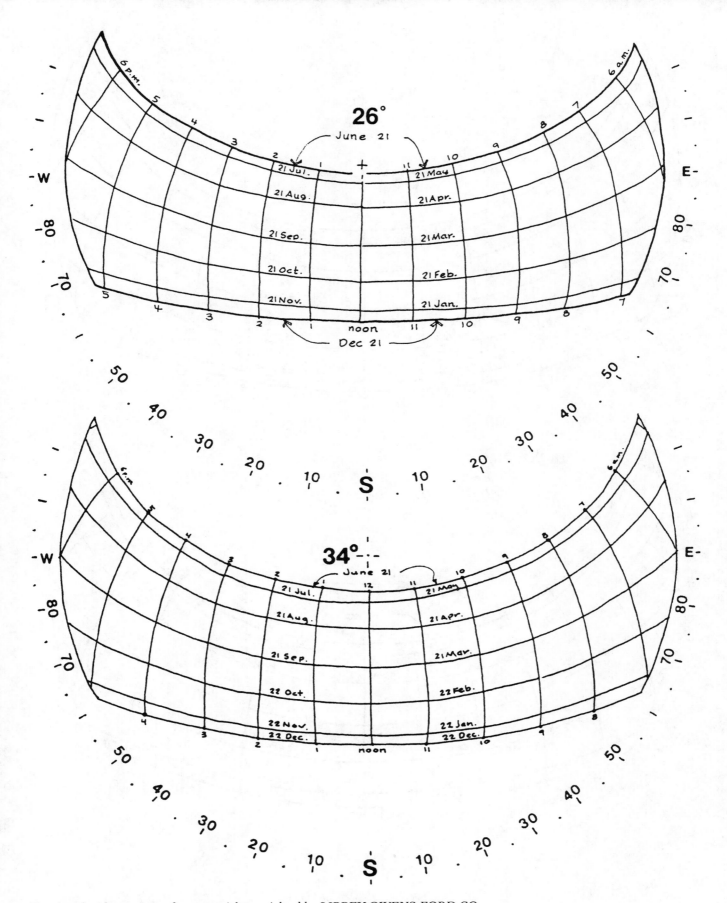

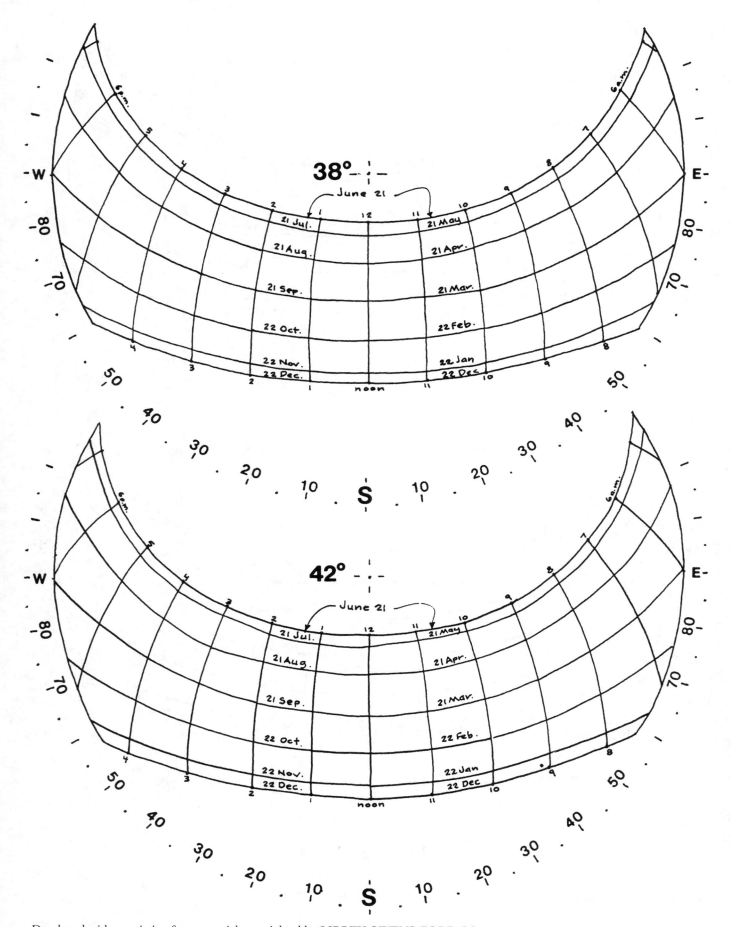

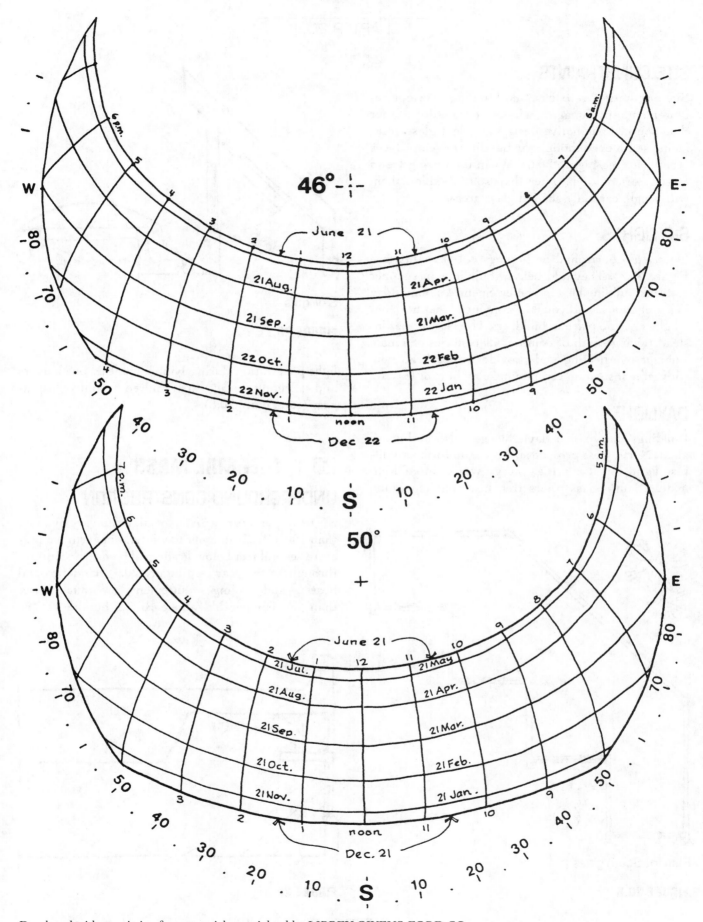

SITE CONSTRAINTS

Site orientation, setbacks, and building envelope requirements may suggest window orientations other than south. Imaginative designers orient glass areas facing south even though the building envelope faces otherwise (see Figure 20.6). When interesting views exist at orientations other than south, exploit them with small, carefully shaded glass areas.

SKYLIGHTS

Do *not* use horizontal or sloping roof skylights (see Figure 20.7). They add heat to buildings in summer and leak building heat in winter. Spend a summer day in a greenhouse and you'll be convinced that horizontal glass has no place in buildings. Use clerestories instead (facing south of course). Clerestories eliminate summer overheating and provide winter heat gain. Moveable insulation can minimize night heat losses.

DAYLIGHT

Buildings that exploit daylight use north and south glass. North glass provides the most uniform and diffuse light, but it is a real energy waster in cold climates. South glass offers free heat and maximum

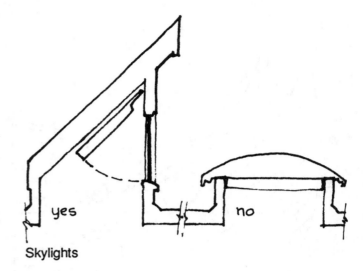

Skylights

FIGURE 20.7

light penetration during winter months, but louvers and reflecting surfaces are required to control brightness (see Figure 20.8).

20.1 THERMAL MASS

UNDERGROUND CONSTRUCTION

Air temperatures at a given location can vary by more than 100°F from summer to winter, but earth temperature several feet below grade will be nearly constant throughout the year (see Figure 20.9). Underground heating and cooling requirements are much lower than for comparable aboveground buildings. The

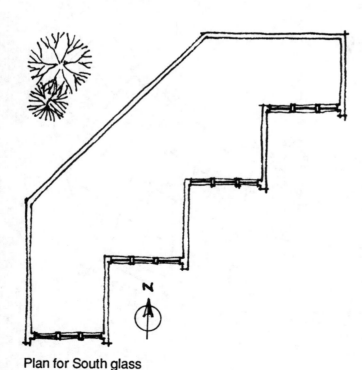

Plan for South glass

FIGURE 20.6

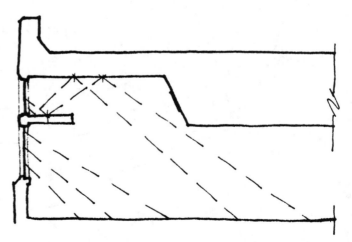

FIGURE 20.8

Underground

FIGURE 20.9

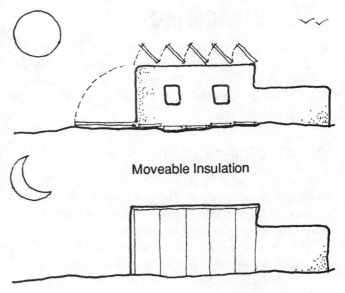

Moveable Insulation

FIGURE 20.11

next time a client asks for an energy-efficient home, watch his or her eyes when you suggest an underground structure.

EARTH-SHELTERED CONSTRUCTION

Where underground construction is not practical, south-facing windows plus earth-sheltered designs can provide similar benefits (see Figure 20.10). Careful detailing is required to prevent water and heat leaks where structural members penetrate the earth envelope. Interested? Read *Gentle Architecture* by M. Wells.

MOVEABLE INSULATION

In desert climates, heavy adobe walls can moderate outdoor temperature change. The sun warms adobe walls during the day, and they release some heat indoors at night. But adobe walls lose lots of heat to the desert night, and moveable insulation can minimize these losses.

An improved desert design might add exterior moveable insulating panels (see Figure 20.11). In winter months, insulation on the east, south, and west walls would be opened to admit solar energy during the day and closed to retain heat during the night. In the summer, wall and roof insulation would be opened at night to maximize radiant heat loss to the night sky, and closed during the day to keep the interior cool. Reflective skins on the insulating panels are a further refinement. Interested? Read *Sunspots* by Steve Baer.

FIXED EXTERNAL INSULATION

Insulation on the outside of a building lets interior mass store some heat and delay interior temperature changes. It may save some energy in climates where outdoor conditions fluctuate around a comfortable temperature, but the savings will be small. External insulation is a worthless investment on south-facing walls and in climates where outdoor conditions are uncomfortable for 24 hours each day.

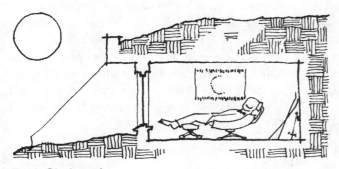

Earth Sheltered

FIGURE 20.10

Heat Capacity* of Materials	Volume (cu. ft.)
Insulation (fiberglass, 1 lb/cu. ft.)	150,000
Wood (pine, 32 lb/cu. ft.)	2,400
Rock (gravel up to 3F3, 100 lb/cu. ft.)	1,250
Concrete (150 lb/cu. ft.)	833
Water (62.4 lb/cu. ft.)	400

*Space required in cubic feet to store 1 million BTU at 40°F above ambient.

20.2 EVAPORATION

In air-conditioning equipment a refrigerant evaporates and condenses to move heat. A lot of heat* is absorbed when liquid refrigerant changes to a vapor (evaporates), and this heat is released by the vapor when it condenses.

EVAPORATIVE COOLERS

In dry climates, evaporative coolers can be used instead of air conditioners. They change hot-dry air into cool-moist air and use less than 10% of the energy required by a conventional air conditioner.

Unfortunately, evaporative coolers don't work in humid climates; their output is cool air, but its moisture content is too high for comfort. Indirect evaporative coolers can provide *limited* cooling in humid climates because they cool without adding moisture to the building air supply (see Figure 20.12).

COOLING TOWERS

Cooling towers are evaporative coolers used to dump large quantities of heat from generation or refrigeration processes. A perfectly efficient cooling tower would cool water to the wet-bulb temperature of surrounding air, but operating towers usually deliver cooled water at about 10°F above the ambient wet bulb.

Large power plants use convective cooling towers to save fan energy (Figure 20.12).

FOUNTAINS AND TREES

A sheltered courtyard with a fountain is cooler than nearby dry spaces because of evaporation. Fountains can be used for outdoor cooling, or they can replace cooling towers as a means of dumping heat to the environment.

Trees also do a fine job of cooling summer air by transpiration. A depressed seating area under a large tree can be very comfortable on a hot summer day.

---

*"Latent heat of vaporization" is the amount of heat required to change a liquid into a vapor. About 1,000 BTU (1,060) are absorbed when a pound of water becomes a pound of water vapor. These BTUs are released when the vapor condenses.

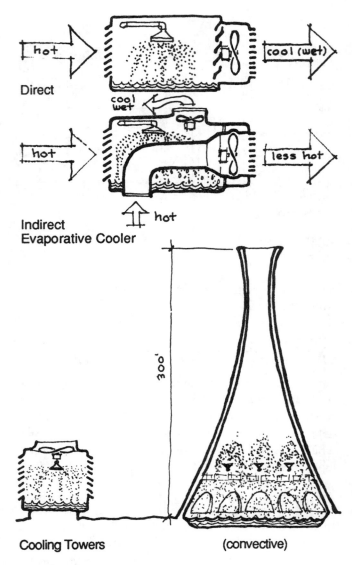

FIGURE 20.12

20.3 AIR MOVEMENT

Building air-conditioning equipment is designed to provide gentle air movement at a speed of about 50 feet per minute (FPM) around people. Increased air speed will increase skin heat losses and permit summer comfort at higher air temperatures (see Figure 20.13). At 200 FPM, air temperature can be increased 5°F. At 700 FPM, air temperature can be increased 10°F.

WIND

Wind can induce indoor air circulation. Designs using natural ventilation exploit breezes with large openings

50 FPM-75°   200 FPM-80°   700 FPM 85°

Velocity - Temperature, relationships

FIGURE 20.13

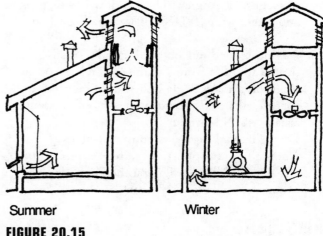

Summer   Winter

FIGURE 20.15

Wind plus convection

FIGURE 20.14

high on leeward walls and small openings on the wind-ward walls at occupant level (see Figure 20.14). Small openings can provide a jet effect, increasing air velocity in the occupied zone. Natural ventilation designs are effective where wind direction and speed are fairly constant, but they're useless on sites where wind speed and direction vary a great deal (most locations)!

CONVECTION

Warm air rises, and cool air settles, due to density difference. In winter, upper stories can enjoy a second use of heat that rises from the floor below. In summer, spaces with high ceilings will be cooler than those with low ceilings because the warmest air stratifies above the occupied zone. Summer designs exploit convective heat flow with high outlets and low inlets for occupied spaces (see Figure 20.15).

Convection can also be used to lower summer ceiling temperatures. Continuous eave and ridge vents wash the under side of a hot roof with air, which limits heat gain and cuts the ceiling temperature.

FANS

In locations with slow or unpredictable summer winds, fans provide continuous air movement that may substitute for cooling equipment. An exhaust fan that draws outdoor air through a building uses much less energy than an air conditioner.

Summer energy savings are also possible using ceiling fans and higher indoor temperature settings for air-conditioning equipment, but air speeds above 250 FPM in the occupied zone can be disturbing (papers move, pages flutter, etc.).

Winter use of ceiling fans is usually not successful. They bring back warm air from the ceiling but the draft voids the warming effect. Fans can be used in winter to recapture the heat at the ceiling and use it to warm lower floors if moving air is kept outside the occupied zone.

REVIEW QUESTIONS

Solar time is used in all the following review questions. Be sure to adjust solar calculations for your location in a time zone and for daylight savings time. Solar noon is midway between the sunrise and sunset times usually found in a local newspaper.

1. Size an overhang at the head of a 4-foot-high south-facing window at 42°N latitude to exclude direct sun from 1 June through 21 August.

2. Size an overhang for the window in question 1 if it is relocated to 30°N latitude.

3. Same window as question 1 but relocated to 38°N latitude and window faces 30° west of south.

4. Size an exterior vertical shade to exclude direct sun from a window between 3:30 PM and

sunset on 21 August. The window is 4 feet high and 3 feet wide. It faces 30° west of south at 38°N latitude.

5. A winery conducts outdoor summer tastings from 4 to 6 PM from 1 June through 31 August. Find the section angle that should be used to space trellis joists above a west-facing tasting area at 42°N latitude.

6. Do horizontal skylights conserve energy?

7. Name the four opportunity areas exploited by efficient building designers; then discuss two example applications of each.

ANSWERS

1. The sun is lowest on 21 August at noon when the section angle is about 60°. Use an adjustable triangle or COT 60° to size the overhang.

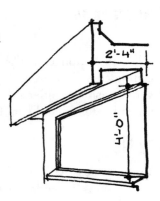

   (4)(0.58) = 2.3 feet

   COT = cotangent

2. The sun is lowest on 21 August at noon when the section angle is 70°. Use an adjustable triangle or COT 70° to size the overhang.

   (4)(0.33) = 1.32 feet

3. The sun is lowest during the afternoon on 21 August. Section angles are:

   3 PM = 48°
   4 PM = 41°
   5 PM = 32°
   6 PM = 18°

It is not practical to use an overhang to shade this window. Consider using an overhang equal to the window height, which will protect the window from direct sun until about 3:30 PM (section angle 45°). Then add a vertical shade to protect the window from 3:30 PM until sunset.

4. The bearing of the sun at 3:30 PM on 21 August is 75° west of south. After 3:30 PM, bearing shifts toward north. Since the window faces 30° west of south, incident solar rays bear 45° or more to the glass surface. The window is 3 feet wide, so specify a vertical shade projecting 3 feet from the west window jamb.

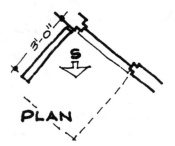 

5. About 28° is the highest section angle (see p. 298).

6. No! Solar gain is maximum in summer and minimum in winter.

7. Solar control, thermal mass, evaporation, and air movement.

CHAPTER 21

Demand, Solar Heating, and First Estimates

My eyes following
until the bird was lost at sea
found a small island.

Basho

Chapters 14 through 20 have developed information to build on as you accept increasing responsibility for the design and construction of efficient buildings. In any field of endeavor there is always more to learn. This chapter explores two further topics: peak electrical demand and active solar heating.

Peak demand is of interest because it is a significant part of a building's monthly electric bills, and the methods used to cut demand are different from the techniques used to reduce heat gains and losses.

"Active" solar-heating systems use some external energy to collect, store, and move heat. The solar designs in Chapter 20 were "passive." They invited or excluded radiation without help from pumps or fans. On sites where sun is plentiful and electricity is expensive, an active solar-heating system may have a reasonable payback. It's a benign energy resource that deserves consideration by designers and builders who have mastered passive solar control.

Finally, this chapter offers "first estimates" that can be helpful before beginning detailed heating and cooling load calculations.

21.0 ELECTRICAL PEAK DEMAND

Commercial buildings have two electric meters. One measures consumption in kilowatt hours (kWh) just like a residential meter, but a second meter records peak demand kilowatts (kW) (see Figure 21.1). The demand meter is like an auto speedometer that registers the maximum speed driven but does not return to zero when the car stops.

Electrical costs for commercial buildings include charges for both consumption and demand. Utilities meter and assess demand charges because they must build and maintain generating capacity that meets their total demand. Generating capacity can be expensive and many utility customers pay more for demand than they pay for energy consumption.

REDUCING COMMERCIAL PEAK

Commercial building designers, constructors, and operators use three methods to reduce demand charges.

1. **Minimize** electrical loads. Efficient buildings are designed and constructed with a continuing commitment to reduce heating, cooling, and lighting loads. Reduced loads and efficient equipment mean less kW demand and lower peak charges.
2. **Delay** electrical loads (thermal storage). Commercial customers can reduce peak demand costs by installing storage capacity and arranging an off-peak rate with the utility. Laundries can heat and store wash water at night when utility demand is low. Churches can make and store ice with a small refrigeration unit that operates seven days a week; ice can provide a lot of air conditioning for a few hours on Sunday morning without the peak electrical demand of a large chiller.

    Many utilities provide substantial cash allowances and rate reductions for large commercial customers who install thermal storage equipment.
3. **Sequence** electrical loads. Many refrigeration, heating, pump, and air-handler loads can be operated sequentially instead of concurrently. Necessary investments in computers, sensors, and switches can show a one-year payback from peak demand savings.

REDUCING RESIDENTIAL DEMAND

Most utilities experience system peak demand during hot summer weather, and some are offering cash incentives to builders who reduce residential peak demand (see Figure 21.2). An effective incentive program pays builders a substantial "allowance" for homes with a maximum air-conditioning capacity of 1 ton for each 800 to 1,000 square feet of conditioned area.

Utilities can also limit system demand by installing radio-controlled switches on residential air conditioners or water heaters. Radio signals can shut off a selected group of air conditioners or water heaters for 5 or 10 minutes each hour. Utilities that install this type of equipment enjoy improved system loading and positive peak demand control that can reduce electric costs.

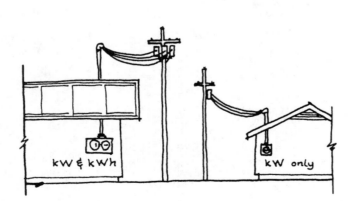

FIGURE 21.1

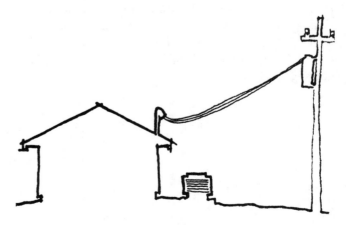

FIGURE 21.2

21.1 **SOLAR HEATING**

A south-facing window is a passive solar collector. Active solar equipment is distinguished from passive because its operation requires a small amount of nonsolar energy. Domestic water heating is the solar application most likely to be cost-effective because hot water is an annual residential need.

ORIENTATION

Solar collectors are positioned to maximize incident insolation.* For maximum annual insolation, collectors are set at an angle equal to the local latitude. To maximize winter insolation, collectors are set at latitude plus 10° (see Figure 21.3).

EFFICIENCY

Solar collectors usually circulate a fluid to absorb energy and store it as heat. Collector efficiency de-

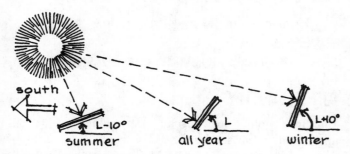

Collector Orientation

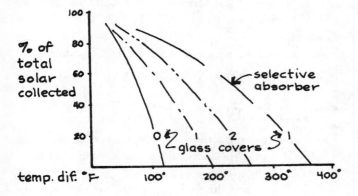

Collector (Efficiency - Temperature)

FIGURE 21.3

---

*Insolation = solar radiation. Verify local values for insolation because the map in Figure 21.4 is very general.

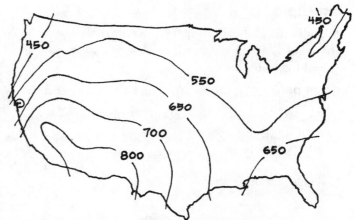

BTU per sqft. per day collected (40% efficiency)

FIGURE 21.4

creases as the temperature of captured heat increases because of losses to the environment. Operating efficiency for solar water-heating equipment can range from 30 to 60%.

EVALUATION

Consider a solar water heater located in San Francisco. The map in Figure 21.4 indicates that each square foot of collector will deliver about 600 BTU per day, or 219,000 BTU per year at 40% collector efficiency.

If solar heat is used to replace electric resistance heat at $0.10 per kWh, each square foot of collector can save about $6.44 per year in electric costs.

$$(219,000)(\$0.10) \div (3,400) = \$6.44$$

If solar heat replaces oil at $1.00 per gallon (and 70% heater efficiency), the annual savings per square foot of collector will be about $2.23.

$$(219,000)(\$1) \div (140,000)(70\%) = \$2.23$$

Projected annual savings can be used to evaluate the rate of return for a solar equipment investment.

SIZING

Sizing collectors to do 100% of the heating task is not cost-effective. As a first guess, size a collector to deliver half of the estimated daily heating need.

Example

A family uses 80 gallons of 140° water each day; supply water is 60°. Estimate the solar collector area required for water heating.

Need $(80)(8.33)(140 - 60) = 53{,}312$ BTU per day.

If the collector will deliver 600 BTU/sq. ft./day, specify a 45 sq. ft. collector.

$$53{,}312 \div 600 = 89 \text{ sq. ft.}$$

Use one-half or 45 sq. ft.

21.2 FIRST ESTIMATES

These estimates are as good as the individuals who use them. Applied with perception and understanding, they can be an excellent starting point for HVAC estimates and projections, but used without comprehension they will embarrass. Values given are for average or typical practice. Capable designers can achieve substantially better results.

HEATING

For residential or commercial buildings in the United States, estimate peak heating loads at 20 to 60 BTUH per square foot of heated area. Use lower values in southern locations and higher values in northern locations.

Example: Estimate the heating capacity required for a 2,000 sq. ft. home in Miami.

$$(2{,}000)(20) = 40{,}000 \text{ BTUH}$$

Example: Estimate required heating capacity for a 20,000 sq. ft. office building in Minneapolis.

$$(20{,}000)(60) = 1{,}200{,}000 \text{ BTUH}$$

Commercial buildings often require less winter heating than residences due to the extra internal heat they receive from lights and occupants. However, they usually need as much peak capacity as a residence to warm up Monday morning after a cold weekend.

COOLING

The range for building peak cooling loads is from 15 to 60 BTUH per square foot of conditioned area. Load variations for air-conditioning are more sensitive to occupancy and internal heat than building location. The cooling season is shorter in Minnesota than in Florida, but a hot day can be as bad in the North as the South. Estimate peak load by building type as follows:

Building Type	Cooling BTUH per sq. ft.
Apartment	15–20
Residence	20–25
School or office	25–35
Retail (furniture or carpet)	25–30
Retail (department store)	30–35
Restaurant	40–50
Hospital	50–60
Dance hall	50–60
Supermarket (why so low?)	10

Exception! For theaters or churches where most of the cooling load is caused by people, just estimate 800 BTUH per seat (range is 600 to 1,000).

Example: Estimate the peak cooling load for a 40,000 sq. ft. school.

$$(40{,}000)(30) = 1{,}200{,}000 \text{ BTUH or 100 tons}$$

Example: Estimate peak cooling load for a 112-seat movie theater.

$$(112)(800) = 89{,}600 \text{ BTUH or 7.5 tons}$$

AIR QUANTITY

Building cooling load usually determines the total amount of air carried by the duct system. Air quantity ranges from 300 to 500 CFM per ton depending on the proportion of sensible and latent load (higher sensible = higher CFM). Use 400 CFM per ton as a first estimate.

Example: Find air quantity for a 20-ton cooling load.

$$(20)(400) = 8{,}000 \text{ CFM}$$

MAIN DUCT SIZE

Low-velocity air handlers use air speeds of 1,500 to 2,000 FPM in main ducts. To estimate main duct size, divide CFM by FPM.

Example: Find the main duct size for a 20-ton installation approximately 1,500 FPM.

$$8{,}000 \div 1{,}500 = 5.33 \text{ sq. ft.}$$

The duct could be 16″×48″; don't forget that duct insulation will increase overall dimensions. With 1″ insulation this duct will be 18″×50″.

ELECTRICAL REQUIREMENTS

One horsepower (hp) requires about 1 kW (actually 746 watts at 100% efficiency).

Fans

Exhaust, not ducted	10,000 CFM/hp
Air handler, low velocity	0.3* hp/ton
Air handler, high velocity	Up to 1 hp/ton

AC and Heat Pumps

Air-cooled AC or heat pump	1 kW/ton
Chiller	0.6 to 1 kW/ton

Cooling Tower

Minimum (pump and fan)	0.5kW/ton

---

*Residential air handlers have much less hp per ton.

WATER QUANTITY

Heating—hot water supply to air handler (hot water at 20°F TD) 1 gpm carries 10,000 BTUH	
Cooling—chill water supply to air handler (chill water at 10°F TD)	3 gpm/ton
Cooling tower (10°F TD)	3 gpm/ton
Cooling tower make up water	6 gal/ton hour
Water source/sink heat pumps	3 gpm/ton

ASHRAE R VALUE TABLES

The following pages of R value information are reprinted by permission of the American Society of Heating, Refrigerating and Air-Conditioning Engineers, Atlanta, Georgia, from the 2001 *ASHRAE Handbook—Fundamentals*. (Copyright 2001, American Society of Heating, Refrigerating and Air-Conditioning Engineers, Inc. [www.ashrae.org]. This text may neither be copied nor distributed in either paper or digital form without ASHRAE permission.)

Publications of the American Society of Heating, Refrigeration, and Air-Conditioning Engineers Inc. are primary HVAC topic references.

Table 1 Surface Conductances and Resistances for Air

Position of Surface	Direction of Heat Flow	Surface Emittance, ε					
		Non-reflective ε = 0.90		Reflective			
				ε = 0.20		ε = 0.05	
		h_i	R	h_i	R	h_i	R
STILL AIR							
Horizontal	Upward	1.63	0.61	0.91	1.10	0.76	1.32
Sloping—45°	Upward	1.60	0.62	0.88	1.14	0.73	1.37
Vertical	Horizontal	1.46	0.68	0.74	1.35	0.59	1.70
Sloping—45°	Downward	1.32	0.76	0.60	1.67	0.45	2.22
Horizontal	Downward	1.08	0.92	0.37	2.70	0.22	4.55
MOVING AIR (Any position)		h_o	R				
15-mph Wind (for winter)	Any	6.00	0.17	—	—	—	—
7.5-mph Wind (for summer)	Any	4.00	0.25	—	—	—	—

Notes:
1. Surface conductance h_i and h_o measured in Btu/h·ft²·°F; resistance R in °F·ft²·h/Btu.
2. No surface has both an air space resistance value and a surface resistance value.
3. For ventilated attics or spaces above ceilings under summer conditions (heat flow down), see Table 5.
4. Conductances are for surfaces of the stated emittance facing virtual blackbody surroundings at the same temperature as the ambient air. Values are based on a surface-air temperature difference of 10°F and for surface temperatures of 70°F.
5. See Chapter 3 for more detailed information, especially Tables 5 and 6, and see Figure 1 for additional data.
6. Condensate can have a significant impact on surface emittance (see Table 2).

Table 2 Emittance Values of Various Surfaces and Effective Emittances of Air Spaces[a]

Surface	Average Emittance ε	Effective Emittance ε_{eff} of Air Space	
		One Surface Emittance ε; Other, 0.9	Both Surfaces Emittance ε
Aluminum foil, bright	0.05	0.05	0.03
Aluminum foil, with condensate just visible (> 0.7 gr/ft²)	0.30[b]	0.29	—
Aluminum foil, with condensate clearly visible (> 2.9 gr/ft²)	0.70[b]	0.65	—
Aluminum sheet	0.12	0.12	0.06
Aluminum coated paper, polished	0.20	0.20	0.11
Steel, galvanized, bright	0.25	0.24	0.15
Aluminum paint	0.50	0.47	0.35
Building materials: wood, paper, masonry, nonmetallic paints	0.90	0.82	0.82
Regular glass	0.84	0.77	0.72

[a]These values apply in the 4 to 40 μm range of the electromagnetic spectrum.
[b]Values are based on data presented by Bassett and Trethowen (1984).

these insulation systems (Hooper and Moroz 1952). Deterioration results from contact with several types of solutions, either acidic or basic (e.g., wet cement mortar or the preservatives found in decay-resistant lumber). Polluted environments may cause rapid and severe material degradation. However, site inspections show a predominance of well-preserved installations and only a small number of cases in which rapid and severe deterioration has occurred. An extensive review of the reflective building insulation system performance literature is provided by Goss and Miller (1989).

CALCULATING OVERALL THERMAL RESISTANCES

Relatively small, highly conductive elements in an insulating layer called thermal bridges can substantially reduce the average thermal resistance of a component. Examples include wood and metal studs in frame walls, concrete webs in concrete masonry walls, and metal ties or other elements in insulated wall panels. The following examples illustrate the calculation of R-values and U-factors for components containing thermal bridges.

These conditions are assumed in calculating the design R-values:

- Equilibrium or steady-state heat transfer, disregarding effects of thermal storage
- Surrounding surfaces at ambient air temperature
- Exterior wind velocity of 15 mph for winter (surface with R = 0.17°F·ft²·h/Btu) and 7.5 mph for summer (surface with R = 0.25°F·ft²·h/Btu)
- Surface emittance of ordinary building materials is 0.90

Wood Frame Walls

The average overall R-values and U-factors of wood frame walls can be calculated by assuming either parallel heat flow paths through areas with different thermal resistances or by assuming isothermal planes. Equations (1) through (5) from Chapter 22 are used.

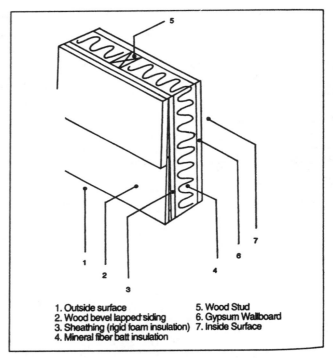

1. Outside surface
2. Wood bevel lapped siding
3. Sheathing (rigid foam insulation)
4. Mineral fiber batt insulation
5. Wood Stud
6. Gypsum Wallboard
7. Inside Surface

Fig. 2 Insulated Wood Frame Wall (Example 1)

The framing factor or fraction of the building component that is framing depends on the specific type of construction, and it may vary based on local construction practices—even for the same type of construction. For stud walls 16 in. on center (OC), the fraction of insulated cavity may be as low as 0.75, where the fraction of studs, plates, and sills is 0.21 and the fraction of headers is 0.04. For studs 24 in. OC, the respective values are 0.78, 0.18, and 0.04. These fractions contain an allowance for multiple studs, plates, sills, extra framing around windows, headers, and band joists. These assumed framing fractions are used in the following example, to illustrate the importance of including the effect of framing in determining the overall thermal conductance of a building. The actual framing fraction should be calculated for each specific construction.

Table 3 Thermal Resistances of Plane Air Spaces[a,b,c], °F·ft²·h/Btu

Position of Air Space	Direction of Heat Flow	Air Space		0.5-in. Air Space[c]					0.75-in. Air Space[c]				
		Mean Temp.[d], °F	Temp. Diff.[d], °F	Effective Emittance ε_{eff}[d,e]					Effective Emittance ε_{eff}[d,e]				
				0.03	0.05	0.2	0.5	0.82	0.03	0.05	0.2	0.5	0.82
Horiz.	Up	90	10	2.13	2.03	1.51	0.99	0.73	2.34	2.22	1.61	1.04	0.75
		50	30	1.62	1.57	1.29	0.96	0.75	1.71	1.66	1.35	0.99	0.77
		50	10	2.13	2.05	1.60	1.11	0.84	2.30	2.21	1.70	1.16	0.87
		0	20	1.73	1.70	1.45	1.12	0.91	1.83	1.79	1.52	1.16	0.93
		0	10	2.10	2.04	1.70	1.27	1.00	2.23	2.16	1.78	1.31	1.02
		−50	20	1.69	1.66	1.49	1.23	1.04	1.77	1.74	1.55	1.27	1.07
		−50	10	2.04	2.00	1.75	1.40	1.16	2.16	2.11	1.84	1.46	1.20
45° Slope	Up	90	10	2.44	2.31	1.65	1.06	0.76	2.96	2.78	1.88	1.15	0.81
		50	30	2.06	1.98	1.56	1.10	0.83	1.99	1.92	1.52	1.08	0.82
		50	10	2.55	2.44	1.83	1.22	0.90	2.90	2.75	2.00	1.29	0.94
		0	20	2.20	2.14	1.76	1.30	1.02	2.13	2.07	1.72	1.28	1.00
		0	10	2.63	2.54	2.03	1.44	1.10	2.72	2.62	2.08	1.47	1.12
		−50	20	2.08	2.04	1.78	1.42	1.17	2.05	2.01	1.76	1.41	1.16
		−50	10	2.62	2.56	2.17	1.66	1.33	2.53	2.47	2.10	1.62	1.30
Vertical	Horiz.	90	10	2.47	2.34	1.67	1.06	0.77	3.50	3.24	2.08	1.22	0.84
		50	30	2.57	2.46	1.84	1.23	0.90	2.91	2.77	2.01	1.30	0.94
		50	10	2.66	2.54	1.88	1.24	0.91	3.70	3.46	2.35	1.43	1.01
		0	20	2.82	2.72	2.14	1.50	1.13	3.14	3.02	2.32	1.58	1.18
		0	10	2.93	2.82	2.20	1.53	1.15	3.77	3.59	2.64	1.73	1.26
		−50	20	2.90	2.82	2.35	1.76	1.39	2.90	2.83	2.36	1.77	1.39
		−50	10	3.20	3.10	2.54	1.87	1.46	3.72	3.60	2.87	2.04	1.56
45° Slope	Down	90	10	2.48	2.34	1.67	1.06	0.77	3.53	3.27	2.10	1.22	0.84
		50	30	2.64	2.52	1.87	1.24	0.91	3.43	3.23	2.24	1.39	0.99
		50	10	2.67	2.55	1.89	1.25	0.92	3.81	3.57	2.40	1.45	1.02
		0	20	2.91	2.80	2.19	1.52	1.15	3.75	3.57	2.63	1.72	1.26
		0	10	2.94	2.83	2.21	1.53	1.15	4.12	3.91	2.81	1.80	1.30
		−50	20	3.16	3.07	2.52	1.86	1.45	3.78	3.65	2.90	2.05	1.57
		−50	10	3.26	3.16	2.58	1.89	1.47	4.35	4.18	3.22	2.21	1.66
Horiz.	Down	90	10	2.48	2.34	1.67	1.06	0.77	3.55	3.29	2.10	1.22	0.85
		50	30	2.66	2.54	1.88	1.24	0.91	3.77	3.52	2.38	1.44	1.02
		50	10	2.67	2.55	1.89	1.25	0.92	3.84	3.59	2.41	1.45	1.02
		0	20	2.94	2.83	2.20	1.53	1.15	4.18	3.96	2.83	1.81	1.30
		0	10	2.96	2.85	2.22	1.53	1.16	4.25	4.02	2.87	1.82	1.31
		−50	20	3.25	3.15	2.58	1.89	1.47	4.60	4.41	3.36	2.28	1.69
		−50	10	3.28	3.18	2.60	1.90	1.47	4.71	4.51	3.42	2.30	1.71

Position of Air Space	Direction of Heat Flow	Air Space		1.5-in. Air Space[c]					3.5-in. Air Space[c]				
Horiz.	Up	90	10	2.55	2.41	1.71	1.08	0.77	2.84	2.66	1.83	1.13	0.80
		50	30	1.87	1.81	1.45	1.04	0.80	2.09	2.01	1.58	1.10	0.84
		50	10	2.50	2.40	1.81	1.21	0.89	2.80	2.66	1.95	1.28	0.93
		0	20	2.01	1.95	1.63	1.23	0.97	2.25	2.18	1.79	1.32	1.03
		0	10	2.43	2.35	1.90	1.38	1.06	2.71	2.62	2.07	1.47	1.12
		−50	20	1.94	1.91	1.68	1.36	1.13	2.19	2.14	1.86	1.47	1.20
		−50	10	2.37	2.31	1.99	1.55	1.26	2.65	2.58	2.18	1.67	1.33
45° Slope	Up	90	10	2.92	2.73	1.86	1.14	0.80	3.18	2.96	1.97	1.18	0.82
		50	30	2.14	2.06	1.61	1.12	0.84	2.26	2.17	1.67	1.15	0.86
		50	10	2.88	2.74	1.99	1.29	0.94	3.12	2.95	2.10	1.34	0.96
		0	20	2.30	2.23	1.82	1.34	1.04	2.42	2.35	1.90	1.38	1.06
		0	10	2.79	2.69	2.12	1.49	1.13	2.98	2.87	2.23	1.54	1.16
		−50	20	2.22	2.17	1.88	1.49	1.21	2.34	2.29	1.97	1.54	1.25
		−50	10	2.71	2.64	2.23	1.69	1.35	2.87	2.79	2.33	1.75	1.39
Vertical	Horiz.	90	10	3.99	3.66	2.25	1.27	0.87	3.69	3.40	2.15	1.24	0.85
		50	30	2.58	2.46	1.84	1.23	0.90	2.67	2.55	1.89	1.25	0.91
		50	10	3.79	3.55	2.39	1.45	1.02	3.63	3.40	2.32	1.42	1.01
		0	20	2.76	2.66	2.10	1.48	1.12	2.88	2.78	2.17	1.51	1.14
		0	10	3.51	3.35	2.51	1.67	1.23	3.49	3.33	2.50	1.67	1.23
		−50	20	2.64	2.58	2.18	1.66	1.33	2.82	2.75	2.30	1.73	1.37
		−50	10	3.31	3.21	2.62	1.91	1.48	3.40	3.30	2.67	1.94	1.50
45° Slope	Down	90	10	5.07	4.55	2.56	1.36	0.91	4.81	4.33	2.49	1.34	0.90
		50	30	3.58	3.36	2.31	1.42	1.00	3.51	3.30	2.28	1.40	1.00
		50	10	5.10	4.66	2.85	1.60	1.09	4.74	4.36	2.73	1.57	1.08
		0	20	3.85	3.66	2.68	1.74	1.27	3.81	3.63	2.66	1.74	1.27
		0	10	4.92	4.62	3.16	1.94	1.37	4.59	4.32	3.02	1.88	1.34
		−50	20	3.62	3.50	2.80	2.01	1.54	3.77	3.64	2.90	2.05	1.57
		−50	10	4.67	4.47	3.40	2.29	1.70	4.50	4.32	3.31	2.25	1.68
Horiz.	Down	90	10	6.09	5.35	2.79	1.43	0.94	10.07	8.19	3.41	1.57	1.00
		50	30	6.27	5.63	3.18	1.70	1.14	9.60	8.17	3.86	1.88	1.22
		50	10	6.61	5.90	3.27	1.73	1.15	11.15	9.27	4.09	1.93	1.24
		0	20	7.03	6.43	3.91	2.19	1.49	10.90	9.52	4.87	2.47	1.62
		0	10	7.31	6.66	4.00	2.22	1.51	11.97	10.32	5.08	2.52	1.64
		−50	20	7.73	7.20	4.77	2.85	1.99	11.64	10.49	6.02	3.25	2.18
		−50	10	8.09	7.52	4.91	2.89	2.01	12.98	11.56	6.36	3.34	2.22

[a]See Chapter 22, section Factors Affecting Heat Transfer across Air Spaces. Thermal resistance values were determined from the relation, $R = 1/C$, where $C = h_c + \varepsilon_{eff} h_r$, h_c is the conduction-convection coefficient, $\varepsilon_{eff} h_r$ is the radiation coefficient $\approx 0.0068\varepsilon_{eff}[(t_m + 460)/100]^3$, and t_m is the mean temperature of the air space. Values for h_c were determined from data developed by Robinson et al. (1954). Equations (5) through (7) in Yarbrough (1983) show the data in this table in analytic form. For extrapolation from this table to air spaces less than 0.5 in. (as in insulating window glass), assume $h_c = 0.159(1 + 0.0016 t_m)/l$ where l is the air space thickness in inches, and h_c is heat transfer through the air space only.

[b]Values are based on data presented by Robinson et al. (1954). (Also see Chapter 3, Tables 3 and 4, and Chapter 36). Values apply for ideal conditions, i.e., air spaces of uniform thickness bounded by plane, smooth, parallel surfaces with no air leakage to or from the space. When accurate values are required, use overall U-factors deter-

mined through calibrated hot box (ASTM C 976) or guarded hot box (ASTM C 236) testing. Thermal resistance values for multiple air spaces must be based on careful estimates of mean temperature differences for each air space.

[c]A single resistance value cannot account for multiple air spaces; each air space requires a separate resistance calculation that applies only for the established boundary conditions. Resistances of horizontal spaces with heat flow downward are substantially independent of temperature difference.

[d]Interpolation is permissible for other values of mean temperature, temperature difference, and effective emittance ε_{eff}. Interpolation and moderate extrapolation for air spaces greater than 3.5 in. are also permissible.

[e]Effective emittance ε_{eff} of the air space is given by $1/\varepsilon_{eff} = 1/\varepsilon_1 + 1/\varepsilon_2 - 1$, where ε_1 and ε_2 are the emittances of the surfaces of the air space (see Table 2).

Table 4 Typical Thermal Properties of Common Building and Insulating Materials—Design Values[a]

Description	Density, lb/ft³	Conductivity[b] (k), Btu·in / h·ft²·°F	Conductance (C), Btu / h·ft²·°F	Resistance[c] (R) Per Inch Thickness (1/k), °F·ft²·h / Btu·in	Resistance[c] (R) For Thickness Listed (1/C), °F·ft²·h / Btu	Specific Heat, Btu / lb·°F
BUILDING BOARD						
Asbestos-cement board	120	4.0	—	0.25	—	0.24
Asbestos-cement board ...0.125 in.	120	—	33.00	—	0.03	
Asbestos-cement board ...0.25 in.	120	—	16.50	—	0.06	
Gypsum or plaster board ...0.375 in.	50	—	3.10	—	0.32	0.26
Gypsum or plaster board ...0.5 in.	50	—	2.22	—	0.45	
Gypsum or plaster board ...0.625 in.	50	—	1.78	—	0.56	
Plywood (Douglas Fir)[d]	34	0.80	—	1.25	—	0.29
Plywood (Douglas Fir) ...0.25 in.	34	—	3.20	—	0.31	
Plywood (Douglas Fir) ...0.375 in.	34	—	2.13	—	0.47	
Plywood (Douglas Fir) ...0.5 in.	34	—	1.60	—	0.62	
Plywood (Douglas Fir) ...0.625 in.	34	—	1.29	—	0.77	
Plywood or wood panels ...0.75 in.	34	—	1.07	—	0.93	0.29
Vegetable fiber board						
Sheathing, regular density[e] ...0.5 in.	18	—	0.76	—	1.32	0.31
...0.78125 in.	18	—	0.49	—	2.06	
Sheathing intermediate density[e] ...0.5 in.	22	—	0.92	—	1.09	0.31
Nail-base sheathing[e] ...0.5 in.	25	—	0.94	—	1.06	0.31
Shingle backer ...0.375 in.	18	—	1.06	—	0.94	0.31
Shingle backer ...0.3125 in.	18	—	1.28	—	0.78	
Sound deadening board ...0.5 in.	15	—	0.74	—	1.35	0.30
Tile and lay-in panels, plain or acoustic	18	0.40	—	2.50	—	0.14
...0.5 in.	18	—	0.80	—	1.25	
...0.75 in.	18	—	0.53	—	1.89	
Laminated paperboard	30	0.50	—	2.00	—	0.33
Homogeneous board from repulped paper....	30	0.50	—	2.00	—	0.28
Hardboard[e]						
Medium density	50	0.73	—	1.37	—	0.31
High density, service-tempered grade and service grade	55	0.82	—	1.22	—	0.32
High density, standard-tempered grade	63	1.00	—	1.00	—	0.32
Particleboard[e]						
Low density	37	0.71	—	1.41	—	0.31
Medium density	50	0.94	—	1.06	—	0.31
High density	62	.5	1.18	—	0.85	—
Underlayment ...0.625 in.	40	—	1.22	—	0.82	0.29
Waferboard	37	0.63	—	1.59	—	—
Wood subfloor ...0.75 in.	—	—	1.06	—	0.94	0.33
BUILDING MEMBRANE						
Vapor—permeable felt	—	—	16.70	—	0.06	
Vapor—seal, 2 layers of mopped 15-lb felt	—	—	8.35	—	0.12	
Vapor—seal, plastic film	—	—	—	—	Negl.	
FINISH FLOORING MATERIALS						
Carpet and fibrous pad	—	—	0.48	—	2.08	0.34
Carpet and rubber pad	—	—	0.81	—	1.23	0.33
Cork tile ...0.125 in.	—	—	3.60	—	0.28	0.48
Terrazzo ...1 in.	—	—	12.50	—	0.08	0.19
Tile—asphalt, linoleum, vinyl, rubber	—	—	20.00	—	0.05	0.30
vinyl asbestos						0.24
ceramic						0.19
Wood, hardwood finish ...0.75 in.	—	—	1.47	—	0.68	
INSULATING MATERIALS						
Blanket and Batt[f,g]						
Mineral fiber, fibrous form processed from rock, slag, or glass						
approx. 3-4 in.	0.4-2.0	—	0.091	—	11	
approx. 3.5 in.	0.4-2.0	—	0.077	—	13	
approx. 3.5 in.	1.2-1.6	—	0.067	—	15	
approx. 5.5-6.5 in.	0.4-2.0	—	0.053	—	19	
approx. 5.5 in.	0.6-1.0	—	0.048	—	21	
approx. 6-7.5 in.	0.4-2.0	—	0.045	—	22	
approx. 8.25-10 in.	0.4-2.0	—	0.033	—	30	
approx. 10-13 in.	0.4-2.0	—	0.026	—	38	
Board and Slabs						
Cellular glass	8.0	0.33	—	3.03	—	0.18
Glass fiber, organic bonded	4.0-9.0	0.25	—	4.00	—	0.23
Expanded perlite, organic bonded	1.0	0.36	—	2.78	—	0.30
Expanded rubber (rigid)	4.5	0.22	—	4.55	—	0.40
Expanded polystyrene, extruded (smooth skin surface) (CFC-12 exp.)	1.8-3.5	0.20	—	5.00	—	0.29

Description	Density, lb/ft³	Conductivity[b] (k), Btu·in / h·ft²·°F	Conductance (C), Btu / h·ft²·°F	Resistance[c] (R) Per Inch Thickness (1/k), °F·ft²·h / Btu·in	For Thickness Listed (1/C), °F·ft²·h / Btu	Specific Heat, Btu / lb·°F
Expanded polystyrene, extruded (smooth skin surface) (HCFC-142b exp.)[h]	1.8-3.5	0.20	—	5.00	—	0.29
Expanded polystyrene, molded beads	1.0	0.26	—	3.85	—	—
	1.25	0.25	—	4.00	—	—
	1.5	0.24	—	4.17	—	—
	1.75	0.24	—	4.17	—	—
	2.0	0.23	—	4.35	—	—
Cellular polyurethane/polyisocyanurate[i] (CFC-11 exp.) (unfaced)	1.5	0.16-0.18	—	6.25-5.56	—	0.38
Cellular polyisocyanurate[i] (CFC-11 exp.) (gas-permeable facers)	1.5-2.5	0.16-0.18	—	6.25-5.56	—	0.22
Cellular polyisocyanurate[j] (CFC-11 exp.) (gas-impermeable facers)	2.0	0.14	—	7.04	—	0.22
Cellular phenolic (closed cell) (CFC-11, CFC-113 exp.)[k]	3.0	0.12	—	8.20	—	—
Cellular phenolic (open cell)	1.8-2.2	0.23	—	4.40	—	—
Mineral fiber with resin binder	15.0	0.29	—	3.45	—	0.17
Mineral fiberboard, wet felted						
Core or roof insulation	16-17	0.34	—	2.94	—	—
Acoustical tile	18.0	0.35	—	2.86	—	0.19
Acoustical tile	21.0	0.37	—	2.70	—	—
Mineral fiberboard, wet molded						
Acoustical tile[l]	23.0	0.42	—	2.38	—	0.14
Wood or cane fiberboard						
Acoustical tile[l] ....0.5 in.	—	—	0.80	—	1.25	0.31
Acoustical tile[l] ....0.75 in.	—	—	0.53	—	1.89	—
Interior finish (plank, tile)	15.0	0.35	—	2.86	—	0.32
Cement fiber slabs (shredded wood with Portland cement binder)	25-27.0	0.50-0.53	—	2.0-1.89	—	—
Cement fiber slabs (shredded wood with magnesia oxysulfide binder)	22.0	0.57	—	1.75	—	0.31
Loose Fill						
Cellulosic insulation (milled paper or wood pulp)	2.3-3.2	0.27-0.32	—	3.70-3.13	—	0.33
Perlite, expanded	2.0-4.1	0.27-0.31	—	3.7-3.3	—	0.26
	4.1-7.4	0.31-0.36	—	3.3-2.8	—	—
	7.4-11.0	0.36-0.42	—	2.8-2.4	—	—
Mineral fiber (rock, slag, or glass)[g]						
approx. 3.75-5 in.	0.6-2.0	—	—	—	11.0	0.17
approx. 6.5-8.75 in.	0.6-2.0	—	—	—	19.0	—
approx. 7.5-10 in.	0.6-2.0	—	—	—	22.0	—
approx. 10.25-13.75 in.	0.6-2.0	—	—	—	30.0	—
Mineral fiber (rock, slag, or glass)[g]						
approx. 3.5 in. (closed sidewall application)	2.0-3.5	—	—	—	12.0-14.0	—
Vermiculite, exfoliated	7.0-8.2	0.47	—	2.13	—	0.32
	4.0-6.0	0.44	—	2.27	—	—
Spray Applied						
Polyurethane foam	1.5-2.5	0.16-0.18	—	6.25-5.56	—	—
Ureaformaldehyde foam	0.7-1.6	0.22-0.28	—	4.55-3.57	—	—
Cellulosic fiber	3.5-6.0	0.29-0.34	—	3.45-2.94	—	—
Glass fiber	3.5-4.5	0.26-0.27	—	3.85-3.70	—	—
Reflective Insulation						
Reflective material (ε < 0.5) in center of 3/4 in. cavity forms two 3/8 in. vertical air spaces[m]	—	—	0.31	—	3.2	—

METALS
(See Chapter 36, Table 3)

ROOFING

Description	Density	k	C	1/k	1/C	Specific Heat
Asbestos-cement shingles	120	—	4.76	—	0.21	0.24
Asphalt roll roofing	70	—	6.50	—	0.15	0.36
Asphalt shingles	70	—	2.27	—	0.44	0.30
Built-up roofing ....0.375 in.	70	—	3.00	—	0.33	0.35
Slate ....0.5 in.	—	—	20.00	—	0.05	0.30
Wood shingles, plain and plastic film faced	—	—	1.06	—	0.94	0.31

PLASTERING MATERIALS

Description	Density	k	C	1/k	1/C	Specific Heat
Cement plaster, sand aggregate	116	5.0	—	0.20	—	0.20
Sand aggregate ....0.375 in.	—	—	13.3	—	0.08	0.20
Sand aggregate ....0.75 in.	—	—	6.66	—	0.15	0.20

Table 4   Typical Thermal Properties of Common Building and Insulating Materials—Design Values[a] (Continued)

Description	Density, lb/ft^3	Conductivity[b] (k), $\frac{Btu \cdot in}{h \cdot ft^2 \cdot °F}$	Conductance (C), $\frac{Btu}{h \cdot ft^2 \cdot °F}$	Resistance[c] (R) Per Inch Thickness (1/k), $\frac{°F \cdot ft^2 \cdot h}{Btu \cdot in}$	Resistance[c] (R) For Thickness Listed (1/C), $\frac{°F \cdot ft^2 \cdot h}{Btu}$	Specific Heat, $\frac{Btu}{lb \cdot °F}$
Gypsum plaster:						
Lightweight aggregate ....0.5 in.	45	—	3.12	—	0.32	—
Lightweight aggregate ....0.625 in.	45	—	2.67	—	0.39	—
Lightweight aggregate on metal lath ....0.75 in.	—	—	2.13	—	0.47	—
Perlite aggregate	45	1.5	—	0.67	—	0.32
Sand aggregate	105	5.6	—	0.18	—	0.20
Sand aggregate ....0.5 in.	105	—	11.10	—	0.09	—
Sand aggregate ....0.625 in.	105	—	9.10	—	0.11	—
Sand aggregate on metal lath ....0.75 in.	—	—	7.70	—	0.13	—
Vermiculite aggregate.......................	45	1.7	—	0.59	—	—
MASONRY MATERIALS						
Masonry Units						
Brick, fired clay	150	8.4-10.2	—	0.12-0.10	—	—
	140	7.4-9.0	—	0.14-0.11	—	—
	130	6.4-7.8	—	0.16-0.12	—	—
	120	5.6-6.8	—	0.18-0.15	—	0.19
	110	4.9-5.9	—	0.20-0.17	—	—
	100	4.2-5.1	—	0.24-0.20	—	—
	90	3.6-4.3	—	0.28-0.24	—	—
	80	3.0-3.7	—	0.33-0.27	—	—
	70	2.5-3.1	—	0.40-0.33	—	—
Clay tile, hollow						
1 cell deep ....3 in.	—	—	1.25	—	0.80	0.21
1 cell deep ....4 in.	—	—	0.90	—	1.11	—
2 cells deep ....6 in.	—	—	0.66	—	1.52	—
2 cells deep ....8 in.	—	—	0.54	—	1.85	—
2 cells deep ....10 in.	—	—	0.45	—	2.22	—
3 cells deep ....12 in.	—	—	0.40	—	2.50	—
Concrete blocks[n, o]						
Limestone aggregate						
8 in., 36 lb, 138 lb/ft^3 concrete, 2 cores	—	—	—	—	—	—
Same with perlite filled cores	—	—	0.48	—	2.1	—
12 in., 55 lb, 138 lb/ft^3 concrete, 2 cores	—	—	—	—	—	—
Same with perlite filled cores	—	—	0.27	—	3.7	—
Normal weight aggregate (sand and gravel)						
8 in., 33-36 lb, 126-136 lb/ft^3 concrete, 2 or 3 cores	—	—	0.90-1.03	—	1.11-0.97	0.22
Same with perlite filled cores	—	—	0.50	—	2.0	—
Same with vermiculite filled cores	—	—	0.52-0.73	—	1.92-1.37	—
12 in., 50 lb, 125 lb/ft^3 concrete, 2 cores	—	—	0.81	—	1.23	0.22
Medium weight aggregate (combinations of normal weight and lightweight aggregate)						
8 in., 26-29 lb, 97-112 lb/ft^3 concrete, 2 or 3 cores..	—	—	0.58-0.78	—	1.71-1.28	—
Same with perlite filled cores	—	—	0.27-0.44	—	3.7-2.3	—
Same with vermiculite filled cores	—	—	0.30	—	3.3	—
Same with molded EPS (beads) filled cores	—	—	0.32	—	3.2	—
Same with molded EPS inserts in cores................	—	—	0.37	—	2.7	—
Lightweight aggregate (expanded shale, clay, slate or slag, pumice)						
6 in., 16-17 lb 85-87 lb/ft^3 concrete, 2 or 3 cores	—	—	0.52-0.61	—	1.93-1.65	—
Same with perlite filled cores	—	—	0.24	—	4.2	—
Same with vermiculite filled cores	—	—	0.33	—	3.0	—
8 in., 19-22 lb, 72-86 lb/ft^3 concrete.......................	—	—	0.32-0.54	—	3.2-1.90	0.21
Same with perlite filled cores	—	—	0.15-0.23	—	6.8-4.4	—
Same with vermiculite filled cores	—	—	0.19-0.26	—	5.3-3.9	—
Same with molded EPS (beads) filled cores	—	—	0.21	—	4.8	—
Same with UF foam filled cores	—	—	0.22	—	4.5	—
Same with molded EPS inserts in cores................	—	—	0.29	—	3.5	—
12 in., 32-36 lb, 80-90 lb/ft^3 concrete, 2 or 3 cores...	—	—	0.38-0.44	—	2.6-2.3	—
Same with perlite filled cores	—	—	0.11-0.16	—	9.2-6.3	—
Same with vermiculite filled cores	—	—	0.17	—	5.8	—
Stone, lime, or sand.............................	180	72	—	0.01	—	—
Quartzitic and sandstone	160	43	—	0.02	—	—
	140	24	—	0.04	—	—
	120	13	—	0.08	—	0.19
Calcitic, dolomitic, limestone, marble, and granite	180	30	—	0.03	—	—
	160	22	—	0.05	—	—
	140	16	—	0.06	—	—
	120	11	—	0.09	—	0.19
	100	8	—	0.13	—	—

Description	Density, lb/ft[3]	Conductivity[b] (k), Btu·in / h·ft[2]·°F	Conductance (C), Btu / h·ft[2]·°F	Resistance[c] (R) Per Inch Thickness (1/k), °F·ft[2]·h / Btu·in	Resistance[c] (R) For Thickness Listed (1/C), °F·ft[2]·h / Btu	Specific Heat, Btu / lb·°F
Gypsum partition tile						
3 by 12 by 30 in., solid	—	—	0.79	—	1.26	0.19
3 by 12 by 30 in., 4 cells	—	—	0.74	—	1.35	—
4 by 12 by 30 in., 3 cells	—	—	0.60	—	1.67	—
Concretes[d]						
Sand and gravel or stone aggregate concretes (concretes	150	10.0-20.0	—	0.10-0.05	—	—
with more than 50% quartz or quartzite sand have	140	9.0-18.0	—	0.11-0.06	—	0.19-0.24
conductivities in the higher end of the range)	130	7.0-13.0	—	0.14-0.08	—	—
Limestone concretes	140	11.1	—	0.09	—	—
	120	7.9	—	0.13	—	—
	100	5.5	—	0.18	—	—
Gypsum-fiber concrete (87.5% gypsum, 12.5%						
wood chips)	51	1.66	—	0.60	—	0.21
Cement/lime, mortar, and stucco	120	9.7	—	0.10	—	—
	100	6.7	—	0.15	—	—
	80	4.5	—	0.22	—	—
Lightweight aggregate concretes						
Expanded shale, clay, or slate; expanded slags;	120	6.4-9.1	—	0.16-0.11	—	—
cinders; pumice (with density up to 100 lb/ft[3]); and	100	4.7-6.2	—	0.21-0.16	—	0.20
scoria (sanded concretes have conductivities in the	80	3.3-4.1	—	0.30-0.24	—	0.20
higher end of the range)	60	2.1-2.5	—	0.48-0.40	—	—
	40	1.3	—	0.78	—	—
Perlite, vermiculite, and polystyrene beads	50	1.8-1.9	—	0.55-0.53	—	—
	40	1.4-1.5	—	0.71-0.67	—	0.15-0.23
	30	1.1	—	0.91	—	—
	20	0.8	—	1.25	—	—
Foam concretes	120	5.4	—	0.19	—	—
	100	4.1	—	0.24	—	—
	80	3.0	—	0.33	—	—
	70	2.5	—	0.40	—	—
Foam concretes and cellular concretes	60	2.1	—	0.48	—	—
	40	1.4	—	0.71	—	—
	20	0.8	—	1.25	—	—

SIDING MATERIALS (on flat surface)

Shingles

Description	Density, lb/ft[3]	Conductivity[b] (k)	Conductance (C)	Resistance Per Inch (1/k)	Resistance For Thickness Listed (1/C)	Specific Heat
Asbestos-cement	120	—	4.75	—	0.21	—
Wood, 16 in., 7.5 exposure	—	—	1.15	—	0.87	0.31
Wood, double, 16-in., 12-in. exposure	—	—	0.84	—	1.19	0.28
Wood, plus ins. backer board, 0.312 in.	—	—	0.71	—	1.40	0.31
Siding						
Asbestos-cement, 0.25 in., lapped	—	—	4.76	—	0.21	0.24
Asphalt roll siding	—	—	6.50	—	0.15	0.35
Asphalt insulating siding (0.5 in. bed.)	—	—	0.69	—	1.46	0.35
Hardboard siding, 0.4375 in.	—	—	1.49	—	0.67	0.28
Wood, drop, 1 by 8 in.	—	—	1.27	—	0.79	0.28
Wood, bevel, 0.5 by 8 in., lapped	—	—	1.23	—	0.81	0.28
Wood, bevel, 0.75 by 10 in., lapped	—	—	0.95	—	1.05	0.28
Wood, plywood, 0.375 in., lapped	—	—	1.69	—	0.59	0.29
Aluminum, steel, or vinyl[p, q], over sheathing						
Hollow-backed	—	—	1.64	—	0.61	0.29[q]
Insulating-board backed nominal 0.375 in.	—	—	0.55	—	1.82	0.32
Insulating-board backed nominal 0.375 in., foil backed	—	—	0.34	—	2.96	—
Architectural (soda-lime float) glass	158	6.9	—	—	—	0.21

WOODS (12% moisture content)[e,f]

Hardwoods

Description	Density, lb/ft[3]	Conductivity[b] (k)	Conductance (C)	Resistance Per Inch (1/k)	Resistance For Thickness Listed (1/C)	Specific Heat
Hardwoods						0.39[s]
Oak	41.2-46.8	1.12-1.25	—	0.89-0.80	—	
Birch	42.6-45.4	1.16-1.22	—	0.87-0.82	—	
Maple	39.8-44.0	1.09-1.19	—	0.92-0.84	—	
Ash	38.4-41.9	1.06-1.14	—	0.94-0.88	—	
Softwoods						0.39[s]
Southern Pine	35.6-41.2	1.00-1.12	—	1.00-0.89	—	
Douglas Fir-Larch	33.5-36.3	0.95-1.01	—	1.06-0.99	—	
Southern Cypress	31.4-32.1	0.90-0.92	—	1.11-1.09	—	
Hem-Fir, Spruce-Pine-Fir	24.5-31.4	0.74-0.90	—	1.35-1.11	—	
West Coast Woods, Cedars	21.7-31.4	0.68-0.90	—	1.48-1.11	—	
California Redwood	24.5-28.0	0.74-0.82	—	1.35-1.22	—	

PART IV
WATER AND PLUMBING

☐

CHAPTER 22

Water Supply and Wastewater

Caressing a rock
Marked with a warrior's moko
The waters of Wai-Kimihia
Warm my bones
As I sing to Hinemoa
The trees whisper
Their reply
The fantails twitter
Their reply
The stitchbirds chorus
Their reply
The water
Brings my chant
Back to me.

Paula Harris

This chapter discusses the properties, sources, collection, and supply systems of water. Primary sources of water supply include rainfall, surface runoff, and underground reservoirs known as aquifers. Some methods of treating water to make it potable are discussed at length. Wastewater is generated from buildings once the supply water has been used. This is discharged from the plumbing fixtures. Methods of removal and treatment of wastewater are also discussed in this chapter.

22.0 WATER AND PLUMBING TERMS

acidity   The ability of a water solution to neutralize an alkali or base.

activated carbon   A material that has a very porous structure and is adsorbent for organic matter and certain dissolved solids.

aeration   The process by which air becomes dissolved in water.

aerobic   An action or a process conducted in the presence of oxygen.

air gap   The unobstructed vertical distance through the free atmosphere between the lowest opening from any pipe or faucet supplying water to a tank, plumbing fixture, or other device and the flood level rim of the receptacle.

alkalinity   The quantitative capacity of water to neutralize an acid. In the water industry, alkalinity is expressed in mg/L of equivalent calcium carbonate.

alum   A common name for aluminum sulfate, used as a coagulant.

anaerobic   An action or a process conducted in the absence of oxygen.

anion   An ion with a negative electrical charge.

aquifer   A natural water-bearing geological formation (e.g., sand, gravel, sandstone) that is found below the surface of the earth.

area drain   A receptacle designed to collect surface or storm water from an open area.

artesian   Water held under pressure in porous geological formations confined by impermeable geological formations. An artesian well is free flowing.

Biochemical Oxygen Demand (BOD)   The amount of oxygen (measured in mg/L) required in the oxidation of organic matter by biological action under specific standard test conditions. It is widely used to measure the amount of organic pollution in wastewater and streams.

biodegradation   Decomposition of a substance into more elementary compounds by the action of microorganisms such as bacteria.

brine   A strong solution of salt(s) (usually sodium chloride) with total dissolved solid concentrations in the range of 40,000 to 300,000 or more milligrams per liter.

BTU   British thermal unit. Quantity of heat. One BTU will increase the temperature of 1 pound of water 1°F (1 Calorie = 4 BTU).

BTUH   British thermal units per hour.

cation   A positively charged ion in an electrolyte solution, attracted to the cathode under the influence of a difference in electrical potential. Sodium ion (Na^+) is a cation.

chlorination   The treatment process in which chlorine gas or a chlorine solution is added to water for disinfecting and control of microorganisms. Chlorination is also used in the oxidation of dissolved iron, manganese, and hydrogen sulfide impurities.

cleanout   An accessible opening in the drainage system used for removal of obstructions.

drain   A pipe, conduit, or receptacle in a building which carries liquids by gravity to waste. The term is sometimes limited to refer to disposal of liquids other than sewage.

effluent   Treated wastewater discharged from sewage treatment plants.

entropy   The capacity of a system or a body to hold energy that is not available for changing the temperature of the system (or body) or for doing work.

fecal matter   Matter (feces) containing or derived from animal or human bodily wastes that are discharged through the anus.

flocculation   The process of bringing together destabilized or coagulated particles to form larger masses or flocs (usually gelatinous in nature), which can be settled and/or filtered out of the water being treated.

flood rim   The edge of a receptacle (such as a plumbing fixture) from which water will overflow.

flow rate   The quantity of water or regenerant that passes a given point in a specified unit of time, often expressed in U.S. gpm (or L/min).

flush tank   The chamber of the toilet in which the water is stored for rapid release to flush the toilet.

flush valve (flushometer)   A self-closing valve used for flushing urinals and toilets in public buildings. This type of valve allows very high flow rates for a few seconds.

free groundwater   Unconfined groundwater whose upper surface is a free water table.

grade   The elevation of the invert of the bottom of a pipeline, canal, culvert, or similar conduit.

gray water   Wastewater other than sewage, such as sink drainage or washing machine discharge.

groundwater   Water below the surface of the ground. It is primarily the water that has seeped down from the surface by migrating through the interstitial spaces in soils and geological formations.

half-life   The time required for half of the substance present at the beginning to dissipate or disintegrate.

hardness   A water quality parameter that indicates the level of alkaline salts, principally calcium and magnesium, and expressed as equivalent calcium carbonate ($CaCO_3$). Hard water is commonly recognized by the increased quantities of soap, detergent, or shampoo necessary to raise lather.

head   The pressure at any given point in a water system, generally expressed in pounds per square inch (psi).

hydrologic cycle   The cyclic transfer of water vapor from the earth's surface through evaporation and transpiration into the atmosphere, from the atmosphere via precipitation back to earth, and through runoff into bodies of water.

interface   The surface that forms a common boundary between two spaces or two parts of matter, such as the surface boundary formed between oil and water.

invert   The lowest point of the channel inside a pipe, conduit, or canal.

ion exchanger   A permanent insoluble material (usually a synthetic resin) which contains ions that will exchange reversibly with other ions in a surrounding solution.

makeup water   Treated water added to the water loop of a boiler circuit or cooling tower to make up for the water lost by steam leaks or evaporation.

mole   The molecular weight of a substance, usually expressed in grams.

Nephelometric Turbidity Unit (NTU)   The standard unit of measurement used in the water analysis process to measure turbidity in a water sample.

osmosis   The natural tendency for water to spontaneously pass through a semipermeable membrane separating two solutions of different concentrations (strengths).

oxidizing agent   A chemical substance that gains electrons (i.e., is reduced) and brings about the oxidation of other substances in chemical oxidation and reduction (redox) reactions.

percolation   Laminar gravity flow through unsaturated and saturated earth material.

permeability   The ability of a material (generally an earth material) to transmit water through its pores when subjected to pressure or a difference in head. Expressed in units of volume of water per unit time per cross-sectional area of material for a given hydraulic head.

pH   A measure of the relative acidity or alkalinity of water. Defined as the negative log (base 10) of the hydrogen ion concentration. Water with a pH of 7 is neutral; lower pH levels indicate increasing acidity, while pH levels above 7 indicate increasingly basic solutions.

psia   Pounds per square inch, absolute. This is the pressure above an absolute vacuum. A 1-inch square column of the earth's atmospheric air exerts a pressure of about 14.7 psia at sea level.

psig   Pounds per square inch, gauge. This is the sum of the pressure indicated by a gauge and the absolute pressure.

pneumatic tank   A pressurized holding tank that is part of a closed water system (such as for a household well system) and is used to create a steady flow of water and avoid water surges created by the pump kicking on and off.

potable water   Water of a quality suitable for drinking.

precipitation   Rain, snow, hail, dew, and frost.

runoff   Drainage or flood discharge that leaves an area as surface flow or as pipeline flow, having reached a channel or pipeline by either surface or subsurface routes.

saturated zone   The area below the water table where all open spaces are filled with water.

soil pipe   A pipe that conveys sewage containing fecal matter.

stack   A general term used for any vertical line of soil, waste, or vent pipe, except vertical vent branches that do not extend through the roof.

stack vent   The extension of a soil or waste stack above the highest horizontal drain connected to the stack.

storm drain   A pipe that conveys rainwater, surface water, condensate, cooling water, or similar liquid wastes.

Total Dissolved Solids (TDS)   The total weight of the solids that are dissolved in the water, given in ppm per unit volume of water.

venturi   A tube with a narrow throat (a constriction) that increases the velocity and decreases the pressure of the liquid passing through it, creating a partial vacuum immediately after the constriction in the tube.

vortex   A revolving mass of water that forms a whirlpool.

viscosity   The tendency of a fluid to resist flowing due to internal forces such as the attraction of the molecules for each other (cohesion) or the friction of the molecules during flow.

waste pipe   A pipe that conveys discharge from any plumbing fixture or appliance that does not contain fecal matter.

water table   The upper boundary of a free groundwater body, at atmospheric pressure.

yield   The amount of product water produced by a water treatment process. It is also the quantity of water (expressed in GPM, GPH, or GPD) that can be collected for a given use from surface or groundwater sources.

zeolites   Hydrated sodium aluminum silicates, either naturally occurring mined products or synthetic products, with ion exchange properties.

zone of aeration   The comparatively dry soil or rock located between the ground surface and the top of the water table.

22.1 WATER SUPPLY

PROPERTIES OF WATER

Water is a basic necessity of life; most of the protoplasm in the human body is comprised of water. In its purest form, water is an odorless, colorless, and tasteless liquid.

Water is a very stable chemical compound. It can withstand a temperature of about 4,900°F before being decomposed into individual atoms. Its boiling point is 212°F, and its freezing temperature is 32°F. The boiling temperature of water drops 1°F for every 500-foot increase in elevation.

A water molecule is formed by covalent bonding between two atoms of hydrogen and one atom of oxygen. This type of bonding occurs when atoms share electrons in order to balance

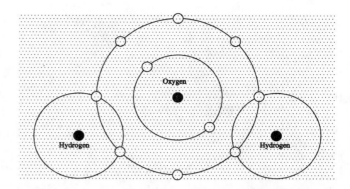

FIGURE 22.1 Chemical Composition of Water

the structure of their outer orbits. An oxygen atom requires two electrons to complete its outer orbit, while a hydrogen atom requires only one. By sharing electrons with two hydrogen atoms, oxygen can balance its outermost orbit. Water's chemical formula is, therefore, H_2O—two hydrogen atoms to one oxygen atom (Figure 22.1).

Water is a universal solvent. It acts as an ideal cleansing agent by promoting dissolution. It is also a remarkably efficient medium for transport of organic waste. The large heat-storage capacity of water makes it an effective means of heat transfer in heating, ventilating, and air-conditioning systems. This particular characteristic also makes it an excellent heat transfer medium for active thermal solar systems.

HISTORY OF WATER SUPPLY

Water has played a great role in the development of humankind throughout the history of the world. It was a determining factor in the triumph and survival of all early human settlements. The first elaborate water supply systems were designed and constructed by the Romans, during the time of Augustus from about 27 B.C. to A.D. 14 (Figure 22.2). They built a system involving nine aqueducts, some of them 50 to 60 miles long, for supply of water to the city of Rome. These aqueducts had a capacity of about 84 million gallons of water per day, to be supplied mainly to the public baths, pools, and fountains.

FIGURE 22.2 A Roman Aqueduct

SOURCES OF WATER

Approximately 70% of the earth's surface is covered by water bodies: oceans, rivers, lakes, and glaciers. Over 97% of this water is either salty (in seas and oceans) or frozen (polar regions). Out of a total of 326,000,000 cubic miles of water, only about 1,000,000 is available to us as fresh water.

One of the primary sources of water supply is **precipitation** in the form of either rain or snowfall. Precipitation is part of the **hydrologic cycle**, a continuous exchange of water among atmosphere, the earth, and the water bodies (Figure 22.3). Water evaporates from the water bodies, soil, and vegetation, changing from a liquid to a gaseous state. The evaporated water goes up into the atmosphere, cools down, and condenses to form very small droplets of water. Clouds are formed due to the concentration of these tiny droplets.

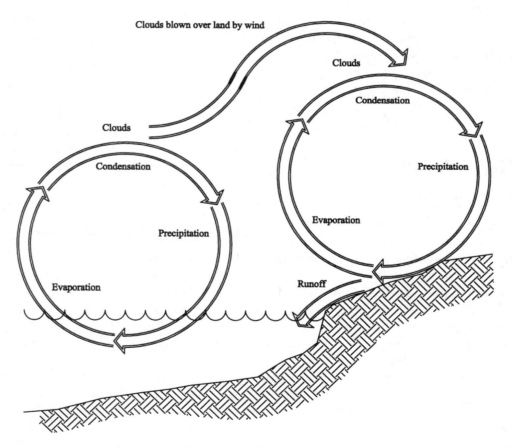

FIGURE 22.3 Hydrologic Cycle

As the water vapor continues to cool and condense, larger droplets of water form around nuclei of dust or ice crystals. Finally, when the droplets coalesce, they fall to the earth as rain or snow.

The United States (excluding some desert areas) receives an average annual rainfall of 40 inches, which yields a total of about 1,080,000 gallons of water per acre. Mean annual rainfall in different regions of the United States is shown in Figure 22.4.

Rainwater is one of the purest sources of water available. Its quality almost always exceeds that of ground or surface water in terms of pollution. The pH-value of rainwater is about 5, which means that it is slightly acidic. A small amount of buffering can, however, neutralize this acid. Rainwater, therefore, can provide clean, safe, and reliable water if it is collected using a clean catchment area and then given appropriate treatment for intended uses (Figure 22.5). Table 22.1 shows quantities of water that can

be obtained per square foot of catchment area for different quantities of rainfall.

Much of the rainwater that hits the ground infiltrates into the earth. Some of it is held in a shallow root zone for use by the plants. The rest of it percolates deeper into the ground through rocks and soil to form a supply of water below the earth's surface. These underground reservoirs of water are known as **aquifers**. A significant quantity of water is available from aquifers, which are geological formations that contain saturated permeable material. The upper level of an aquifer is called the water table (Figure 22.6). The impermeable rock layer below an aquifer is called an aquiclude. Water obtained from an aquifer by drilling wells is relatively pure and, therefore, requires less elaborate treatment.

Water that does not infiltrate to the ground is known as surface **runoff**. It flows across the ground and is collected in rivers, streams, lakes, or other depressions. These

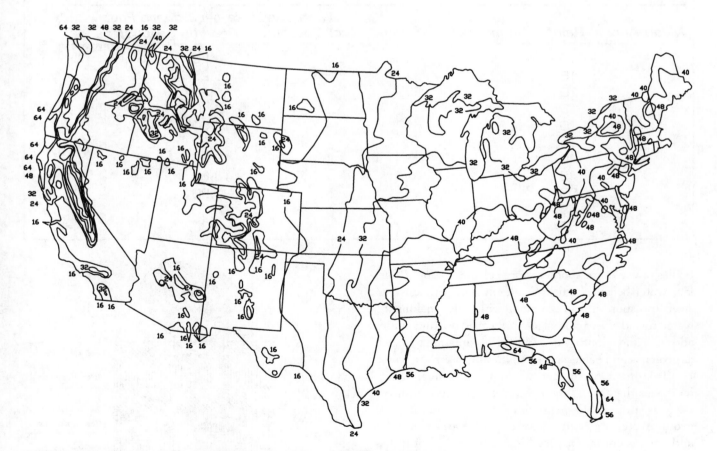

FIGURE 22.4 Mean Annual Rainfall (in inches) in the United States

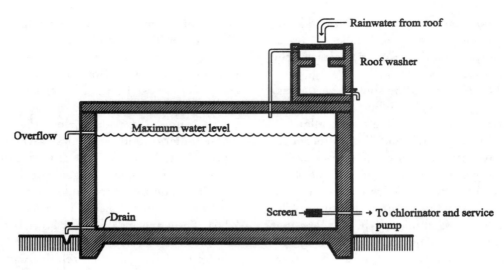

FIGURE 22.5 Rainwater Collection

TABLE 22.1

Quantity of Rainwater Provided per Square Foot of Catchment Area

Mean Annual Rainfall (in inches)	Water Available per Square Foot of Catchment Area (in gallons)
10	4.20
15	6.20
20	8.30
25	10.40
30	12.50
35	14.50
40	16.60
45	18.60
50	20.80
55	22.90
60	25.00

bodies of water are another source of water supply. Surface water becomes quickly contaminated. Streams and rivers transport silt and clay along with water, making the water turbid. Decaying algae and other organic matter add color and odor to water. Biological contaminants such as protozoa and bacteria occur naturally in surface water.

Treatment of surface water is extensive, which increases the cost of such water. Surface water, however, is preferable in regions where pumping of groundwater may cause subsidence. Water from seas and oceans contain a very high quantity of **total dissolved solids (TDS)**. It is about 35,000 ppm, which is 70 times higher than the maximum allowable quan-

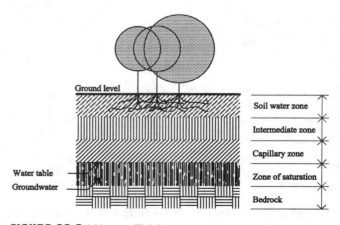

FIGURE 22.6 Water Table

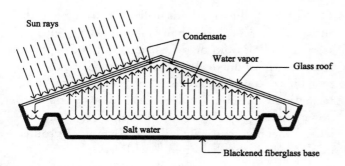

FIGURE 22.7 Solar Distillation

tity for continual human consumption. (TDS in rain-water ranges from 100 to 150 ppm.) If water from such sources is to be used for consumption, then the TDS must be removed by distillation, electrodialysis, or ion exchange methods. Seawater reverse osmosis (SWRO) plants may also be constructed for the purpose. Such methods used to be quite expensive. But the development of new technologies has resulted in cost of production of desalinized water.

Low-cost systems, utilizing solar energy, can be designed for desalination of brackish water. A solar still consists of a basin with a sloping glass cover and low walls, resembling a house with a glass roof (Figure 22.7). The base is made of blackened fiber-glass with insulation at the bottom. Gutters run on the inside, at the bottom edge of the glass. Salty water, about 12 to 16 inches deep, is introduced in the basin and exposed to the sun. As the water evaporates, the air inside the basin becomes saturated, water condenses on the glass surface, flows down the inside of the glass, and collects in the gutter. The water is then directed to a storage tank.

GROUNDWATER SUPPLY SYSTEM

Groundwater is usually collected by digging either shallow or deep wells. When the depth of the well is less than 25 feet, it is called a shallow well. Such wells can be dug, driven, or drilled. Wells deeper than 25 feet are called deep wells. They are either bored or drilled through the earth and rocks.

Dug wells are quickly contaminated by surface-water flow or by seepage of polluted groundwater. Being generally shallow, these wells tend to fail during periods of drought or whenever water is rapidly withdrawn.

Driven wells are very simple and inexpensive to construct. Such wells are driven into the ground

using a well point attached to a steel pipe. The well point, having a diameter of $1\frac{1}{4}$ to 2 inches, is also made of steel (Figure 22.8).

Bored wells are similar to dug wells. They are dug using earth augers to a depth not exceeding 100 feet. These wells are usually lined with vitrified tile, concrete, or metal. The diameter of a bored well may range from 2 to 30 inches (Figure 22.9).

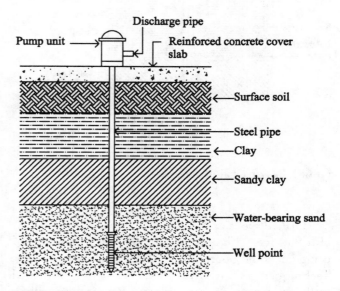

FIGURE 22.8 Driven Well

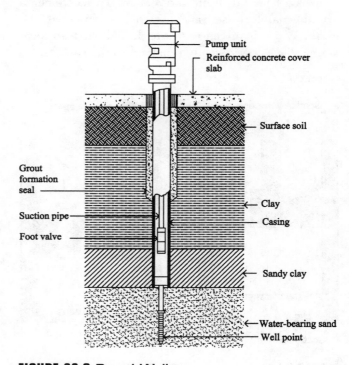

FIGURE 22.9 Bored Well

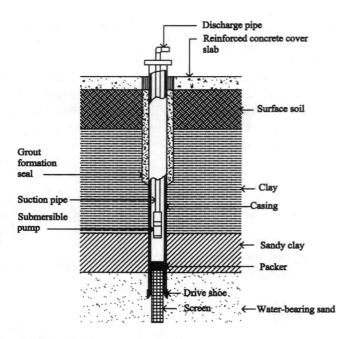

FIGURE 22.10 Drilled Well

Drilled wells are very deep, sometimes in excess of 1,000 feet (Figure 22.10). They are constructed using machine-operated drilling equipment or rigs. Techniques employed for the construction of drilled wells are percussion and rotary. Percussion drilling utilizes a heavy drill bit and a stem. These elements are alternately raised and lowered to pulverize the earth. Water is mixed with the drilled earth to form slurry that is periodically removed to the surface.

Rotary drilling methods utilize a cutting bit at the end of a drill pipe, a revolving table for drill pipe passage, a drilling fluid or pressurized air, and a power source to drive the drill. As drilling proceeds, the cutting bit rotates and advances to break up the rock formation. Drilling fluid is continuously pumped down the drill pipe for removal of cuttings to the surface. A metal case is installed, and sometimes grouted in place, after the drill stem is withdrawn.

PUMPS

Two categories of pumps are used for collection of water from wells: positive displacement and dynamic. In positive displacement pumps, energy is periodically added by application of force to the movable parts of a pump to increase the water velocities. In dynamic pumps, energy is continuously added to increase the water velocities within the machine.

There are two major classes of positive displacement pumps: reciprocating and rotary. A reciprocating pump works on a two-stroke principle (Figure 22.11). It consists of a piston within a cylinder, which draws in water on an ingoing stroke and delivers the same on the outgoing stroke. It is equipped with check valves (one-way valves) on both the suction and delivery sides.

A rotary pump has a helical or spiral rotor within the pumping chamber that rotates during the opera-

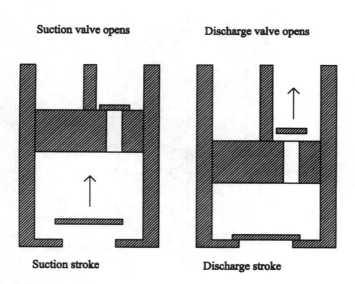

FIGURE 22.11 Operation of a Reciprocating Pump

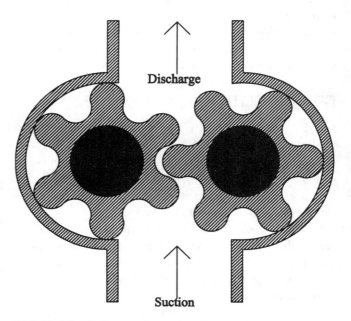

FIGURE 22.12 Operation of a Rotary Pump

tion of the pump. Water entering the suction side is forced to the delivery side due to the rotation of the rotor (Figure 22.12).

The major classes of dynamic pumps used for water supply are centrifugal and jet pumps. Centrifugal pump is a general description for equipment that utilizes an impeller mounted on a rotating shaft for suction and delivery of water. An impeller together with its casing is called a stage. Pumps with more than one stage are called multistage pumps. Two common types of centrifugal pumps that are used for collecting well water are turbine and submersible. The assembly of a turbine pump consists of a suction head, an impeller or a set of impellers, discharge bowl, intermediate bowl or bowls, and a vertical impeller shaft along with various bearings (Figure 22.13). The driving motor of the pump is located over well casing at grade.

Pump and motor in a submersible pump are coupled as one unit. The unit is completely submerged in the well along with the impellers. The need for lengthy pump shaft is thus eliminated.

A jet or ejector pump is a special-effect dynamic pump that consists of a nozzle, a diffuser, and a suction chamber. It utilizes the motive power of a high-pressure stream of water directed through a nozzle

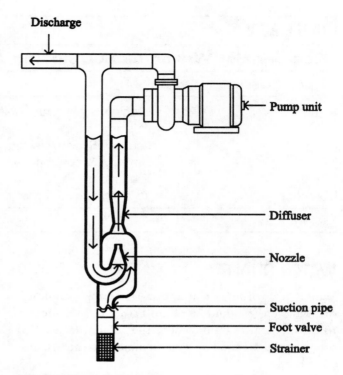

FIGURE 22.14 Jet Pump

designed to produce high velocity. The high-velocity jet stream creates a low-pressure area in the mixing chamber that causes the suction water to flow into this chamber. The diffuser entrains and mixes the water and converts velocity to pressure energy. The increased pressure pushes the suction water toward the discharge port (Figure 22.14).

The vertical distance traversed by water from the point of intake to the level where it is supplied is called the head of a pump. It is dependent on the total pump pressure that includes suction lift, static head, and friction loss plus pressure head. The relationship between pressure (P) and head (H) can be expressed as follows:

$$P \text{ (lb/in}^2) = [62.4 \text{ (lb/ft}^3) \cdot H \text{ (ft)}]/144$$
$$= H \text{ (ft)}/2.31 \qquad (1)$$

WATER QUANTITY

The supply water is usually estimated in terms of gallons per person per day (gppd). Domestic water consumption can be broken down into two categories: indoor use and outdoor use. An average residential water consumption pattern in the United States is given in Table 22.2.

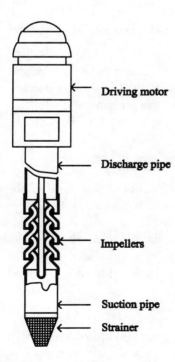

FIGURE 22.13 Turbine Pump

TABLE 22.2

Residential Water Consumption Pattern in the United States

Type of Use		Quantity of Water (in gppd)	
Indoor	Bathing & personal hygiene	21	
	Flushing toilets	30	
	Laundry & dishwashing	15	
	Drinking & cooking	4	
	Total indoor		70
Outdoor	Yard irrigation, car wash, etc.		70
Total			140

WATER QUALITY

Water used for consumption (bathing, cooking, drinking, etc.) should be **potable**, that is, safe enough to drink. It also should meet the minimum level of other qualities based on several major characteristics, as follows:

Turbidity

The clarity of water is measured by the presence of insoluble suspended solids, that is, its turbidity. Turbidity is more common in surface than groundwater. Clarity of water is determined by the amount of light scattered by solid particles in a sample of water. Potable water should not contain more than 5 turbidity units (TU). One turbidity unit equals 1 mg of suspended matter per liter of water sample.

pH-value

pH refers to "potential hydrogen," and is a measure of hydrogen ion concentration in water. It is expressed as follows:

$$pH = \log(1/H^+) \qquad (2)$$

where:

H^+ = number of hydrogen ions.

The higher the concentration of hydrogen ions in water, the lower is the pH-value, indicating a high concentration of acid. Addition of alkali reduces the number of free hydrogen ions, causing an increase in pH-value. It is measured on a scale ranging from 0 to 14. The pH-value is 0 when water is extremely **acidic**; the value is 14 when water is extremely **alkaline**. At a pH-value of 7, water is considered to be neutral. Recommended pH-value of water ranges from 6.5 to 8.5.

Hardness

Hardness is a measure of detergent-neutralizing ions present in water. A hardness of about 300 mg/L in public water supplies is considered to be excessive. It results in hindering the cleansing action of detergents and in forming scales in cooking utensils and supply pipes. Calcium and magnesium salts predominantly cause hardness. A level of hardness between 60 and 120 mg/L is acceptable in potable water.

Biochemical Oxygen Demand (BOD)

BOD is the amount of oxygen, measured in mg/L, required in the oxidation of organic matter in water by biological action. BOD is thus a measure of pollution in wastewater and streams. Supply water should not contain more than 4 mg/L of BOD.

WATER TREATMENT

A supply system should provide potable water that is safe from chemical and bacteriological points of view. Domestic water should also be free from unpleasant taste and smell, and improved for human health. Water should be conditioned by utilizing different treatment processes before it is supplied.

Oxidation

Oxidation is a method of breaking down organic pollutants or chemicals present in supply water by addition of oxygen. This process also helps remove such minerals as iron and manganese, thereby improving the taste and color of water.

Sedimentation

This process involves the removal of suspended solids from water by allowing the particles to settle out. The water has to be kept still in a pond or a basin for at least 24 hours for sedimentation to take place. This method helps to remove the turbidity of water. Adding a chemical such as hydrated aluminum sulfate (commonly known as **alum**) can increase the efficiency of the method. This chemical causes fine suspended particles in water to coagulate and settle down to the bottom of the sedimentation pond.

Filtration

Filtration is a process of removing suspended matter from water by passing it through beds of porous materials such as sand, gravel, porous stone, diatomaceous earth, activated charcoal, or filter paper. Filtration not only removes suspended matter but also helps remove some bacteria and improves turbidity, color, taste, and odor of water to a certain extent.

Disinfecting

Disinfecting may be a simple or complex matter depending on the source of water supply. Chlorine is extensively used as a disinfecting agent to treat water for municipal and individual supplies. Chlorine has a strong oxidizing power on the chemical structure of bacterial cells; it destroys the enzymatic processes of the cells required for their existence.

Ozone is also used as a disinfectant. The use of ozone helps remove color, taste, and odor that may be present in water. It does not, however, provide a lasting residual like chlorine in treated water. Therefore, a secondary **chlorination** is recommended in order to provide a protective residual in the distribution system.

Softening

Softening is the process of removing hardness from water. Hardness occurs when calcium and magnesium carbonates dissolve in groundwater. The process begins when rainwater dissolves carbon dioxide to form carbonic acid:

$$H_2O + CO_2 \Rightarrow H_2CO_3 \qquad (3)$$

This weakly acidic water dissolves limestone when it comes into contact with the material and

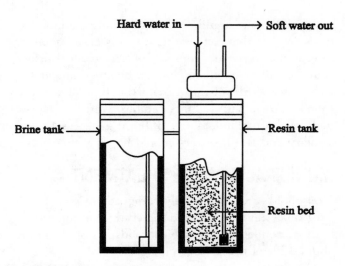

FIGURE 22.15 Water Softener

forms solutions of calcium and/or magnesium bicarbonate as follows:

$$H_2CO_3 + CaCO_3 \Rightarrow Ca(HCO_3)_2 \qquad (4)$$

$$H_2CO_3 + MgCO_3 \Rightarrow Mg(HCO_3)_2(5) \qquad (5)$$

The hardness may be removed utilizing various methods of water softening. One of the most popular methods is the **zeolite** system. The process involves an exchange of calcium or magnesium ions with sodium ions. The softener consists of a resin tank filled with ion-exchange resins, a group of hydrated sodium aluminosilicates (Figure 22.15). Another component of the system is a brine tank containing sodium solution for regenerating the resins. When water is passed through the sodium-saturated resins, the resins remove calcium and magnesium ions from water and release sodium ions into it. Hardness of water is thus removed.

Use of a deionizing system can also provide water softening. This method is capable of removing hardness caused not only by calcium and magnesium ions but also by other types of alkali metal ions.

22.2 WASTEWATER

SEWER

Wastewater generated from buildings either has to be transported by sewers to central facilities for treatment and disposal, or may be treated using individual waste systems. Drainage in buildings has two major components: sanitary waste and storm water. Sanitary waste consists of liquid discharged from plumbing fixtures. A pipe that carries this waste is called sanitary sewer. Storm water consists of rainwater, surface water, condensate, cooling water, or other similar liquid wastes. A **drain** that carries this type of waste is called a **storm sewer**. Since sanitary waste contains a high level of pollutants and BOD, it is necessary that sanitary and storm sewers should be separate.

MUNICIPAL SEWAGE TREATMENT

Household sewage is largely composed of water with very small amounts of solid material. The quantity of both suspended and dissolved solids in waste is only about 0.1% of the total volume.

Sanitary sewers in urban areas carry sanitary waste to a central treatment plant before it is discharged into streams or rivers. The treatment is provided in four stages:

Sedimentation

The waste is first diverted to a grit chamber where the larger solids and heavy objects settle out. It then

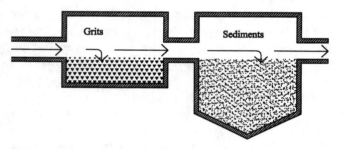

FIGURE 22.16 Sedimentation Tank

moves on to a sedimentation tank where lighter suspended materials settle to the bottom and grease floats on top (Figure 22.16). The mass of sediments is called sludge.

The sludge and scum from the sedimentation tank is usually pumped to a sludge thickener for hydration. The thickened sludge then may be transferred to a digester where it is anaerobically decomposed. Methane gas produced during the process is a good source of energy for operating the treatment plant. The decomposed sludge from the digester may be used as a fertilizer.

Aeration

The waste is pumped to an **aeration** tank where it mixes with oxygen in air for several hours (Figure 22.17). During this time, colonies of **aerobic** bacteria break down the organic matter present in waste. Aeration of the sewage results in the removal of about 90% of organic matter.

Chlorination

The **effluent** from the aeration tank is carried to a chlorination chamber where it is treated using chlorine gas for about 15 to 30 minutes. This process kills the remaining organisms in the effluent.

Dechlorination

After the effluent has been disinfected, it is given a further treatment to remove the chlorine. The process involves either the use of some chemicals such as sodium metabisulfate or the use of oxygen in air. Once chlorine is removed, the effluent is released into streams or

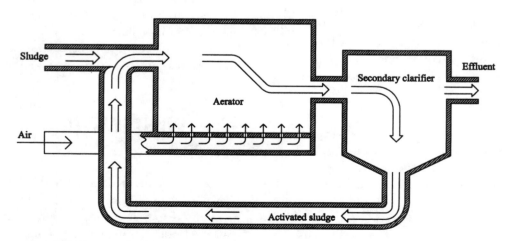

FIGURE 22.17 Aeration Tank

rivers, or is used for irrigation. It has to be ensured that the effluent coming out from the final stage of treatment contains a low BOD.

SEPTIC TANK SYSTEMS

A septic tank is an individual waste system for households in areas that are not connected to a municipal treatment system. It consists of a tank for primary treatment of waste and a method of filtration of the effluent as secondary treatment.

The system works by allowing wastewater to separate into layers and decomposing the organic matter while it is contained within the tank.

Anaerobic bacteria, naturally present in the septic systems, digest the solids that settle out at the bottom of the tank. The outflow, or effluent, from the tank is then given a secondary treatment.

Construction of Septic Systems

Septic tanks are usually constructed underground using brick, stone, concrete blocks, precast concrete, fiberglass, or steel. Use of precast concrete, however, is very common.

The tank may either be rectangular or circular in configuration (Figure 22.18). It is divided into two chambers using a baffle wall with an opening 18 inches

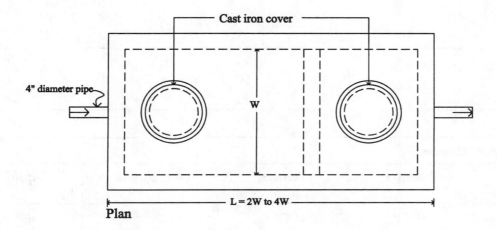

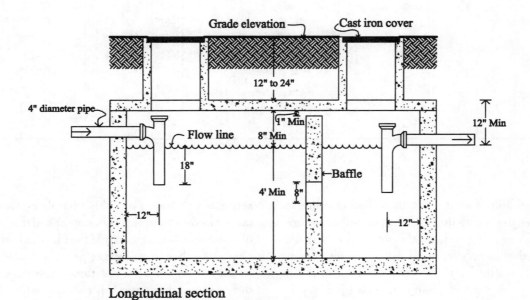

FIGURE 22.18 Rectangular Septic Tank

TABLE 22.3

Septic Tank Capacity

Single-Family Dwellings: Number of Bedrooms	Multiple Dwelling Units or Apartments: One Bedroom Each	Other Uses: Maximum Fixture Units Served	Minimum Septic Tank Capacity (in gallons)
1–3		20	1,000
4	2	25	1,200
5 or 6	3	33	1,500
	4	45	2,000
	5	55	2,250
	6	60	2,500
	7	70	2,750
	8	80	3,000
	9	90	3,250
	10	100	3,500

TABLE 22.4

Septic Tank Sizes

Capacity (in gallons)	Length	Width	Air Space	Liquid Depth
1,000	8'–0"	4'–0"	1'–0"	4'–0"
1,200	9'–0"	4'–6"	1'–0"	4'–0"
1,500	9'–6"	4'–9"	1'–0"	4'–6"
2,000	10'–6"	5'–3"	1'–3"	4'–9"
2,250	11'–0"	5'–6"	1'–3"	5'–0"
2,500	11'–6"	5'–9"	1'–3"	5'–0"
2,750	12'–0"	6'–0"	1'–3"	5'–3"
3,000	12'–6"	6'–3"	1'–3"	5'–3"
3,250	13'–0"	6'–6"	1'–3"	5'–3"
3,500	13'–6"	6'–9"	1'–3"	5'–3"

below the flow line of waste. A 4-inch-diameter house sewer delivers the waste into the first chamber where the grits and organic solids settle out at the bottom. Grease and other lighter material rise to the top, forming scum. The baffle wall stops the sludge and scum from entering the second chamber. The finer solids that enter the second chamber along with the liquid settle at the bottom of this chamber. The sludge retained in the chambers is decomposed due to the action of anaerobic bacteria, while the clarified effluent is discharged either to a seepage pit or a drainfield for secondary treatment. Septic tank capacities needed for different buildings and representative tank dimensions are given in Tables 22.3 and 22.4, respectively.

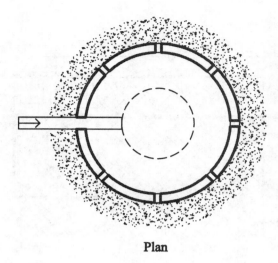

Plan

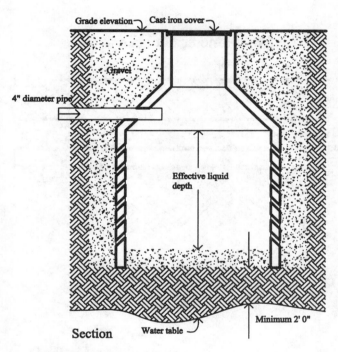

Section

FIGURE 22.19 Seepage Pit

Seepage Pit

A **seepage pit** or leeching pool is also commonly constructed using precast concrete (Figure 22.19). The pit has a circular wall with perforations that allow the effluent to seep into the gravel surrounding it before being absorbed by the soil. A seepage pit cannot be constructed in a place with

a high water table. The bottom of the pit should be placed at least 2 feet above the water table.

Drainfield

A **drainfield** is an area where the effluent from a septic tank is distributed by horizontal underground piping in order to expedite the process of natural leeching and percolation through the soil. The drainage system consists of 4-inch-diameter clay tiles or perforated plastic pipes buried in trenches (Figure 22.20). Clay tiles are laid with a $\frac{1}{4}$-inch separation between the ends to allow the effluent to run out. The tiles or pipes are covered on all sides by clean stone or gravel. The tops of the trenches are backfilled with earth. Drainfield areas required for residential building are given in Table 22.5.

Percolation Test

It is necessary to determine the absorption capacities of soil before constructing a drainfield. This can be done using a method known as **percolation test**. In order to perform percolation tests, several straight-sided holes, each at least 4 inches in diameter, are dug down to the level where drainfield pipes are to be laid. These holes should be at least 14 inches in depth. Smeared soil surfaces are then removed from the holes, and 2 inches of coarse sand or fine gravel are placed at the bottom. The purpose of this filling is to protect the holes from scouring. They are now filled with clear water to a minimum depth of 12 inches. This water level is maintained for at least 4 hours for all types of soils except sands.

After soaking, 6 inches of water are added over the sand or gravel filling. The drop in this water level is measured from a fixed reference point at ground level, at approximately 10-minute intervals for one hour. The drop that occurs during the final 30-minute period is used for calculating the absorption rate. This rate is expressed in number of minutes required for water to drop an inch.

LAGOONS

Sewage lagoons are shallow ponds constructed to hold sewage while aerobic bacteria decompose the waste. Lagoon floors and sides should be relatively

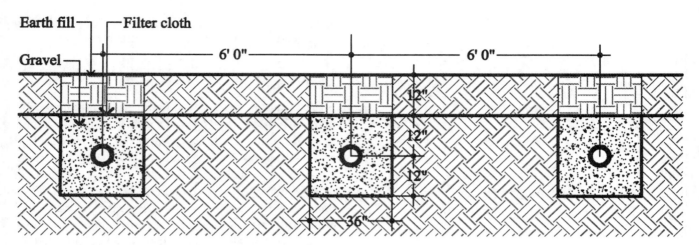

Cross section

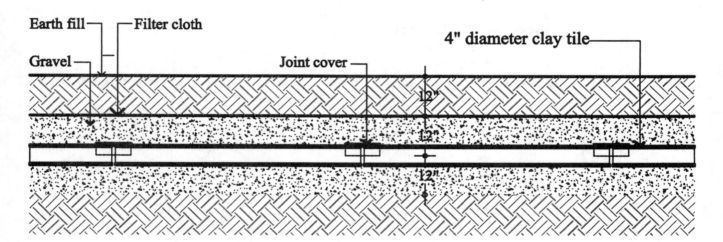

Longitudinal section

FIGURE 22.20 Drainfield Trench

impervious, in order to minimize seepage and contamination of groundwater. A lagoon is required to provide a seven-day retention of sewage. Soils that contain a large amount of organic matter are not suitable for the floor of an aerobic lagoon.

STORM WATER

Storm sewers, commonly designed as gravity flow conduits, usually carry storm water. Infiltration of storm water into sanitary sewers should be avoided as far as possible. Excessive clean water inputs in sanitary sewer result in overloading the treatment plants.

Storm water should be disposed of in such a manner that it does not create damages to property or people during periods of heavy rainfall and subsequent runoff. Surface drainage has to be considered as a function of grading.

TABLE 22.5

Drainfield Area per Bedroom

Average Percolation Rate (minutes/inch)	Area of Trench per Bedroom (sq. ft.)	Length of Trench in Feet		
		18″ Wide Trench	24″ Wide Trench	36″ Wide Trench
5	125	84	63	42
10	165	110	83	55
15	190	127	95	64
20	215	144	108	72
30	250	167	125	84
45	300	200	150	100
50	315	210	158	105
60	340	227	170	113
70	360	240	180	120
80	380	254	190	127
90	400	267	200	134

REVIEW QUESTIONS

1. What is the approximate temperature at which a water molecule is decomposed into individual atoms of hydrogen and oxygen?
2. What is surface runoff?
3. What is the maximum recommended limit of turbidity units in potable water?
4. What is sedimentation?
5. Hardness in water is predominantly caused by the presence of _____.
6. What is indicated by the pH-value of water?
7. What is a drainfield?
8. What is the method used to determine absorption capacities of soil?
9. What is a storm sewer?
10. Name the term used to indicate the height up to which water can be lifted by a pump from the point of intake to the level of supply.

ANSWERS

1. About 4,900°F.
2. Water that does not infiltrate into the ground.
3. 5
4. It is the process of removing suspended materials from water.
5. Calcium and magnesium ions.
6. pH-value is a measure of hydrogen ion concentration in water. A low pH-value indicates a high concentration of hydrogen ions and high acidity.
7. It is a horizontal area used for leeching and percolation of effluent from a septic tank through the soil.
8. Percolation test
9. It is the drainpipe that carries rainwater, surface water, condensate from HVAC systems, and other similar liquid wastes.
10. Head of a pump

CHAPTER 23
Site and Roof Water Drainage

Marvels are there in every land,
Where nature shows her wondrous hand;
Rivers are there in East and West,
Where tourists throng for change and rest.

J. A. O'Reilly

This chapter discusses the principles of site and roof drainage. Proper site planning and development is essential for efficient site drainage. Capably designed, constructed, and maintained drainage systems, for both the site and roof, promote the comfort and durability of a building. Site development includes landscaping. The chapter elaborates on the water requirement for maintenance of landscaping and techniques of irrigation. It also deals with the design of parking areas as part of site development.

Apart from practical purposes, water may also be used to enhance the aesthetic qualities of a site. Issues related to the design of pools and fountains are discussed in this chapter.

23.0 SITE DRAINAGE

OBJECTIVE OF SITE DRAINAGE

The principal objective of site drainage is to keep storm water from entering the buildings located on the site. Storm water should be disposed of in such a way that it does not create any damage either to the properties or to people during periods of heavy rainfall and subsequent runoff.

SUBSURFACE SYSTEMS

Subsurface systems are the primary techniques of removing rainwater from site. The basic objectives of subsurface systems are collection, transfer, and disposal of surface runoff. In order to ensure that water does not enter the buildings, it is necessary that they be positioned at a higher elevation than the street level. Proper grading is an integral part of positive site drainage. Surface drainage has to be considered as a function of grading. Storm water should be diverted from the buildings by grading the site to provide at least 6 inches of vertical fall in the first 10 feet of horizontal distance toward the street level. It should be disposed of in such a manner that it does not create damages to property or people during periods of heavy rainfall and subsequent runoff. Since water always runs downhill and perpendicular to the contour line, it may be necessary to revise the site contours for easy removal of surface water. See Figures 23.1(a) and 23.1(b).

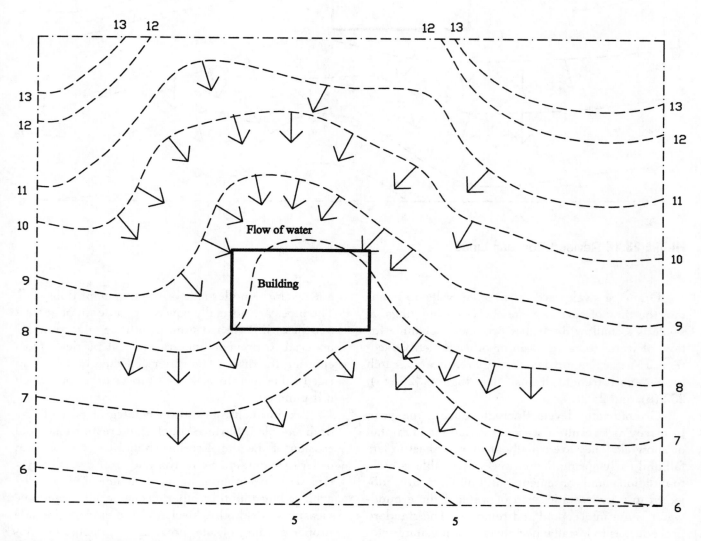

FIGURE 23.1A Original Contour Lines

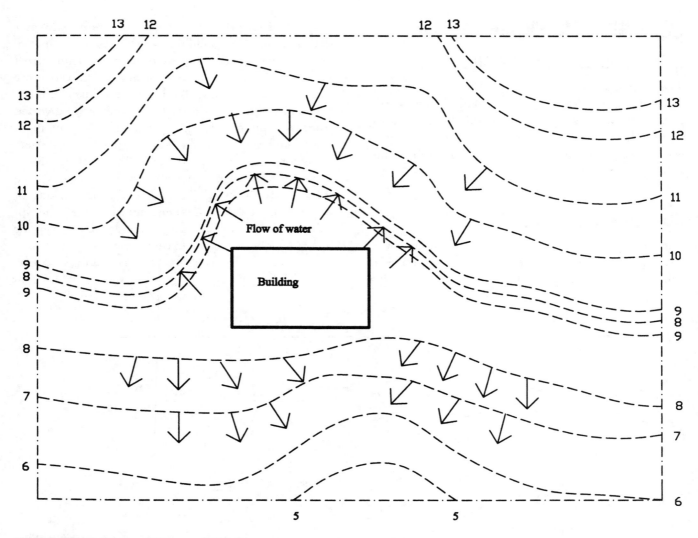

FIGURE 23.1B Revised Contour Lines

Drainage swales are required in order to maintain the flow of surface water between buildings. A swale is a shallow ditch that provides a channel for flow of water to catch basin or other outlet. Collection of surface runoff is done by **area drains**, catch basins, and trench drains. See Figures 23.2(a), 23.2(b), and 23.2(c).

Groundwater levels fluctuate during the year, primarily as a result of seasonal variations in rainfall or snowfall. They are usually at their highest in late fall and early spring. When the water table is high, foundations and basements of buildings are subjected to a hydrostatic head of water if the groundwater is not intercepted and removed. The standard procedure is to install a peripheral drain using perforated pipes around the basement so that water does

not reach the under side of the basement floor slab. The pipes are normally covered to a depth of at least 6 inches with crushed stone or other course granular material to prevent fine-grained soil particles from entering the drain. The water collected by the drain is led either to a storm sewer or to a sump from which it is pumped.

Erosion control is also an integral part of any drainage plan. Surface runoff can create substantial erosion of the site. Erosion should be controlled in order to maintain an effective and clear drainage system with a minimum of maintenance. This can be achieved by adopting either mechanical or vegetative measures, or both. Mechanical measures include proper grading of site, creating diversions, or constructing sediment basins. Vegetative measures include preserving trees and shrubs on site, planting of turf or ground cover immediately after grading, using straw mulch to protect constructed slopes, and using fibrous materials directly on soil to protect newly seeded channels.

Once the water is collected, it is carried by storm sewers and disposed of in streams or rivers.

Water collected by peripheral drains may sometimes have to be led to a sump from which it is pumped to storm sewers. Storm sewers, designed commonly as gravity flow conduits, usually carry storm water. Infiltration of storm water into sanitary sewers should be avoided as far as possible. Excessive clean water inputs in sanitary sewer re-

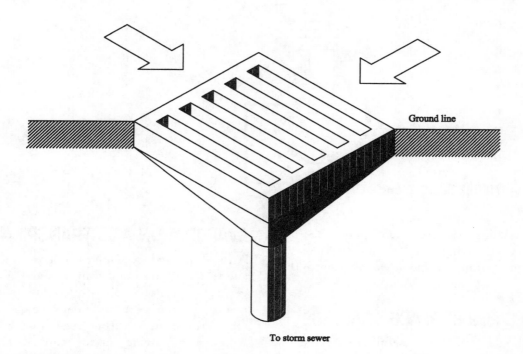

FIGURE 23.2A Area Drain

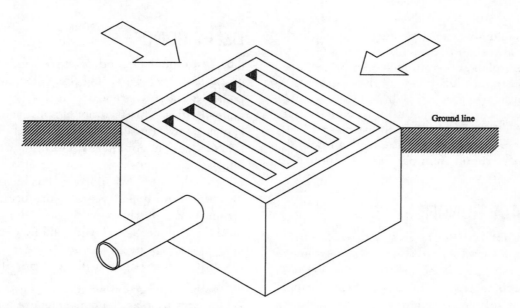

FIGURE 23.2B Catch Basin

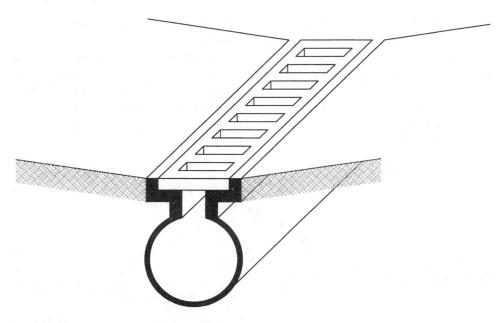

FIGURE 23.2C Trench Drain

sult in overloading the treatment plants. Straw bales may be used during installation of storm sewers to prevent sediments from washing into the gutters.

GUIDELINES FOR SURFACE DRAINAGE

- Runoff should not be purposefully redirected from its natural course on one's own property such that it creates a problem on another property.
- A method of reducing the velocity of surface runoff should always be considered so that it might be absorbed in the soil.
- A drainage plan and grading should be so devised as to take advantage of the existing natural systems.
- Drainage of large paved areas across pedestrian paths should be avoided. Catch basins and trench drains can be used to collect substantial quantities of runoff from parking lots and paved pedestrian areas.

SUBSURFACE RUNOFF

Subsurface runoff is the secondary method of removal of water from a site. This process allows the water to permeate the surface, be absorbed by the soil, and eventually become part of the groundwater supply and the aquifers. It is a very important contributor to the recharge of groundwater.

EVAPORATION AND TRANSPIRATION

A third method by which water is removed from a site is evaporation and transpiration. A site that receives plenty of sunlight will lose surface water through evaporation and reinforce the natural systems. Transpiration from vegetation at site is also helpful in removing surface water from site. Plants and vegetation absorb water during photosynthesis and transpire it to the atmosphere.

DELAY RUNOFF

The traditional method of storm water disposal in city areas aims at draining the water away as fast as possible, via gutters and storm sewers, to the natural water bodies. This traditional approach, however, is reported to contribute to increased flooding and stream erosion. An alternative solution to this problem is local disposal of storm water at its source of runoff. Although this approach has been getting a lot of attention in recent years, it has been used in different parts of the world for a long time. This method, known as delayed runoff, can be used to control storm water from individual lots.

Advantages cited for delayed runoff include:

- recharge of groundwater;
- preservation and enhancement of natural vegetation;

- reduction of pollution to the natural water bodies;
- smaller storm sewers at low cost; and
- reduction of downstream flow peaks.

Disadvantages of the system include:

- eventual sealing of soil, leaving the property owners with a failed disposal system; and
- rise in groundwater level, causing flooding of basements or damage to building foundations.

23.1 SITE IRRIGATION

WATER FOR IRRIGATION

The amount of water required for landscaping depends on the variety and location of plants in a site. On an average, about 70 gallons of water per person are used each day for outdoor use. Most of this water is utilized to maintain outdoor landscaping.

The quantity of water required for irrigation depends on type of soil and climatic conditions, as well as variety and location of plants. In order to optimize outdoor water use, it is a good idea to design a zone plan for the site based on the water requirements of the different types of trees, plants, and grass. Vegetation with similar water requirements may be grouped under a single irrigation zone. Such zoning will not only simplify the landscaping plan, but will also help to conserve water.

SPRAY IRRIGATION

The most effective way to irrigate a lawn is by spray irrigation using automatic sprinklers. A spray irrigation system works very efficiently for a well-defined site with simple configuration. Narrow lawn areas with less than 10 feet width are not recommended because of a possibility of overspray.

Types of Sprinklers

Sprinklers for spray irrigation come in two basic types: pop-up and shrub. Pop-up sprinklers are installed below ground (Figure 23.3); part of the sprinkler rises above grade when the sprinkler is in operation and retracts back when it is shut off. Shrub sprinklers are installed above ground on top of a riser (Figure 23.4). It is advantageous to use pop-up sprinklers from the point of view of safety. It is easy to trip

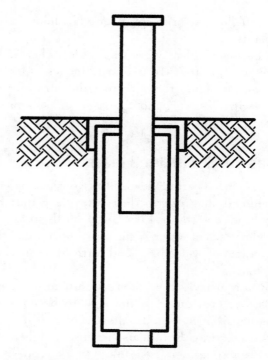

FIGURE 23.3 Pop-up Sprinkler

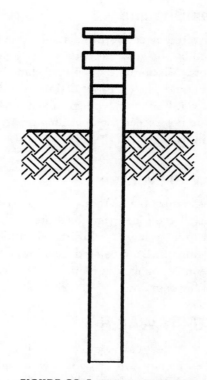

FIGURE 23.4 Shrub Sprinkler

over, or fall on, a shrub sprinkler while taking care of the plants.

The height of a pop-up sprinkler when it rises above ground ranges from 3 to 12 inches. It includes a spring retraction for operating the sprinkler riser and a wiper seal around the riser stem to prevent leakage.

Types of Sprinkler Heads

Sprinkler heads are divided into two types based on the method they use to distribute the water: fixed spray heads and rotors. Fixed spray heads spray a fan-shaped pattern of water, while rotors operate by rotating streams of water back and forth over the shrubs and plants.

Rotors usually cover more area than fixed spray heads, but cost more. If the area to be covered is more than 18 feet, then it is more economical to use a fixed head spray. Rotors also require a higher water pressure to operate. They do not work properly at a pressure less than 40 psi.

Sprinkler Spacing

The area covered by each sprinkler should overlap the area covered by the adjacent sprinkler to avoid creating dry spots. Maximum on-center distance between sprinklers should not exceed 1.2 times the radius of the spray head. For rotor-type sprinklers, the spacing in feet should not exceed the operating pressure of the sprinklers in psi.

Backflow Preventer

Local codes in most places require a backflow preventer to be installed on the irrigation system. This equipment helps to keep lawn fertilizers, weed killers, and other contaminants from being drawn into the potable water system. Manual or automatic valves cannot be substituted for backflow preventers.

USE OF GRAY WATER

Using gray water can almost double residential water-use efficiency and provide a water source for landscape irrigation. It is the wastewater from facilities such as lavatory sinks and showers. Community-wide use of this recycled water could allow a reduction in the size of water-purification and sewage-treatment facilities. It should be very

carefully separated from black water (WC and urinal flush water). Local plumbing codes should be checked before installing a gray water recovery system.

As a general rule, gray water does not require extensive chemical or biological treatment before it can be used for irrigation. Soaps and detergents are the components in gray water that could adversely affect plants the most. Therefore, the water most preferred for irrigation is from the shower or lavatory sink since this wastewater generally contains only a small amount of soap and has very few solid residues.

23.2 SITE WATERSCAPE

POOLS AND FOUNTAINS

Water is often used for decorative or ornamental purposes in site design. Pools are usually constructed to evoke the impression of quiet water found in natural lakes; constructing fountains may simulate the essence of moving water such as waterfalls and geysers. The reflective quality of a pool adds tranquility to the environment, and the moving quality of water imparted by a fountain enhances the visual interest of any landscape. Water fascinates all humans, regardless of their culture, ethnicity, and religious belief. It soothes and relaxes, inspires reflection, and is a source of pleasure.

Apart from aesthetic reasons, fountains also serve practical purposes. It may be effectively used for noise mitigation. The sounds of splashing, flowing, or moving water can mask adjacent irritating sources of noise in an urban environment. A fountain's ability to overcome noise is directly proportional to the sound produced by its water. A fountain may also serve as a source of evaporative cooling for an area with high ambient temperature and low humidity.

FOUNTAIN AND POOL EQUIPMENT

Mechanical systems of pools and fountains consist of three components: water effects system, water level control and drainage system, and filtration system. The water effects system of a fountain requires a pump, a discharge line to supply water, and a nozzle through which water will be discharged.

Nozzles

There are four basic types of nozzles: aerating, smooth-bore, spray head, and formed. Aerating nozzles produce white, frothy water visible from a significant distance (Figure 23.5). This is accomplished by installing the nozzle under the water level and drawing a mixture of air and water through a perforated cover at the bottom of the nozzle. This type of nozzle requires a larger horsepower pump than other types of nozzles.

Clear and smooth water columns characterize a smooth-bore nozzle (Figure 23.6). The first part of the trajectory from such a nozzle is a solid-stream jet of water, which gradually breaks up into fine sprays as it reaches maximum height and descends downward.

A spray head nozzle produces delicate combinations of clear thin water jets (Figure 23.7). The distribution head of such a nozzle is usually circular or fan-shaped, with multiple exposed tube nozzles or holes in the face.

A formed nozzle spurts out thin sheets of water (Figure 23.8). Size and shape of the water sheets

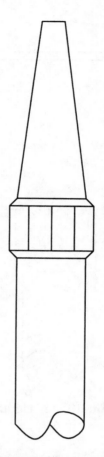

FIGURE 23.6 Smooth-bore Nozzle

vary according to the design of the nozzle. Unlike other nozzles, formed nozzles are not dependent on the level of water in pools; they can be located any distance above the water level.

Pumps

Fountains nowadays are predominantly designed as closed systems; water held in a basin is continually recirculated from basin to nozzle and back again to the basin. A pump mechanism is required to generate pressure to recirculate the water. Submersible pumps are used for small fountains requiring a discharge rate of 100 gpm or less; centrifugal pumps are used for fountains where large water volumes or high pressures are required.

FILTER SYSTEMS

Fountains are usually equipped with a filter system, which prevents damage to the nozzle and pump, and thereby reduces maintenance costs. A filter system

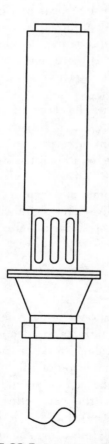

FIGURE 23.5 Aerating Nozzle

consists of a pump and motor, a filtration medium, and a discharge line. A pump and motor for filter systems is a separate unit from the water effects system. These allow the filter system to operate independent of the water effects system. Cartridge filters, diatomaceous-earth filters, or permanent-media sand filters are most often used for filtration.

A filter system should be capable of recirculating the entire volume of pool water in 6 to 10 hours. If total volume of water of a basin containing 9,000 gallons is to be recirculated in 10 hours, the filter system must circulate 900 gallons per hour or 15 gallons per minute.

WATER LEVEL CONTROL

Water in a pool may be lost due to evaporation and splash, causing a fall in water level. This fall may cause damage to nozzles and pumps that require minimum water protection for their continued operation. Water level controls are therefore required in fountains to replenish the lost water.

Appropriate water levels may be maintained using float controls or probe-type water level controls. A float control consists of a plastic float and a brass valve. The float, attached to an arm, operates the valve. The float is depressed when the water level falls, causing the valve to open and fresh water to flow into the fountain. This device is commonly used for small fountains.

Probe-type water level controls are used for large fountains. These devices consist of an electronic probe box for sensing water level, a relay control panel, and a solenoid-operated valve. The probe sends a signal to the relay panel when the water level drops below a minimum, which in turn sends a signal to the solenoid valve. This signal causes the valve to open, allowing fresh water to enter the fountain.

SIZING

Fountain Basin

Fountain basins should be large enough to contain splashes from the nozzles and deep enough to ensure smooth and efficient operation of the equipment. The horizontal dimension of a fountain basin, with a single nozzle at the center, should not be less than twice the height of the water effect above the water surface (Figure 23.9). For windy locations, this dimension should be four times the height of the water effect.

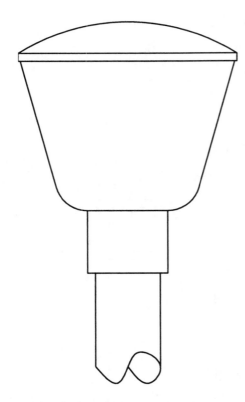

FIGURE 23.7 Spray Head Nozzle

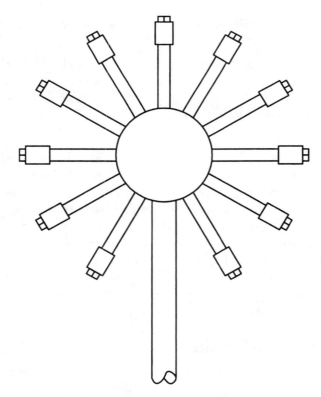

FIGURE 23.8 Formed Nozzle

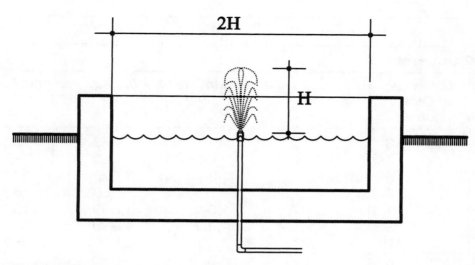

FIGURE 23.9 Sizing of Fountain Basins

Pump Size

The water capacity of fountain pumps should be equal to the total capacity of all nozzles installed on the discharge line. A factor of safety to account for decrease in pump performance due to aging is usually added to the capacity of nozzles when pumps are specified. A common practice is to add 10% to the total nozzle capacity.

23.3 ROOF DRAINAGE

ROOF DRAINS

Sloping Roof

Roof drains are required for disposal of rainwater collected on roofs of buildings. Rainwater from sloped roofs is first collected in gutters, which, in turn, direct the water to vertical leaders. The leader inlets should preferably be installed with leaf screens to prevent clogging of the system. If the vertical leader is not connected to a storm sewer, then it should be ensured that water is released at least 3 feet away from the foundation.

Rainwater may be drained without the use of gutters from a sloping roof if it has an overhang of 3 feet or more. A system without gutters requires that ground surface where drainage water strikes the ground be treated with gravel. The ground must also slope away from the structure on all sides at a minimum of 2 to 5% to carry water immediately away from the foundation.

Flat Roof

For collection of rainwater from a flat roof, a gentle slope should be provided on the built-up roof so that the water can be carried to a vertical roof drain. Tables 23.1, 23.2, and 23.3 can be used for sizing gutters, drains, leaders, and horizontal storm-water pipes, respectively.

Storm sewers are typically used for carrying the rainwater discharged from the roof. These pipes normally have multiple storm-water collection points (catch basins or area basins). They are interconnected through a series of pipes that carry water to a main collector line. The main collector transports the water to a natural creek or other water bodies, if permitted by local code. Otherwise, a dry-well system has to be designed and constructed in accordance with local codes. Table 23.3 can also be used for sizing storm sewers.

Roofs in Cold Climates

Ice buildup on a sloped roof is a major problem in cold climatic conditions. This ice ridge on the edge of a roof is known as an ice dam. The formation of an ice dam is the result of a complex interaction between amount of heat loss from a house, snow cover, and outside air temperature. Snow will melt on the part of the roof through which there is a heat loss. As the melted snow reaches the portion of the roof that is below 32°F, it freezes and forms an ice dam (Figure 23.10). It grows in size as it is fed by more and more melting snow from above. Melted snow will back up behind the ice dam

TABLE 23.1

Sizing of Gutters

Slope of Gutter	Diameter of Gutter (inches)	Maximum Rainfall (inches per hour)				
		2	3	4	5	6
1/16-in	3	340	226	170	136	113
	4	720	480	360	288	240
	5	1,250	834	625	500	416
	6	1,920	—	960	768	640
	7	2,760	1,840	1,380	1,100	918
	8	3,980	2,655	1,990	1,590	1,325
	10	7,200	4,800	3,600	2,880	2,400
1/8-in	3	480	320	240	192	160
	4	1,020	681	510	408	340
	5	1,760	1,172	880	704	587
	6	2,720	1,815	1,360	1,085	905
	7	3,900	2,600	1,950	1,560	1,300
	8	5,600	3,740	2,880	2,240	1,870
	10	10,200	6,800	5,100	4,080	3,400
1/4-in	3	680	454	340	272	226
	4	1,440	960	720	576	480
	5	2,500	1,668	1,250	1,000	434
	6	3,840	2,560	1,920	1,536	1,280
	7	5,520	3,680	2,760	2,205	1,840
	8	7,960	5,310	3,980	3,180	2,655
	10	14,400	9,600	7,200	5,750	4,800
1/2-in	3	960	640	480	384	320
	4	2,040	1,360	1,020	816	680
	5	3,540	2,360	1,770	1,415	1,180
	6	5,540	3,695	2,770	2,220	1,850
	7	7,800	5,200	3,900	3,120	2,600
	8	11,200	7,460	5,600	4,480	3,730
	10	20,000	13,330	10,000	8,000	6,660

Note: Figures in the field indicate horizontal projected roof areas in square feet.

and remain as a liquid. This stagnant water will find cracks and openings on the exterior roof surface and will eventually find its way inside the building.

In order to prevent the buildup of ice dams, the ceiling/roof insulation has to be increased to cut down on heat loss by conduction. Some state codes require insulation with an R-value of 38 above the ceiling. The ceiling also has to be made airtight so that no warm air can flow from inside the house to the attic space.

Immediate actions for preventing the formation of ice dams include removal of snow from the roof using a roof rake. If an ice dam has already formed and it cannot be removed immediately, the least that can be done is to make channels through the ice dam so that the trapped water drains off. Hosing the dam with tap water will make the channels.

Drainage Below Grade

A sump is required to be installed in a building with a subsurface storm drain in the basement. Storm water received by the sump is pumped out, to be eventually discharged either to the storm sewer or to a drywell.

TABLE 23.2

Sizing of Roof Drains and Leaders

Rainfall (inches per hour)	Diameter of Drain or Leader (inches)					
	2	3	4	5	6	8
1	2,880	8,800	18,400	34,600	54,000	116,000
2	1,440	4,400	9,200	17,300	27,000	58,000
3	960	2,930	6,130	11,530	17,995	38,660
4	720	2,200	4,600	8,650	13,500	29,000
5	575	1,760	3,680	6,920	10,800	23,200
6	480	1,470	3,070	5,765	9,000	19,315
7	410	1,260	2,630	4,945	7,715	16,570
8	360	1,100	2,300	4,325	6,750	14,500
9	320	980	2,045	3,845	6,000	12,890
10	290	880	1,840	3,460	5,400	11,600
11	260	800	1,675	3,145	4,910	10,545
12	240	730	1,530	2,880	4,500	9,660

Note: Figures in the field indicate horizontal projected roof areas in square feet.

TABLE 23.3

Sizing of Horizontal Storm-Water Pipes

Slope of Pipe	Diameter of Pipe (inches)	Maximum Rainfall (inches per hour)				
		2	3	4	5	6
1/8-in	3	1,644	1,096	822	647	548
	4	3,760	2,506	1,880	1,504	1,253
	5	6,680	4,453	3,340	2,672	2,227
	6	10,700	7,133	5,350	4,280	3,566
	8	23,000	15,330	11,500	9,200	7,600
	10	41,400	27,600	20,700	16,580	13,800
	12	66,600	44,400	33,300	26,650	22,200
	15	109,000	72,800	59,500	47,600	39,650
1/4-in	3	2,320	1,546	1,160	928	773
	4	5,300	3,533	2,650	2,120	1,766
	5	9,440	6,293	4,720	3,776	3,146
	6	15,100	10,066	7,550	6,040	5,033
	8	32,600	21,733	16,300	13,040	10,866
	10	58,400	38,950	29,200	23,350	19,450
	12	94,000	62,600	47,000	37,600	31,350
	15	168,000	112,000	84,000	67,250	56,000
1/2-in	3	3,288	2,295	1,644	1,310	1,096
	4	7,520	5,010	3,760	3,010	2,500
	5	13,360	8,900	6,680	5,320	4,450
	6	21,400	13,700	10,700	8,580	7,140
	8	46,000	30,650	23,000	18,400	15,320
	10	82,800	55,200	41,400	33,150	27,600
	12	133,200	88,800	66,600	53,200	44,400
	15	238,000	158,800	119,000	95,300	79,250

Note: Figures in the field indicates horizontal projected roof areas in square feet.

23.4 **PARKING LOT**

DESIGN

A major component of building design and site improvement is to make provision for parking vehicles.

In order to organize parking efficiently and to provide adequate spaces, one must be aware of the dimensions of the vehicles.

Parking lot sizes are measured in numbers of vehicles to be accommodated. Lots that accommodate 100 to 200 cars are efficient and practical. For effi-

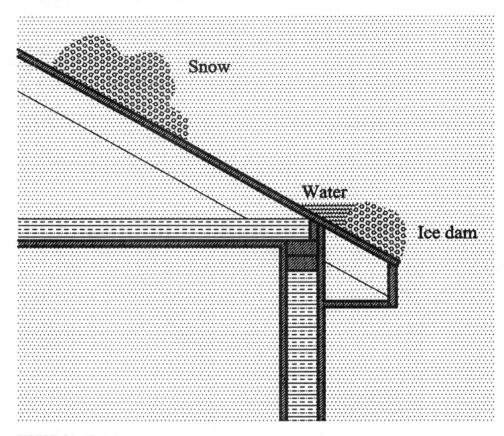

FIGURE 23.10 Ice Dam

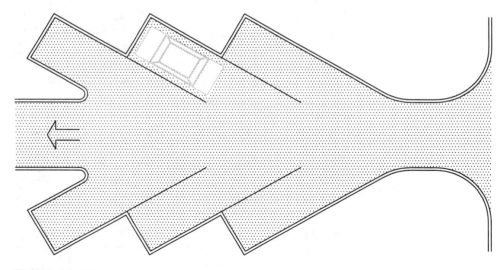

FIGURE 23.11 Parking at 30° Angle

cient land use, a self-parking lot should provide about 300 square feet for each vehicle space. The size and capacity of a parking lot, however, should be tailored to the actual requirements.

Maximum capacity of a parking lot can be achieved by developing parking modules running parallel to the longer dimension of the site. A parking module includes a drive path with parking on either side. Cars may be parked at angles ranging from 0 to 90 degrees (Figures 23.11–23.14). Width of a parking module becomes progressively smaller as the angle of parking decreases. Width of parking stall

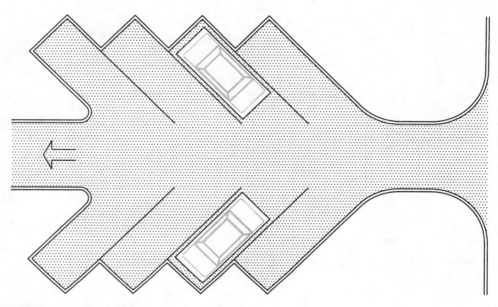

FIGURE 23.12 Parking at 45° Angle

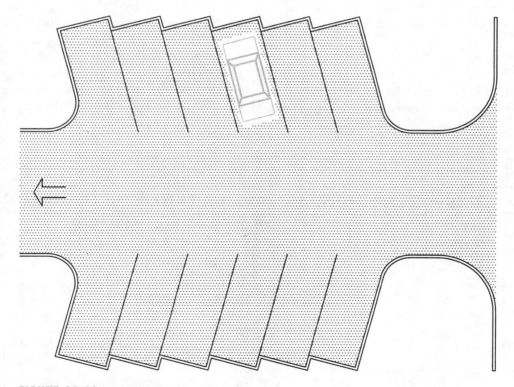

FIGURE 23.13 Parking at 75° Angle

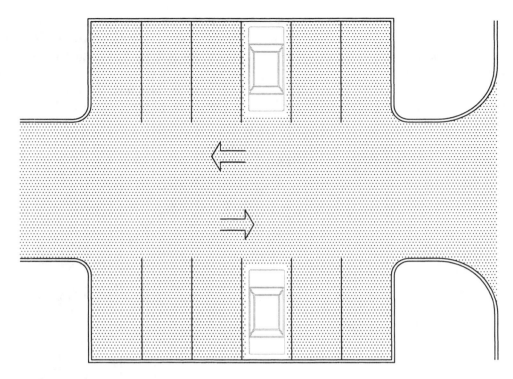

FIGURE 23.14 Parking at 90° Angle

sizes ranges from a minimum of 8′–6′ to a maximum of 12″–0″. Typical parking lot capacity figures at different angles of parking are given in Table 23.4.

DRAINAGE

Adequate drainage of parking lots is important for preventing damage or inconvenience to abutting property and/or public streets and alleys. Parking lots are usually required to be improved with a compacted gravel base surfaced with an all-weather material. The maximum desirable grade of a parking lot in any direction should not exceed 5%. Steeper slopes make it difficult to maneuver cars in slippery weather. Slope, however, should not be less than 1% in all directions to develop a positive drainage pattern.

TABLE 23.4

Maximum Capacity of Parking Lots at Different Parking Angles

Parking Angle (degrees)	Curb Length per Car (ft.)	Depth of Stall (ft.)	Width of Aisle (ft.)	Gross Area per Car (sq. ft.)	Unit Parking Depth (ft.)	Approx. Number of Cars per Acre
0	22	8	12	308	28	141
30	17	16.4	12	380.8	44.8	114
45	12	18.7	12	296.4	49.4	147
60	9.8	19.8	14.5	265.1	54.1	164
75	8.8	19.6	23	273.7	62.2	159
90	8.5	18	24	255	60	171

Some local codes require the use of storm-water retention systems for large parking lots. It helps to manage storm-water runoff and prevent down-stream flooding. Storm water can be retained in catch basins and eventually directed to the storm sewers.

REVIEW QUESTIONS

1. What are the basic objectives of a subsurface drainage system?
2. What is delay runoff?
3. Maximum recommended on-center distance between irrigation sprinklers is _____.
4. What is an ice dam?
5. Rainwater from a vertical leader should be released at least _____ feet away from the foundation of a building.

ANSWERS

1. Collection, transfer, and disposal of surface water at site.
2. It is the disposal of surface water at its source of runoff.
3. 1.2 times the radius of the spray head of a sprinkler.
4. It is an ice ridge on the edge of the roof of a building.
5. 3.

CHAPTER 24

Building Plumbing

A mountain village lost in
snow . . . under the drifts of a sound of water.

Shiki

This chapter deals with water demand, distribution, and drainage. Piping is needed for both water supply and drainage. The chapter elaborates on the different types of materials for each purpose and the fittings, fixtures, and other accessories required for efficient installation and functioning of the systems. Issues related to the sizing of supply and drainage pipes are discussed at length. The chapter ends with some thoughts on conservation and efficient use of water.

24.0 **WATER DEMAND**

The first step in selecting a suitable water supply source is deciding the demand that will be placed on it. Water demand refers to the quantity of potable water required by the inhabitants of a building. There are two different types of demand: average daily demand and peak demand.

Average daily water demand is the total quantity of water required by a building during the period when it is occupied. It depends on the functions served by the building, the types of fixtures used, and the total hours of operation of a facility. Table 24.1 presents a summary of average daily water demand by different types of establishments. An average pattern of home end uses in the United States ranges from about 4 gppd for drinking and cooking to 30 gppd for flushing toilets (refer to Table 22.2 in Chapter 22).

Peak demand refers to the rate of water use during critical periods of the day when water consumption is the highest in a building. This is also known as maximum momentary demand, or maxmo. This rate must be determined for the process of sizing water-supply piping. The total quantity of water used by a household may be distributed over only a few hours of the day, during which period actual water use is higher than the mean rate shown in Table 24.1.

HOT WATER DEMAND

End uses related to bathing, washing, laundry, and dishwashing require hot water. The average usage of hot water in the United States is assumed to be 20 gppd for a family of two persons. Another 15 gppd will have to be added for an increase in the number of family members. Hourly demand for domestic hot water is considered to be 2.1 gallons per person. Table 24.2 presents average usage rates of residential hot water.

Hot water is generated using a wide variety of devices and different types of energy sources. Water is defined as hot when it attains a temperature of 110°F or higher. Energy sources used for generating hot water include electricity, natural gas, oil, and solar energy. Water heaters that use electricity may be either the resistance or heat pump type. Heat pump water heaters use electricity to move rather than create heat. The heat source is the outside air or the air where the unit is located. Refrigerants and compressors transfer heat into an insulated storage tank.

Water heaters may either be storage type or instantaneous. Storage-type water heaters consist of an insulated storage tank and a device for heating. The heating device heats up a fixed quantity of water that is stored in the tank and used when required. An instantaneous water heater converts cold water instantly to hot water as it passes through it. The process continues as long as a fixture draws the hot water. It eliminates standby losses that occur in storage tanks and piping.

Water heaters are rated by their tank capacity in gallons and recovery rate in gallons per hour. The recovery rate is simply the number of gallons per hour that a water heater can produce. This rate is usually based on a cold water inlet temperature of about 50°F and a final temperature ranging from 110°F to 140°F. For a given load, the recovery rate of a water heater is inversely proportional to the hot water storage tank size.

The usable quantity of hot water available from a storage tank is usually assumed to be 70%, unless mentioned otherwise. This means that if the actual capacity of a water heater tank is 100 gallons, the quantity available at a desired temperature would be about 70 gallons. Table 24.3 presents a guideline for sizing residential water heater tanks, considering both household size and the hot water plumbing fixtures.

Water heating typically represents about 15 to 25% of the total household energy use in the United States. It is, therefore, important to ensure that the heater is operated using the most efficient energy source. Annual quantity of fuel required for operating a water heater can be calculated using the following formula:

$$F = Q/(V \cdot n) \qquad (1)$$

where F = total quantity of fuel required per year, Q = total quantity of heat required to produce hot water per year in British thermal units (BTU), V = heat content per unit (gallon, therm, or kWh) of the fuel, and n = percent efficiency of the water heater.

TABLE 24.1

Average Daily Water Demand

Type of Facility	Water Demand (in gallons per day)
Airport (per passenger)	3–5
Camps:	
Construction (per worker)	50
Day without meal (per camper)	15
Luxury (per camper)	100–150
Resorts, day and night, with limited plumbing (per camper)	50
Tourist with central bath and toilet facilities (per person)	35
Cottages with seasonal occupancy (per resident)	50
Country club (per resident member)	100
Country club (per nonresident member present)	25
Courts, tourist, with individual bath units (per person)	50
Dwellings:	
Boardinghouses (per boarder)	50
Luxury (per person)	100–150
Multifamily (per resident)	60
Rooming houses (per resident)	60
Single-family, indoor use (per resident)	70
Single-family, outdoor use (per resident)	70
Estates (per resident)	100–150
Factories (per person per shift)	15–35
Highway rest areas (per person)	5
Hotels with private baths (two persons per room)	60
Hotels without baths (per person)	50
Institutions (per person)	75–125
Hospitals (per bed)	250–400
Laundries, self-service (per customer)	50
Livestock (per animal):	
Cattle (drinking)	12
Dairy (drinking and servicing)	35
Goat (drinking)	2
Hog (drinking)	4
Horse (drinking)	12
Mule (drinking)	12
Sheep (drinking)	2
Steer (drinking)	12
Motels with bath, toilet, and kitchen facilities (per bed space)	50
Motels with bath and toilet (per bed space)	40
Parks:	
Overnight with flush toilets (per camper)	25
Trailers with individual baths, no sewer connection (per trailer)	25
Trailers with individual baths, connected to sewer (per person)	50

TABLE 24.1 (Continued)

Average Daily Water Demand

Type of Facility	Water Demand (in gallons per day)
Picnic facilities:	
With bathhouses, showers, and flush toilets (per person)	20
With toilet facilities only (per person)	10
Poultry:	
Chickens (per 100)	5–10
Turkeys (per 100)	10–18
Public baths (per bather)	10
Restaurants with toilet facilities (per patron)	7–10
Restaurants without toilet facilities (per patron)	2.5–3
Restaurants with bars and cocktail lounge (additional quantity per patron)	2
Schools:	
Boarding (per student)	75–100
Day, with cafeteria, gymnasiums, and showers (per student)	25
Day, with cafeteria, but no gymnasiums or showers (per student)	20
Day, without cafeteria, gymnasiums, or showers (per student)	15
Service stations (per vehicle)	10
Stores (per rest room)	400
Swimming pools (per swimmer)	10
Theaters, movie (per seat)	5
Workers, day (per person per shift)	15

Unit of measurement of heat is BTU. Specific heat of water (i.e., heat required to raise the temperature of one pound of water 1°F) is 1 BTU. Therefore, total quantity of heat required per year to produce hot water can be calculated using the following formula:

$$Q = G \cdot 8.33 \cdot \text{TD} \qquad (2)$$

where Q = BTU per year, G = gallons of hot water produced per year, and TD = temperature difference between cold supply water and hot water.

Total quantity of fuel required per year to produce hot water is calculated using the following formula:

$$E = Q/(F \cdot e) \qquad (3)$$

where E = quantity of fuel (in gallons, therm, or kWh), F = fuel heat content in BTU, and e = efficiency of the heater. Fuel heat content of different types of energy sources is shown in Table 24.4.

Once the total quantity of fuel required per year to produce hot water has been established, the total cost of operation can be calculated using the unit

TABLE 24.2

Residential Hot Water Consumption

Usage	Consumption (in gppd)
Bath	25
Shower (5 minutes)	6
Wash (using lavatory)	3
Dishwasher	10
Clothes washer	36
Kitchen sink	6

TABLE 24.3

Sizing of Domestic Water Heater Tank

Hot Water Load	Number
Size of household	Indicate total number persons in the family
Bathtubs or showers	Indicate total number of fixtures
Washing machines	Indicate total number of fixtures
Automatic dishwashers	Indicate total number of fixtures
Total load	**Sum of the above numbers**

Total Load	Capacity of Gas Water Heater Tank (in gallons)	Total Load	Capacity of Electric Water Heater Tank (in gallons)
3 or less	30	4 or less	40
4 or 5	40	5 to 7	60
6	50	8 or more	80
7 or more	60		

cost of the fuel type. The annual cost of production of hot water can be calculated using different types of fuel. This will form a simple basis for choosing the type of water heater to be installed according to fuel type.

The most economical choice of fuel is usually natural gas. Natural gas costs about two-thirds less per million BTUs than electricity. For locations where natural gas is not available, electricity or oil is commonly used for water heating. Initial cost of **heat pump water heaters is high, but the efficiency of such a unit is much higher than an electrical resistance water heater. Solar water heaters use a renewable resource. This system is the most economical way to produce hot water during the period**

TABLE 24.4

Fuel Heat Content

Fuel	Unit	Heat Content (in BTU)
Coal	Pound	14,000
Electricity	kWh	3,400
Gasoline	Gallon	125,000
Natural gas	MCF (1000 cu. ft.)	1,000,000
Oil #2	Gallon	140,000
Propane	Gallon	95,500

when it is abundantly available. Solar energy can meet most of the summertime demand for domestic hot water throughout the United States. The system needs to be protected against freezing in winter months, especially in the northern part of the country. Solar water heaters may be either active or passive. The active system relies on a pump or controller to move the water or collector fluid from the roof collector to a storage tank located in the house; in the passive system the storage tank rests directly upon the roof.

24.1 WATER DISTRIBUTION

MUNICIPAL STREET MAIN

Potable water is distributed in urban areas through municipal street mains. These are large pipes that usually run underground below the streets. Supplied by the water companies, the water flows under pressure that normally ranges between 50 and 70 pounds per square inch (psi) by the time it gets to a building. The supply main pipe of each building must be buried below frost line in cold climates so that it does not freeze. This depth can vary from about 2 to 7 feet depending on the climate.

The pressure of water supply to a building must be great enough to overcome the frictional resistance offered by the distribution system and the static pressure (static head) of water. Moreover, there must be adequate pressure left for the plumbing fixtures to operate properly. The water pressures in the municipal street main are usually adequate for supply to the highest and remotest fixtures in low-rise buildings.

Static Pressure

Static pressure (static head) is pressure exerted by water standing in vertical piping. The weight of a cubic foot of water is 62.4 pounds. Therefore, downward pressure exerted by a 1-square-inch column of water, 1 foot in height, is 0.433 psi (Figure 24.1). If the supply pressure in a municipal street main is 50 psi, it can withstand the pressure exerted by a column of water more than 115 feet in height. Considering the pressure necessary to operate the fixtures on the top floor of a building, a 50-psi pressure is powerful enough to supply water to fixtures more than 50 feet above the supply main level.

Upfeed Distribution

When water is fed to fixtures in a building by the incoming pressure of the water, it is called upfeed distribution. As stated earlier, this method is good for buildings up to five or six stories high. For upfeed distribution to work in taller buildings, additional pumps have to be installed to increase pressure. The augmented system is known as pumped upfeed distribution. This method is good for medium-size buildings that cannot rely on street main pressure.

Upfeed distribution for medium-size buildings can also be provided using hydro-pneumatic tanks. In this system, water is forced into hermetically sealed vessels, compressing the air within. This captive, compressed air maintains the required pressure and forces the water to flow into the distribution network.

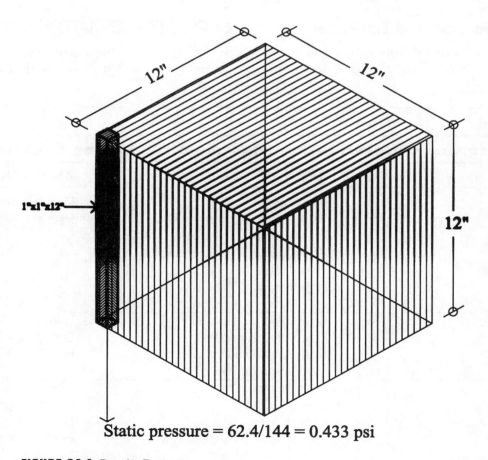

Static pressure = 62.4/144 = 0.433 psi

FIGURE 24.1 Static Pressure

Downfeed Distribution

Downfeed distribution systems may be designed for buildings more than six stories in height. In this system, water from a street main or suction tank is pumped to the roof of the building to storage tanks. The water from the storage tanks serves the floors below due to the force of gravity.

The water pressure required to serve the top floor of a building is usually 25 psi. Maximum pressure exerted on the fixtures of the bottom floor is recommended not to exceed 50 psi. A water pressure above 80 psi can damage the fixture valves. In order to keep the water pressure within proper limits, buildings higher than 15 stories in height are divided into more than one zone. Each zone should have a reservoir placed approximately 35 feet above the floor of the top level to produce a minimum pressure required to operate the fixtures. Use of pressure-reducing valves on the water main at bottom-floor level of each zone is recommended in order to avoid exceeding the maximum pressure limit.

Cross-Connection and Backflow

Discussions about water distribution would be incomplete without introducing the concepts of cross-connection and backflow. A cross-connection is any temporary or permanent connection between a public water system or consumer's potable water system and any source or system containing nonpotable water or other substances. It may occur due to backflow, which is an undesirable reversal of flow of nonpotable water or other substances into the potable water system. A reversal of flow in a building supply system can be created by any change of system pressure wherein the pressure at the supply point becomes lower than the pressure at the point of use.

Plumbing codes require the installation of cross-connection control devices. There are four distinct types of piping or mechanical assemblies, which are considered to be backflow prevention assemblies: (1) air gap, (2) atmospheric vacuum breaker, (3) pressure vacuum breaker, (4) double check valve assembly, and (5) reduced pressure principle assembly. Some of these devices are discussed in details under the Water Supply Accessories and Controls and Plumbing Fixtures sections in this chapter.

SUPPLY PIPING MATERIALS

Water pipes and fittings may be of brass, black steel, copper, galvanized steel, or plastic (see Table 24.5).

TABLE 24.5

Comparison of Different Pipe Materials for Water Supply

Material	Major Advantages	Major Disadvantages
Copper	Long lasting Easy to put together and dismantle Resists attacks by most acids Thin-walled Lightweight Low frictional resistance	Very expensive Requires soldering
Galvanized steel	Strong Relatively inexpensive Resistant to rough handling High pressure rating	Heavy Susceptible to corrosion High frictional resistance
Plastic	Inexpensive Lightweight Easy to install Very low frictional resistance Corrosion resistant	High thermal expansion Low strength Brittle when cold Easily scratched

However, the local plumbing code may specify the type of materials that may be used for each particular piping system.

Steel and Galvanized Steel

Steel pipe may be used for supply when water is non-corrosive. It is made from mild carbon steel as either welded or seamless pipe. Unprotected steel pipe rusts upon exposure to the atmosphere and moisture. In order to provide a protective coating, the steel pipe is dipped in a hot bath of molten zinc. This process is known as galvanizing. Nominal sizes of galvanized steel pipe range from 1/8 inch to 12 inch, in several wall thicknesses. The pipe wall thickness is usually described using the terms Schedule 40, for standard wall, and Schedule 80, for extra strong wall. Schedule 40 is normally used for typical plumbing applications.

Steel and galvanized steel pipes are connected with threaded fittings. For this reason, the pipes are normally manufactured with threads on both ends and supplied with a coupling threaded onto one end. Steel pipes in sizes larger than 4 inches are usually welded or connected by bolted flanges.

Steel and galvanized steel pipes tend to become clogged with corrosion over time. They are also more likely to develop leaks with aging. In addition, the rough interior walls of these pipes offer high frictional resistance to water supply. Because of these reasons, steel or galvanized steel pipes are seldom used in new construction today.

Copper

Copper pipe is substantially resistant to the chemical attack of many acids, salts, and bases. It is a durable pipe that handles high water pressure loads and is relatively easy to work with. However, copper is expensive and can cost much more than either steel or plastic pipe.

The pipes are manufactured from pure copper by drawing a heated billet through a die. There are four different types of copper pipe or tubing available according to wall thickness: K, L, M, and DWV. Type K tubing has the heaviest wall, and type DWV has the thinnest wall. Types K, L, and M are used for water supply while DWV is used for drainage. K and L types are available both as rigid and soft copper tubes, while M and DWV types are available only as rigid copper tubes. They are marked with color codes: green for type K, blue for type L, red for type M, and yellow for type DWV.

Copper pipes fit together with lead-free, solid-core solder using solder-type fittings. The soldering process involves heating the surfaces and then applying molten solder (usually a tin-antimony alloy) between the surfaces of the pipe and the fitting. Solder joints depend on capillary action drawing free-flowing molten solder into the narrow clearance between the fitting and the pipe. Molten solder metal is drawn into the joint by capillary action regardless of whether the solder flow is upward, downward, or horizontal. Solder type fittings are used above ground for both rigid and soft copper tubes. Shapes and patterns of these fittings are similar to those used for galvanized pipes.

Mechanical joints involving flared tube ends are frequently used for soft copper tubes or when the tubing is subject to vibration. The joint is made by slipping a flare-type nut over the end of the copper tube, flaring the end of the tube, and then screwing the flare nut onto a threaded fitting. A compression fitting may also be used to join copper tubes.

Plastic

Plastic pipes are available in different varieties. They are produced from synthetic resins derived from fossil fuels such as coal and petroleum. There are seven types of plastic: (1) polyvinyl chloride (PVC), (2) chlorinated polyvinyl chloride (CPVC), (3) acrylonitrile butadiene styrene (ABS), (4) polyethylene (PE), (5) styrene rubber (SR), (6) polybutylene (PB), and (7) polypropylene (PP). Out of these seven types, four are commonly used for plumbing pipes and fittings: (1) acrylonitrile butadiene styrene (ABS), (2) polyvinyl chloride (PVC), (3) chlorinated polyvinyl chloride (CPVC), and (4) polyethylene (PE). ABS is commonly used for drainage piping; PVC is used to distribute cold water (DWV-type PVC is also used for drainage piping); CPVC is used for distribution of hot and cold water; and PE is used for cold water supply piping, below ground, from a municipal water main to a building.

Plastic piping is lightweight, inexpensive, and resistant to corrosion and weathering. Smooth interior walls of plastic piping offer low frictional resistance to water supply. However, it has a low resistance to heat, a high coefficient of expansion, and low crush resistance. Pressure ratings of a plastic pipe are also lower than metal pipes. The plastic piping that can withstand higher temperature is CPVC. It is a rigid, high-strength thermoplastic polymer (polyvinyl dichloride) that is practically inert toward water, inorganic reagents, hydrocarbons, and alcohol over a broad temperature range. The CPVC pipe is suitable for carrying hot water limited to temperatures not greater than 180°F.

The fittings for plastic pipes are similar to those used for galvanized steel pipes. Pipes are joined with the fitting using a process called solvent welding. The material used for the purpose is known as solvent cement. It softens the piping material on the outside and the fitting material on the inside. When joined together under proper conditions, the surfaces run together and fuse, producing a joint that is as strong as the pipe itself.

Joints in PE pipes are made with tapered and notched inserts of either polystyrene or galvanized steel. After the insert has been slipped into the ends, the pipe is compressed against it by means of metal steel clamps.

Heat fusion (popularly known as butt fusion) is also used for joining PE pipes. The joint is made by melting the ends of the pipes and then holding the melted ends together, allowing them to cool. This is a very strong, reliable, leak-free, and seamless connection that provides the same strength as any other part of the pipe.

Threaded joints are also available for plastic pipes. They are usually used for Schedule 40 PVC pipes.

WATER SUPPLY ACCESSORIES AND CONTROLS

Valve

The valve is a fitting used on a piping system to control the flow of fluid within that system in one or more ways. It is desirable to install a valve to control individual fixtures, branch supply lines that serve bathrooms and kitchens, and every riser (i.e., vertical supply line). A gate valve is the most commonly used device that can obstruct the flow of water by means of a gate-like wedge disk fitted within the valve body (Figure 24.2). It is used in locations where the valve is required to be left completely open most of the time. The gate valve mainly performs shut-off duty. It is not intended for flow regulations.

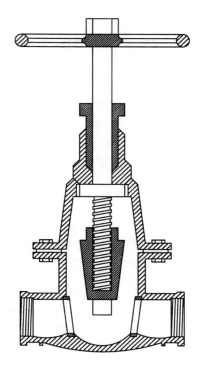

FIGURE 24.2 Gate Valve

A globe valve is installed when it is necessary to regulate the flow of water. It is a compression-type valve that controls the flow of water by means of a circular disk installed within the valve body (Figure 24.3). The globe valve usually has small ports, an "S" flow pattern, and relatively high pressure drop. Globe valves provide tight, dependable seals with minimum maintenance.

A check valve is a device that prevents the flow of water in a direction reverse to the normal flow. It is used to direct the flow of water in only one direction. Any reversal of flow closes the valve (Figure 24.4). The closure element in the valve can be internally spring-loaded to promote rapid and positive closure.

A double check valve assembly is suitable for protection only against low-hazard pollutants in supply water due to backflow. It consists of two internally loaded and independently operating check valves, along with two shut-off valves. The assembly is also equipped with pressure-test fittings to ensure the tightness of the valves.

In order to ensure against high-hazard contaminants in supply water due to backflow, a reduced pressure assembly has to be used. This is similar to a

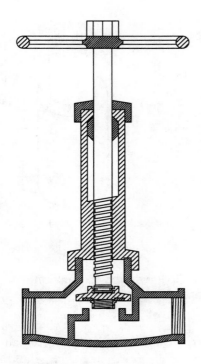

FIGURE 24.3 Globe Valve

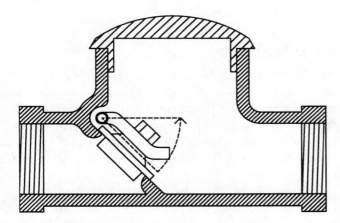

FIGURE 24.4 Check Valve

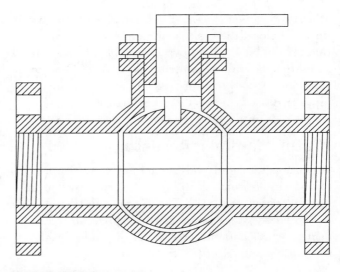

FIGURE 24.5 Ball Valve

the on position. The valve fits tightly against a resilient seat in its body. Primarily used as an isolation valve, it tends to be very reliable and trouble-free. It is usually used in pipes smaller than 3 inches in size.

A butterfly valve has a rotating disk that controls the water flow (Figure 24.6). Most butterfly valves have a single, round disk that rotates on an axle.

double check valve assembly with a mechanically independent, hydraulically dependent relief valve located between the check valves. The relief valve is designed to maintain a zone of reduced pressure between the two check valves at all times.

A ball valve controls the water by means of a rotating ball with a cylindrical hole through its center (Figure 24.5). When the hole is aligned with the water flow, the water flows freely through the valve with almost no friction loss. The flow is completely shut off when the ball is rotated 90 degrees from

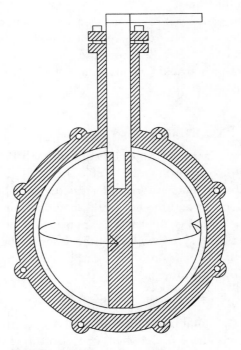

FIGURE 24.6 Butterfly Valve

When fully open, the disk is aligned with the water flow. To close, the disk is rotated at a right angle so that it fully blocks the flow. Butterfly valves are used as both isolation and control valves. They tend to be very reliable and trouble free and are used mostly on pipes that are 3 inches or larger in size.

Water Hammer Arrestor

When a water supply valve or a fixture in a supply system is closed quickly, the force exerted by the fast-flowing water causes the pipe to shake and rattle. This is known as water hammer. It can be controlled by using a water hammer arrestor at the end of each fixture in the system.

Insulation of Pipes

Pipes carrying cold water usually have a surface temperature of about 60°F. If the ambient air temperature rises to 85°F with a relative humidity of 40%, condensation will occur on the pipe surfaces. The higher the dry-bulb temperature and relative humidity, the more will be the condensation. Deterioration of finished surfaces will occur due to the dripping of the moisture from the supply pipes. All cold water pipes and fittings should be insulated to prevent condensation. Material commonly used for the purpose is fiberglass, 1/2 to 1 inch in thickness.

Hot water pipes should also be insulated using the same material to prevent heat loss to the surrounding air. Parallel hot and cold water pipes should be separated by a minimum of 6 inches to prevent heat interchange.

Pipe Expansion

The temperature of water flowing through hot water supply pipes can range from 60°F to 180°F. It causes a significant temperature difference to exist between the indoor air and the hot water in supply pipes. Depending on the coefficient of expansion of the material, the piping system will experience some elongation. The actual expansion of pipe can be calculated using the following formula:

$$\text{Total expansion in feet} = L_0 \cdot \Delta t \cdot \alpha \qquad (4)$$

where L_0 = total pipe length in feet, Δt = temperature difference, and α = coefficient of expansion of the piping material.

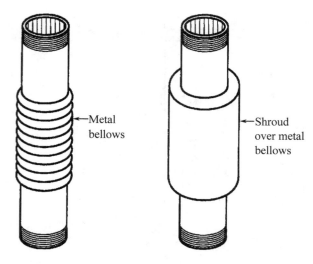

FIGURE 24.7 Bellows-type Expansion Compensator

Pipe expansion is negligible for small buildings but may be quite significant for large and tall buildings. Expansion joints, as illustrated in Figure 24.7, are used to prevent pipes from buckling as a result of expansion.

Pipe Support

One cubic foot of water weighs 62.4 pounds. Water supply pipes are therefore quite heavy and require adequate support. Different types of anchors and hangers are used to provide support to the pipes (Figure 24.8). Horizontal pipes should be supported at 10-foot intervals. Closer spacing of 8 to 10 feet is preferred for

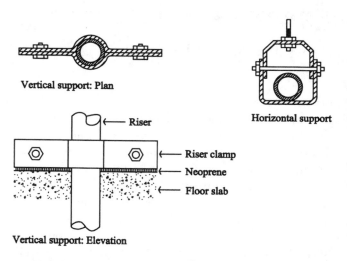

FIGURE 24.8 Pipe Hangers

metal pipes 1/2 inch and smaller in size. Plastic pipes 1 1/2 inch and smaller in size should be supported at 5-foot intervals. Vertical run of 1-inch pipes should be supported at every floor level; larger pipe sizes may be supported at every two floors.

PLUMBING FIXTURES

The plumbing fixtures are the visible part of a water supply system. They are the devices that receive water from a distribution system or that receive waterborne wastes and discharge them into the drainage system.

Minimum Requirements

All nationally recognized plumbing codes normally include the minimum number of plumbing fixtures required to be installed in a building. The number is based on the use of the building and its population. Table 25.1 in Chapter 25 lists such requirements.

General Fixture Characteristics

Fixtures are made of dense, impervious, and abrasion-resistant materials. Commonly used materials are enameled cast iron, stainless steel, fiberglass-reinforced plastic, and vitreous china. They are installed with fittings such as faucets or control valves, waste pipes, and other accessories as necessary.

Air Gap

The connection between a water supply component and the relevant plumbing fixture must have an unobstructed vertical separation through the free atmosphere. This separation is known as an **air gap** (Figure 24.9). The purpose of maintaining this air gap is to prevent the possibility of contaminating supply water with wastewater. It is measured between the lowest opening from any pipe or faucet conveying potable water and the **flood-level rim** of the fixture. Minimum air gap is affected by the size of opening of a faucet and the side walls of a fixture. It ranges from 1 to 3 inches.

Vacuum Breaker

A vacuum breaker is installed at the branch connection to a plumbing fixture that lacks an air gap (e.g., dishwasher) in order to prevent backsiphonage. If the

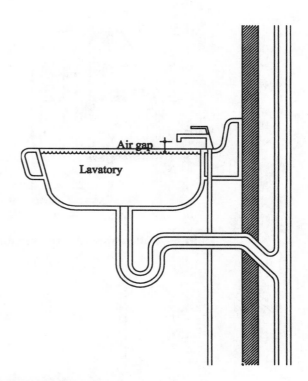

FIGURE 24.9 Air Gap

water pressure in a supply system drops below the atmospheric pressure level, then it is possible for the foul water in the fixture to be siphoned into the supply system. Under such conditions, the device will close the pipe automatically to prevent backflow.

A vacuum breaker may be either the atmospheric or pressure type. An atmospheric vacuum breaker consists of a checking member and an air inlet (Figure 24.10). A pressure vacuum breaker includes a check valve that is designed to close with the aid of a spring when flow stops (Figure 24.11).

Supply Fixture Units (SFU)

Demand for water by a plumbing fixture varies according to its type and the occupancy category (i.e., either public or private) of the building in which it is installed. An index called supply fixture unit is used for measuring the demand load of a plumbing fixture. Table 25.2 in Chapter 25 provides the supply fixture ratings of standard plumbing fixtures. The table also lists the minimum water pressure required for proper operation of those fixtures. The supply fixture unit for a separate maximum demand for either cold or hot water supply by a fixture is 75% of the total supply fixture unit of that fixture.

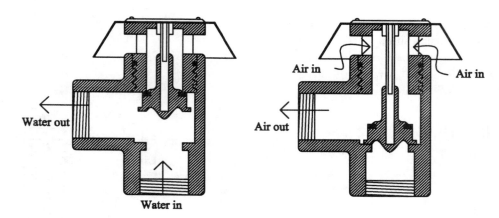

FIGURE 24.10 Atmospheric Vacuum Breaker

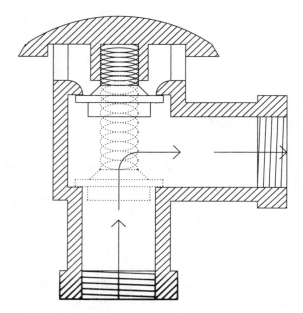

FIGURE 24.11 Pressure Vacuum Breaker

SIZING OF SUPPLY PIPES

Total Water Demand

In order to determine the size of water supply main to a building and the subsequent branch sizes, it is necessary to determine the maximum momentary load that the supply main should carry. This demand is calculated in terms of gallons of water per minute. Once the total supply fixture units for all the plumbing fixtures installed in a building have been calculated, the total water demand can be found out using one of the demand curves shown in Figure 25.4 of Chapter 25. These curves are based on a study by R. B. Hunter, who established a curvilinear correlation between the supply fixture units and the peak demand flow rate of the water supply distribution systems. These curves indicate that the demand rate does not increase proportionately to the increase in supply fixture units. This is because in most installations, the total number of plumbing fixtures is not expected to be used concurrently.

Curves labeled "valve" in Figure 25.4 (Chapter 25) are used for systems having predominantly flush-valve-operated water closets or urinals. Curves labeled "tank" are used for systems consisting of predominantly flush-tank-operated water closets or urinals. According to most building codes, all public buildings must use flush valves.

Water Pressure

Sufficient pressure must be maintained in the building water supply system to ensure adequate flow of water through the pipes and the fixtures. The pressure components in a building supply system are the following:

- **Water pressure in municipal supply main (P_{SM}):** This is the available pressure at the point in the municipal supply main from which it is connected to the building supply main. This pressure may range from 50 to 70 psi. The local water supply authorities should be contacted to get the correct figure. The sum of the pressure losses in the distribution system described hereafter should be equal to the water pressure in the municipal supply main.
- **Pressure required for fixture (P_F):** This is the pressure needed to operate the most remote fixture on the highest floor of an upfeed distribution system. The figure can be obtained by consulting Table 25.3 in Chapter 25.

- **Pressure lost due to height (static pressure) (P_{HT}):** This is the pressure required to overcome the loss when water travels vertically from the level of the municipal supply main to the highest fixture in the building. It is a product of the total vertical distance traversed by water and unit static pressure.
- **Pressure loss due to flow of water through meter (P_M):** The water supply to a building is routed through a meter in order to measure the quantity of water consumed. The loss in flow through meter is a function of the demand rate and the size of the meter. This value is estimated, usually based on a meter size of 2 inches for residences and small offices. After the supply main size has been determined, the value must be rechecked, and, if necessary, a recalculation should be made.
- **Pressure required for friction loss in piping (P_{FLH}):** Loss of pressure occurs because of the resistance offered to the flow of water by the pipe walls. It is also affected by the **total equivalent length** of the piping system. **Total equivalent length (TEL)** is the sum of **developed length (DL)** (i.e., the total linear distance of water travel from municipal main to the remotest and highest fixture) and **equivalent length (EL)** (i.e., length of the piping system equivalent to the fittings used for construction). The pipe length equivalent to the fittings is commonly estimated to be equal to 50% of the developed length. Therefore, the **TEL** of the supply piping of a building is equal to 1.5 times its **DL**. Total pressure loss in the distribution system is equal to the difference between municipal supply main pressure and pressure losses generated by other pressure components in the system. Unit pressure loss due to friction, also known as friction loss in head, is calculated per 100 feet of total equivalent length. The following formula is used for the calculation:

$$P_{FLH} = [P_{SM} - (P_F + P_{HT} + P_M)] \cdot 100/TEL \quad (5)$$

Sizing of Building Supply Main

Once the demand rate in gallons per minute and friction loss in head per 100 feet of pipe length have been determined, the size of the building supply main can be obtained using the flow graphs shown in Figures 25.5 and 25.6 of Chapter 25. The horizontal lines in the chart represent demand rate, the vertical lines represent friction loss in head, the diagonal lines rising from left to right represent pipe sizes, and the diagonal lines rising from right to left represent the velocity of water flowing through the pipes. Examples of how to size supply pipes are given in Chapter 25.

Water Velocity

Water flowing through supply pipes tends to produce noise due to friction. The higher the velocity, the greater the noise. Moving water can be heard within the pipes if the water velocity is higher than 10 feet per second. It may sometimes be necessary to install a higher pipe size in order to bring down the water velocity within this limit.

24.2 BUILDING DRAINAGE

BASIC ELEMENTS OF DRAINAGE SYSTEM

What comes in must go out; half of building plumbing is for getting rid of wastes. A drainage system consists of the least visible and least glamorous elements of a building's plumbing, but it is just as important as the supply system. It safely removes wastes for treatment and provides a critical barrier that keeps sewer gases, insects, and rodents from entering the building. Commonly known as DWV (drainage, waste, and venting), the system consists of drainage pipes, traps, and vents.

Drainage Pipes

All plumbing fixtures receiving water from the supply system must discharge the used water to the drainage system through soil or waste pipes. **Soil pipes** convey wastes containing **fecal matter**, while **waste pipes** convey discharge free of any fecal matter. Vertical pipes are called either **soil** or **waste stacks**.

Traps

One of the basic principles of drainage is that every plumbing fixture must be installed with a trap. It provides a water barrier against the infiltration of sewer gases into the building without materially affecting the flow of sewage or wastewater through it (Figure 24.12). The water barrier in a trap is termed the trap seal, and may be defined as the column of water retained between the crown weir and the dip of the trap (Figure 24.12). The depth of a trap seal should not exceed a minimum of 2 inches and a maximum of 4 inches.

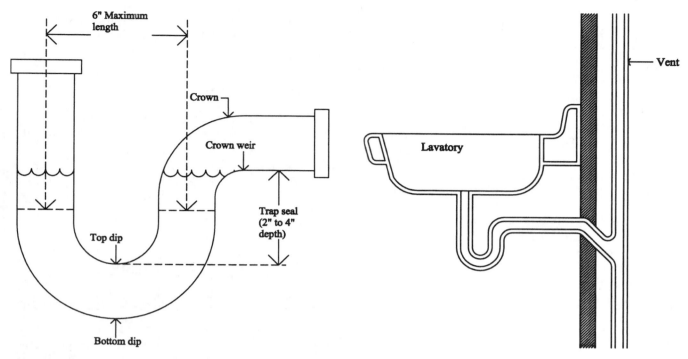

FIGURE 24.12 Fixture Trap

FIGURE 24.13 Fixture Vent

All plumbing fixtures must be equipped with a trap. They are usually placed within 2 feet of the fixture. The most common form of trap is constructed in the form of letter P; hence it is called a P-trap. All traps should be capable of being completely flushed each time a plumbing fixture is operated. Many are accessible for cleaning through a bottom opening closed by a plug.

Vents

A drainage system will not function properly without the installation of vent pipes. Vents provide free circulation of air within the drainage system. They equalize pressure to aid drainage and allow sewer gases to escape to the atmosphere. Without venting, high pressure in the drains may force sewer gas out through traps and toilets. Also, low pressure in the drains may cause siphoning in the traps whenever the fixtures are drained. If the trap seals are broken, sewer gas will enter directly into the house (Figure 24.13).

A vertical pipe installed to provide circulation of air to and from the drainage system is called a **vent stack**. When a soil or waste stack is extended above

the highest horizontal drain for a group of fixtures, it is called a **stack vent**.

In order to provide efficient ventilation to the drainage system, it is necessary to use individual or continuous vents. However, plumbing codes allow the use of circuit and loop vents, which do not provide an individual vent for every plumbing fixture.

A **circuit vent** serves more than two traps and extends from the front of the last fixture connection of a horizontal branch to the vent stack (Figure 24.14). This may be used on intermediate floors of a multistoried building. A circuit vent that serves more than three water closets should have a relief vent in front of the first fixture that is connected to the circuit.

A **loop vent** is similar to a circuit vent except that it loops back and connects with a stack vent instead of a vent stack. Both these types of vents are dependent on the air space at the top of fixture branch pipes for delivery and removal of air from individual plumbing fixture connections. Therefore, the relevant branch drains should be conservatively sized. Even though they are legal, the use of circuit and loop vents are not recommended in best practice.

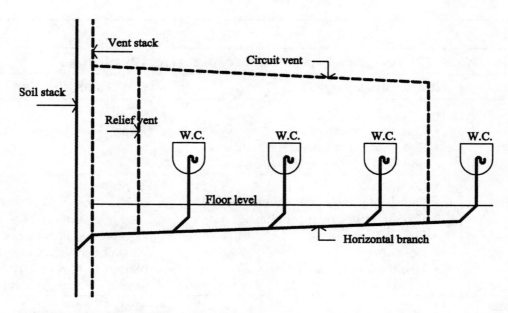

FIGURE 24.14 Circuit Vent Detail

PIPING AND FITTING MATERIALS

The principle materials used for piping different components of the drainage system are cast iron, copper, plastic (ABS and PVC DWV), and galvanized steel (Table 24.6). Extra-strength vitrified clay is allowed to be used for underground sewer connections. Joining methods of pipes and fittings must be appropriate for the material used. Codes limit the installation of approved plastic piping to three floors above grade. Galvanized steel drainage pipes are not allowed to be installed underground. Different types of joining methods are illustrated in Figure 24.15.

CHANGE OF DIRECTION

All direction changes in the flow line of a drainage system should be made with easy bends to avoid clogging of the pipes. Therefore, a branch drain should always be connected to a main drain at a 45° angle using a Y fitting or a combination of Y fitting and 1/8 bend (Figure 24.16). Right-angle connections are not used for the purpose.

CLEANOUTS

A **cleanout** is an accessible opening in a drainage system used for removal of obstructions. It is an essential part of the drainage system and should be provided in the following locations:

- at the outside wall of the building where the building drain connects to the house drain;
- at the base of all soil and waste stacks;
- at the upper terminal of all horizontal branch drains;
- every 50 feet on horizontal piping that is 4-inch size or smaller;
- every 100 feet on horizontal piping that is larger than 4 inches; and
- at all direction changes that are greater than 45°.

OTHER ACCESSORIES

Many other accessories are required for efficient operation and maintenance of a drainage system. Some of the common devices are as follows.

Floor Drain

This is a receptacle installed on the floor to receive water that is to be drained from the floor into the drainage system. A floor drain is provided with a strainer and a trap, and is suitably flanged to provide a watertight joint in the floor. It is necessary in areas where floors are washed after food preparation and cooking, mechanical rooms, and rest rooms.

Backwater Valve

This is a type of check valve installed to prevent the backflow of sewage from flooding the basement or

TABLE 24.6

Comparison of Different Pipe Materials for Drainage

Material	Major Advantages	Major Disadvantages
Cast iron	• Long lasting • Corrosion resistant • Suitable for both above-ground and underground applications	• Heavy • Rough inner surface
Copper	• Long lasting • Easy to put together and dismantle • Resists attack by most acids • Thin-walled • Lightweight • Low frictional resistance	• Very expensive • Time consuming to install • Does not work very well in underground service because of high conductivity of the material (only K or L type is allowed for underground use)
Galvanized steel	• Strong • Relatively inexpensive • Resistant to rough handling	• Heavy • Susceptible to corrosion • Susceptible to clogging • Not allowed for underground use
Plastic (ABS and PVC)	• Inexpensive • Lightweight • Easy to install • Very low frictional resistance • Corrosion resistant • Suitable for both above- and below-ground applications	• High thermal expansion • Low strength • Brittle when cold • Easily scratched • Not allowed to be used in buildings having more than three floors above grade • ABS pipes become brittle over time when exposed to sunlight

lower levels of a building. It is not capable of protecting the entire drainage system and should be used only when necessary.

Interceptor

This is a device installed to separate and retain deleterious or hazardous matter from normal wastes, while allowing the normal wastes to discharge into the drainage system by gravity. It includes devices to intercept hair, grease, plaster, lubricating oil, glass grindings, or such other unwanted materials. A common interceptor installed in a residential drainage system is the grease trap.

SIZING OF DRAINAGE PIPES

Drainage pipes (mains and branches) are sized using an index called the **drainage fixture unit (DFU)**. It is a measure of the probable discharge into the drainage system by various types of plumbing fixtures. The drainage fixture unit rating of a particular fixture depends on its rate of drainage discharge, on the time of duration of a single drainage operation, and on the average time between successive operations. One drainage fixture unit is approximately equal to 7 1/2 gallons of waste discharge per minute. DFU values of various plumbing fixtures are given in Table 25.3 in Chapter 25. Branch and main drainpipe sizes are selected directly from a table (see Tables 25.5, 25.6, and 25.7 in Chapter 25) in accor-

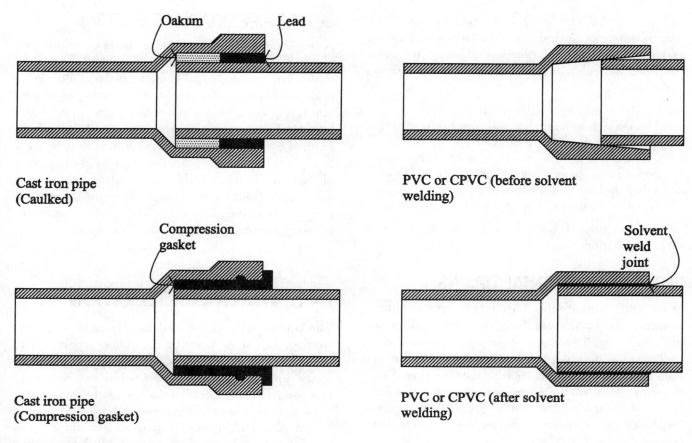

Oakum Lead

Cast iron pipe
(Caulked)

PVC or CPVC (before solvent
welding)

Compression
gasket

Solvent
weld
joint

Cast iron pipe
(Compression gasket)

PVC or CPVC (after solvent
welding)

FIGURE 24.15 Pipe Joints

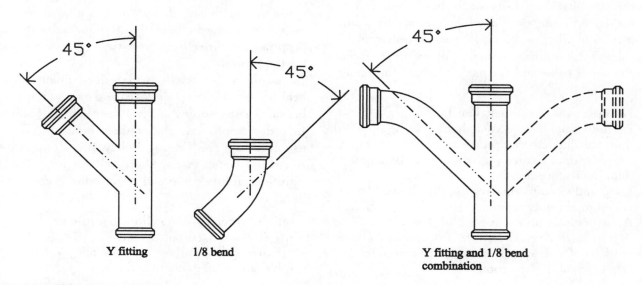

45°

45°

45°

Y fitting

1/8 bend

**Y fitting and 1/8 bend
combination**

FIGURE 24.16 Drainage Fittings

dance with the load in DFU carried by the drain. Size of the pipes for individual fixtures are the same as their respective trap sizes. Following are some supplemental rules and recommendations for sizing of drains:

- No drain may be smaller than 1 1/4 in.
- The maximum load on a 1 1/4-in. drain should not exceed one fixture unit.
- Codes allow a water closet to be connected to a 3-in. drain, but it is highly recommended to use a 4-in. minimum connection.
- A house drain may not be smaller than 4 in.

An example of how to size drainage pipes is given in Chapter 25.

SLOPE OF HORIZONTAL DRAINS

Water and waste products discharged from plumbing fixtures into horizontal drainpipes flow under the pull of gravity. Therefore, these pipes have to be installed at a certain slope. It is expressed as the fall per foot of horizontal pipe length. Factors that affect slope or grade of drainpipe are pipe size and connected drainage fixture unit. Table 25.6 in Chapter 25 shows the relationship between slope and these factors.

SIZING OF VENT PIPES

Vent pipes are also sized using drainage fixture units. The other factors that affect the size of a vent pipe are the size of soil or waste stack that is required to be vented and the developed length of the vent. Table 25.5 in Chapter 25 may be used to determine vent sizes for a drainage system. Following are some supplemental rules and recommendations for sizing of vents:

- No vent may be smaller than 1 1/4 in.
- An individual fixture vent should not be smaller than half the fixture drain size.
- An individual fixture vent should not be larger than the fixture drain.
- Loop and circuit vents are usually allowed by code but are not recommended in best practice.
- A water closet must have a 2-in. vent.
- A vent stack or stack vent that penetrates a roof must be of a size not less than 4 in. and must rise 12 in. above the roof.

An example of how to size vents is given in Chapter 25.

SINGLE-PIPE DRAINAGE SYSTEM

A single-pipe drainage system consists of only drainpipes. It is a specially engineered method, known as the **sovent system**, suitable for high-rise buildings. It is an all copper or plastic system consisting of an aerator fitting at each floor and a de-aerator fitting at the bottom of each stack. The effluent from the plumbing fixtures is mixed with air in the aerators to produce a foamy substance that lacks the stack-filling tendency of the liquid effluent. It does not produce pressures of more than 1-inch water gauge. Therefore, a normal trap seal is safe against being siphoned out or penetrated. It eliminates the need for vent pipes.

24.3 WATER CONSERVATION

The United States, in general, has abundant water resources. But some parts of the country, particularly the western region, have an emerging problem with water supplies adequate to meet the current consumption patterns. The rate of consumption of groundwater exceeds the rate at which it is replenished by recharge. Recent drought in some areas has accentuated the need to balance water demand with available supply. The future health and economic welfare of the nation's population are dependent upon a continuing supply of fresh, uncontaminated water.

One of the ways to maintain this balance is through water conservation. Simply stated, water conservation means doing the same with less, by using water more efficiently or reducing where appropriate, in order to protect the resource now and for the future.

Using water wisely will reduce pollution and health risks, lower water costs, and extend the useful life of existing supply and waste treatment facilities. In order to promote the conservation and efficient use of water, the Energy Policy and Conservation Act was enacted in the United States in 1992. The statute establishes national water conservation standards for:

- Showerheads—2.5 gallons per minute
- Toilets—1.6 gallons per flush
- Faucets—2.5 gallons per minute
- Urinals—1.0 gallon per flush

Over 42% of our indoor water use is for flushing water closets. The conventional toilet uses at least

3.5 gallons per flush. If low-flush toilets using 1.6 gallons per flush replace them, the overall indoor use is reduced by over 20%. The low-flush toilets are proven technology. They work efficiently and without problem.

The shower accounts for approximately 20% of indoor water use. The typical showerhead allows a water flow of 4 to 8 gallons per minute. Installing a low-flow showerhead with a flow rate of 2.5 gallons per minute will reduce consumption by one-half, yet most people will not notice the effects of the reduction.

Water-conserving faucet aerators are available in sizes ranging from approximately 0.5 to 2.5 gallons per minute. Low-flow aerators mix air with the water to make an effective spray pattern. Therefore, by installing a low-flow aerator, one can save a lot of water.

Washing clothes accounts for approximately 25% of residential water use. Front-loading horizontal-axis machines use one-third less water than top-loading vertical-access machines. The standard top-loader uses 35 to 55 gallons per load, whereas a front loader will use 25 to 30 gallons per load. As well as saving water, the front-loading machines also save energy. Front-loading machines, however, are still more expensive than the standard top-loading models in the United States.

REVIEW QUESTIONS

1. The recovery rate of a water heater is indicated to be 5 gph. What does it mean?
2. Water pressure in a municipal street main is 60 psi. How many stories (where floor-to-floor height is 10 feet) can a building have in order to receive water supply on all the floors? Assume that a minimum pressure of 25 psi is required to operate all the plumbing fixtures on the top floor.
3. What type of plastic piping would you recommend for hot water supply?
4. What type of copper tube has the heaviest wall?
5. What does a check valve do?
6. It is recommended that the supply pipes should be insulated. Why?
7. What is a trap?
8. What is the minimum recommended depth of a trap seal?
9. What is a vent stack?
10. What is a drainage fixture unit?

ANSWERS

1. It indicates that the water heater is capable of producing hot water, at a specified temperature, at the rate of 5 gallons per hour.
2. Pressure available to traverse vertical distance = 60 psi − 25 psi = 35 psi
Vertical distance traversed by water having a pressure of 35 psi = 35/0.433 = 80 feet
Number of floors of the building = 80/10 = 8
3. Polyvinyl dichloride
4. K-type
5. It prevents backflow in a water supply system.
6. To prevent condensation on pipe surfaces; to prevent heat exchange between supply water and the surrounding air.
7. It is a fitting, installed in a drainage system, that provides a water barrier against the infiltration of sewer gases into a building.
8. 2 inches
9. It is a vertical pipe that provides circulation of air to and from a drainage system.
10. It is an indication of the probable discharge of waste by a plumbing fixture into the drainage system.

CHAPTER 25
Plumbing Examples

hot water left~cold water to the right, shake hands judiciously

This chapter begins with a discussion of the sequential plumbing activities that meet a construction schedule. Architectural drawings locate plumbing fixtures with appropriate clearances for accessibility, and plumbing drawings detail the piping and connections needed to serve each fixture.

Next a series of tables and graphs explains the estimates and calculations involved in sizing a building's plumbing system. Complete the examples that follow each table or graph to develop your ability to number and size plumbing components.

Two example buildings, a 1,600-sq. ft. home and a 2-story 22,000-sq. ft. office building, are used to illustrate supply and DWV piping layouts. Plans and drawings define the components of plumbing installations for both examples, and high-rise plumbing system components are also discussed. Review these examples in detail, and then prepare plumbing drawings for a building you have designed. Be sure to verify fixture requirements and plumbing tables in the applicable *local* building and plumbing codes. Tables here have been developed from various codes and will NOT apply in all localities.

25.0 PLUMBING SEQUENCE

Three sequential plumbing activities must fit the construction schedule. Underground work begins after the building corners are set, and must be completed before work on the floor slab or floor framing begins. "Rough in" work begins after the wall framing and roof deck are in place, and must be completed before insulation and sheetrock work can start. "Set and finish" work can begin when interior sheetrock, tile, and base cabinets are complete.

UNDERGROUND

Piping below the ground floor (Figure 25.1) is installed and capped before placing concrete or setting floor joists. Drain lines are tested by filling the pipe with water and looking for leaks before backfilling trenches. A water test is done, after expelling all air, by subjecting the system to 10 feet of hydrostatic pressure for at least 15 minutes. All water distribution piping and fittings are pressure tested with air or water to reveal possible leaks. Tests have to be performed, obtaining water from a potable source of supply, under a pressure not less than the maximum working pressure for which the system has been designed. The piping has to withstand the test without leaking for a period of at least 15 minutes. Tests are not usually required for loop copper water supplies because all joints are made above the floor.

Drain piping below the slab falls 1/8" or 1/4" per foot, so long runs to the city sewer can require deep trenches. Flow velocities in horizontal drain pipes decrease with reductions in downward slopes. All site piping is kept at least 18" below grade, and water supply lines must be set below the frost line. Site trenches should be filled with sand prior to backfill to protect pipes and warn future excavators.

Natural gas piping is NOT run below the ground floor, for leaks could follow the pipe into the building.

ROUGH IN

When the roof deck and wall sheathing are complete, the plumbing crew "roughs" for fixtures and hose bibs. Water supply pipes and drainpipes are stubbed at the proper height for individual fixtures, and DWV trees are completed through the roof (Figure 25.2).

In high-rise buildings, DWV trees for each floor can be shop fabricated to minimize site labor. Horizontal vent branches are sloped to drain, and the plumbing contractor provides flashings for vents that penetrate the roof. Codes require at least one large VTR (vent through roof), and thicker walls are used

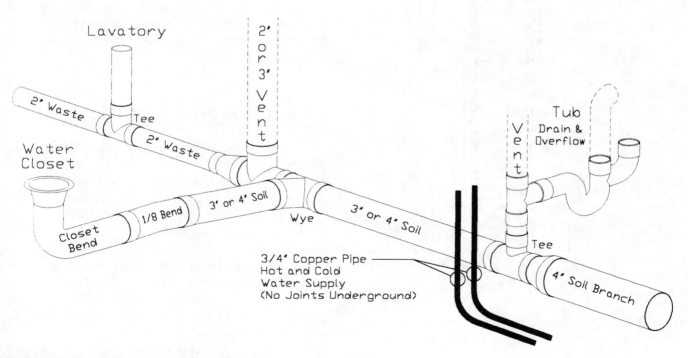

FIGURE 25.1 Underground Piping Schematic

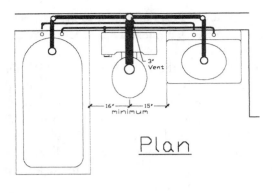

Plan

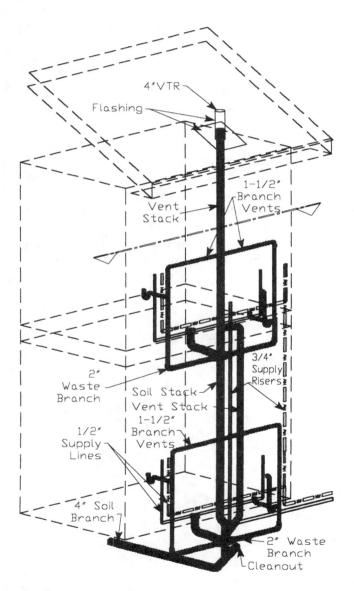

FIGURE 25.2 Plumbing Tree

behind water closets to enclose large vent and soil stacks. Wherever possible, small individual vents are connected to large vents in the wall or above the ceiling to minimize floor and roof penetrations.

Built-in fixtures like one-piece tub-showers are set now, but sinks and toilets are not installed until tile work, sheetrock, and counters are complete.

Water piping in vented roofs or attics is insulated to prevent freezing, and in very cold climates drain valves or freeze-proof hose bibs are specified. Black steel pipe is used for interior gas piping because natural gas reacts with galvanized steel.

When "rough in" is complete, piping is leak tested again, before building insulation and sheetrock work begins. DWV trees are capped and filled with water, hot and cold water supply lines are checked for leaks under pressure, and natural gas piping must maintain air pressure for 24 hours.

After "rough in," the building insulation contractor seals plumbing and electrical penetrations with expansive foam to minimize air infiltration.

SET AND FINISH

When interior walls, ceilings, and cabinets are complete, the plumbing crew returns to set toilets, sinks, water heaters, and other fixtures, and to install the valves and trim (Figure 25.3).

After final inspection, connections are completed to the city water and sewer mains, supply piping is flushed with chlorinated water, and all fixtures are checked for proper operation.

25.1 PLUMBING REQUIREMENTS

Local codes set minimum requirements for building plumbing systems. Most cities and towns adopt one of the National Codes and then modify it to suit particular local requirements. The International Association of Plumbing and Mechanical Officials (IAMPO) is currently updating the Uniform Plumbing Code and Uniform Mechanical Code through a consensus-based process in association with the National Fire Protection Association (NFPA).

Work the examples on each of the following pages to develop your understanding of plumbing system requirements.

Use Table 25.1 to work the following examples before checking your answers.

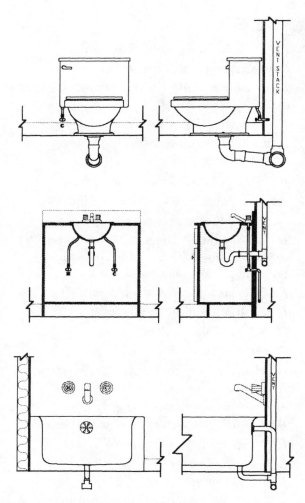

FIGURE 25.3 Setting Fixtures, Valves, and Trim

TABLE 25.1

Occupancy Estimates

WARNING! Use the following APPROXIMATE values ONLY for first estimates. Codes vary, and many exceptions apply. *Local codes govern.*

Floor Area In Square Feet per Occupant

Church–Theater–Restaurant–Stadium–Arena etc.
–fixed chairs—allow 1 occupant per seat
–fixed pews or benches—(allow 18" per seat)

–moveable chairs	7 sq. ft.
–moveable chairs and tables	15 sq. ft.
–standing room only	3 sq. ft.
Office	*150 sq. ft.
Retail ground floor and basement	30 sq. ft.
–other floors	60 sq. ft.
School–classroom	20 sq. ft.
Library–reading areas	*50 sq. ft.
–stack areas	100 sq. ft.
Hospital	*240 sq. ft.
Residential	*200 sq. ft.

Gross building floor area enclosed by exterior walls. Values without asterisk are *Net>* occupied room areas.

EXAMPLES

1. An office building has a gross floor area of 44,000 sq. ft. Estimate occupancy.
2. A church sanctuary has 200 pews, each 12' long. Estimate occupancy.
3. A 4,400-sq. ft. night club includes 4,000 sq. ft. with moveable chairs and tables, plus 400 sq. ft. of standing room. Estimate the number of occupants.
4. A school includes ten classrooms; each is 600 sq. ft. Estimate occupancy.
5. A retail store has a total floor area of 15,000 sq. ft. Estimate occupants.
6. A 10,000-sq. ft. library includes 4,000 sq. ft. of reading area and 6,000 sq. ft. of stacks. Estimate occupants.

ANSWERS

1. Office = $293 - 44{,}000 \div 150$
2. Church = $1{,}600 - (200)(12) \div 1.5$
3. Night club = $400 - (4{,}000 \div 15) + (400 \div 3)$
4. School = $300 - (600)(10) \div 20$
5. Grocery store = $500 - 15{,}000 \div 30$
6. Library = $140 - (4{,}000 \div 50) + (6{,}000 \div 100)$

HOW MANY FIXTURES?

Building codes use occupancy estimates to set the required minimum number of WC's (water closets), Lav's (lavatories), and DF's (drinking fountains). The following brief tables are satisfactory for preliminary estimates, but verify fixture requirements with the applicable *local* code authority before preparing contract drawings.

When the building occupants include equal numbers of males and females, codes allow for 60% of total occupancy by each gender.

EXAMPLE ESTIMATES

Estimate the number of water closets (WC's), lavatories (Lav's), urinals (UR's), and drinking fountains (DF's), required in a one-story 44,000-sq. ft. office building.

Solution

Estimate 293 building occupants (refer to preceding example #1). Estimate 60% men and 60% women, say 176 of each (if one gender dominates, adjust the percentages of males and females as appropriate).

- Eight WC's are required to serve 176 women (1 for the first 15, plus 6.44 for the next 161 @ 1 for 25 = 7.44; say 8).
- Eight more WC's are required to serve 176 men. Conserve space by replacing half these WC's with urinals.
- Eight lavatories are required in the building—half of the total WC requirement of 16.
- Three drinking fountains are required to serve 293 occupants at 1 DF per 100. In multistory buildings, codes also require at least one DF on *each floor*.

ANSWER

Minimum fixture requirements include:
12 WC's, 4 UR's, 8 Lav's, and 3 DF's.

MORE EXAMPLES

Use Table 25.2 to estimate the minimum number of WC's, UR's, Lav's, and DF's required to serve the following occupancies. Make your estimates before checking the solutions in the column at right.

1. A theater* seating 1,600
2. A 4,400-sq. ft. night club for 400
3. An elementary school classroom wing for 600 students
4. A store with a total of 500 customers and employees
5. A library for 140

---

*In some commercial and institutional occupancies like theaters and prisons, facilities for employees (actors or guards) must be calculated separately.

ANSWERS

1. A theater* seating 1,600 requires:
   estimate 960 of each sex (60%)
   WC—1 for first 50 plus 6 for the remaining 810, or 7 for each sex = 14.
   UR—replace 50% = 3 men's WC.
   LAV—half of required WC = 7.
   DF—1 per 100 = 16.
   WC = 11, UR = 3, LAV = 7, DF = 16

2. A night club for 400 occupants requires:
   estimate 240 of each sex (60%)
   WC—1 for first 40 men and women, plus 5 for the remaining 200 men, plus 10 for the remaining 200 women = 17.
   UR—replace up to 50% = 3 men's WC.
   LAV half of required WC = 9.
   DF—1 per 75 = 5.33, say 6.
   WC = 14, UR = 3, LAV = 9, DF = 6

3. An elementary school classroom wing for 600 students requires:
   estimate 360 boys and 360 girls
   WC—1 for 25 = 30, half for boys and half for girls.
   UR—replace 50% = 7 boys' WC.
   LAV—half of required WC = 15.
   DF—1 per 40 (600 ÷ 40) = 15
   WC = 23, UR = 6, LAV = 15, DF = 15

4. A store serving 500 requires:
   allow 300 men and 300 women (60%)
   WC—1 for first 15, plus 12 for the remaining 285 of each sex = 26.
   UR—replace up to 50% = 6 men's WC.
   LAV—half of required WC = 13.
   DF—1 per 100 = 5.
   WC = 20, UR = 5, LAV = 13, DF = 5

5. A library for 140 requires:
   estimate 84 men and 84 women (60%)
   use the table values for "college"
   WC—1 for each 40 men = 3, plus 1 for each 30 women = 3, total = 6.
   UR—replace 50% = 1 men's WC.
   LAV—half of required WC = 3.
   DF—1 per 100 = 1.4 say 2
   WC = 5, UR = 1, LAV = 3, DF = 2

TABLE 25.2

Fixtures Required (minimum)

Private	WC	Lav	Bath
Hotel/Motel–room	1	1	1
Hospital–private room	1	1	1
Residence	1	1	1

WARNING! Use the following APPROXIMATE values ONLY for first estimates. Codes vary and many exceptions apply. *Local codes govern.*

Public	WC*	Lav	DF
maximum number of occupants per fixture			
Church–theater			
–first 50	50	1/2 WC	100
–over 50	150	1/2 WC	100
Dormitory			
8 dorm occupants per shower			
Office & Retail			
–first 15	15	1/2 WC	100
–over 15	25	1/2 WC	100
Educational–nursery	15	1/2 WC	30
–elementary school	25	1/2 WC	40
–secondary school	30	1/2 WC	50
–college*	40/30*	1/2 WC	100
Hospital Ward	8	1/2 WC	100
20 patients/bath			
occupants per fixture: male–female			
College, male/female	40/30	1/2 WC	100
Restaurant			
–first 150, male or female	50	1/2 WC	200
–over 150, male/female	200/100	1/2 WC	200
Night club			
–first 40, male or female	40	1/2 WC	75
–over 40, male/female	40/20	1/2 WC	75
Stadiums & arenas			
–first 150, male or female	50	1/2 WC	200
–over 150, male/female	300/150	1/2 WC	200

*Up to 50% of required male WC's may be replaced one-for-one with urinals.

25.2 FIXTURE UNITS

Fixture units are numbers that indicate water quantity. An *SFU* (supply fixture unit) is a flow of about 1 GPM, and a *DFU* (drainage fixture unit) is a flow of about 0.5 gpm. Fixture units are used to determine pipe sizes.

TABLE 25.3

Fixture Units

Private	SFU	DFU	psi*
Bathroom group (gravity tank)	6	6	10
Bathroom group (pressure tank)	5	5	25
Bathroom group (flush valve)	8	8	25
Lavatory	1	1	10
Tub or shower	2	2	10
Water closet (gravity tank)	3	4	10
Water closet (pressure tank)	2	2	25
Water closet (flush valve)	6	6	25
Kitchen sink	2	2	10
Washer (clothes—8 lb.)	2	3	10
Dishwasher	1	2	10
Hose bib	4	—	10+

Public	SFU	DFU	psi
Lavatory	2	1	10
Tub or shower	4	2	10
Urinal (gravity tank)	3	2	10
Urinal (flush valve)	5	4	15
Water closet (gravity tank)	5	4	10
Water closet (pressure tank)	2	2	25
Water closet (flush valve)	10	6	25
Kitchen sink	4	3	10
Service sink	3	3	10
Drinking fountain	1/4	1/2	10
Hose bib	4	—	10+

psi = minimum fixture supply pressure

EXAMPLE

Use Table 25.3 to calculate SFU and DFU for a public building with 12 water closets, 4 urinals, 8 lavatories, and 3 drinking fountains.

ANSWER

Supply	SFU
Public, flush valve WC's—12 @ 10	= 120
Public, flush valve urinals—4 @ 5	= 20
Public lavatories 8 @ 2	= 16
DF's 3 @ 1/4 say	= 1
Hose bibs, allow for 6 @ 4	= 24
Total SFU	= 181

Drainage	DFU
Public, flush valve WC's—12 @ 6	= 72
Public, flush valve urinals—4 @ 4	= 16
Public lavatories 8 @ 1	= 8
DF's 3 @ 1/2, say	= 2
Total DFU	= 98

Answer: 181 SFU and 98 DFU

EXAMPLES

Calculate total SFU and DFU for the following:

1. A *hotel* wing with 400 private bathrooms using pressure tank water closets.
2. A 100-unit *apartment* with 25% 1 BR, 50% 2 BR, and 25% 3 BR. One-bedroom units have 1 bathroom, 2- and 3-bedroom units have 2 bathrooms, and all water closets have gravity tanks.
3. A *school* with 23 flush valve water closets, 6 urinals, 15 lavatories, 2 service sinks, 15 drinking fountains, and 6 hose bibs.

ANSWERS

1. The hotel has 2,000 SFU and 2,000 DFU.

Supply	SFU
Private bathroom group 400 @ 5	= 2,000

Drainage	DFU
Private bathroom group 400 @ 5	= 2,000

SFU and DFU for drinking fountains, hose bibs, service sinks, etc., would be added for a typical hotel wing.

2. The apartment has 1,666 SFU and 1,250 DFU.

Supply	SFU
175 bathroom groups @ 6	= 1,050
100 kitchen sinks @ 2	= 200
Washroom, 8 clothes washers @ 2	= 16
allow 100 hose bibs @ 4	= 400*
Total SFU	= 1,666

Drainage	DFU
200 bathroom groups @ 6	= 1,050
100 kitchen sinks @ 2	= 200
Total DFU	= 1,250

3. The school has 324 SFU and 191 DFU.

Supply	SFU
Public, flush valve WC's—23 @ 10	= 230
Public, flush valve urinals—6 @ 5	= 30
Public lavatories—15 @ 2	= 30
service sinks—2 @ 3	= 6
DF's 15 @ 1/4 say	= 4
hose bibs—6 @ 4	= 24
Total SFU	= 324

Drainage	SFU
Public, flush valve WC's—23 @ 6	= 138
Public, flush valve urinals—6 @ 4	= 24
Public lavatories—15 @ 1	= 15
Service sinks—2 @ 3	= 6
DF's 15 @ 1/2 say	= 8
Total SFU	= 191

*When a landscape sprinkler system is used, include its SFU and reduce the hose bib allowance (diversity).

SUPPLY GPM

The graphs in Figure 25.4 link total SFU to water supply demand in gallons per minute (supply GPM). Read solid-line curves for residential and commercial occupancies; use the dashed curves for large-assembly occupancies* and supply risers in multi-riser occupancies.

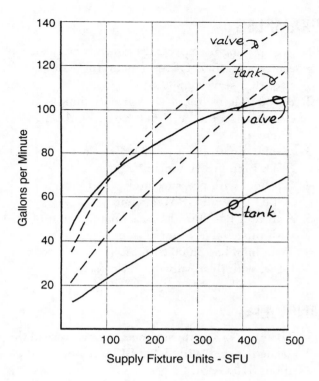

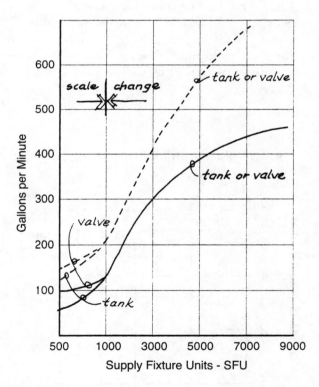

FIGURE 25.4

---

*Large-assembly examples: stadium, theater, arena.

EXAMPLES

1. A *hotel* wing uses pressure tank water closets in each room. Find GPM demand for a supply riser that serves 2,000 SFU.
2. A 100-unit *apartment* with gravity tank water closets has a total of 1,666 SFU. Find GPM demand.
3. A school with flush valve water closets has 324 SFU. Find GPM demand.
4. A school with pressure tank water closets has 324 SFU. Find GPM demand.
5. A theater has 400 SFU. Estimate demand GPM with flush valve water closets.
6. A stadium has 3,000 SFU. Estimate demand GPM with flush valve water closets.

ANSWERS

1. The hotel is a multiriser occupancy, so read the dashed curve. With 2,000 SFU, the demand is about 310 GPM.
2. Read the solid-line curve for this residential occupancy. With 1,666 SFU, the demand is about 200 GPM.
3. Read the solid-line curve for flush valve water closets. With 324 SFU, the demand is about 97 GPM.

4. Read the solid-line curve for flush tank water closets. With 324 SFU, the demand is about 50 GPM.
5. Read the solid-line curve for flush valve water closets. With 400 SFU, the demand is about 102 GPM.
6. Read the dashed curve for flush tank water closets. With 3,000 SFU, the demand is about 410 GPM.

25.3 SIZE SUPPLY

SIZE WATER SUPPLY PIPE

Gallons per minute (GPM) delivered by a main, branch, or riser depends on flow velocity in feet per second (fps). Faster flow means more GPM, more pressure loss due to friction (psi/100′), and more noise. Use Figure 25.5 to find flow velocity (fps) and friction loss (psi/100′) for a 1/2″ pipe delivering 4 GPM. Notice that if flow is cut 50% (to 2 GPM), velocity drops from 4 to 2 fps, but friction loss drops from 7 to 2 fps.

A building's water supply piping can be sized with flow velocities as high as 10 feet per second (fps), but to minimize noise and pressure losses due to friction this text assumes a very conservative maximum flow velocity of *6 feet per second* for in-building piping and 8 fps for underground piping.

Look at Figure 25.5 again; when the required flow is 4 GPM, 1/2″ or 3/4″ pipe could be used. While 1/2″ pipe is cheaper, 3/4″ pipe will make less noise when water is flowing and generate less pressure drop due to pipe friction. Using a maximum flow velocity of 6 fps, 1/2″ pipe would be specified.

EXAMPLES

Study Figure 25.5 for a moment and fill in the underlined values:

Size	GPM	fps	psi/100′
1/2″	2	?	?
3/4″	10	?	?
3/4″	?	?	3

Check your answers on the following page after you use Figure 25.6 to answer the following questions.

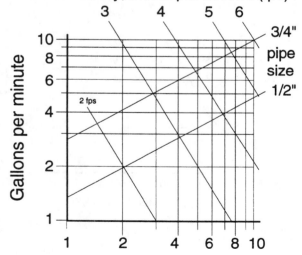

Flow velocity, in feet per second (fps)

Friction loss, in psi per 100' of pipe

FIGURE 25.5 GPM and Friction Loss

1. Select a riser to deliver 310 GPM in a large hotel. Limit flow velocity to 6 feet per second and check the friction loss.
2. Select a water main to deliver 200 GPM to a 100-unit apartment. Limit flow velocity to 8 feet per second and check the friction loss.
3. Select a water main to deliver 100 GPM to a school. Limit flow velocity to 8 feet per second and check the friction loss.
4. Select a water main to deliver 50 GPM to a store. Limit flow velocity to 8 feet per second and check the friction loss.

ANSWERS

Size	GPM	fps	psi/100"
1/2"	2	2	2
3/4"	10	6	10
3/4"	5	3	3

1. Select 5" pipe to deliver 310 GPM. Actual friction loss will be about 1 psi/100'. Many engineers would specify 4" riser for economy, even though flow will be noisy with a velocity of 8 fps.
2. Select 3" pipe to deliver 200 GPM. Actual friction loss will be about 4 psi/100', and velocity will be a little more than 8 fps.
3. Select 2½" pipe to deliver 100 GPM. *Actual* friction loss will be about 3 psi/100'.
4. Select 1½" pipe to deliver 50 GPM. Actual friction loss will be about 8 psi/100'.

SIZE RISERS AND BRANCHES

The building main is sized to meet expected demand for hot *and* cold water, but individual risers and branches are sized for hot *or* cold water demand. The actual friction loss (psi/100') for the pipe size selected for the main is used to size *all* branches and risers.

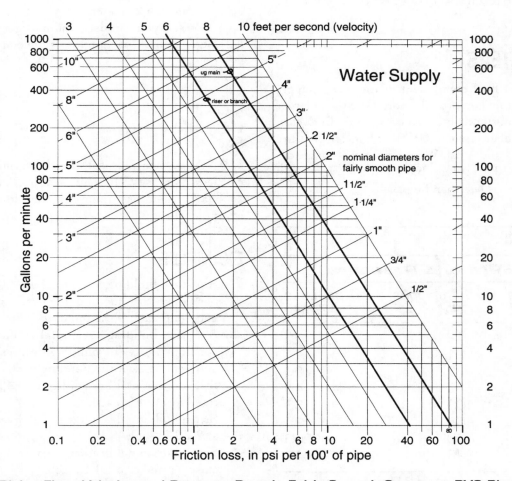

FIGURE 25.6 Piping Flow, Velocity, and Pressure Drop in Fairly Smooth Copper or PVC Pipe

Fixtures connected to both hot and cold water are rated at 75% for each—for example, a lavatory rated at 4 SFU is counted as 3 SFU for cold water and 3 SFU for hot water.

EXAMPLE

A gymnasium main calculation established a friction loss of 5 psi/100′ for all pipe runs. Size hot and cold supply branches for a locker room with 1 flush valve WC, 1 Urinal, 2 Lav's, and 4 Showers.

Fixtures	SFU	CW	HW
1 WC @ 10	10	10	
1 Urinals @ 5	5	5	
2 Lav's @ 4	8	6	6
5 Showers @ 4	20	15	15
		SFU 36	SFU 21
Supply GPM required.		51	12

See Figure 25.4; read CW from valve curve—HW from tank curve
Branch sizes @ 5 psi/100′—Figure 25.6 *CW 2″, HW 1″*

25.4 METERS

Meters are the final component of the water supply system. Larger meters provide more GPM, and larger meters add substantially to the system's first cost. Table 25.4 tabulates flow in GPM with meter pressure drops of 10 psi and 4 psi. Where city pressure is adequate, use the 10 psi values, but use the 4 psi values if supply pressure is less than 50 psi.

TABLE 25.4

Meter GPM and Pressure

Meter Size	10 psi loss GPM	4 psi Loss GPM
5/8″	12	8
3/4″	21	14
1″	33	20
1½″	63	40
2″	100	63
3″	200	125
4″	350	200
6″	625	440

EXAMPLES

1. Select a meter to deliver 60 GPM if a 10-psi pressure drop is acceptable.
2. Select a meter to deliver 60 GPM if low-supply pressure limits meter pressure drop to 4 psi.
3. A building with a flush valve system has a total of 350 SFU. Find the main size (Figures 25.4 and 25.6) if flow velocity is limited to 6 fps, and find the meter size (Table 25.4) if a 10-psi drop through the meter is acceptable.

ANSWERS

1. $1\frac{1}{2}$″
2. 2″
3. Select a $2\frac{1}{2}$″ main (from Figure 25.4 GPM is about 95, from Figure 25.6, @ 95 GPM select $2\frac{1}{2}$″ @ 6fps).

   Select a 2″ meter.

   (Yes, you can use a main larger than the meter to limit friction losses, velocity, and noise.)

25.5 SIZE DWV

DWV pipe size increases as the number of DFU carried increases. Review Tables 25.5 through 25.8 and then use them to complete the sizing examples.

EXAMPLES

1. A 5-story building has a total of 400 DFU (80 per floor). Size the building drain, soil stack, and vent stack if the building is 80′ tall.
2. A 4-story building has a total of 400 DFU (100 per floor). Size the building drain, soil stack, and vent stack if the building is 60′ tall.
3. Size a horizontal branch to carry 4 lavatories rated at 2 DFU each.
4. Size a building drain to carry 4 lavatories rated at 2 DFU each.
5. Size a horizontal branch to carry 3 water closets rated at 3 DFU each.

TABLE 25.5

DWV Minimums

Drains

Drainpipe size	1¼″
Branch size below the floor	2″

Allow 1/4″ per foot fall for waste and soil branches

Water closet outlet	4″

A 3″ drain can serve 2 WC's, but 4″ preferred

Vents

Vent pipe size	1¼″
Individual vent	1/2 size of trap served
Circuit vent	1/2 size of drain branch served
Vents > 40′ long	increase to next std. pipe size
Vents > 100′ long	increase another std. pipe size

Std. pipe sizes 1¼″–1½″–2″–2½″–3″–4″–5″–6″–8″–10″

TABLE 25.7

Soil and Waste Branches and Stacks

Maximum DFU for *Branches** & Stacks* condensed from the National Plumbing Code
BI = Branch Interval (each building story with drain branches is counted as one BI)

Pipe Size	Branch	Stack-3BI or Less	Stack Total	> 3BI Each BI
2″	6	10	24	6
3″	20*	48*	72*	20*
4″	160	240	500	90
6″	620	960	1900	350
8″	1400	2200	3600	600
10″	2500	3800	5600	1000

*Not more than 2 water closets
**Except branches of the building drain

TABLE 25.8

Vent Stacks

Maximum *Vent* length*—set by soil stack DFU condensed from the National Plumbing Code

Soil-Waste Stack (dfu)		Vent Pipe Size			
		3″	4″	6″	8″
3″	(60)	400′			
4″	(500)	180′	600′		
6″	(1900)	20′	70′	700′	
8″	(3600)		25′	250′	800′
10″	(5600)			60′	250′

*Field numbers are the maximum vent length in feet.

TABLE 25.6

Building Drains

Maximum DFU for *Building Drains* condensed from the National Plumbing Code

Pipe Size	Fall in inches per foot of run			
	1/16″	1/8″	1/4″	1/2″
2″			21	26
2½″			24	31
3″		36*	42*	50*
4″		180	216	250
6″		700	840	1,000
8″	1,400	1,600	1,920	2,300
10″	2,500	2,900	3,500	4,200

*Not more than 2 water closets

ANSWERS

1. Building drain = 6″ @ 1/8″ per foot fall, soil stack = 4″, vent stack = 3″.
2. Building drain = 6″ @ 1/8″ per foot fall, Soil stack = 6″, Vent stack = 4″.
   (Soil stack is 6″ because a 4″ stack will carry a maximum of 90 DFU per floor.)
3. 3″ branch

4. 2″ building drain @ 1/4″ fall if all lavatories are connected to the building drain, but 3′ if lavatory branch is not part of the building drain.

5. 4″ branch (A 3″ branch will carry 20 DFU, but not more than 2 water closets.)

25.6 **EXAMPLE RESIDENCE**

The example residence in Figure 25.7 is a 1,568-sq. ft. slab on grade structure used in companion texts to illustrate lighting, electrical, and HVAC work. Plumbing work for the house will consume 8 to 10% of the construction budget if municipal water and sewer are available. A municipal plumbing permit costs hundreds of dollars and usually includes inspections and connections to city water and sewer lines. On building sites outside the city limits, a well and septic system will add thousands of dollars to the project budget.

FIXTURE LOCATION

Codes require a minimum of one bathroom and one kitchen sink in each residence. The example will include plumbing fixtures and connections for two bathrooms, kitchen, laundry, and exterior hose bibs (Figure 25.8). Where handicapped access is required, designers consult Uniform Federal Accessibility Standards for dimensions, clearances, and grab bar requirements. Experienced designers locate fixtures on interior partition walls so that DWV trees don't interfere with foundations, structure, and fascia, but the example house has a window above the kitchen sink, so wall framing there must be reinforced.

PLUMBING PLAN

Complete plumbing plans and isometrics are typical in commercial construction, but for many residential projects the plumber develops a piping layout working from the fixture locations shown on the architectural plans. Plumbers consult the city inspector to find water and sewer lines that will serve the home, and then locate site trenches offering the shortest route to avoid existing tree roots and driveways (Figure 25.9).

UNDERGROUND

Fixture waste and soil connection points are carefully set to meet the architectural plans. Waste and soil pipe runs fall 1/4″ per foot as they dodge foundations, grade beams, and piers (Figure 25.10). Floor drains must be set to elevation and anchored before placing concrete (2″ floor drains must be vented within 6 ft. of the trap).

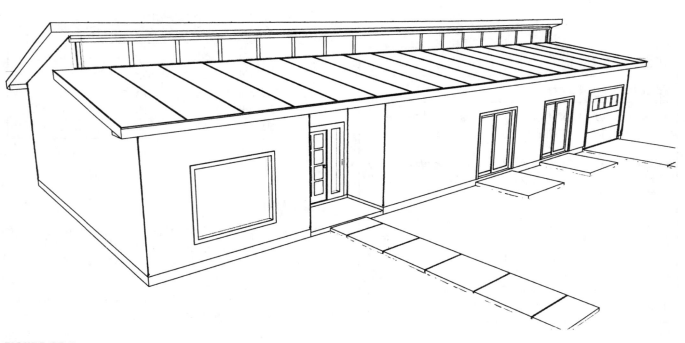

FIGURE 25.7

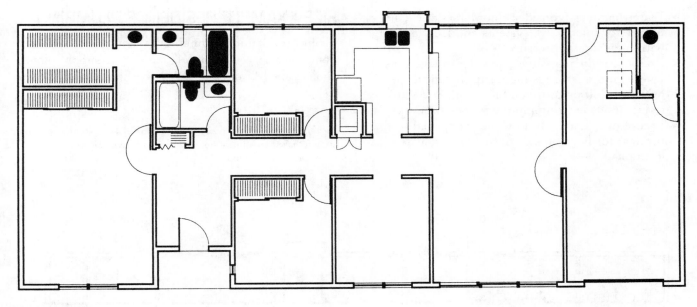

FIGURE 25.8

All joints below grade are leak tested before backfilling. Underground hot and cold water supply piping is often run with soft temper copper loops because no fittings are required below the floor. Supply loops rise along the center lines of future wall framing, where the plumber sleeves, seals, and identifies each loop.

ROUGH IN

The plumbing crew begins "rough in" when the slab, roof deck, and exterior sheathing are complete. Water supply and drain pipes are stubbed at the proper height for individual fixtures, and DWV trees are completed through the roof. A 3″ VTR (vent through roof) located in a 2 × 6 stud partition wall serves the bathroom DWV tree; vent flashing is usually provided by the plumber. Smaller VTR trees serve the kitchen and laundry (Figure 25.11).

Two bathtubs are set in alcoves faced with water-resistant sheetrock. Access panels for tub-shower valves are located in the entry closet and behind the medicine cabinet and below the lavatory in the guest bath.

A laundry supply-waste box is installed in the wall behind the washer, and water lines are run up the laundry wall to serve the washer and the water heater. Pipe stub-outs at each fixture are securely anchored to permit installation of threaded fittings for fixture service valves.

Black steel natural gas outlets are stubbed at the water heater and the furnace, and all piping is leak tested again before insulation and sheetrock work begins.

SET AND FINISH

When interior tile, walls, ceilings, and cabinets are complete, the plumbing crew returns to set toilets, sinks, hose bibs, and the water heater. Valves, trim, and the garbage disposer are installed, and a final inspection is scheduled.

City personnel usually make or supervise connections to city water and sewer mains, supply piping is flushed with chlorinated water, and all fixtures are checked for proper operation.

CHECKLIST

The following items are usually covered in the plumbing specifications.

* Condensate drain for AC unit(s)
* Drain pan under AC (if in attic)
* Drain pan under water heater (in attic)
* Water heater pressure relief valve (check local codes — some codes require PRV discharge outdoors; other codes allow discharge via drain line)

- Water hammer (consider eliminators at each fixture for flush valve systems and locations with high supply pressure)
- Pressure reducing valve (consider in locations where city pressure is high; supply landscape outlets at full pressure and locate the pressure reducing valve to supply the indoor fixtures)
- Freeze protection (drain valves or special fittings) for hose bibs or fixtures in unheated locations

SIZE EXAMPLE RESIDENCE PLUMBING

Sizing residential plumbing is not difficult. Most homes are served by a 3/4″ meter,* so supply piping choices are 1/2″ or 3/4″. DWV size choices range from

---

*5/8″ meters are used by some water supply districts, and 1″ meters are purchased by affluent homeowners with swimming pools and lawn sprinkler systems.

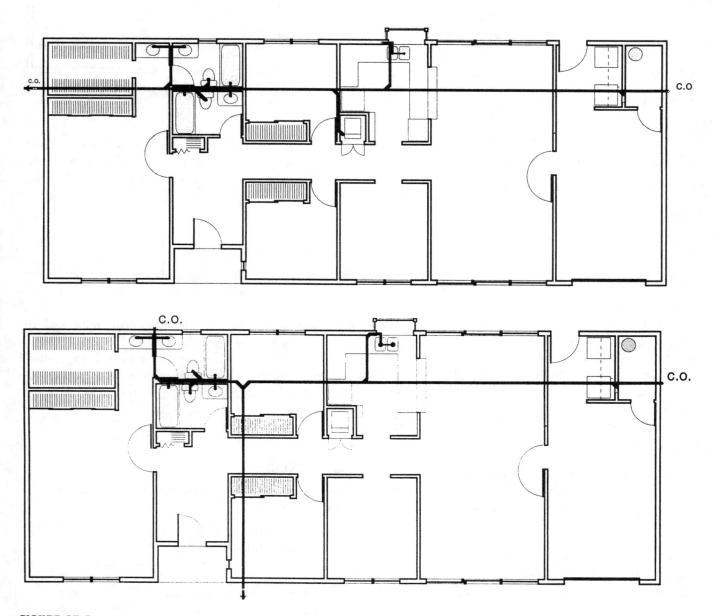

FIGURE 25.9

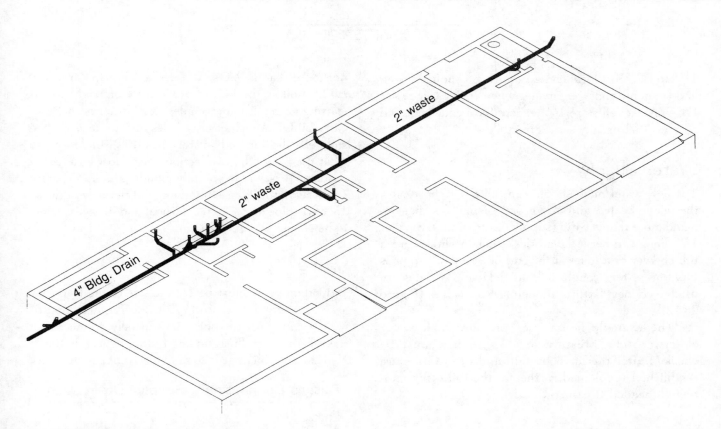

FIGURE 25.10 Underground Drains

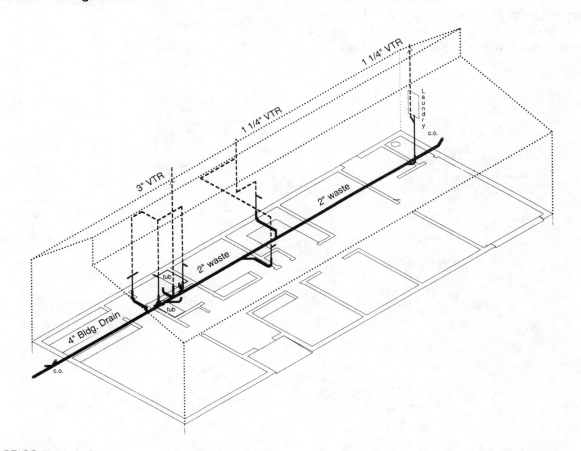

FIGURE 25.11 Rough In

$1\frac{1}{4}''$ to 4″. Most local codes require a 4″ house sewer, and branch drains get smaller when they carry fewer DFU. Most vents are $1\frac{1}{4}''$ except for a code-required 3″ VTR that usually serves the toilets.

Water Supply

The 3/4″ water main should run full size to hose bibs, the water heater, and any group of fixtures that demand more than 4 SFU (supply fixture units). Smaller 1/2″ lines can serve lavatories, kitchen sinks, and a tub-shower or a toilet at the end of a run. 3/8″ pipe is NOT used for "rough in," but flexible 3/8″ lines are used to connect fixtures to their service valves (Figure 25.12).

The example house uses pressure tank water closets, so the pressure at these toilets must be checked after the main friction and head losses are established (explained further in the following sub-section headed Pressure Check).

Waste and Soil

In a single-story residence, all waste and soil lines are part of the building drain. A 2″ drain will carry 20 DFU (drain fixture units); 3″ will carry 30 DFU but not more than 2 WC's; and a 4″ drain can carry 160 DFU. Follow the house sewer line from right to left starting at the 2″ laundry inlet. A 2″ drain branch can carry 20 DFU, so it easily carries the laundry, kitchen sink, tub-shower, and lavatory (8 DFU). Increase it to 4″ at the first toilet connection since most local codes require a 4″ sewer beyond the building line. A 3″ drain could serve this house, but the cost difference between 15′ of 3″ pipe and 4″ pipe is trivial and 4″ is less likely to be clogged (Figure 25.13).

Vents

Individual vents must be 1/2 the size of the trap they serve but not less than $1\frac{1}{4}''$. Code requires one 3″ VTR (vent through roof) that's usually connected at the toilet outlets. Surrounding fixture vents join the 3″ VTR in the wall or attic to minimize roof penetrations.

Fixture Connections	Supply	Drain	Vent
Lavatory	1/2″	$1\frac{1}{4}''$	$1\frac{1}{4}''$
Laundry box	1/2″	2″	$1\frac{1}{4}''$
Tub–Shower	1/2″	$1\frac{1}{2}''$	$1\frac{1}{4}''$
Water Closet(s)	1/2″	4″	3″*

*One 3 vent required by code

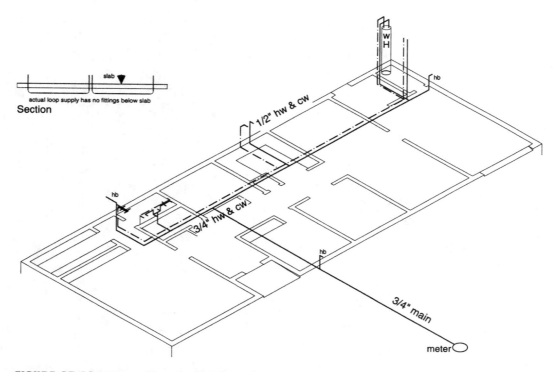

FIGURE 25.12 Water Supply Schematic

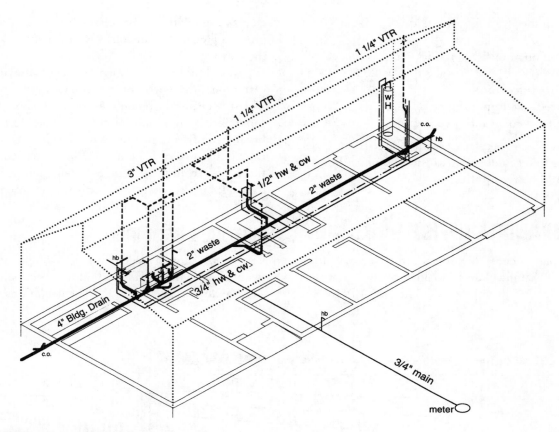

FIGURE 25.13 DWV and Supply

PRESSURE CHECK

Pressure tank toilets specified for the example residence need 25 psi to flush effectively. Pressure losses between the municipal water supply and the toilet inlet must be tabulated to verify adequate residual pressure (Figure 25.14).

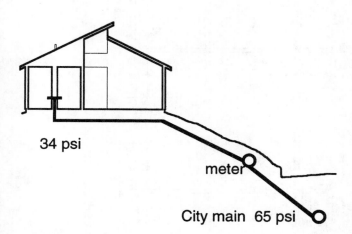

34 psi

meter

City main 65 psi

FIGURE 25.14 Pressure Losses

Given: The pipe run from city main to most distant toilet measures 80′, but allow 50% extra for friction losses in the pipe fittings. Total friction losses will be calculated for a 120′ run.

The City main flows @ 65 psi, and its elevation is 9′ below the pressure tank toilet inlets.

From sections 4.3 and 4.6:
Maximum home demand—about   14 gpm
(15 SFU excluding hose bibs*—Table 25.4)
Friction pressure loss—about   19 psi/100′
(3/4″ pipe,19 psi/100′ @ 14 GPM—Table 25.6)

Calculate pressure losses:
head (9)(0.434) = 3.9, say 4   4 psi
(meter elevation is 9′ below toilet tank)
meter   4 psi

---

*House fixtures total 15 SFU plus 16 SFU for hose bibs. If you calculated the house main to carry 31 SFU (17 GPM), you would select a 1″ main and pay more for a 1″ meter (or do a challenging calculation for flow through a 3/4″ meter with a 1″ main, or schedule hose bib use when your showers and laundry are inactive).

(3/4″ meter @ 14 GPM—allow 4 psi; see Table 25.7).

pipe friction (19 psi/100′)(120′) = 22.8   <u>23 psi</u>
(3/4″ pipe @ 14 GPM—see Figure 25.6)

Total pressure loss	31 psi
Required toilet operating pressure	25 psi
Pressure at toilet inlets (65 − 31)	34 psi

The available pressure is adequate. If the pressure at the toilet inlet was less than 25 psi, a booster pump would be required.

25.7 EXAMPLE OFFICE BUILDING

The example office is a 21,600-sq. ft. two-story structure used in Volumes 1 and 2 to illustrate lighting, electrical and HVAC installations (Figure 25.15). Where city water and sewer connections are available, plumbing costs for this office will consume 4 to 6% of the construction budget.

Most building plumbing is concentrated in the four rest rooms adjacent to the two-story entry lobby (Figure 25.16). Men's and women's rest rooms on both floors are separated by the main circulation corridors, and the kitchens and drinking fountains (not shown) are located near individual office entries. Pipe sizing calculations that follow include allowances for all building plumbing fixtures, but drawings detail only the rest room area.

FIXTURES REQUIRED

Table 25.1 shows 1 occupant per 150 sq. ft., or 144 people, but allow for a total building occupancy of 160 as estimated in previous HVAC calculations.

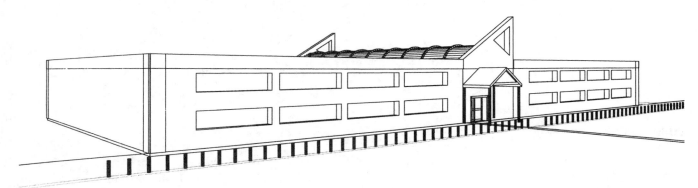

FIGURE 25.15

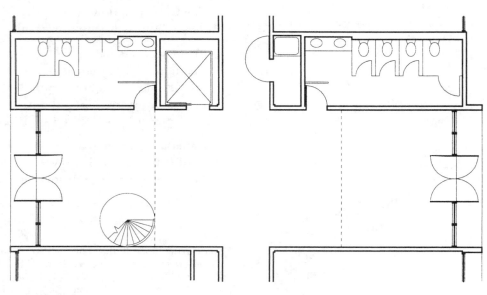

FIGURE 25.16

Minimum fixtures required to serve the building population are 5 WC's and 3 Lav's for women, 3 WC's, 2 UR's, and 3 Lav's for men, and 2 DF's. These fixture requirements are developed from Table 25.2, allowing for 96 women and 96 men (60% of 160).

The building owner orders more fixtures—12 WC's, 8 Lav's, 4 UR's, 4 Kitchen Sinks, 2 Service Sinks, and 4 DF's—to accommodate occupants of a planned future addition (Table 25.9).

Plumbing the office requires three separate labor activities: underground work, rough in, and set and finish. These activities were described in detail for the example house on preceding pages, so only differences for the office are covered.

Underground

Underground DWV piping is similar to the example house whether the first floor is slab on grade construction or spans over a crawl space. If a full basement is used, a sump pump and a sewage ejector pump may be required depending on the elevation of the city sewer.

The water supply main will enter the building below the rest room plumbing walls. Two risers, sized for quiet flow, will serve the rest rooms unless the building is sprinklered. Mains for a sprinklered building are sized for fire protection; they enter the building at a location where firefighters can access control valves, and the smaller domestic water main is connected there.

TABLE 25.9

SFU and DFU from Table 25.3 (fixture requirement)

Fixture	SFU	DFU	psi
WC's, flush valve (12)	120	72	25
Lav's (8)	16	8	10
UR's, flush valve (4)	20	16	15
DF's (4)	1	2	10
Kitchen Sinks (4)	16	12	10
Service Sinks (2)	6	6	10
Hose Bibs (6)	*	—	10+
Landscape water	*	—	10+
totals		179	116

*SFUs are not counted for hose bibs or the building's landscape watering system because their operation will be scheduled when fixture water demand is low.

Rough In

DWV trees repeat on each floor, so separate soil and vent stacks are required. The vent stack connects to the soil stack above the highest vent branch, and in high-rise buildings relief vents connect vent branches and soil branches on all except the top floor.

Supply branches run horizontally in the rest room walls, and piping that serves remote fixtures is usually run above the ceiling. Natural gas piping in multistory commercial buildings is run in fire-resistant vertical chases with natural ventilation openings above the roof.

Set and Finish

When interior walls, ceilings and cabinets are complete, the plumbing crew returns to set toilets, sinks, water heater, and the like to install the valves and trim.

Supply piping is flushed with chlorinated water, all fixtures are checked for proper operation, and after final inspection, connections to the city water and sewer mains are completed.

PLUMBING PLAN AND ISOMETRIC

The plan and isometric shown in Figure 25.17 and Figure 25.18 define the plumbing work in the example office rest rooms. The plan illustrates ground floor piping and the isometric details pipe sizes. Pipe size calculations can be found on the following pages.

DWV

The building drain runs from one end of the building to the other. It carries waste from the kitchens and drinking fountains (not shown) and picks up the rest room drains below the lobby floor. Building drain DFUs and sizes are shown on the plan; stack and branch piping sizes are on the isometric. Circuit vents are shown in the women's rooms, and individual vents are illustrated in the men's rooms. Some local codes prohibit circuit venting.

Supply

The building supply main splits below the lobby floor to serve the men's and women's rest rooms. The women's room supply riser carries all building hot water and both service sinks. Water lines for kitchens, DF's and remote hose bibs are run overhead.

DWV PIPE SIZES

A 4" drain sloped 1/8" per foot will be used to carry the building's 116 DFU (Table 25.6). 4" branches and

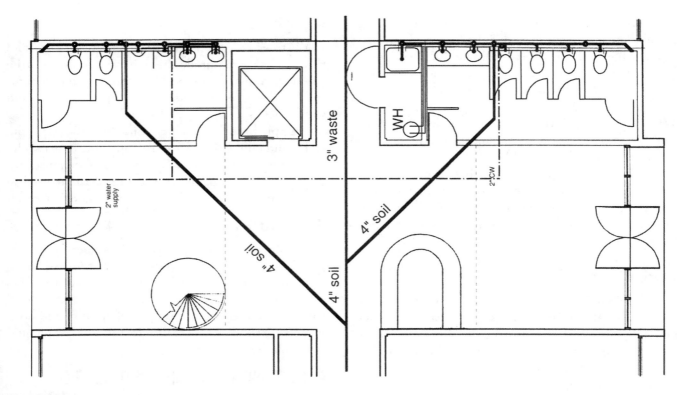

FIGURE 25.17 Ground Floor Plan

soil stacks will also serve each rest room because 3″ pipe cannot carry more than 2 WC's (Table 25.7). A 3″ branch must be used to carry 2 kitchen sinks and 2 DF's (7DFU). 3″ vent stacks are required to serve each restroom soil stack (Table 25.8).

The remaining DWV piping is sized using Table 25.8.

SUPPLY PIPE SIZES

GPM

Figure 25.4 shows a demand of about 80 GPM for the building's 179 SFU.

Main and Meter Size

Use Table 25.6 to select a 2″ main with a friction loss of 5 psi/100′. Flow velocity in the main is just under 8 fps, and by splitting the main as it enters the building the branches and risers can be sized at a quiet 6 fps.

Check Operating Pressure

The most demanding fixture in the building is the second-floor flush valve WC that is most distant from the city main. If there is a residual pressure of 25 psi at this WC, pressure will be more than adequate at all other fixtures.

Given: The pipe run from City main to the most distant second-floor WC is 200′, but allow 50% extra for friction losses in the pipe fittings. Total friction losses will be calculated using a developed pipe length of 300′.

The City main flows at 60 psi. Its elevation is 20′ below the flush valve WC's on the second floor.

From Tables 25.3 and 25.6:

Maximum demand	80 GPM
(179 SFU—Table 25.3)	
Pipe friction pressure loss	5 psi/100′
(2″ pipe, 5 psi/100′ @ 80 GPM—Table 25.6)	

Calculate pressure losses:

head (20)(0.434) = 8.7, say	9 psi
(main elevation is 20′ below toilet tank)	
meter	7 psi
(2″ meter @ 80 GPM—interpolate	
—Table 25.7).	
pipe friction (5 psi/100′)(300′) = 15	15 psi
total pressure loss	31 psi.
Required toilet operating pressure	25 psi
Pressure at toilet inlets (60 − 31)	29 psi

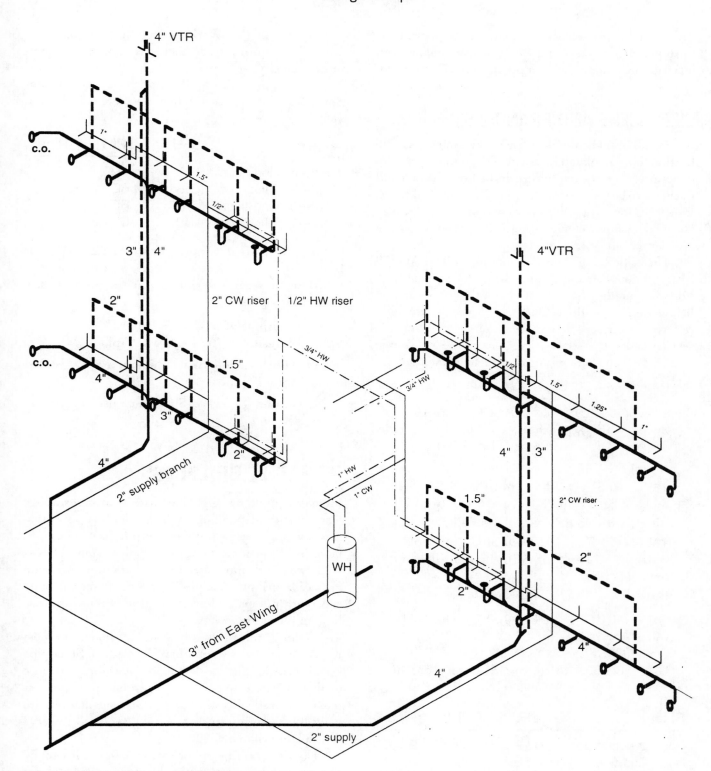

4" VTR

c.o.

1"

1.5"

1/2"

3" 4"

2"

c.o.

4"

3"

4"

2"

2" CW riser

1/2" HW riser

1.5"

3/4" HW

2" supply branch

4"VTR

3/4" HW

1/2"

1.5"

1.25"

1"

4" 3"

2" CW riser

1.5"

1" HW

1" CW

WH

2"

2"

4"

3" from East Wing

4"

4"

2" supply

FIGURE 25.18 Isometric Plan with Pipe Sizes
Women's rooms below right (circuit vents), men's rooms below left (individual vents).
DWV soil and waste = heavy solid lines, vents = heavy dashed lines. Sizes are bold numbers.
Cold water = light solid lines, hot water = light dashed lines. Sizes are light numbers.
Vertical lines at each fixture are water hammer eliminators.

The available pressure is adequate. If the pressure at the toilet inlet was less than 25 psi, a larger main (with less friction loss) or a booster pump would be required.

SIZE RISERS AND BRANCHES

Risers and branches are sized with the same 5 psi/100′ friction loss calculated for the main. This assures adequate pressure at each fixture and allows branch and riser size to be reduced as each supplies fewer fixtures. Pipe sizes for in-building risers and branches are also checked for flow velocity; if velocity exceeds 6 fps, the next larger pipe size is selected.

The 2″ building main supplies up to 80 GPM (179 SFU). It connects to the landscape water system outside the building and then divides into two risers below the ground floor. One riser serves the women's rest rooms on two floors, the water heater, and the service sinks; the second riser serves the men's rest rooms, DF's, hose bibs, and the kitchen sinks.

Riser One — 112 SFU @ ground floor

8 WC (80SFU), 4 Lav's (8 SFU), 2 Service Sinks (6 SFU), plus hot water* for 4 Lav's (6* SFU) and 4 Kitchen Sinks (12* SFU).

Riser One — 53 SFU @ second floor

4 WC (40 SFU), 2 Lav's (4 SFU), 1 Service Sink. (3 SFU), and 2 Kitchen Sinks (6 SFU).

Riser Two — 78 SFU @ ground floor

4 WC (40 SFU), 4 UR (20 SFU), plus cold water* for 4 Lav's (6* SFU), and 4 Kitchen Sinks (12* SFU).

Riser Two — 39 SFU @ second floor

2 WC (20 SFU), 2 UR (10 SFU), plus cold water* for 2 Lav's (3* SFU), and 2 Kitchen Sinks (6* SFU).

Use Table 25.4 to convert SFU to GPM; then size risers at 5 psi/100′ using Table 25.6.

Riser	SFU	GPM	Pipe
One—ground floor	112	70	2½″
One—2nd floor	58	60	2″
Two—ground floor	78	65	2″
Two—2nd floor	39	55	2″

---

*HW & CW SFU = 3/4 of total fixture SFU

Wow! All that work and no smaller pipe, but what about noise? Redo these risers, limiting maximum velocity to 6 fps.

Use Figure 25.6 and revise riser size to limit flow noise. Size risers at 6 feet per second maximum velocity.

Riser	SFU	GPM	Pipe
One—ground floor	112	70	2½″
One—2nd floor	58	60	2″
Two—ground floor	78	65	2″
Two—2nd floor	39	55	2″

At the end of supply runs, where a riser or branch serves just a few fixtures, you cannot read GPM on the demand curves. In these cases, size supply piping as follows:

One flush valve WC—minimum 1″—two or three flush valve WC's, use 1¼″ pipe.

One flush valve Urinal—minimum 3/4″—two to four, use 1″ pipe.

Lavatories, Showers, Sinks, etc., use 1/2″ up to 4 SFU and 3/4″ up to 12 SFU.

25.8 TALL BUILDINGS

Increasing building height increases head losses; above 4 or 5 stories (depending on the available supply pressure), water storage tanks and/or pumps will be components of the domestic water supply system. Downfeed systems pump water to a roof tank and rely on gravity distribution. Upfeed systems maintain pressure with pumps and are typical in recent high-rise buildings (Figure 25.19).

Codes require standpipes and sprinklers in most high-rise buildings (more than 75 feet tall). Standpipe systems are designed to deliver lots of water to fire hose cabinets located on each floor. The domestic water system is metered and previous pipe size calculations are valid, but a large separate main is required to serve standpipes (Figure 25.20).

High-rise buildings store water to fight a fire for 30 minutes, while firefighters make pumper and hydrant connections. A standpipe must deliver 500 GPM at 65 psi minimum for 30 minutes. For an installation with multiple standpipes, allow for 1,250 GPM, which is 150 tons of water over 30 minutes, so large storage tanks are required (Figure 25.21).

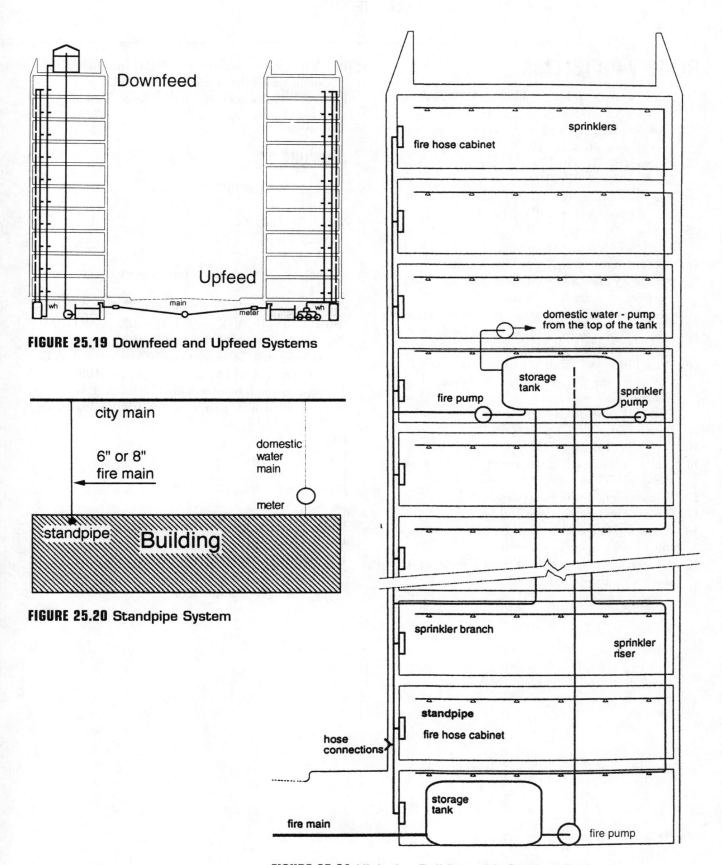

FIGURE 25.19 Downfeed and Upfeed Systems

FIGURE 25.20 Standpipe System

FIGURE 25.21 High-rise Building with Storage Tanks

REVIEW PROBLEMS

1. Find the friction loss in a 200' run of $2\frac{1}{2}$" pipe delivering 80 GPM.
2. How many GPM will a $1\frac{1}{2}$" pipe deliver with a friction loss of 10 psi/100'?
3. How many GPM will a $1\frac{1}{2}$" pipe deliver with a friction loss of 1 psi/100'?
4. A 2" pipe must deliver 30 GPM. What is the friction loss over a 600' pipe run?
5. A 2" pipe must deliver 30 GPM. Find the water velocity in feet per second.
6. A 700' length of $1\frac{1}{2}$" pipe is delivering 40 GPM. Find the water velocity in feet per second and the *total* pressure loss due to friction.
7. If water velocity is limited to 6 feet per second, find the maximum GPM that can be delivered by 1", 2", and 8" pipe.
8. If friction loss is limited to 1 psi/100" feet, find the maximum GPM that can be delivered by 1", 3", and 6" pipe.
9. Can a 1" pipe deliver 50 GPM? Why is a 2" pipe a better choice at 50 GPM?
10. A service sink has a 3" trap. Size its vent.
11. Size a building drain to carry 800 DFU with a slope of 1/4" per foot.
12. Size a soil branch to carry 600 DFU.
13. Size a soil stack to carry 400 DFU.
14. A 6-story soil stack carries 400 DFU. What is the maximum DFU at any floor?
15. A soil stack carries 400 DFU. Find the vent stack size if the stacks are 150' tall.
16. A soil stack carries 400 DFU. Find the vent stack size if the stacks are 200' tall.
17. A 2" vent branch will be 50' long. Should the vent size be increased?

ANSWERS

1. 4 psi — (2 psi/100')(200') = 4
2. 60 GPM
3. about 11 GPM
4. about 6 psi
5. 3 fps
6. velocity = 6 fps; the total pressure loss is about 31 psi.
7. 1", = 16 GPM; 2", = 64 GPM; 8", = 900 GPM
8. 1", = 5+ GPM; 3", = 90 GPM; 6", = 550 GPM
9. Yes a 1" pipe can deliver 50 GPM, *but* water velocity will be more than 10' per second and friction loss will be about 55 psi per 100'. 2" pipe is a much better choice because flow velocity is less than 5 fps and friction losses are reduced by 95%.
10. $1\frac{1}{2}$"
11. 6"
12. 6"
13. 6" for 3 stories or less, 4" if more than 3 stories.
14. 90
15. 3"
16. 4"
17. Yes, increase to $2\frac{1}{2}$"

CHAPTER 26

Water Supply for Fire Protection and HVAC

The steaming river has washed the hot round red sun down under the sea.

Basho

This chapter deals with water supply systems for two other important uses of water in buildings— fire protection, and heating and cooling. Water is a very economical and readily available medium for suppression of fire. The capability of water to carry large quantities of heat also makes it a popular fluid for use in HVAC—heating, ventilating, and air-conditioning systems.

26.0 WATER FOR FIRE PROTECTION

BASIC FACTORS OF FIRE

Fire or combustion is a chemical reaction involving fuel, oxygen, and high temperature (Figure 26.1). In this reaction, molecules of a fuel are combined with the molecules of oxygen in a reaction that also results in an evolution of light and energy. If any of the three factors involved in the reaction is removed, a fire will be extinguished.

CLASSIFICATION OF FIRE HAZARDS

The National Fire Protection Association (NFPA) has categorized fire hazards into four different classes: A, B, C, and D. Class A fires involve solid combustibles (such as wood or paper), which are best extinguished by water or dry chemicals. Fires in flammable liquids are categorized as Class B, which are best extinguished by foam, carbon dioxide, or dry chemicals. Class C fires involve live electrical equipment, which must be extinguished by a non-conductive extinguishing agent, such as carbon dioxide or dry chemicals. Class D fires involve combustible metals such as magnesium or titanium, which must be extinguished by special dry chemical extinguishers.

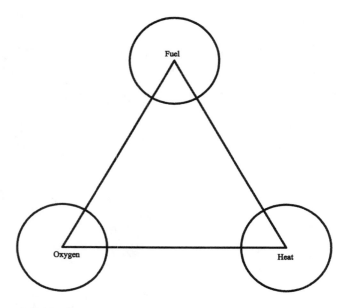

FIGURE 26.1 The Fire Triangle

CLASSIFICATION OF OCCUPANCY HAZARDS

The occupancy hazard rating is a way to classify an occupancy or a building to determine the extent of sprinkler system requirements.

According to NFPA, there are three categories of hazard occupancies:

- **Light hazard:** Occupancies where the quantity and combustibility of contents are low. Fires in this category of occupancies tend to develop at a relatively low rate with low rates of heat release. Light hazard occupancies are typically institutional, educational, religious, residential, and commercial properties.
- **Ordinary hazard:** Ordinary hazard occupancies are divided into two types: Group 1 and Group 2.

  Group 1 includes occupancies consisting of materials low in combustibility, moderate quantity of combustibles, and stockpiles of combustibles not exceeding 8 feet. Occupancies in this group include automobile showrooms, bakeries, canneries, electronic plants, laundries, parking garages, and restaurant serving areas.

  Group 2 includes occupancies where quantity and combustibility of materials are moderate to high, with the stockpiles of combustibles not exceeding 12 feet. Examples of occupancies in this group include cereal mills, confectioners, distilleries, feed mills, libraries, machine shops, paper mills, and wood product assembly plants.

- **Extra hazard:** Occupancies where the quantity and combustibility of materials are very high are classified as extra hazard. Fires in this category of occupancies develop rapidly with a high rate of heat release. Extra-hazard occupancies are also divided into two types: Group 1 and Group 2.

  Group 1 includes occupancies having hydraulic systems with flammable or combustible hydraulic fluids under pressure. Properties with process machinery that use flammable or combustible liquids in closed systems and those having dust and lint in suspension are also included in this group. Aircraft hangars, plywood manufacturing plants, rubber vulcanizing plants, and sawmills belong to this group.

Group 2 type contains larger amounts of flammable or combustible liquids than Group 1 type. Examples of occupancies in this group include asphalt saturating, open oil quenching, plastics processing, and varnish and paint dipping.

METHODS FOR FIRE DETECTION

The National Fire Protection Association has developed NFPA No. 74, Standard for Household Fire Warning Equipment, in order to protect the customers and establish criteria for the manufacture and installation of residential fire detection systems.

There are two types of residential fire detection systems: smoke detectors and heat detectors. The basic residential detection system relies primarily on the use of smoke detectors.

Flame detectors are used to detect the direct radiation of a fire. They are used mainly in industrial processes and for the protection of combustion equipment.

Smoke Detectors

There are two kinds of smoke detectors: ionization and photoelectric. The **ionization smoke detector** operates on the principle of changing the conductivity of air. It uses a radioactive source to ionize the air between two charged surfaces. The ionization of air causes a small flow of electrical current. When smoke from a fire enters the detection chamber, its presence causes a reduction in the current's flow. The electronic circuitry senses the reduced flow and triggers an alarm. An ionization detector is capable of sensing microscopic particles of combustion and is the best for detecting fire at an incipient stage.

A **photoelectric smoke detector** operates on the principle of the scattering of light. It consists of an LED (light-emitting diode) light source, a supervisory photocell directly opposite the light source, and an alarm photocell within the detection chamber. When smoke is present in the chamber, the alarm photocell, which is located at right angles to the light beam, senses the light scattered off the smoke particles. This activates the electric circuit to sound an alarm. This device is the best for detecting fire at the smoldering stage, when smoke particles are visible to the naked eye.

Heat Detectors

The NFPA standard does not require the use of heat detectors as part of the basic protection scheme, but it recommends that heat detectors be used to supplement the basic smoke detector system. There are two types of heat detectors: fixed temperature and rate of rise.

Fixed-temperature heat detectors may be either the self-restoring or the nonrestoring type. The self-restoring type consists of an open contact held by a bimetallic element. When the ambient temperature reaches a fixed setting (usually 135°F or 185°F), the contact is closed, thereby triggering an alarm. The contact will return to open position when the ambient temperature returns to normal. The nonrestoring (i.e., nonresettable) type utilizes a special alloy retainer designed to melt at a specific temperature. When the ambient temperature reaches the fixed setting, the fusible alloy melts to close the electrical contacts and to initiate an alarm.

A **rate-of-rise type heat detector** reacts when the temperature in the immediate vicinity rises higher than the preset rate per unit of time. It incorporates a sealed but slightly vented air chamber and a flexible metal diaphragm. Air within the chamber expands and contracts due to the fluctuation of temperature under normal conditions. The vent, which is calibrated, releases or admits air to compensate for the changes in pressure. When the temperature within the vicinity of the device rises quickly, air within the chamber expands faster than it can be vented. The resulting pressure distends the diaphragm and closes a set of normally open contacts, thereby activating an electrical circuit.

Flame Detectors

Flame detectors are of two types: infrared and ultraviolet. Infrared radiation is present in most flames. Therefore, an **infrared (IR)** detector can detect the presence of fire instantaneously. A relatively long IR wavelength allows it to penetrate smoke, making detection possible. Because there are many sources of IR other than fire in an industrial setting, it is possible that a simple IR detector may cause false alarms.

Ultraviolet (UV) flame detectors operate by detecting UV radiation produced by fire. These detectors are not plagued by as many possible sources of false alarms. Sources of false alarms for UV detectors are well defined. The most common are lightning, X rays, and arc welding.

Dual spectrum flame detectors are also available that employ UV sensors in combination with a narrowband IR detector. To prevent false alarms caused by nonfire sources, the UV-IR detector utilizes an IR detector sensitive to wavelengths in the range of 4.1 to 4.6 microns. All hydrocarbon fires produce radiation within this range.

METHODS OF FIRE CONTROL

Fire has a need for oxygen, fuel, and heat. When deprived of any of these needs, there will be no fire. The most common method of fire suppression is the use of water. A universal medium for extinguishing fire, water is available in large quantities and is less expensive than any other fire-suppression medium. It works as a cooling agent by lowering the ignition temperature of combustion and also deprives the source of combustion from oxygen supply. Water supply for fire suppression may be provided by either manual or automatic systems. The manual systems use standpipes and hose, and the automatic systems use sprinklers.

Automatic Sprinkler Systems

Automatic sprinkler systems consist of a network of pipes connecting water supply to a series of sprinkler heads. They provide automatic fire suppression in areas of a building where fire temperature has reached a predetermined level. Automatic sprinkler systems are the most effective in the suppression of Class A fire that contains wood, paper, and plastics.

The sprinklers are generally installed in a gridiron pattern near the ceiling. There are four basic types of sprinkler systems: wet pipe, dry pipe, preaction, and deluge.

Wet pipe type is the most common sprinkler system. In this system, the piping network contains water at all times. Water is immediately discharged onto the fire when a sprinkler activates (Figure 26.2). This system is very simple, reliable, and relatively inexpensive to install and maintain. The main disadvantage of the system is that it is not suitable for subfreezing environments.

A **dry pipe** sprinkler piping system is filled with pressurized air or nitrogen (Figure 26.3). The system is connected to a dry pipe valve that remains closed under normal conditions. When fire causes one or more sprinklers to operate, air within the system es-

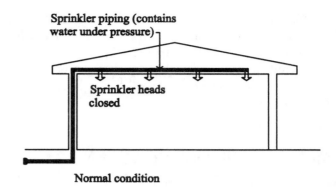

Normal condition

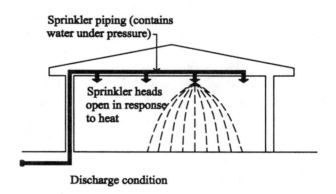

Discharge condition

FIGURE 26.2 Wet Pipe System

capes and the dry pipe valve is released. Water is thus allowed to enter the piping network and flow through the open sprinkler heads. The main advantage of this system is its capability to provide protection to spaces that are subject to freezing temperature conditions. It is more complex than a wet pipe system and requires greater attention in design, installation, and maintenance. It may produce greater fire damages because of a higher response time. A dry pipe system must be thoroughly drained and dried following operation in order to prevent the corrosion of pipes.

A **preaction** system employs the basic concept of a dry pipe system. It consists of a preaction valve that controls the flow of water in the sprinkler piping network (Figure 26.4). Discharge in the system is initiated by two independent events. When the detection device identifies a developing fire, the preaction valve is opened, allowing water to enter the pipes. The sprinklers also open independently in response to the heat from fire and discharge water. A preaction system with double interlock allows water

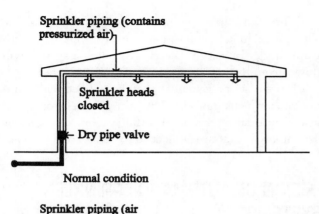

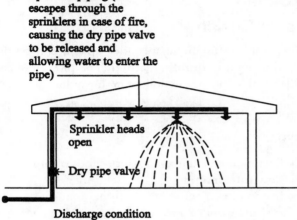

FIGURE 26.3 Dry Pipe System

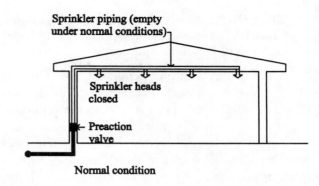

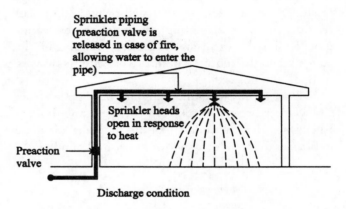

FIGURE 26.4 Preaction System

to enter the sprinkler piping when both the detection device and the sprinklers identify a developing fire. This system provides an added level of protection against any inadvertent discharge and is suitable for areas where water damage is a serious problem.

A **deluge** system is a variation of preaction, equipped with open-type sprinklers (Figure 26.5). Actuation of the fire detection systems releases a control valve, which in turn causes immediate water flow through all sprinklers installed in the network. Since all sprinkler heads are open, every sprinkler on a deluge system will discharge water simultaneously when the control valve is released. This system is recommended for buildings where the spread of fire is anticipated to be rapid and for very high-hazard areas.

Standpipe Systems

A standpipe system is required for high-rise buildings where the hose from firefighting equipment cannot reach the upper floors. It provides fire hose stations for manual application of water to fires in buildings. The system consists of piping, valves, hose racks, hose connections, and auxiliary equipment necessary to provide sprays of water to suppress fire. Standpipes are classified into three categories.

- **Class I:** This category of standpipe supplies $2\frac{1}{2}''$ hose outlets at each floor level for use by the fire department. The $2\frac{1}{2}''$ outlet hose valves furnish the firefighters with a water hydrant and adequate water supply for using hose stream during fire. The minimum water supply requirement for Class I category is 500 gpm for the first standpipe and 250 gpm for each additional standpipe.
- **Class II:** This category of standpipe is designed to be used by the occupants of the building until the arrival of the firefighters. It supplies $1\frac{1}{2}''$ hose stations along with a hose rack. The minimum water supply requirement for this category is 100 gpm. Hose stations should be located so that all spaces on a floor are within 30 feet of a nozzle attached to a 100-foot-long hose.

◆ **Class III:** This is a combination of Class I and Class II standpipes. It supplies both $2\frac{1}{2}''$ hose outlets and $2\frac{1}{2}''$ hose stations.

A fire department pumper connection, known as the Siamese connection, is required to be provided outside the building. It is a two-way connection having two $2\frac{1}{2}''$ outlets supplied by a $4''$ pipe.

Maximum height for a standpipe system is 275 feet. For buildings higher than this limit, separate standpipe systems are required for each additional 275 feet or less.

The water pressure requirement for all categories of standpipes is a supply that will provide 65 psi of residual pressure at the highest outlet on the standpipe. Residual pressure is the pressure reading on the gauge when the required quantity of water is flowing through the hose.

Water supply for standpipes may be supplied from city water main, gravity tank, or pressure tank. It must be ensured that the capacity of the source is adequate to furnish the total demand for at least 30 minutes.

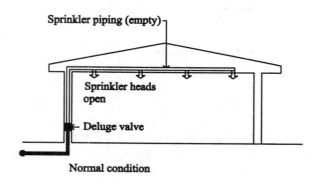

Normal condition

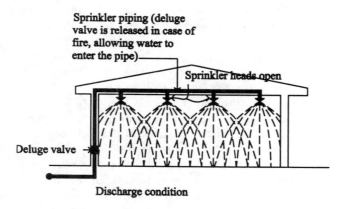

Discharge condition

FIGURE 26.5 Deluge System

The standpipes may either be wet or dry types. The pipes in a wet system are always filled with water under pressure. Water from such a system will flow through the hose as soon as it is activated.

Pipes in a dry system are generally filled with compressed air instead of water. A valve has to be operated in order to admit water into the system. Dry pipe is used in areas that are subject to freezing conditions.

DESIGN OF AUTOMATIC SPRINKLER SYSTEMS

Piping Design

Piping for an automatic sprinkler system can be designed using a **hydraulic method**. The pipe sizes can be determined using the following formula:

$$p_f = (4.52 \cdot Q^{1.85})/(C^{1.85} \cdot \partial^{4.87}) \qquad (1)$$

where:

p_f = frictional resistance, in psi per foot of pipe
Q = flow rate, in gpm
C = friction loss coefficient of piping material (100 for cast iron; 120 for black steel, galvanized steel, and plastic; 140 for cement-lined cast iron; and 150 for copper and stainless steel)
∂ = internal diameter of pipe in inches

The orifice size of a sprinkler and the residual pressure of the water supply affect the **flow rate** of a sprinkler in any sprinkler system. It is calculated using the following formula:

$$Q = K \cdot \sqrt{p} \qquad (2)$$

where:

Q = flow rate in gpm
K = flow constant, per unit; varies from 1.3 to 1.5 for a $\frac{1}{4}''$ orifice, 5.3 to 5.8 for a $\frac{1}{2}''$ orifice, and 13.5 to 14.5 for a $\frac{3}{4}''$ orifice
p = residual pressure in psi

A simple method of piping design for small automatic sprinkler systems is the use of **piping schedule**. The sizes of branch pipes and risers are determined from a pipe schedule (see Table 26.1), assuming that residual pressure and flow rate are in compliance with code. The method is very useful for preliminary design and cost estimates.

TABLE 26.1

Pipe Schedule for Number of Sprinklers Allowed in an Automatic Sprinkler System

Pipe Size	Hazard Classification					
	Light		Ordinary		Extra	
	Steel	Copper	Steel	Copper	Steel	Copper
1″	2	2	2	2	1	1
1¼″	3	3	3	3	2	2
1½″	5	5	5	5	5	5
2″	10	12	10	12	8	8
2½″	30	40	20	25	15	20
3	60	65	40	45	27	30
3½″	100	115	65	75	40	45
4	a	a	100	115	55	65
5			160	180	90	100
6			275	300	150	170
8			b	b		

a One 4-inch system may serve up to 52,000 sq. ft. of floor area.
b One 8-inch system may serve up to 52,000 sq. ft. of floor area.

Sprinklers on a Branch Line

A maximum of eight sprinklers are allowed to be installed on a branch line, on either side of a cross main of a sprinkler system designed for either a light or an ordinary hazard occupancy. The limit for extra hazard occupancy is six sprinklers.

Protection Area

The maximum floor area that can be protected by a single sprinkler system shall not exceed 52,000 sq. ft. for light or ordinary hazards. The maximum area protected by a single system for extra hazard occupancy shall not exceed 25,000 sq. ft. if the piping is designed using the pipe schedule method, and 40,000 sq. ft. if it is designed using the hydraulic method.

The actual number of sprinklers to be installed and their spacing depend on the hazard classification of the building. Table 26.2 gives the maximum sprinkler protection area and spacing between sprinklers.

WATER DEMAND FOR FIRE PROTECTION

A small part of the sprinkler system needs to be operated during the early stage of a fire. Therefore, fire protection codes require only a small area of the building to be taken into consideration for simultaneous water demand. This space is called the area of sprinkler operation. The area selected should be the one that is the most remote from the source of water supply. The size of the area of sprinkler operation varies from 1,500 sq. ft. for light hazard to 5,000 sq. ft. for extra-hazard occupancies. A minimum density of water flow has to be maintained for different areas of sprinkler operation. The relationships between area of operation and water flow density is shown in Figure 26.6.

The minimum water requirement for fire protection is determined independently of the actual size of a building. This is because a fire can usually be expected to start in only one area. Therefore, total water demand is calculated using the area of sprinkler operation. Water demand of hoses is also included if the sprinkler systems are supported by standpipe systems. Table 26.3 shows water demand

TABLE 26.2

Maximum Spacing Between Sprinklers and Coverage per Sprinkler

Construction Type		Classification of Hazard		
		Light	Ordinary	Extra
Unobstructed	Spacing (ft.)	15	15	12
	Area (sq. ft.)	200[a]–225[b]	130	90[a]–130[b]
Obstructed, noncombustible	Spacing (ft.)	15	15	12
	Area (sq. ft.)	200[a]–225[b]	130	90[a]–130[b]
Obstructed, combustible	Spacing (ft.)	15	15	12
	Area (sq. ft.)	130[c]–225[d]	130	90[a]–130[b]

[a]To be used when pipe sizing is based on pipe schedule.
[b]To be used when pipe sizing is based on hydraulic method.
[c]To be used when framing members are spaced less than 3 feet on center.
[d]To be used when framing members are spaced 3 feet or more on center.

by hose pipes and the required duration of the supply. The total water demand for combined sprinkler and standpipe systems can be calculated using the following formula:

$$TWD = (ASOP \cdot D \ OVF) + HSD \qquad (3)$$

where:

TWD = total water demand in gpm
ASOP = area of sprinkler operation in sq. ft.
D = density of water flow in gpm/sq. ft.
OVF = overage factor (usually 1.1)
HSD = hose stream demand in gpm

Supply pipe sizes for standpipe connections considering the total water demand are shown in Table 26.4.

OTHER METHODS OF FIRE SUPPRESSION

Alternative methods of fire suppression are available in situations where the use of water may not be a practical solution. These methods include the use of carbon dioxide, dry chemicals, foam systems, and halogenated gas.

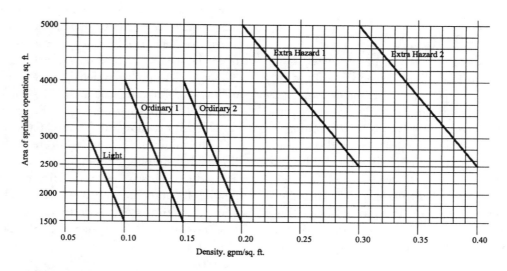

FIGURE 26.6 Area/Density Curves

TABLE 26.3

Water Requirements by Hose Pipes and Duration of Supply

Hazard Type	Inside Hose (gpm)	Inside + Outside Hose (gpm)	Duration (minutes)
Light	0, 50, or 100	100	30
Ordinary	0, 50, or 100	250	60–90
Extra	0, 50, or 100	500	90–120

Carbon Dioxide

Carbon dioxide is a colorless and odorless gas. It can extinguish a fire without leaving any residue. The introduction of CO_2 into a fire displaces oxygen in the atmosphere, thereby retarding the combustion process. At the same time, the extinguishing process is aided by a reduction in the concentration of gasified fuel in the fire area. CO_2 also provides some cooling in the combustion zone, completing the extinguishing process. The evaporation of the liquefied gas absorbs about 120 BTU (British thermal unit) of heat per pound of the gas.

CO_2 is effective in confined and unventilated spaces. Such spaces include display areas, electrical and mechanical chases, and unventilated areas above suspended ceilings that are not inhabited by people or other living creatures.

Dry Chemicals

The best candidates for dry chemicals that suppress fire are industrial applications. They are extremely effective firefighting agents that suppress fire by covering the surface of the fuel. The coating separates the fuel from the oxygen supply and thus retards fire. The dry chemicals usually contain bicarbonates, chlorides, and phosphates.

Foam Systems

Foams are gas-filled bubbles, produced by using a generator that mixes water with detergent or other chemicals. It is possible to produce about 1,000 gallons of foam using one gallon of water. Foam systems are most effective for fires involving flammable liquids. Being lighter than the combustible liquids, they float on the surfaces of the liquids, thereby insulating them from oxygen. Moreover, the vaporization of water contained in the films of the foam provide a certain measure of cooling by heat absorption. A good foam blanket strongly resists the heat and flame of a fire.

Halogenated Agents

Halogenated agents are gases that contain halogen atoms (bromine, chlorine, fluorine, or iodine). The most commonly used halogenated gas is Halon 1301. It is a very effective fire suppressant that extinguishes

TABLE 26.4

Supply Pipe Sizes for Standpipe Connections

Total Accumulated Flow in GPM	Total Distance of Piping from Furthest Outlet		
	Less Than 50 Feet	50–100 Feet	Over 100 Feet
100	2"	2½"	3"
101–500	4"	4"	6"
501–750	5"	5"	6"
751–1250	6"	6"	6"
Over 1250	8"	8"	8"

fire by inhibiting the chemical reaction of fuel and oxygen. But scientific evidence indicates that the agent is an ozone-depleting chemical. Therefore, the use of halogenated agents has been banned; they are being replaced with a number of alternatives such as IG-541 (commonly known as INERGEN) and HFC-227 (commonly known as FM-200). These agents have either zero or negligible ozone-depleting potential. Their effectiveness as fire suppressants is still being evaluated.

26.1 WATER FOR HVAC

WATER AS A MEDIUM FOR HEATING AND COOLING

Water is a chemically stable, nontoxic, and inexpensive fluid. It also has a remarkable capability of carrying large quantities of heat. Compared to air, it is a much less bulky medium for conveying heat or cooling. One pound of air is about 14 cubic feet and can carry 0.24 BTU of heat per 1°F temperature difference; one pound of water, on the other hand, has volume of only about 0.016 cubic foot and is capable of carrying 1 BTU of heat per 1°F temperature difference. Therefore, a system of water piping takes up much less space of building structure than ductwork for air required for an equivalent amount of thermal distribution. Water is, therefore, a very popular fluid used in HVAC systems, particularly for large buildings.

Water can be used for both heating and cooling. For heating purposes, the temperature of water is raised to 160°F to 250°F or more in a boiler. It is lowered to 40°F to 50°F in a chiller for cooling purposes. Pipes are used to convey the product from the boiler or chiller to an air-handling unit, which transfers the heat or cooling to an airstream for ducted delivery to different spaces in the building. In small buildings, heated or chilled water is piped to a terminal device that transfers the heat or cooling directly to the room air.

WATER DEMAND FOR HEATING AND COOLING

The water demand for heating or cooling is dependent on the total quantity of heat required to be added to or removed from a building. It also depends on the designed temperature difference of the flow of water entering and leaving the heat exchanger. In other words, it is the function of the peak heating or cooling load of a building and the required rise or drop in temperature of water used in the process.

Flow rate or demand of water is calculated in terms of gallons per minute (gpm). One gallon per minute equals 60 gallons per hour. Since a gallon of water weighs 8.33 pounds, one gallon per minute equals $8.33 \times 60 \approx 500$ pounds per hour. Using this relationship, we can calculate water demand for heating or cooling as follows:

$$FL = Q/500 \cdot (t_1 - t_2) \qquad (4)$$

where:

FL	=	flow rate of water, in gpm
Q	=	heat required to be added or removed, in BTUH (BTU per hour)
$t_1 - t_2$	=	difference between supply and return water temperatures

For heating systems, a temperature drop of 20°F between supply and return water temperatures is an industry standard. Therefore, heat transferred by 1 gpm of water equals $500 \cdot 20 = 10,000$ BTUH. A building with a heating load of 50,000 BTUH will require 5 gpm of water at 20°F drop in temperature between supply and return.

PIPING

Many types of materials are used for HVAC piping. Most common are Schedule 40 steel pipe, copper pipe (types K, L, and M), and Schedule 80 CPVC (thermoplastic) pipe. Pipe size is based on either velocity or head loss, both of which provide roughly equal results. Head or friction loss is the reduction of water pressure due to the friction that the walls of the pipe impose on a liquid—that is, a measure of the resistance of the piping system to the flow of the water through it. It is dependent on the viscosity of the fluid and the turbulence of the flow. Head or friction loss in feet of water is equal to 2.31 feet/psi. A friction loss of 2.5 feet per 100 feet of pipe length is used as design criterion.

Pipe sizes should be large enough to minimize the noise produced by water flow. Larger pipe sizes also help in reducing the amount of friction that the pump has to overcome. Water velocities of 3 to 5 feet per second are generally recommended for piping design. Once the flow rate or demand of water has been established, the size of HVAC pipes can be determined using sizing charts (Figures 26.7 and 26.8).

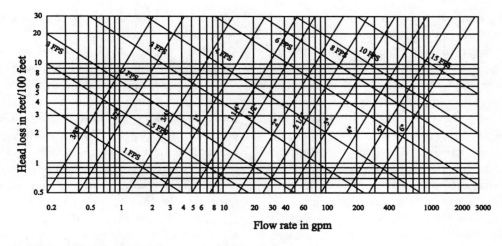

FIGURE 26.7 Sizing Chart for Copper Pipes

Water-bearing HVAC pipes have to be insulated to minimize the transfer of heat to the surrounding unconditioned air before the water reaches the terminal device. Insulation can increase the pipe diameter by 2 to 8 inches. Added precaution has to be taken in insulating chilled water pipes to prevent condensation on the pipe surfaces.

HVAC WATER QUALITY

Water quality is important for both heating and cooling purposes in order to maintain the reliability and efficiency of the equipment, such as boilers, chillers, and cooling towers. Mineral scale deposit, corrosion, and growth of microorganisms may cause failure of the systems.

Boiler Water

Scale-forming minerals such as calcium and magnesium, introduced by makeup water, primarily cause scale and sludge deposit in a boiler system. The primary means for controlling boiler scale caused by minerals in the makeup water is to treat the water, using ion exchange softening or demineralization. This treatment should remove the majority of the scale-causing minerals.

Oxygen that enters a boiler system may contribute to accelerated corrosion of the feedwater system, boiler, and condensate return system. De-aeration of makeup water is helpful in controlling this type of corrosion. Once the makeup water is de-aerated, various organic and inorganic oxygen-scavenger (40 to 60 mg/L of sulfite, for example) compounds are added to

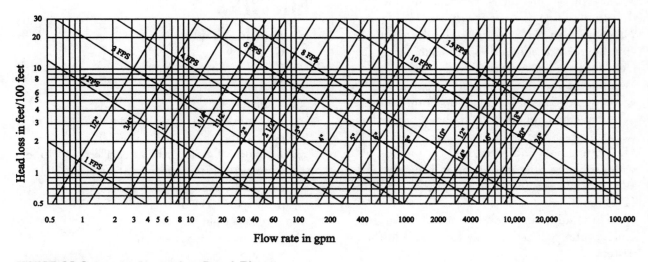

FIGURE 26.8 Sizing Chart for Steel Pipes

complete oxygen removal. Oxygen scavengers should be added to the boiler makeup water stream immediately after de-aeration so as to protect all downstream components of the system.

Inner surfaces of a boiler system must also be protected against corrosion caused by contact with water. The best method of protection is to simply maintain the boiler water at a pH-value between 8 and 8.5. This will make the water noncorrosive to steel boiler internals. It is also required to control carbonate and phosphate scale formation that occurs at higher pH-value levels.

Despite the treatment of makeup water, some dissolved minerals may enter the boiler system. This material remains in the drum as solids. Excessive concentration of these solids will eventually lead to unsatisfactory operation of the system. Boiler blowdown is necessary under such circumstances. It means completely or partially removing boiler water and replacing it with treated water.

Chiller and Cooling Tower Water

Water for cooling is also required to be treated to remove dissolved solids and to prevent scale formation. Growth of microorganisms has to be controlled to prevent tube fouling. Chlorine and ozone work well on microorganisms. Since the chilled water loop is usually a closed one, it needs to be treated as often as makeup water is added.

Open cooling tower systems are subject to deposits of airborne dust and debris. Fine particles of these foulants tend to collect in the condenser-water system, get deposited on the heat transfer surfaces in the form of sticky mud, and interfere with efficient operation of the system. These suspended solids are removed by filtration using strainers and filters on cooling tower systems. Makeup water for cooling systems, containing turbidity or suspended matter, should be treated before use by coagulation and filtration.

Microbiological growths such as algae, slime-forming bacteria, and mold take place in cooling towers. Algae and fungi can cause serious problems by blocking the air in cooling towers and plugging the distribution systems. Deposits of bacterial slime on heat transfer surfaces will affect the efficiency of the cooling system. Traditional biocides, such as gaseous or liquid chlorine, are quite effective for removal of these organic growths. Use of ozone is also a common practice for treatment of cooling tower water.

Cooling Tower blowdown is periodically required in order to control the concentration of dissolved solids by systematic drainage of a portion of the circulating water. Water thus removed from the system has to be replaced with treated water.

REVIEW QUESTIONS

1. What are the basic factors of fire?
2. What category of fire involves flammable liquids?
3. What type of detector is the most suitable for detection of fire at a very early stage?
4. Name an automatic fire sprinkler system that is suitable for protection of spaces subject to freezing temperature conditions.
5. What is the maximum number of sprinklers recommended to be installed on branch pipe?
6. Why have halogenated agents been banned as fire suppressants?
7. How much heat can be carried by a gallon of water when temperature drop is 1°F?
8. What is the recommended velocity of water in HVAC piping?
9. Name the minerals, the presence of which in boiler water will cause scale formation in the system.
10. What is the recommended pH-value of water for a boiler system

ANSWERS

1. Fuel, heat, and oxygen
2. Class B
3. Ionization smoke detector
4. Dry pipe system
5. 8 sprinklers
6. Halogenated agents have high ozone-depleting potential
7. 1 gallon = 8.33 lbs.; Heat carried by 1 lb. of water @ 1°F temperature drop = 1 BTU; Heat carried by 8.33 lbs. of water @ 1°F temperature drop = 8.33 BTU.
8. 3 to 5 feet per second
9. Calcium and magnesium.
10. 8 to 8.5.

REFERENCES

Read the following sources for a more comprehensive coverage of energy, design, and HVAC topics.

ENERGY

Cook, Earl. *Man, Energy, Society*. W.H. Freeman, San Francisco, CA, 1976. *Thoughtful analysis of energy resources and end use efficiency.*

Daniels, Farrington. *Direct Use of the Sun's Energy*. Yale University Press, New Haven, CT, 1964 (also Ballantine Books, New York, 1974). *Explicit coverage of most solar topics.*

DESIGN

Baer, Steve. *Sunspots*. Cloudburst Press, Seattle, WA, 1979. *Solar design, moveable insulation, thermal storage.*

Brown, Reynolds, Ubbelohde. *Insideout*. John Wiley & Sons, New York, 1985. *Solar design workbook.*

Olgyay, Victor. *Design with Climate*. Princeton University Press, Princeton, NJ, 1963. *Passive design.*

Wells, Malcolm. *Gentle Architecture*. McGraw-Hill, New York, 1981. *Earth-sheltered construction.*

HVAC

Air Conditioning and Refrigeration Institute. *Refrigeration and Air-Conditioning*. Prentice Hall, Upper Saddle River, NJ, 1979.

Ambrose, E.R. *Heat Pumps and Electric Heating*. John Wiley & Sons, New York, 1966.

ASHRAE Handbook Series; *2001 Fundamentals*, and current *Systems and Applications*. American Society of Heating, Refrigeration, and Air-Conditioning Engineers Inc., Atlanta, GA, 2001.

Egan, M. David. *Concepts in Thermal Comfort*. Prentice Hall, Upper Saddle River, NJ, 1975.

HVAC PLUS

Lewis, Jack. *Support Systems for Buildings*. Prentice Hall, Upper Saddle River, NJ, 1986.

Stein, Benjamin, and John S. Reynolds. *Mechanical and Electrical Equipment for Buildings*. John Wiley & Sons, New York, 1992.

Tao, William, K.Y., and Richard R. Janis. *Mechanical and Electrical Systems in Buildings*. Prentice Hall, Upper Saddle River, NJ, 1997.

INDEX